JAVA Programming Fundamentals

Problem Solving Through Object Oriented Analysis and Design

Premchand S. Nair

CRC Press
Taylor & Francis Group
Boca Raton London New York

CRC Press is an imprint of the
Taylor & Francis Group an **informa** business
A CHAPMAN & HALL BOOK

CRC Press
Taylor & Francis Group
6000 Broken Sound Parkway NW, Suite 300
Boca Raton, FL 33487-2742

Printed in the United States of America on acid-free paper
10 9 8 7 6 5 4 3 2 1

International Standard Book Number-13: 978-1-4200-6547-3 (Softcover)

Library of Congress Cataloging-in-Publication Data

Nair, Premchand S., 1956-
Java programming fundamentals : problem solving through object oriented analysis and design / by Premchand S. Nair.
p. cm.
Includes bibliographical references and index.
ISBN-13: 978-1-4200-6547-3
ISBN-10: 1-4200-6547-5
1. Java (Computer program language) 2. Object-oriented programming (Computer science) I. Title.

QA76.73.J38N345 2008
005.1'17--dc22 2008017335

Visit the Taylor & Francis Web site at
http://www.taylorandfrancis.com

and the CRC Press Web site at
http://www.crcpress.com

Dedication

To five people in my life,

M.K. Krishna Pillai
Grandfather

S. Sukumaran Nair
Dad

A. Sarada Devi
Mom

Suseela Nair
Wife

Meera Nair
Daughter

Contents

Preface

Programming is an art. Although traditional art imitates life, programming simulates life. Every abstract concept in programming, and to a great extent in the field of computer science, has its roots in our daily life. For example, humans and possibly all other living forms were *multiprocessing* long before the term entered into computer science lingo. Therefore, any concept in programming can in fact be illustrated through examples from our day-to-day life. Such an approach not only enables the student to assimilate and internalize the concept presented in a programming situation but also provides a solid foundation for the very process of programming, namely, the simulation of the real world. Unfortunately, textbooks currently on the market do not exploit this fact through examples or meaningful discussions. Thus, for many students, an abstract concept remains abstract. This is especially true in the case of object-oriented programming. The "wow moment" one gets by seeing programming as a simulation of the real-world situation is never realized.

This book on Java programming teaches object-oriented design and programming principles in a completely integrated and incremental fashion. This book allows the reader to experience the world we live in as object-oriented. From the very outset the reader will realize that everything in this world is an object. Every concept of object-oriented design is first illustrated through real-life analogy. Corresponding Java language constructs are introduced in an integrated fashion to demonstrate the programming required to simulate the real-world situation. Instead of compartmentalizing all the object-oriented concepts into one chapter, this book takes an incremental approach.

The pedagogy of this book mirrors the classroom style the author has developed over the years as a teacher of computer science. In particular, every programming concept is introduced through simple examples followed by short programming examples. Case studies at the end of each chapter illustrate various design issues as well as the usefulness of many new concepts encountered in that chapter.

Java has emerged as the primary language for software development. From a software engineering perspective, object-oriented design has established itself as the industry standard. Thus, more and more teaching institutions are moving toward a CS1 course that teaches Java programming and object-oriented design principles. A common approach followed in many textbooks on the market is to introduce object-oriented concepts from the very beginning and ignore many traditional programming techniques completely. The objective of this book is to present object-oriented programming and design without compromising the training one needs on traditional programming constructs and structures.

OUTSTANDING FEATURES

- *Object first approach and unified modeling language (UML).* The object-oriented design principles and UML notation are introduced from the very beginning. Case studies at the end of each chapter take the reader through a journey that starts at requirement specification and ends at an object-oriented program.
- *Incremental approach to topic presentation.* Object-oriented concepts are introduced in an incremental fashion. This book does not compartmentalize all object-oriented principles into one chapter; rather, new concepts are introduced and used throughout the book.
- *In-depth treatment of topics.* Concepts on object-oriented design and programming are presented in an in-depth fashion. The reader could easily master all concepts by working through various examples. Topics that can be skipped in an introductory course are labeled as Advanced Topic and can be omitted.
- *Numerous examples drawn from everyday life.* This book contains many fully developed programming examples. In addition, each concept is illustrated through simple examples drawn from everyday life. Examples do not depend on mastery in mathematics.
- *Notes on common pitfalls and good programming practice.* Notes on common pitfalls and good programming styles appear throughout this book.

PEDAGOGICAL ELEMENTS

Every chapter begins with a list of objectives. This list in a way summarizes the theme of the chapter.

This book uses examples at four different levels. First, simple examples are presented. To follow those examples, the reader need not know anything about programming or material covered in the book until then. Second, examples are provided to illustrate the proper and appropriate usage. Third, examples illustrate the new concept through a simple program. Fourth, case study examples are employed to demonstrate the need and effectiveness of the concept in a broader problem-solving context.

Introspection, a unique pedagogical element of this book, is a thought-provoking technique that will empower the instructor with ample materials to start a discussion on the major concepts discussed in each section. This technique will enable the student to internalize the concepts in a meaningful way.

Self-check questions are presented at the end of each subsection. It not only checks the understanding of the subject matter presented in the subsection but also highlights the major concepts the reader is expected to know from that point on.

The quick review presented at the end of each chapter provides a summary of the chapter. The aim of the quick review is to highlight major points explained in the chapter. Thus, quick review works as a checklist for the student as well.

Each chapter presents major constructs of Java language along with topics that can be covered depending on the availability of time and the student's level of comprehension. Those topics are labeled as Advanced Topic for easy identification.

The case study at the end of the chapter serves two important purposes. First, it allows the reader to see the application of the new concepts learned in a practical situation. Second, we have used two major themes throughout the book. The same theme is used at different levels of complexity to illustrate the application and usefulness of the new concepts.

Solved exercises at the end of each chapter provide enough challenges and further consolidate the concepts introduced in the chapter.

SUPPLEMENTS AND COMPANION WEBSITE

The companion website www.premnair.net contains many useful links, documents, and programs. The instructor can obtain the instructor's manual that contains solutions to all problems presented at the end of the chapter, including the programming exercises. Both the instructor and the students can access all programs presented in this book as well as PowerPoint presentations of each of the chapters.

CHAPTER DEPENDENCY

Chapters in this book can be taught in the sequence presented. However, the instructor has the liberty to tailor the course on the basis of the needs as long as the following dependency diagram is followed:

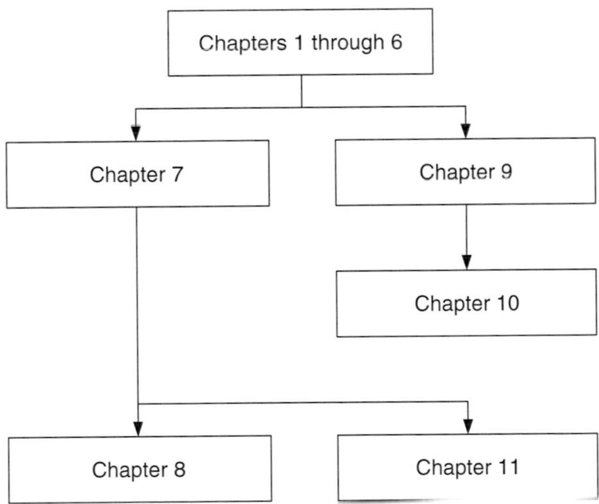

Acknowledgments

I am extremely thankful to the following reviewers whose valuable suggestions and corrections transformed my manuscript into this book in its present form: Dr. Mark Meysenburg (Doane College) and Dr. Charles Riedesel (University of Nebraska–Lincoln). I have a wonderful person, Randi Cohen, as my acquisition editor at Taylor & Francis. She is one of the most kind-hearted and efficient persons I have ever known. I am so lucky to work with her on this project. I would like to thank Amber Donley, the project coordinator, Editorial Project Development, for the successful and timely completion of this project.

I would like to thank my wife Dr. Suseela Nair, who patiently read the entire manuscript several times and gave me many valuable suggestions and corrections. I am grateful to my daughter Meera Nair for all her love. My thanks are also due to my parents, S. Sukumaran Nair and A. Sarada Devi, and grandfather M.K. Krishna Pillai for all the wonderful things in my life. Thanks are also due to George and Susan Koshy, who are my best friends. They have been my cheerleaders and advisors for the past two decades.

I am so grateful to Creighton University, which has supported all my professional efforts throughout my career. I would, in particular, like to thank Dr. Robert E. Kennedy, dean of the College of Arts and Science, one of the most decent, honest, and fair-minded persons this great institution has ever had. It is my privilege and honor to work under his leadership.

I welcome your comments and suggestions. Please contact me through e-mail at prem@premnair.net.

Premchand S. Nair

Author

Dr. Premchand S. Nair is a professor of computer science at Creighton University, Omaha, Nebraska, where he has been teaching programming for the past 19 years. He has two PhD degrees: in mathematics and computer science (Concordia University, Montreal, Canada). This is his seventh book and his first undergraduate book as the sole author. He has published his research work in many areas, including algorithm design, computer vision, database, graph theory, network security, pattern recognition, social network analysis, and soft computing.

CHAPTER 1

Object Model of Computation

In this chapter you learn

- Object-oriented concepts
  - Object model of computation and use case analysis
- Java concepts
  - Compiler, interpreter, virtual machine, bytecode, Unicode character set, and file naming conventions

INTRODUCTION

Human history is punctuated by inventions of various useful machines. Machines either make our day-to-day life more comfortable or enable us to perform some new tasks. Invention of the air conditioner has greatly improved our comfort level during the hot summer months, and the invention of the television added a whole new dimension to our experience.

Along came computer, another human-made machine. Computers revolutionized the world well beyond human imagination. Computers are becoming smarter and smaller day-by-day. A computer can be used to perform many different tasks:

1. To write a poem using your favorite word processor
2. To send an instant message to your friends
3. To show your creativity on a web page
4. To listen to the music

You may be well aware of the fact that each of the above tasks is accomplished with the help of different *software*. Software is created using *programming languages*. Java is one of the latest programming languages. Since Java is a recently developed language, its design is guided by wisdom, insights, and experience gained over past half a century of software development. In particular, Java is designed as a programming language that naturally promotes object-oriented software development. This text teaches you Java programming in an object-oriented way.

Self-Check

1. Name a popular software for word processing.
2. Name a programming language other than Java.

OBJECT MODEL OF COMPUTATION

Object-oriented analysis and design is the latest paradigm for software development. Therefore, it is imperative that you have a good understanding of some of the fundamental concepts involved. In this section, we look at the world we live in with a new perspective, the object-oriented way.

You may be surprised to hear me say that objects are everywhere. In fact, we live in an object-oriented world! Consider a very simple situation. You are entering your home at night after a long day at the library. Due to the darkness, you may turn on a lamp switch. In this situation, the lamp in your room is an *object*. The *switch on* is an *operation* of the lamp. After a while, when you are ready to retire for the day, you switch off the lamp. Again, *switch off* is another operation of the lamp.

Your lamp is an object. We have identified at least two operations: turn on the switch and turn off the switch. The *formal specification* or the *template* of an object is called a *class*. Thus, lamp is a class. Your lamp is just an *instance* of the lamp class.

Example 1.1

The DVD player you own is an object or an instance of the DVD player class. The DVD player I own is another instance of the DVD player class. The DVD player owned by Mr. Smith is yet another instance of the DVD player class. Thus, each individual DVD player is an instance of the DVD player class. Observe that all DVD players have a set of operations such as play, next, previous, pause, and eject.

Example 1.2

The concept of an object is so encompassing even your dog cannot escape from it! Sorry! Your dog, Mr. Boomer, is an object. Mr. Boomer is an instance of the dog class. Your apartment is an object and is an instance of the apartment class. This book is an object and it is an instance of the book class. The chair you sit on is an object. The notebook you are currently using to take notes is an object. Thus, everything in this world is an object.

Observation 1.1

Objects are everywhere and everything is an object.

All objects of a class have identical *behavior*. For example, all lamps behave in the same way. If you turn on the switch, it lights up, and if you turn off the switch, it no longer provides any light. In other words, every instance of the lamp class has at least two operations: `switch on` and `switch off`. Thus, every lamp behaves identically and all DVD players have identical behavior. This leads to Observation 1.2.

Observation 1.2

Behavior of all objects of a class is identical.

Note that to use your lamp, you need not know any internal details. No prior knowledge of electricity, physics, or electrical engineering background is required to use the lamp. In fact all you need to know is what are the available operations and how will the object behave under those operations. For example, as long as you understand the behavior of the lamp object under switch on and switch off operations, you can use your lamp very efficiently.

Example 1.3

Consider the following two features of a DVD player.

To use your DVD player,

- You need not know how it is built, what are its internal components, and what is the underlying technology.
- All you need to know is the behavior of the DVD under each of its operations.

These two features are not specific to DVD players. Suppose you have a checking account in Great Friendly Bank. Any time you want to know your current balance, you can request that service. Your bank will promptly comply with your request. As far as you are concerned, you need not know how the bank keeps track of data relevant to your account such as account balance, account number, name, address, and so on. In fact, you do not have to know anything about the internal workings of a bank to carry out any transaction. However, you need to know the valid operations that can be performed. In the case of checking account, some of the valid operations are depositing an amount, withdrawing an amount, and checking the current balance.

Observation 1.3 (Encapsulation)

An object has data along with a set of valid operations.

Observation 1.4 (Information Hiding)

An object keeps its own data. The data maintained by the object is hidden from the user. To use an object all that the user needs to know is the specification of its operations.

Example 1.4

Have you ever thought of yourself as an object? You are an object. You are an instance of the student class. By the same token, I am an object of the professor class.

You being a student, I know, you can perform the following operations:

- Attend lectures
- Submit assignments
- Take examinations
- Ask probing questions

I can expect you to perform all of the above operations. And I being a professor, you can expect the following *services* from me:

- Deliver lectures during the allocated class time
- Explain concepts through examples or analogies
- Create and assign class assignments and examinations
- Grade your work
- Answer your questions
- Assign a letter grade for your work
- Write recommendation letters

Example 1.5

Mr. Boomer is a member of the Golden Retriever class. One of the operations of Mr. Boomer is fetch. Mr. Boomer can understand the message fetch. So if you want Mr. Boomer to fetch an item, you request the fetch service of Mr. Boomer by *sending the message* "fetch" to Mr. Boomer. However, "fly" is not an operation of Golden Retriever class. Therefore, fly is not an operation of Mr. Boomer. If you send the message "fly" to Mr. Boomer, he cannot carry out the service you requested.

In general, objects request the service of other objects through message passing. The receiver of the message can carry out the service so long as there is a corresponding operation in the class it belongs.

Observation 1.5 (Client–Server Paradigm)

Objects communicate through message passing. The object that requests a service from another object is known as the *client* and the object that provides the service is called the *server.*

There are many other important concepts in object-oriented programming. You will be introduced to those in later chapters.

Self-Check

3. In the case of a teacher–student relationship, teacher is a _________ and student is a _________.
4. The book you are currently reading is an _________.

TABLE 1.1 Binary Representations of Numbers 0 through 31

0	1	10	11	100	101	110	111
1000	1001	1010	1011	1100	1101	1110	1111
10000	10001	10010	10011	10100	10101	10110	10111
11000	11001	11010	11011	11100	11101	11110	11111

DATA REPRESENTATION

Data in a computer is represented through physical quantities. One of the options is to represent data by means of the electrical potential difference across a transistor. The presence of the potential difference can be treated as 1 and the absence of the potential difference can be treated as 0. The smallest unit of data that can be represented is a *binary digit* or *bit*, which has just two possible values: 0 and 1. The term binary means having two values. It is customary to represent these two values using the symbols 0 and 1. Every data value is represented as a sequence of these two values.

Table 1.1 summarizes the first 32 binary numbers. First row contains numbers 0 through 7, the second row contains numbers 8 through 15, and so on.

We could perform a similar encoding for characters. The American Standard Code for Information Interchange (ASCII code) is one of the widely used coding schemes for characters. The ASCII character set consists of 128 characters numbered 0 through 127. In the ASCII character set, the first character is at position 0 and is represented using 8 bits (or a *byte*) as `00000000`. The second character is at position 1 and is represented as `00000001`. The character `A` is at position 65 and thus `A` is represented as `01000001`. The character `B` is at position 66 and thus `B` is represented as `01000010`. The character `a` is at position 97 and the character `b` is at position 98. Every ASCII character has a character code and this in turn determines an ordering of characters. For example, character `B` is smaller than character `C`. Similarly, character `F` is smaller than character `f`. The ordering of characters in a character set based on their character code is called the *collating sequence*. A complete list of ASCII character set can be found in Appendix B.

Unicode is the character set used by Java. Unicode consists of 65,536 characters and uses 2 bytes to represent a character. The first 128 characters of Unicode and ASCII are identical. Thus, Unicode is a superset of ASCII. This is all you need to know about Unicode to learn Java programming language. You can get more information on Unicode characters at www.unicode.org.

Self-Check

5. The binary representation of number 32 is ________.
6. The character at position 100 in ASCII character set is ________.

HARDWARE OVERVIEW

Your programs and data reside in secondary storage such as hard disk, compact disk, or flash memory. These are permanent storage devices. The data stored in permanent storage devices will not be lost even if there is no power supply. However, information kept in memory is lost once the power is turned off. The memory is much faster compared to secondary storage devices. Therefore, the program as well as the data is first placed (or *loaded*) in the memory as a prelude to execution.

The central processing unit (CPU) is responsible for carrying out various operations and it communicates with the memory for instructions and data. Note that all arithmetic calculations and logical decisions are made by the CPU. In fact, CPU contains a subcomponent called arithmetic and logical unit (ALU) to carry out all arithmetic and logical operations. To perform an ALU operation, the operands have to be present in the memory. To carry out an ALU operation, CPU *fetches* operands from the memory. Once the operation is performed, the result is *stored* in the memory.

The *memory* is connected to the CPU through a bus. The memory can be visualized as an ordered sequence of cells, called *memory words*. Each word has a unique number called the *memory address*. The CPU uses these addresses to fetch and store data. At this point all you need to know about memory is that each word can keep only one data at any time. As soon as CPU stores a new data at a memory word, the old data is lost forever.

Introspection

Each memory cell is like a pigeonhole. At any time only one pigeon can occupy a pigeonhole.

Self-Check

7. The CPU uses memory address to ________ and ________ data from the memory.
8. All arithmetic operations are carried out in the ________ of the computer.

BRIEF HISTORY OF PROGRAMMING LANGUAGES AND MODELS OF COMPUTATION

Every computer has a *machine language*. If you want to program in machine language, you need to write every *instruction* as a sequence of 0s and 1s. In fact early programmers used machine language to program their computers. Thus, their model of computation directly reflected the machine's organization itself. However, this approach had many drawbacks. First, the process of writing those programs was difficult. Further, once a program is written, it was quite difficult and time-consuming to understand or modify the program. Even though every computer has instructions to perform operations such as addition, multiplication, and so on, the system architects of the computer may choose binary codes of their choice to perform those operations. Thus, machine languages of any two machines are not identical. Programs written for one machine will not run on another machine. In other words, you need to rewrite your program for each and every machine.

By mid-1950s symbolic languages came to the rescue of the machine language programmer. Symbolic languages were developed to make the programmer's job easier. The model of computation no longer mirrors the machine. Rather, a program is conceived as a sequence of simple operations such as addition, multiplication, subtraction, division, comparison, and so on. It is much easier to write instructions in symbolic language. Above all, it is much easier to maintain a symbolic language program than a machine language program. However, a computer can understand its machine language only. Therefore, a computer cannot execute symbolic language programs directly. The symbolic language instructions need to be translated into the machine language. A program called *assembler* is written to translate a symbolic language program into a machine language program. The symbolic language of a machine is also known as its *assembly language.*

Assembler. The software that translates an assembly language program into equivalent machine language program.

Assembly language. The symbolic language of a machine.

The advent of assembly language is a major leap in the history of programming languages. However, the programmer was forced to know the inner details and working of the machine, and also to assign memory locations for the data and manipulate them directly. Then came the *high-level languages* such as Basic, FORTRAN, Pascal, C, and C++. The model of computation became that of a series of tasks that had to be carried out one after another. With the arrival of high-level languages came the need for another translator program that can convert a high-level language program such as one written in C++ to corresponding symbolic language program, and then ultimately translate it into the machine language program. Such a program is known as a *compiler.* Thus, each high-level language required a compiler. However, for the compiler to do translation, your source code must obey all the grammatical rules of the high-level language. Compiler is a very strict grammarian and does not allow even a very simple mistake in your program, which can be as simple as a colon or a period in place of a semicolon. All such grammatical errors are known as *syntax errors.* Thus, a compiler checks for correctness of syntax, gives helpful hints on syntax errors, and translates a syntax error–free high-level language program into the equivalent machine language program.

Compiler. The software that translates a high-level language program into an equivalent machine language program.

An alternate approach was to translate each line of a high-level language program into a machine language and execute it immediately. In this case, the translator program is called an *interpreter.* Thus, the main difference between a compiler and an interpreter is that in the case of a compiler, a high-level program is translated into an equivalent machine language program only once. The machine language program is stored in a file and consequently can be executed any number of times without compiling. However, an interpreter does not keep a machine language equivalent of a source program in a file and as such interpretation takes place each time you execute the program. The programming language LISP used an interpreter.

A program written in a high-level language is known as a *source program* or *source code.* The file containing a source program is called a *source file.* A *source file* is a plain text file and can be created using any text editor that can create a plain text file. With the introduction of compilers and interpreters, a source program no longer needs to be written for a specific machine. For example, once you write a certain program in C++, all that is required to execute it in different machines is to compile the source program on the machine of your choice. This seemed like a perfect solution.

Once Internet became popular, it was necessary to execute the same program on different machines without compiling on each one of those machines. This need resulted in the introduction of the concept of a *virtual machine*. Designers of Java introduced a common symbolic language called *bytecode*. The virtual machine that can understand the bytecode is termed as *Java virtual machine (JVM)*. Thus, it became possible to compile a Java source program into *bytecode* and execute it on any machine that has JVM without compiling again. JVM performs the enormous task of interpreting the bytecode into equivalent machine language instruction. Thus, Java language introduced the concept of *platform independence* to the world of computing.

Self-Check

9. The Java program you write is an example of_________ code.
10. A Java compiler translates a source program into its equivalent _________.

CREATING AND EXECUTING JAVA PROGRAM

There are two types of Java programs: applications and applets. In this book you will be introduced to application programs first. We shall introduce applets in Chapter 8. In this section, we outline the steps involved in creating and executing a Java application or an applet.

Step 1. Create Java Source File

Every Java program, whether it is an application or an applet, is a Java class that needs to be created. You will see your first Java class in Chapter 2. You can use any text editor to type in a Java program. However, do not use any word processor. Once you have typed in your program, you must save your file as a `.java` file. That is, the extension of the file must be `.java`. Further, the name of the file should be the same as the name of the class you have created. For example, a file containing a Java class named `HiThere` must be saved as `HiThere.java`.

Common Programming Error 1.1

Giving a file name different from the class name is a common error.

Common Programming Error 1.2

Saving a file with extension other than `.java` is a common error.

Step 2. Compile Source Code into Bytecode

In this step you translate your source code program into equivalent bytecode. The Java compiler supplied by the Sun Microsystems is known as `javac` (see Step 3 for more details). You may have to go back to Step 1 and correct errors before you can successfully complete this step. It is quite common to have syntax errors in your program. All you need is some patience. If no error is found, the compiler will produce the equivalent bytecode program and save it in a file with the same name but with an extension `.class`. For instance, the compiler will create the bytecode for `HiThere.java` in a new file `HiThere.class`.

Step 3. Execute Java Program

To execute a Java application, first, the corresponding `.class` file needs to be placed in main memory. However, to execute a Java applet one must use a web browser or an `appletviewer`. An `appletviewer` is a simpler version of a web browser capable of executing applets. More details on executing an applet are presented in Chapter 8.

The software supplied by Sun Microsystems to execute the Java application is known as `java`. In fact, the software you used in Step 2 to compile your Java language program also contains many programs that are useful in creating a Java program. Therefore, the software is quite often known as a software development kit (SDK). In the case of Java, Sun Microsystems calls it Java development kit (JDK). The Sun Microsystems provides three versions of JDK: enterprise edition, standard edition, and microedition. The one you need is the standard edition. The latest version, JDK 6u10 (Java development kit 6 update 10), can be obtained free at java.sun.com/javase/downloads/index.jsp. Keep in mind that website addresses change quite frequently. If the above address does not work, you can search for the correct page starting with Java home page java.sun.com. Some of the freely available integrated development environments (IDEs) for Java are Eclipse (www.eclipse.org), NetBeans (www.netbeans.org), and Jdeveloper (www.oracle.com/technology/products/jdev/index.html). IDEs provide an environment for editing, compiling, debugging, executing, documenting, and so on through a user-friendly graphical interface.

Many Java programs may need various mathematical functions. Similarly, almost all Java programs you write need to communicate to the user by receiving input and producing output. Thus, to reduce the burden of a typical programmer, Java provides many precompiled programs organized as various *libraries* also known as *packages*. Your program may be using many of those programs from various packages. A system software called *linker links* or connects the bytecode program you created in Step 2 with necessary precompiled programs from the packages.

> *Linker.* A system software that links a user's bytecode program with necessary other precompiled programs to create a complete executable bytecode program.

Once a complete executable bytecode program is created, a system software called *loader loads* the bytecode program into memory and starts the execution of the first bytecode instruction.

> *Loader.* A system software that loads a linked bytecode program into main memory and starts the execution of the first bytecode instruction.

Once the program is executed, you can observe the result produced. If the behavior of the program is different from what you expected, there are *logical errors.* For example, if you add 2 and 3 you expect 5 as answer. Anything other than 5 is an error. This type of error is due to some logical error on the part of the programmer. Programmers call it a *bug* in the program and the process of finding and eliminating bugs from the program is known as *debugging*. If the program has a logical error, you need to go back to Step 1 and make corrections to your program using the editor.

As a programmer, the process of linking and loading is more or less hidden from you. Therefore, the above discussion can be summarized as follows.

There are three steps in creating a Java application:

Step 1. Create a Java file with `.java` extension using a text editor.

Step 2. Compile Java program created in Step 1 using the Java compiler `javac` to create the corresponding `.class` file. If there are any syntax errors, `.class` file would not be created and you need to go back to Step 1 and make corrections. Thus, steps 1 and 2 need to be repeated until there is no syntax error.

Step 3. Execute the class file created in Step 2 using `java`. If there are any logical errors, you need to go back to Step 1 and correct your program. Steps 1 through 3 need to be repeated until there are no logical errors.

Figure 1.1 illustrates the editing, compiling, and executing of a Java application program.

Self-Check

11. To create a Java class `FirstTrial`, the file must be named _________.
12. The error detected during compilation is known as _________ error.

INTRODUCTION TO SOFTWARE ENGINEERING

Software development is an engineering activity and as such has lot in common to building a house. Suppose you want to construct a house. You approach the local building construction firm with your ideas and dreams. You may start out exploring the requirements and the feasibility of the project. How much money you have and what type of house you would like to build are to be discussed with an architect of the firm. Let us call this the analysis phase. Your architect will prepare a blueprint of the house. Note that it is quite easy to make any changes on the blueprint without tearing down any wall. Once you and the architect finalize the blueprint, as a customer, you have very little to do with the actual building of the house.

The construction company makes a detailed design of your house for framers, plumbers, electricians, and so on. We call this the design phase. Next, the implementation phase results in the construction of the house. During the construction many tests are done to verify the correct functioning of various systems. Once the house is handed over to you, it enters the maintenance phase.

The software development process mirrors the above five phases:

1. Analysis
2. Design
3. Implementation
4. Testing
5. Maintenance

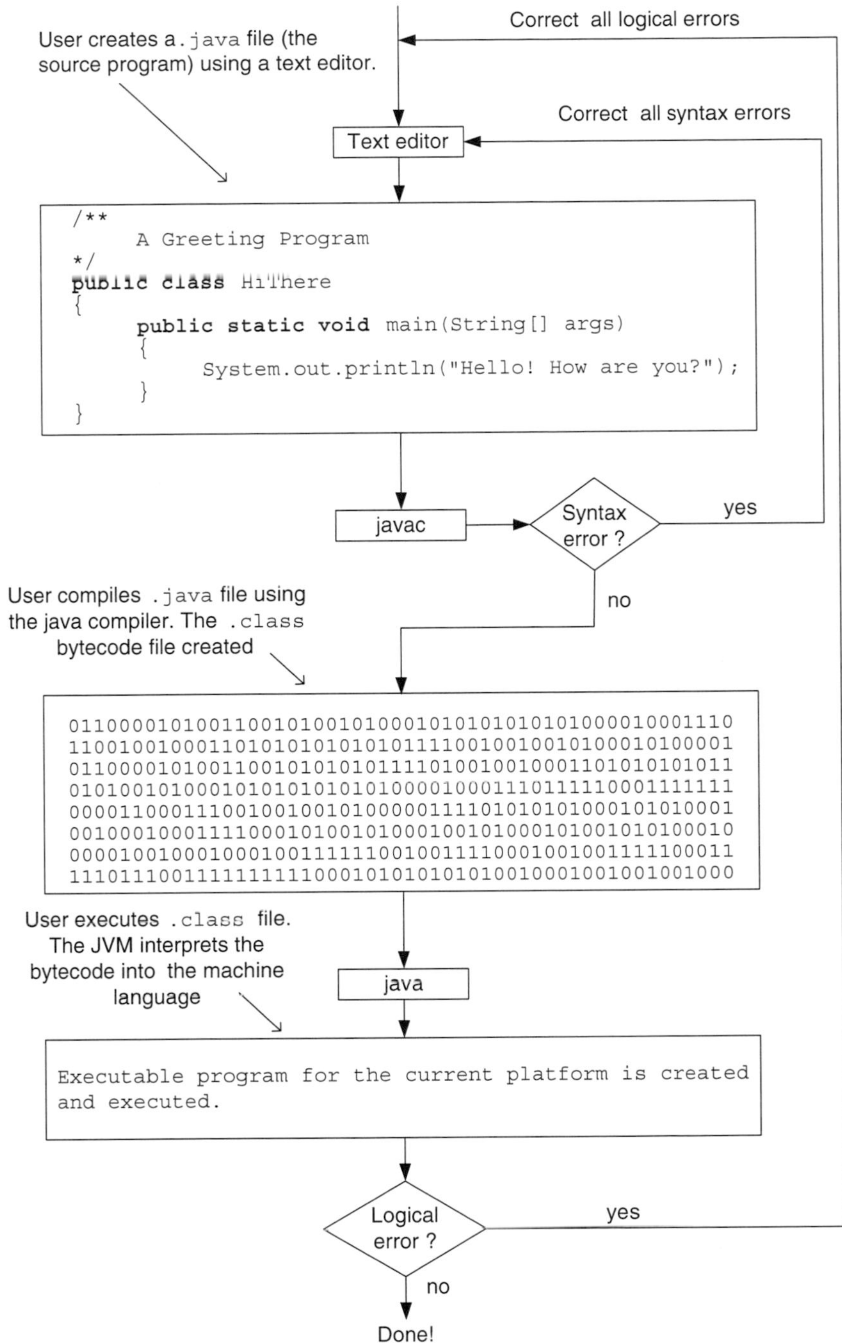

FIGURE 1.1 Editing, compiling, and executing HiThere.java.

Phases 2 through 4 involve understanding of many concepts and as such cannot be fully explained at this point. However, phase 1 can be introduced with remarkable clarity even at this point. Further, the main focus of this book is on phases 2 through 4. Therefore, it makes perfect sense to discuss phase 1 now and focus on phases 2 through 4 in the rest of this book. Therefore, we shall concentrate on the analysis phase in the rest of this chapter. Phases 2 through 4 will be explained in Chapter 3.

Analysis and Use Case Diagram

A use case diagram specifies the functional requirements of the system to be developed. Thus, a use case diagram clearly identifies the boundary of the system to be developed. Graphically, we show the boundary by drawing a rectangle. Everything inside the rectangle is a functional requirement and as such needs to be developed. However, anything shown outside the rectangle is not part of the system. As the name suggests, use case diagrams contain use cases. Each *use case*, denoted by an oval symbol, stands for a functional requirement of the system to be developed. Users of the system as well as other external systems with which the system needs to interact are called *actors*. Actors are shown using stick figures. Note that each actor stands for a group of users or external systems with identical behavior. An actor and a use case are connected by arrows starting from the initiator of the action.

We will now illustrate these concepts through a series of simple examples.

Example 1.6

Problem statement. Write a program to create a digital dice.

We are all familiar with a dice. A dice is a cube with numbers 1–6 on its six faces. You may roll the dice at any time. Each time you roll the dice, it is supposed to *show* one of the six faces.

Observe that all users belong to one category, say *user.* Thus, our software has only one actor. The user can always roll the dice. Thus, the dice has a use case *roll.* In fact roll the dice is the only functionality of a dice. The digital dice has exactly one use case and one actor. In other words, digital dice is a class with one service: *roll* (see Figure 1.2).

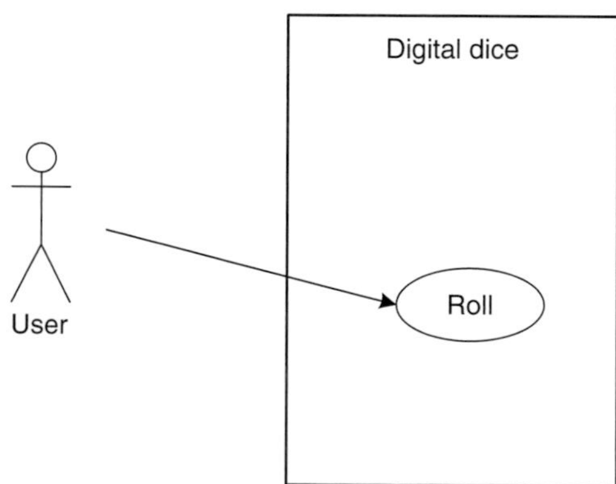

FIGURE 1.2 Use case diagram for a digital dice.

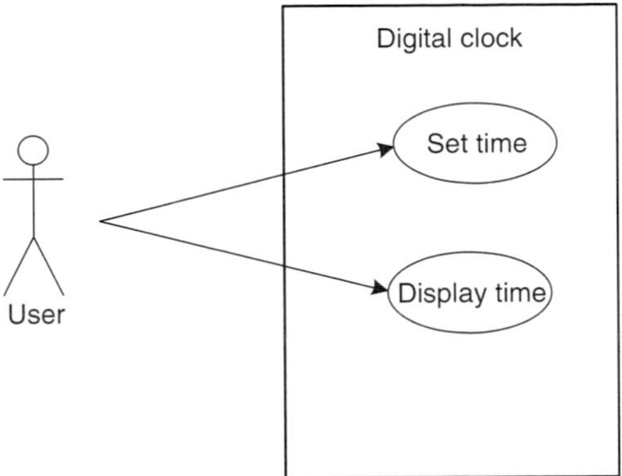

FIGURE 1.3 Use case diagram for a digital clock.

Example 1.7

Problem statement. Write a program to create a digital clock.

Every clock must have, at a minimum, the ability to set current time and display current time. Once again, all users of the clock fall under one category. Thus, we have the use case diagram shown in Figure 1.3. Observe that both arrows start from the actor to indicate the fact that it is the actor who initiates the function.

Example 1.8

Problem statement. Write a program to simulate an ATM machine.

An ATM machine should be capable of the following services:

- Deposit cash or check by a customer
- Withdraw cash by a customer
- Check balance by a customer
- Print receipt by a customer
- Display messages by the ATM
- Dispense cash by the ATM
- Maintenance by a service employee

Each of these services can be modeled as a use case. Having decided on the use cases, let us examine the possible actors. Clearly, customer is an actor. Unlike the dice or clock illustrated in the previous examples, an ATM machine cannot function by itself. It is part of a network and it must communicate to the network for verifying customer information as well as for updating the customer account balance whenever a deposit or withdrawal takes place. A bank's authorized employee is required to carry out various maintenance operations of the ATM. The use case diagram is shown in Figure 1.4.

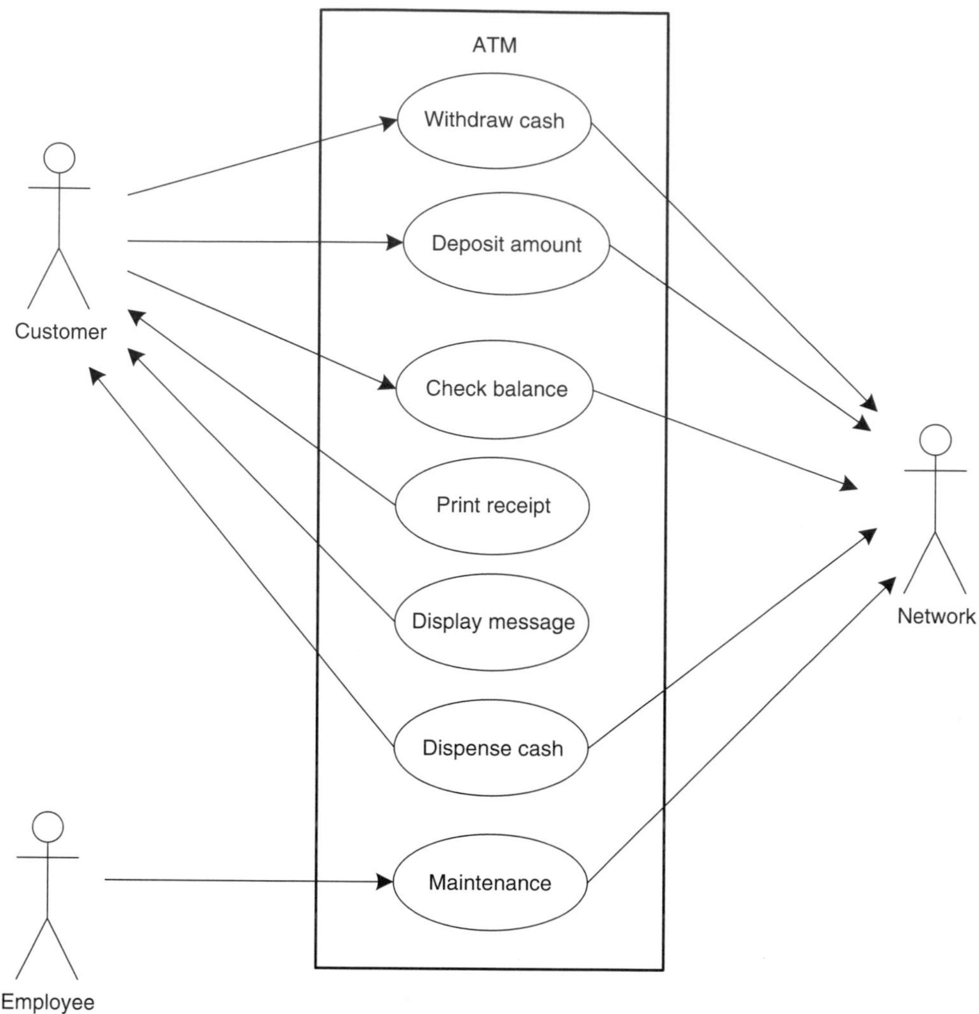

FIGURE 1.4 Use case diagram for an ATM.

Self-Check

13. If the proposed software uses an external database, in the use case diagram the database is represented as a/an __________.
14. Consider the digital clock in Example 1.7. How will you accommodate the additional functional requirement that the user must be capable of setting the alarm?

REVIEW

1. Software is created using programming languages. Java is one of the latest programming languages.
2. The formal specification or the template of an object is called a class.
3. All objects of a class have identical behavior.

4. An object has data along with a set of valid operations.
5. The data maintained by the object is hidden from the user.
6. To use an object all that the user needs to know is the specification of its operations.
7. Objects communicate through message passing.
8. The object that requests a service from another object is known as the client and the object that provides the service is called the server.
9. The smallest unit of data that can be represented is a *binary digit* or *bit*, which has just two possible values: 0 and 1. A sequence of 8 bits is known as a byte.
10. Unicode is the character set used by Java. Unicode consists of 65,536 characters and uses 2 bytes to represent a character.
11. The program as well as the data is first placed (or loaded) in the memory as a prelude to execution.
12. The central processing unit (CPU) is responsible for carrying out various operations and it communicates with memory for instructions and data.
13. The memory can be visualized as an ordered sequence of cells, called words.
14. Every computer has a machine language.
15. The symbolic language of a machine is also known as its assembly language.
16. An assembler translates a symbolic language program into a machine language program.
17. A compiler translates a high-level language program into a machine language program.
18. The virtual machine that can understand the bytecode is termed as Java virtual machine (JVM).
19. Java provides many precompiled programs organized as various libraries also known as packages.
20. The system software linker *links* a user's bytecode program with other necessary pre-compiled programs to create a complete executable bytecode program.
21. The system software loader *loads* a linked bytecode program into main memory and starts the execution.
22. A use case diagram specifies the functional requirements of the system to be developed. Thus, a use case diagram clearly identifies the boundary of the system to be developed.
23. Graphically, we show the boundary by drawing a rectangle.
24. Each use case, denoted by an oval symbol, stands for a functional requirement of the system to be developed.
25. Users of the system as well as other external systems with which the system needs to interact are called actors. Actors are shown using stick figures.
26. An actor and a use case are connected by arrows starting from the initiator of the action.

EXERCISES

1. Indicate true or false.
   a. Corresponding to a class there can be only one instance.
   b. An object can have any number of operations.
   c. When you request your friend, "could you please pass the book," you are passing a message.
   d. Assembly language is the language of 0s and 1s.
   e. The grammatical error in a program is known as syntax error.
   f. ALU is part of memory.
   g. The value of 17 in binary number system is 100001.
   h. In ASCII system, character `A` is 65.
   i. ASCII stands for American Standard Character Interchange for Information.
   j. Every high-level language needs a compiler.
   k. Each memory word has a unique memory address.
   l. It is possible to have a use case diagram with 0 use cases.
2. For each of the following objects, list at least two operations:
   a. Microwave oven
   b. Oven
   c. Thermostat
   d. Savings account
   e. Telephone
3. Explain the following terms:
   a. Class
   b. Instance
   c. Encapsulation
   d. Information hiding
   e. Client
   f. Server
   g. Interpreter
   h. Compiler
   i. Linker
   j. Loader
4. Fill in the blanks.
   a. The binary representation of number 41 is _________.
   b. The binary representation of number 45 is _________.

c. The decimal value of the binary number 100100 is _________.

d. The decimal value of the binary number 100111 is _________.

e. The collating sequence value of character `D` in ASCII set is _________.

f. The collating sequence value of character `e` in ASCII set is _________.

g. The largest integer you can represent using 6 bits is _________.

h. The largest integer you can represent using 8 bits is _________.

i. You save a file containing Java class `HelloThere` as _________.

j. When you compile a file containing Java class `HelloThere`, a new file _________ is created if there are no syntax errors.

k. Having a comma in place of a semicolon is an example of a _________.

l. _________ is the name of a CPU in the market.

5. Why do you need memory? Why cannot CPU use secondary storage instead of memory?
6. Name two system software you may use in connection with your program execution.
7. How did designers of Java achieve platform independence?
8. What are the advantages of using a high-level programming language over assembly language?
9. List one functional requirement and one nonfunctional requirement for each of the objects listed in Exercise 2.
10. Draw a use case diagram for a proposed software development project for course enrollment. Students, faculties, and administrators need to access the system. Students can enroll or drop a course, and can also view their grades. Faculties can view their course assignments as well as assign grades to students in their course. Administrator can add a new course or remove a new course. All the relevant information is kept in a database external to the system.
11. Draw a use case diagram for a proposed software development project for online shopping. There are two types of users of the system: customers and administrator. A customer can order an item or search for an item. They can also rate an item and post comments. The administrator is responsible for maintaining the system. All the relevant information is kept in a database external to the system.
12. Draw a use case diagram for a proposed software development project for library checkout system. There are two types of users for the system: patrons and staff. All the relevant information is kept in a database external to the system.

ANSWERS TO SELF-CHECK

1. Microsoft (MS) Word
2. C++
3. server, client

4. object
5. 100000
6. `d`
7. store, fetch
8. ALU
9. source
10. bytecode
11. `FirstTrial.java`
12. syntax
13. actor
14. Include a use case “set alarm” and draw an arrow from the user to it.

CHAPTER 2

Class and Java Fundamentals

In this chapter you learn

- Object-oriented concepts
  - Class, object, instance, attribute, operation, method
- Java concepts
  - Named constants, variables, primitive data types, and operators, `String` class, reference variables, literals, assignment statements, `String` operators, operator precedence, evaluation of mixed expressions, input/output statements
- Programming skills
  - Create, compile, execute, and test an application program

Today you went to your bank and deposited $1000.00 to your savings account. Your savings account is an *object* that keeps track of many closely related data items such as your account number, name, address, phone number, e-mail address, current balance, and so on. If I have a savings account in your bank, my savings account is another object that will keep track of many closely related data items such as my account number, name, address, phone number, e-mail address, current balance, and so on. In fact, anyone who has a savings account with your bank will have a similar savings account object. Recall that the *formal specification* or the *template* of an object is called a *class*. Thus savings account is a class. Your savings account is just an *instance* of the savings account class. Instances are called objects and two objects are identical except for the values.

The *behavior* of all objects of a class is identical. Depositing $1000.00 will result in adding $1000.00 to the current balance. Withdrawing $200.00 in turn results in subtracting $200.00 from the current balance. Thus, a class is a collection of objects having identical behavior.

The object-oriented analysis and design is the latest methodology for software development. To understand what is meant by object-oriented, you need to understand many concepts. In this chapter you have been introduced to the following four concepts: class, object, instance, and behavior. Examples 2.1 through 2.3 will further illustrate concepts introduced so far and introduce two new concepts: operation and attribute.

Example 2.1

All DVD players have certain identical behaviors. They all have a play button, a pause button, and a stop button. In fact, you expect every DVD player to have at least these three buttons. Thus, DVD players form a class. Play, pause, and stop are *operations* of the class DVD player. The DVD player you have is an object of the DVD player class. If Mr. Jones has a DVD player, then that specific DVD player is another object or instance of the DVD player class.

Example 2.2

What are the common behaviors of clocks? You expect every clock to "display the current time." There has to be an operation to "set the current time." Thus, clock is a class with at least two operations: display the current time and set the current time. The Big Ben tower in London, the watch you wear every day, the clock that is part of your DVD player or your car, and the system clock of your computer are all various instances or objects of the clock class.

Example 2.3

Every clock needs to keep the current time. The current time is in turn made up of hours, minutes, and seconds. Thus, a clock needs three *attributes*: hours, minutes, and seconds. In a mechanical clock such as Big Ben, these data values are kept through mechanical components. However, in a digital clock similar to the one in your car, these values are kept in three memory locations.

Self-Check

1. Consider the class student. Suggest three attributes and at least one operation for this class.
2. Consider the class rectangle. Suggest at least two attributes and two operations for this class.
3. Consider the class circle. Suggest attribute(s) and operation(s) for this class.
4. Consider the class rocket. Suggest at least two attribute(s) and two operation(s) for this class.

JAVA APPLICATION PROGRAM

Every programming language before Java was designed to create stand-alone programs. Java, however, has two categories of programs. Apart from the traditional stand-alone programs, Java introduced applets. Applets are small applications intended to run on a web browser. In this book, application programs are introduced first. All the Java language fundamentals, good programming skills, and software engineering principles are taught through application programs. Java applets are discussed in Chapter 8.

A Java program is a collection of one or more classes. So a very simple application program must have at least one class. The basic syntax of class can be shown using a syntax template as follows:

```
[accessModifiers] class className [classModifiers]
{
    [members]
}
```

An item inside a pair of square brackets in a syntax template is optional. In the syntax template, `className` is a name you would like to give to the class. The optional `members` consist of attributes and operations of the class. The optional `accessModifiers` determines, among other things, the accessibility of the class to other classes and Java virtual machine (JVM). For example, the access modifier `public` makes a class accessible to JVM. Because JVM needs access to your program, every application must have one class with `public` or `package` access. We will have more to say about this topic later. Both `public` and `class` are reserved words or keywords. A reserved word is a word with special meaning. For example, the reserved word class indicates the beginning of a class definition.

In a syntax template, all reserved words are shown in bold. The optional `classModifiers` are not used in the first six chapters of this book and it will be explained in Chapter 7. Following the reserved word class, you must supply a name for the class. Note that the name *identifies* a class. Similarly, you need to provide a name to attributes and operations of a class. All such names are collectively known as *identifiers.*

Self-Check

5. In a syntax template, how is an optional item shown?
6. In a syntax template ________ is shown in bold.

Identifier

A Java identifier can be of any length. However, it must obey the following three rules:

1. Every character in an identifier is a letter (A to Z, a to z), or a digit (0 to 9), or the underscore character (_), or the dollar sign ($).
2. The first character of an identifier must be a letter or the underscore character or the dollar sign.
3. An identifier cannot be a reserved word.

Common Programming Error 2.1

Using a space or blank character as part of an identifier is an error.

Common Programming Error 2.2

Using a hyphen as part of an identifier is an error.

Note 2.1 Java is a case-sensitive language. In other words, uppercase letter A is not the same as lowercase letter a. Thus, `Welcome`, `welcome`, `WelCome`, and `WELCOME` are four different identifiers.

Good Programming Practice 2.1

Create meaningful identifiers. An identifier must be self-explanatory.

Good Programming Practice 2.2

Java system has many predefined identifiers. Most frequently encountered identifiers are `System`, `print`, `println`, `main`, and `next`. Even though predefined identifiers can be redefined, it is wise to refrain from it.

Example 2.4

The following identifiers are legal:

```
Carrot
Tiger
bonusSalary
number_of_lines_per_page
_0              //legal but bad choice
$1              //legal but bad choice
```

Even though identifiers `_0` and `$1` are legal, they do not convey the purpose or the meaning of the identifier and as such should be avoided.

Example 2.5

Table 2.1 provides a list of five illegal identifiers.

It is a well-accepted convention to keep an identifier used to name a class in all letters and the first letter of each word in uppercase. Thus `HiThere`, `FirstJavaProgram`, and `DigitalClock` obey the convention followed by class names.

TABLE 2.1 Illegal Identifiers

Identifier	Explanation
`Yearly bonus`	The space character cannot appear in an identifier
`British#Sterling`	The pound sign cannot appear in an identifier
`Jim&Mary`	The symbol & cannot be used in an identifier
`9thValue`	A digit cannot be the first character of an identifier
`public`	Reserved word cannot be used as an identifier

Note that a class need not have any members. Therefore, you can create a class named HiThere as follows:

```
public class HiThere
{
}
```

Note 2.2 You must always keep the opening and closing pair of braces ({ and }) in the class definition.

Note 2.3 As a matter of style, we always align a pair of braces vertically and any Java statements appearing inside a pair of braces will be indented.

Once you have created this class, you must save it in a file HiThere.java. If you compile this class, Java will create a new file HiThere.class without any error messages. However, if you try to execute HiThere.class, Java will give you an error message similar to the following one:

```
Exception in thread "main" java.lang.NoSuchMethodError: main
```

In other words, to execute an application, there needs to be a *method* named main. The term *method* stands for Java code that implements an operation of the class.

The basic syntax of the method main is

```
public static void main(String[] args)
{
    [statements]
}
```

You need not be concerned about the details of the syntax at this point. Instead, note that statements are optional. The simplest method main one could write is

```
public static void main(String[] args)
{
}
```

We have the following simple application program:

```
public class HiThere
{
    public static void main(String[] args)
    {
    }
}
```

You may compile and execute this program without any error. However, this program produces no output at all. All application programs in this chapter will be created out of the above template.

So far you have encountered identifiers, reserved words, methods, classes, and a pair of opening and closing braces. You have also mastered the rules for forming identifiers. Let us now have a closer look at reserved words.

Self-Check

7. List four distinct identifiers that spell the word "fine."
8. Why using the identifier `abc123` a bad choice?

Reserved Word

You have already seen the reserved words `public`, `static`, `void`, and `class`. In Java, all *reserved words* are made up of lowercase letters. Each reserved word is treated as a single symbol and must be typed exactly as shown. For example, there are two reserved words `throw` and `throws`. Both have different intended usage and the reserved word `throw` is not the plural of the reserved word `throws`. A complete list of reserved words can be found in Appendix C. Recall, from the rules of identifiers, that you cannot create an identifier that matches a reserved word. *Reserved words* are also known as *keywords*.

Note 2.4 In this book, all reserved words appearing in the code are highlighted in bold. Many Java program editor software display the keywords in a different color. This feature is known as *syntax coloring*.

Self-Check

9. Is Class a reserved word in Java?
10. True or false: In Java, all reserved words are made up of lowercase letters.

Comment Lines

One of the most important qualities of a good program is its readability. Programmers are notorious for creating programs that are very difficult to understand not only by others, but also by themselves after a short period of time. Since software maintenance is quite costly, importance of documentation in software creation cannot be overemphasized.

As a programmer, you can place comment lines in your programs to fully explain the purpose as well as the logic of each of the important steps. Java provides two types of comment lines: single line format and multiple lines format. A *Single line comment* starts with // and can begin at any place in a line. The compiler ignores everything that appears after two consecutive / characters. A *Multiple line comment* starts with /* and ends with */. Everything that appears between /* and */ is ignored by the compiler.

Example 2.6

In this example, we illustrate both single line and multiple line comments:

```
//  This is a single line comment
```

and is equivalent to the following:

```
/*  This is a single line comment */
```

Similarly, six lines of comment

```
//  This is a comment. Next five lines are also comments
//  Line 1
//  Line 2
//  Line 3
//  Line 4
//  Line 5
```

is equivalent to the following:

```
/*  This is a comment. Next five lines are also comments
    Line 1
    Line 2
    Line 3
    Line 4
    Line 5
*/
```

Note 2.5 Every Java program is a collection of one or more classes and every class has one or more methods. Since you have not seen many classes or methods, the material covered in this note may not be completely clear. From a pedagogical perspective, it is better to develop the habit of including comments in every program. In Java, there is a standard for documenting classes and methods. If you follow these guidelines, you can use a Java utility called javadoc to generate documentation in HTML format from the comments in the source code. The documentation comment is placed before the class or method definition and it starts with /** and ends with */. The javadoc utility copies the first line of each comment into a table. Therefore, compose your first lines in each comment with some care. Here is a sample comment you may place before a class:

```
/**
    A student class tracks gpa, major and advisor
*/
```

The additional conventions you need to follow in the case of methods will be discussed in Chapter 3. Comments in this book will follow conventions consistent with javadoc.

Self-Check

11. True or false: In Java, every comment starts with /*.
12. True or false: Including comments in a program is optional.

JAVA GREETINGS PROGRAM

In this section, you will learn how to display a welcome message on the monitor. Suppose you want to display a message such as

```
Hello! How are you?
```

All we need to do is add the following statement inside the method main:

```
System.out.println(("Hello! How are you?");
```

Thus, we have the following Java application program:

```
/**
    A Greeting Program
*/
public class HiThere
{
    public static void main(String[] args)
    {

        System.out.println("Hello! How are you?");
    }
}
```

Since the program has changed, you must first save and compile the file `HiThere.java`. If you execute this program, it will display the following line on the monitor:

```
Hello! How are you?
```

Note that double quote characters appearing before the character H and after the character ? are not displayed on the monitor. In Java, a sequence of zero or more characters enclosed within a pair of double quotes is a `String literal` or `String` for short; and the `println` operation will display it on the monitor. A `String` containing zero characters is called an *empty*. Thus, `""` is an empty `String`. The empty `String` is different from a *null* `String`. The `null` is a keyword in Java and it signifies the absence of the object. Thus in the case of a `String`, the null signifies the absence of a `String` object.

Example 2.7

String Literal

String	Explanation
`null`	Absence of a `String` object
`""`	Empty `String` or a `String` of length 0
`" "`	A `String` containing a single space character
`"P"`	A `String` containing single character `P`
`"Peter"`	A `String` containing 5 characters
`"Peter, Paul and Mary"`	A `String` containing 20 characters

Example 2.8

Consider the Java statement:

```
System.out.println("Hello! How are you?");
```

The above statement can be replaced by the following two statements:

```
System.out.print("Hello! ");
System.out.println("How are you?");
```

Further, the following five statements will also produce the same output:

```
System.out.print("Hello! ");
System.out.print("How ");
System.out.print("are");
System.out.print(" you?");
System.out.println();
```

Observe that the statement

```
System.out.println(); // statement 1
```

is used to position the cursor at the beginning of the next line.

Common Programming Error 2.3

Omitting the pair of parentheses in statement 1 above is a syntax error.

You can include extra characters known as escape sequences in a `String` to control the output and to print special characters. An escape sequence in Java consists of two characters and starts with the backslash character. For example, the escape sequence `\n` can be included in the `String` to move the cursor to the beginning of the next line.

Self-Check

13. True or false: A `String` in Java is enclosed within a pair of single quotes.
14. Write the Java statement that outputs the message `All is well in the eastern border!`

Advanced Topic 2.1: Frequently Used Escape Sequences

Table 2.2 summarizes some of the most frequently used escape sequences.

TABLE 2.2 Frequently Used Escape Sequences

Purpose	Escape Sequence	Name
Move cursor to the beginning of the current line	`\r`	Return
Move cursor to the beginning of the next line	`\n`	Newline
Move cursor to the beginning of the next line of next page	`\f`	Form feed
Move cursor to the next tab stop	`\t`	Tab
Move cursor one space to the left	`\b`	Backspace
Print backslash character	`\\`	Backslash
Print single quotation character	`\'`	Single quotation
Print double quotation character	`\"`	Double quotation

Example 2.9

Consider the following Java application to illustrate various escape sequences introduced in this section:

```
/**
    Demonstration of Escape Sequences
*/
public class EscapeSequence
{
    public static void main(String[] args)
    {
        System.out.println("A return\r<- character");
        System.out.println("A newline\n<- character");
        System.out.println("A tab stop\t<- character");
        System.out.println("A backspace\b<- character");
        System.out.println("A backslash\\<- character");
        System.out.println("A single quotation\'<- character");
        System.out.println("A double quotation\"<- character");
    }
}
```

Output

```
<- character
A newline
<- character
A tab stop <- character
A backspac <- character
A backslash\<- character
A single quotation'<- character
A double quotation"<- character
```

Observe that the two character sequence <- (less than followed by a negative sign) is used to indicate the position of the cursor after the escape sequence. Note that the return character placed the cursor at the beginning of the current line, and consequently, all we can observe is a space followed by the word character. Compare the effect of return character with the newline character. In particular, in the case of the newline character, the cursor moved to the next line. Clearly the tab stop character inserted extra space. Notice that due to the backspace character, the letter e at the end of the word backspace is replaced by the next character in the String.

Advanced Topic 2.2: Details on println Method

Consider the Java statement:

```
System.out.println("Hello! How are you?");
```

The System class has an attribute out that in turn is an instance of the PrintStream class. The PrintStream class has an operation println that takes String as an *argument* or *parameter*. The println operation of the PrintStream class is used to print the argument and to move the cursor to the beginning of the next line.

Therefore, the Java statement

```
System.out.println("Hello! How are you?");
```

prints the String "Hello! How are you?" and moves the cursor to the beginning of the next line. If you do not want to move the cursor to the beginning of the next line, you could use the operation print of the PrintStream class.

Concatenation and the length of Strings

The concatenation is a binary operation that appends the second String after the first one. Java uses + as the concatenation operator.

Example 2.10

The following Java application illustrates the concatenation operator. Note that the space between a String and the operator + has no impact on the String produced. However, space between a String and the operator + improves the readability.

```
/**
    Demonstration of String concatenation
*/
```

```
public class ConcatOperator
{
    public static void main(String[] args)
    {
        System.out.println("\n\n\tHello");
        System.out.println("This line concatenates"+" two
                                                   Strings");
        System.out.println("This " + "line " + "concatenates"
                                                 + " 4 Strings");
        System.out.println("\n\n\n\tBye Now");
    }
}
```

Output

```
 Hello
This line concatenates two Strings
This line concatenates 4 Strings

 Bye Now
```

Good Programming Practice 2.3

Always leave space before and after an operator such as + for better readability.

The concatenation operator + can also be used to join a `String` and a numeric value or a character. In that case, Java first converts the numerical value or character to a `String` and then uses the concatenation operator +. We will discuss this topic in detail, after introducing other data types, in Advanced Topic 2.6.

Self-Check

15. Concatenating `"North"` and `"America"` will produce the `String` ______.
16. Concatenating `"  South"` and `"Africa"` will produce the `String` ______.

Positional Value of a Character in `String`

In a `String`, every character has a positional value. The positional value of the first character in a `String` is 0. The positional value of the second character is 1, and so on. The number of characters in a `String` is called the length of the `String`. A `String` of length 10 has 10 characters and they are at positions 0 through 9.

For example, consider the `String`

```
"God bless America"
```

The characters and their positional values in the `String` can be summarized as follows. In Java, a character literal is enclosed within a pair of single quotations.

`'G'` at 0	`'o'` at 1	`'d'` at 2	`' '` at 3	`'b'` at 4
`'l'` at 5	`'e'` at 6	`'s'` at 7	`'s'` at 8	`' '` at 9
`'A'` at 10	`'m'` at 11	`'e'` at 12	`'r'` at 13	`'i'` at 14
`'c'` at 15	`'a'` at 16			

The `String` is of length 17. It is worth noticing that in determining the length of a `String`, spaces are included. In fact, within a `String` the space character is treated just like any other character.

Note 2.6 In Java, `"A"` and `'A'` are different. While `"A"` is a `String` of length 1, `'A'` is a character.

Self-Check

17. Consider the `String` "Sea to shinning sea!" What is the positional value of the character !
18. What is the length of the `String.` "Sea to shinning sea!"

DATA TYPES

A computer has to process various types of data. One such data type is a sequence of characters called a `String`, which you have already encountered. Another data type you have encountered is character. You may use a `String` to manipulate names, addresses, book titles, and so on. But a computer, as its name suggests, is a computing or calculating machine as well. Today computers are used in a wide variety of applications. Therefore, the data it needs to process also varies widely. Observe that the set of possible values associated with a person's name is not the same as the set of values associated with a person's salary. Further, there is no need to multiply or divide names or addresses. However, various numeric calculations are required to determine the tax to be withheld. In other words, the set of operations that can be performed on a `String` is not the same as the set of operations that can be performed on a number. Based on these facts, high-level programming languages classify data into different sets called data types. The designers of Java incorporated the above feature as well. In fact, Java is a *strictly typed language*. In other words, if you try to multiply two `String`s or multiply a `String` by a number and so forth, the built-in error checking mechanisms will alert the user during the compilation time.

Data type is a set of values along with a set of meaningful basic operations on those data values.

Introspection

Can you use apples in place of oranges or vice versa? How about using oranges to bake an apple pie?

The data type determines how the data is represented in the computer. Data types in Java can broadly be classified into two categories: primitive data types and reference types.

Self-Check

19. The name of a company is of the type _______.
20. The legal operations that can be performed are determined by their _______.

Primitive Data Types

Many types of data are used quite frequently by almost all Java programs. Therefore, Java provides them as built-in data types. Java has eight built-in data types that are implemented not as objects. We call them primitive data types. They are `boolean`, `char`, `byte`, `short`, `int`, `long`, `float`, and `double` (Figure 2.1).

The eight primitive data types of Java fall under three disjoint categories:

1. *Boolean type.* There is only one primitive data type that falls under this category, `boolean`. This data type deals with true or false values.
2. *Integral* or *integer type.* There are five primitive data types under this category: `byte`, `char`, `short`, `int`, and `long`. This data type deals with integers or whole numbers.
3. *Floating point type.* There are two primitive data types under this category: `float` and `double`. This data type deals with values that are not necessarily whole numbers. In other words, numbers can have decimal part.

There are five integral data types and two floating point data types. Java has its roots in C++. The programming language C++ in fact is an enhancement of the programming language C. Thus, Java came to support many data types that are most commonly used in C and C++. The programming language C was designed to replace assembly language

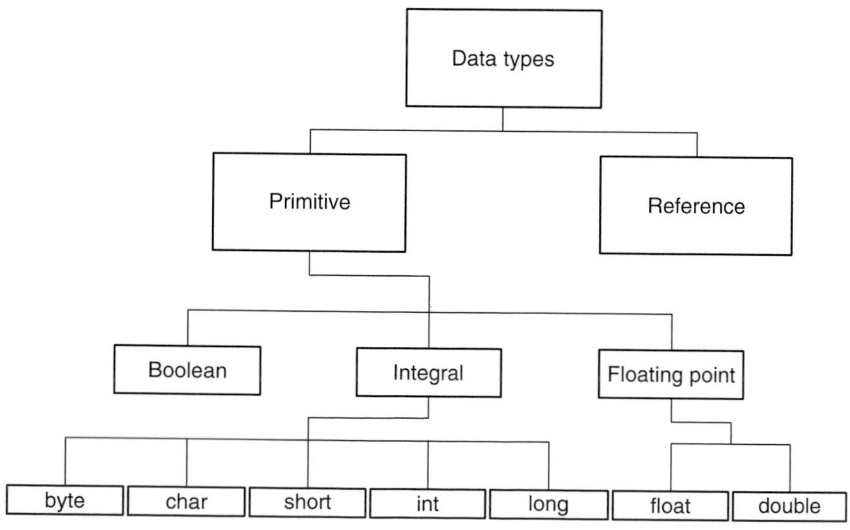

FIGURE 2.1 Java data types.

programming and as such provided mechanisms for efficient storage and manipulation of data. Thus, Java also comes with the capability to write memory-efficient programs. For example, the `byte` data type can be used for any value between −128 and 127. However, the `int` data type can be used to represent any value between –2147483648 and 2147483647, and `long` data type is used to represent integers between −9223372036854775808 and 9223372036854775807. If optimization of memory usage is not an issue, you can choose `int` to represent integers. Similarly, data type `float` can be used to represent any real number between −3.4E+38 and 3.4E+38, and data type `double` can be used to represent any real number between −1.7E+308 and 1.7E+308. Once again if optimization of memory usage is not an issue, you can choose `double` to represent real numbers.

Range of Values and Memory Requirements of Data Types

Data Types	Values	Bytes Required
Boolean type		
`boolean`	`False` and `true`	(1 bit)
Integer type		
`char`	0 to 65535	2 (16 bits)
`byte`	−128 to 127	1 (8 bits)
`short`	−32768 to 32767	2 (16 bits)
`int`	−2147483648 to 2147483647	4 (32 bits)
`long`	−9223372036854775808 to 9223372036854775807	8 (64 bits)
Floating point type		
`float`	−3.4E+38 to 3.4E+38, 6 or 7 significant digits	4 (32 bits)
`double`	−1.7E+308 and 1.7E+308, 15 significant digits	8 (64 bits)

`boolean` *Data Type*

An expression that is either `false` or `true` is called a *logical* expression or *boolean expression*. Computers make decisions on the basis of logical expressions. Thus, computer programs have to evaluate many logical (Boolean) expressions. Logical (Boolean) expressions will be formally introduced and explained in detail in Chapter 4.

The data type `boolean` has two values: `false` and `true`. In Java, `boolean`, `false`, and `true` are reserved words. The memory allocated for the `boolean` data type is 1 bit.

Note 2.7 The `boolean` value `false` is not equal to integer 0 and `boolean` value `true` is not equal to 1. Therefore, `boolean` value cannot be part of any arithmetic expression.

`char` *Data Type*

The `char` data type represents single characters. A single character can be any of the letters (a to z, A to Z), digits (0 to 9), and special characters (!, @, #, $, &, *, {, }, [,], and so on). In particular, the `char` data type can be used to represent any character you see on the keyboard. A `char` literal is represented using a pair of single quotations.

Examples of char literals are

```
'J', 'a', 'v', '@', '-', '_', '%','7','0'.
```

It may be noted that a blank space is `char` literal and is written as `' '`, with a space between a pair of the single quotations. Note that '`Java`' is not a `char` literal.

TABLE 2.3 Some of the Special Symbols

Special Symbol	Meaning
<=	Less than or equal to
>=	Greater than or equal to
==	Equal to
!=	Not equal to

Note 2.8 Java uses certain character combinations as *special symbols*. Some of the special symbols and their intended meanings are shown in Table 2.3.

Just as an identifier or a keyword is treated as one *token*, Java treats special symbols also as one token. Similarly, an identifier such as `HiThere` or a keyword such as `class` or a special symbol such as <= is not a `char` literal.

Java uses the Unicode character set containing 65536 characters. Therefore, `char` data type needs 2 bytes or 16 bits. The 65536 characters are represented using values 0 to 65535. In other words, the first character has integral value 0 and the second character has an integral value 1 and the last character has an integral value 65535. As mentioned in Chapter 1, the first 128 characters of Unicode are the same as the American Standard Code for Information Interchange (ASCII) character set and thus Unicode is a superset of ASCII character set (see Appendix B).

As mentioned in Chapter 1, an important consequence of each character having an associated integral value is that of an implicit ordering of all characters. This implicit ordering of characters is called the collating sequence. Based on the collating sequence, any two characters can be compared. For instance, character `'a'` has value 97 and character `'A'` has value 65. Thus `'A'` is less than `'a'` in the collating sequence. In fact all uppercase letters are less than `'a'` or any lowercase letter in the collating sequence. Further, it is possible to compute the difference between any two characters, as shown in the following example.

Example 2.11

The following Java application illustrates collating sequence:

```
/**
    Illustration of collating sequence
*/
public class CollatingSequence
{
    public static void main(String[] args)
    {
        System.out.println('a' - 'A');
        System.out.println('A' - 'a');
        System.out.println('q' - 'Q');
        System.out.println('Q' - 'q');

    }
}
```

Output

```
32
-32
32
-32
```

Note that the difference between a lowercase letter and the corresponding uppercase letter is always 32.

Note 2.9 The escape sequences characters `'\r'`, `'\n'`, `'\f'`, `'\t'`, and `'\b'` you have encountered in Advanced Topic 2.1 are treated as a single character.

Good Programming Practice 2.4

Use `char` for characters.

Advanced Topic 2.3: Unicode Character Specification

You can specify any character using the format `'\uXXXX'`, where each `X` can be any hexadecimal digit: `0, 1, 2, 3, 4, 5, 6, 7, 8, 9, A, B, C, D, E, F`. Here `A` stands for 10, `B` stands for 11, and so on. Further, in this context, you can use lowercase or uppercase letters. Thus `A` or `a` will represent 10. For example, `'\uFa1B'` stands for the character $15 \times (16)^3 + 10 \times (16)^2 + 1 \times (16) + 11 = 64027$. As another example, `'\u0046'` is $4 \times 16 + 6 = 70$. The 0th character is known as the null character. The null character is represented as `'\0'` (backslash followed by zero) and is the same as `'\u0000'`.

Integer Data Type

Integers are whole numbers. In Java, integers are a sequence of digits starting with a nonzero digit preceded by a negative sign or an optional positive sign. In other words, in the case of positive integers, you need not include the sign. Further, both positive and negative integers must begin with a nonzero digit and must not contain any commas.

Example 2.12

Examples of Legal and Illegal Integers

Integer	**Explanation**
`-34552`	All integers are legal
`34`	
`0`	
`+79`	
`12,340`	Comma not allowed
`-34489.0`	Decimal part not allowed
`0989`	A decimal integer cannot have a leading 0

Note 2.10 An integer literal is always treated as type `int` and as such Java will store it using 4 bytes. To force a literal to be `long`, you must add either letter L or letter l

immediately following the last digit. Since letter l can easily be confused with digit 1, we will use only L in this book.

Good Programming Practice 2.5

Use `int` for integers. In the rare event where you need to deal with larger numbers, use `long`.

Good Programming Practice 2.6

Always use uppercase letter L in the context of `long` literal.

Advanced Topic 2.4: Various Integer Representations

Integer literals may be written in decimal (default), octal, or hexadecimal notation. A leading 0 indicates that the integer is in octal format. Similarly, a leading 0x or 0X indicates that the integer is in hexadecimal format.

Examples of Legal Integers

Integer	Explanation	Equivalent Decimal Value
`071`	Octal number	$= 7 \times (8) + 1 = 57$
`71`	Decimal number	$= 7 \times (10) + 1 = 71$
`0X71`	Hexadecimal number	$= 7 \times (16) + 1 = 113$

Good Programming Practice 2.7

Use decimal system for better readability.

Floating Point Data Type

Floating point types are used to deal with real numbers. Floating point numbers have an integer part and a fractional part separated by a decimal point. Either the integer part or the fractional part can be omitted. However, decimal point along with either the integer part or the decimal part must be present. Here is a list of floating point numbers:

```
123.456
123.0
123.
0.456
.456
```

Good Programming Practice 2.8

Avoid using `float` data type. Use `double` data type. If you use `float` to manipulate floating point numbers in a program, you may encounter a warning message such as “truncation from double to float” or “possible loss of data.” You will understand the reason for these error messages later in this chapter.

Note 2.11 In Java, the compiler takes a floating point literal as `double` by default. You can specify a floating point literal as `float` by adding an f or F at the end of the floating point number. Thus, 123.456 is a `double` and 123.456F is a `float`.

Self-Check

21. Area of a property is of the type ________.
22. The number of completed years of service is of the type ________.

Advanced Topic 2.5: Floating Point Notation

The leading 0 in a floating point number such as 0.456 will not make it an octal number. A floating point number is always in decimal format.

In Java, you could also use either the scientific notation or the floating point notation. Table 2.4 illustrates the use of scientific notation by Java in printing real numbers. In floating point notation, the letter E stands for the exponent.

As mentioned before, Java has two data types to represent real numbers:

1. The data type `float` can be used to represent any real number between −3.4E+38 and 3.4E+38. Further, the maximum number of significant digits or the number of decimal places for `float` values is 6 or 7. The memory required for the `float` data type is 4 bytes.
2. The data type `double` (meaning double precision compared to `float`) can be used to represent any real number between −1.7E+308 and 1.7E+308. The maximum number of significant digits for `double` values is 15. The memory required for the `double` data type is 8 bytes.

OPERATIONS ON NUMERIC DATA TYPES

There are five operations on numeric data types. They are

1. Addition (symbol used is +)
2. Subtraction (symbol used is −)
3. Multiplication (symbol used is *)
4. Division (symbol used is /)
5. Modulus (symbol used is %)

TABLE 2.4 Real Numbers Printed in Java Floating Point Notation

Real Number	Scientific Notation	Floating Point Notation of Java
123.456	$1.23456\times(10)^{2}$	1.234560E2
.456	$4.56\times(10)^{-1}$	4.560000E-1
0.00000456	$4.56\times(10)^{-6}$	4.560000E-6
-1.234	$-1.23\times(10)^{0}$	-1.234000E0
1230000.	$1.23\times(10)^{6}$	1.230000E6

The following example illustrates addition, subtraction, multiplication, division, and modulus operations on integer data types.

Example 2.13

Operations on Integer Data

Expression	Result	Explanation
`-12 + 51`	39	
`-51 - 12`	-63	
`3 * 8`	24	
`19 / 5`	3	Dividing `19` by `5` yields `3` as quotient. Thus `19/5 is 3`
`20 / 5`	4	Dividing `20` by `5` yields `4` as quotient. Thus `20/5 is 4`
`19 % 5`	4	Dividing `19` by `5` yields `3` as quotient and `4` as remainder. In the case of modulus, remainder is the result. Thus `19 % 5 is 4`
`-19 % 5`	-4	Dividing `-19` by `5` yields `-3` as quotient and `-4` as remainder. In the case of modulus, remainder is the result. Thus `-19 % 5 is -4`
`19 % -5`	4	Dividing `19` by `-5` yields `-3` as quotient and `4` as remainder. In the case of modulus, remainder is the result. Thus `19 % -5 is 4`
`-19 % -5`	-4	Dividing `-19` by `-5` yields `3` as quotient and `-4` as remainder. In the case of modulus, remainder is the result. Thus `-19 % -5 is -4`
`5 % 19`	5	Dividing `5` by `19` yields `0` as quotient and `5` as remainder. In the case of `modulus`, remainder is the result. Thus `5 % 19 is 5`

The following program illustrates all five integer operators:

```
/**
    Demonstration of numeric operations in integer mode
*/
public class IntegerOperator
{
    public static void main(String[] args)
    {
        System.out.println(" 12 + 51 = "+ (12 + 51));
        System.out.println("-12 + 51 = "+ (-12 + 51));
        System.out.println(" 51 - 12 = "+ (51 - 12));
        System.out.println("-51 - 12 = "+ (-51 - 12));
        System.out.println(" 3 * 8 = "+ (3 * 8));
        System.out.println(" 19 / 5 = "+ (19 / 5));
        System.out.println(" 20 / 5 = "+ (20 / 5));
        System.out.println(" 19 % 5 = "+ (19 % 5));
        System.out.println("-19 % 5 = "+ (-19 % 5));
        System.out.println(" 19 % -5 = "+ (19 % -5));
        System.out.println("-19 % -5 = "+ (-19 % -5));
        System.out.println(" 5 % 19 = "+ (5 % 19));
    }
}
```

Output

```
 12 + 51  = 63
-12 + 51  = 39
 51 - 12  = 39
-51 - 12  = -63
  3 * 8   = 24
 19 / 5   = 3
 20 / 5   = 4
 19 % 5   = 4
-19 % 5   = -4
 19 % -5  = 4
-19 % -5  = -4
  5 % 19  = 5
```

All five operations can be used on two integers or two floating point numbers or one integer and one floating point number.

Example 2.14 illustrates four operations on floating point numbers. Modulus operator is rarely used in connection with floating point operands and hence omitted.

Example 2.14

The following program computes all the expressions given in Table 2.5:

```
/**
    Demonstration of numeric operations in floating point mode
*/
public class FloatingPointOperator
{
    public static void main(String[] args)
    {
        System.out.println(" (12.0 + 49.2) = "+ (12.0 + 49.2));
        System.out.println(" (12.3 + 49.2) = "+ (12.3 + 49.2));
        System.out.println(" (12.3 - 49.2) = "+ (12.3 - 49.2));
        System.out.println(" (12.3 * 49.2) = "+ (12.3 * 49.2));
        System.out.println(" (12.3 / 49.2) = "+ (12.3 / 49.2));
    }
}
```

Output

```
(12.0 + 49.2) = 61.2
(12.3 + 49.2) = 61.5
(12.3 - 49.2) = -36.900000000000006
(12.3 * 49.2) = 605.1600000000001
(12.3 / 49.2) = 0.25
```

TABLE 2.5 Operations on Floating Point Data

Expression	Mathematical Result
`12.0 + 49.2`	`61.2`
`12.3 + 49.2`	`61.5`
`12.3 - 49.2`	`-36.9`
`12.3 * 49.2`	`605.16`
`12.3 / 49.2`	`0.25`

Note that (12.3 – 49.2) is not −36.9, as you might have expected; rather, the value computed by the above program is –36.900000000000006. Again note that multiplication also produced a result different from what you might have expected. In other words, floating point arithmetic is not exact.

So far, we have used all operators with two operands. If an operator has two operands, it is called a *binary operator*. Thus addition, subtraction, multiplication, division, and modulo. are binary operations. In other words, +, –, *, and % are binary operators. However, we also use symbols + and – as unary operators or operators with one operand. For example, in the expressions +12.5 and −51.5, the operators + and – are unary operators. Note further that in the expression +12.5, the symbol + does not stand for addition. Similarly, in the expression −51.5, the symbol – does not stand for subtraction.

Self-Check

23. What is the value of 17/18?
24. What is the value of 17 % 18?

Operator Precedence Rules

Consider the following arithmetic expression:

```
7 + 3 * 2
```

If you add 7 and 3 first and then multiply by 2 the expression is equal to 20 and if you first multiply 3 and 2 and then add 7, it is equal to 13. Which is the correct answer? From your high school mathematics, you know that multiplication needs to be done before addition. Therefore, in this case, the correct answer is 13 and not 20. In other words, the operator * is of higher precedence than the operator +.

Whenever more than one operator appears in an expression, there needs to be a rule to determine the order in which these operators are evaluated. Every programming language provides such rules and Java is no exception. These rules are called *operator precedence rules*. The operator precedence rules for arithmetic operators can be summarized as follows:

- Unary operators + and – have the same precedence, and they are evaluated from right to left.

TABLE 2.6 Operator Precedence

Operator	Operand Types	Operation	Associativity
+, −	Number	unary plus, unary minus	Right to left
*, /, %	Number	Multiplication, division, modulus	Left to right
+, −	Number	addition, subtraction	Left to right

- Binary operators *, /, and % have the same precedence, and they are evaluated from left to right. Further, Unary operators + and − have higher precedence than the operators *, /, and %.
- Binary operators + and − have the same precedence, and they are evaluated from left to right. The operators *, /, and % have higher precedence than the binary operators + and −.

The property that an operator is evaluated from left to right or from right to left is called the *associativity* of that operator. Thus, note that the associativity of unary arithmetic operators is from right to left and that of binary arithmetic operators is from left to right.

Introspection

What is 13 − 7 − 2? Is it (13 − 7) − 2 = 4 or 13 − (7 − 2) = 8?

It is customary to state these rules in the form of a table similar to Table 2.6. Observe that operators in the same row have the same precedence and an operator in a "higher row" has a higher precedence over an operator in a "lower row."

In an arithmetic expression, parentheses can be used to modify the order of execution. Parentheses can also be used for better readability.

Example 2.15

Consider the following expression:

```
-4 * 2 + 8 - 9 / 3 * 5 + 7
```

The order of evaluation based on precedence rules is as follows:

```
= (-4) * 2 + 8 - 9 / 3 * 5 + 7     (unary minus is evaluated)
= ((-4) * 2) + 8 - 9 / 3 * 5 + 7       (leftmost mul./div. is
                                                  evaluated)
= (-8) + 8 - (9 / 3) * 5 + 7           (leftmost mul./div. is
                                                  evaluated)
= (-8) + 8 - (3 * 5) + 7               (leftmost mul./div. is
                                                  evaluated)
= ((-8) + 8) - 15 + 7                  (leftmost add /sub. is
                                                  evaluated)
```

```
= (0 - 15) + 7                  (leftmost add /sub. is
                                          evaluated)
= (- 15 + 7)                    (leftmost add /sub. is
                                          evaluated)
= -8
```

Example 2.16

In the expression

```
6 + 3 / 3
```

the division operator / is evaluated first. Thus, the above expression is equivalent to 6 + (3 / 3) = 7. However, parentheses can be used to modify the order of execution as follows:

```
(6 + 3) / 3 = 3.
```

Good Programming Practice 2.9

The intended use of the `char` data type is to manipulate characters. However, since `char` is an integral data type, it is legal to perform arithmetic operations on `char` data type. Use integral types other than `char` for computational purposes.

Note 2.12 From Appendix B, it can be seen that character `'6'` has collating sequence value `54`. Thus, `'6'+'6'` evaluates to `108`.

Self-Check

25. What is 27 − 8 / 11 + 4?
26. What is (27 − 8) / 11 + 4?

Rules for Evaluating Mixed Expressions

An expression that has operands of more than one numerical data types is called a *mixed expression*. Thus, a mixed expression may contain `int` and `long` values or `int` and `double` values, and so on.

Example 2.17

Consider Table 2.7.

TABLE 2.7 Mixed Expressions

Mixed Expression	Explanation
`31L / 2`	`long and int values`
`32 + 12.5`	`int and double values`
`18.723F * 234L`	`float and long values`
`353.35 - 0.001F`	`double and float values`

The rules for evaluating mixed expressions can be stated as follows:

Step 1. Determine the next operation based on operator precedence rules.

Step 2. Consider the operands involved. If both of them are of the same data type there is nothing to be done at this step. If they are of different type, Java converts the data type of lesser range to the other data type for the purpose of this expression evaluation. That is, data values are converted as follows:

```
byte → short → int → long → float → double
```

or any combination of these in the same direction.

Step 3. Evaluate the expression.

Thus, evaluation of a mixed expression is done on one operator at a time using the rules of precedence. If the operator to be evaluated has mixed operands, Java cannot evaluate as such. Therefore, the operand with lesser range of values is *promoted* (or *implicitly converted* or *coerced*) to the data type of the other operand before the evaluation. For example, if one operand is `int` and the other is `double`, the `int` data value is promoted to `double` value. In this case, the floating point operation takes place and the result is of type `double`. Similarly, if one operand is `int` and the other is `long`, the `int` data value is promoted to `long` value and the result is of type `long`. Thus an operator with integer operands will yield an integer result and if one of the operands is a floating point, the result is a floating point. The following example illustrates the evaluation of mixed expressions.

Example 2.18

Evaluation of Mixed Expressions

Mixed Expression	Step 1 (Operator)	Step 2 (Operands)	Evaluation
`17 / 5 + 34.0`	/	`17` and `5` (no change)	`= 3 + 34.0` (`17 / 5 = 3`; integer division)
	+	`3.0` and `34.0` (`int 3` is promoted to `double 3.0`)	`= 3.0 + 34.0 = 37.0`
`12 / 5.0 + 10`	/	`12.0` and `5.0` (`int 12` is promoted to `double 12.0`)	`= 2.4 + 10` (`12.0 /5.0 = 2.4`; floating point division)
	+	`2.4` and `10.0` (`int 10` is promoted to `double 10.0`)	`2.4 + 10.0 = 12.4`
`9 % 5 + 6 * 7 - 3.19`	%	`9` and `5` (no change)	`= 4 + 6 * 7 - 3.19 (9 % 5 = 4)`
	*	`6` and `7` (no change)	`= 4 + 42 - 3.19 (6 * 7 = 42)`
	+	`4` and `42` (no change)	`= 46 - 3.19 (4 + 42 = 46)`
	-	`46.0` and `3.19` (`int 46` is promoted to `double 46.0`)	`= 42.81`

Self-Check

27. What is 27 / 11 + 4.0?
28. What is 27 / 11.0 + 4?

Advanced Topic 2.6: Mixed Expressions Involving String

As noted before, concatenation operator + can also be used to join a String and a numeric value or a character. In that case, Java first converts the numerical value or character to a String and then uses the concatenation operator +. The following example illustrates the use of concatenation operator.

Strings play a very important role in Java. So we will address issues involving Strings in a later chapter. However, at this point you need to know the behavior of the concatenation operator when other data types are involved.

Example 2.19

The following Java application illustrates the concatenation operator on Strings and numbers, and Strings and characters:

```
/**
    Demonstration of concatenation operation in mixed mode
*/
public class ConcatOperatorNumber
{
    public static void main(String[] args)
    {

        System.out.print
                ("(1)\t 1600 + \" Pensylvania Avenue\" is ");
        System.out.println(1600 + " Pensylvania Avenue");
        System.out.print
                ("(2)\t \"Pensylvania Avenue \" + 1600 is ");
        System.out.println("Pensylvania Avenue " + 1600);
        System.out.print("(3)\t 563 + 34 is ");
        System.out.println( 563 + 34);
        System.out.print
                 ("(4)\t \"Victoria, NE \" + 563 + 34 is ");
        System.out.println("Victoria, NE " + 563 + 34);
        System.out.print
                ("(5)\t 563 + 34 + \" Victoria, NE \" is ");
        System.out.println(563 + 34 + " Victoria, NE ");
        System.out.print
                ("(6)\t \"Victoria, NE \" + (563 + 34) is ");
        System.out.println("Victoria, NE " + (563 + 34));
        System.out.print
        ("(7)\t \"Victoria, \" + \'N\' + \'E\' + \' \' + 56334 is ");
        System.out.println
                   ("Victoria, " + 'N' + 'E' + ' ' + 56334);
        System.out.print("(8)\t 563 * 34 is ");
```

```
        System.out.println( 563 * 34);
        System.out.print
                ("(9)\t \"Victoria, NE \" + 563 * 34 is ");
        System.out.println("Victoria, NE " + 563 * 34 );
        System.out.print
               ("(10)\t 563 * 34 + \" Victoria, NE \" is ");
        System.out.println(563 * 34 + " Victoria, NE ");
        System.out.print
              ("(11)\t \"Victoria, NE \" + (563 * 34) is ");
        System.out.println("Victoria, NE " + (563 * 34));

    }
}
```

Output

```
 (1) 1600 + " Pensylvania Avenue" is 1600 Pensylvania Avenue
 (2) "Pensylvania Avenue " + 1600 is Pensylvania Avenue 1600
 (3) 563 + 34 is 597
 (4) "Victoria, NE " + 563 + 34 is Victoria, NE 56334
 (5) 563 + 34 + " Victoria, NE " is 597 Victoria, NE
 (6) "Victoria, NE " + (563 + 34) is Victoria, NE 597
 (7) "Victoria, " + 'N' + 'E' + ' ' + 56334 is Victoria, NE 56334
 (8) 563 * 34 is 19142
 (9) "Victoria, NE " + 563 * 34 is Victoria, NE 19142
(10) 563 * 34 + " Victoria, NE " is 19142 Victoria, NE
(11) "Victoria, NE " + (563 * 34) is Victoria, NE 19142
```

Explanation

(1) and (2) (A `String` and a number case): The number 1600 is converted to the `String` `"1600"`. Then the concatenation takes place and thus produces the `String` `"1600 Pennsylvania Avenue"` and `"Pennsylvania Avenue 1600"`, respectively. (3) (Numbers only case): In this case, all numerical computations will be done first. Thus, 563 and 34 are added together to produce the number 597. (4) (Numbers after `String` case) : Java converts the `int` value 563 to the `String` "563" and appends at the end of the `String` `"Victoria, NE"`. This produces a `String` `"Victoria, NE 563"`. Now `int` value 34 is converted to the `String` `"34"` and concatenated to `"Victoria, NE 563"`, thus producing `"Victoria, NE 56334"`. (5) (Numbers before `String` case): In this case, the first + symbol is an addition symbol. Both `563` and `34` are numbers. Therefore, + is not considered as a concatenation operator. In other words, for + to work as a concatenation symbol, at least one of the two operands must be a `String`. Therefore, `597` is computed first. Now, `597` is a number and `"Victoria, NE"`

is a `String`, and therefore the `String "Victoria, NE"` gets appended to the `String "597"`. (6) (Use of parenthesis): Due to parenthesis, Java adds `563` and `34` first. Consequently, the number `597` is converted to the `String "597"` and appended to `"Victoria, NE"` as in cases (1) and (2). The output line (7) shows how characters can also be appended to a `String` using the concatenation operator. Note that each character gets appended to the existing `String`. (9), (10), and (11) (Arithmetic operator with a higher precedence): Since multiplication has higher precedence, it is done before the concatenation operation.

NAMED CONSTANTS AND VARIABLES

The data required for your program is kept in memory during the execution of the program. The memory is a collection of cells with individual addresses. You have seen that each primitive data type has its own storage requirements. For example, an `int` data type is always stored using 4 bytes. Similarly, a `double` is always stored in 8 bytes. To store a value in memory, it is important to know the data type for two reasons: First, data type determines the size of memory that needs to be allocated. Second, data type determines the set of legal operations that can be applied on the stored data. All high-level languages allow the programmer to assign an identifier to a memory location so that the programmer need not directly deal with memory addresses. The programmer uses the identifier to store and retrieve data from memory.

There are certain values in your program that need not be modified during the program execution. For example, we all pay 7.5% of our salary as Social Security Tax. Similarly, to work, one must be at least 16 years old. A conversion formula that converts pounds into kilogram must make use of the fact that 1lb is 0.454 kg. During the program execution, these data values need to be guarded against accidental changes. Java allows you to allocate memory using a *named constant* and store a data value. The value of a named constant remains fixed during the program execution.

In Java, you can declare a named constant in two different contexts: at the class level and at the method level. The Java syntax to declare a named constant at the class level is

```
[public] [static] final dataType IDENTIFIER = literal;
```

and that at the method level is

```
final dataType IDENTIFIER = literal;
```

Note that both `static` and `final` are reserved words. The word `static` specifies that only one memory location will be allocated to all instances of a class and such a memory location will be available even if there is no object belonging to the class. Observe that the word `static` is optional. However, from a programming point of view, there is no need to have multiple copies of a constant value. The reserved word `final` indicates that the literal stored in the `IDENTIFIER` is final, or cannot be changed.

Introspection

Date of birth is a named constant of our life. It never changes.

Good Programming Practice 2.10

Any literal value that must remain constant throughout the execution of the program is to be declared as a named constant.

Good Programming Practice 2.11

Use uppercase letters and underscore in a named constant identifier. Use underscore character at the beginning of each new word.

Example 2.20

Consider the following Java declarations.
Class level:

```
public static final char PERCENTAGE = '%';
public static final int MAXIMUM_ALLOWED_WEIGHT = 4000;
```

Method level:

```
final float ERROR_ALLOWED = 0.01F;
final double POUND_TO_KILOGRAM = 0.454;
```

The first statement instructs the compiler to allocate 2 bytes in memory and store the character '%'. Further, the memory location will be known as PERCENTAGE at the class level. Therefore, inside the class you can access the constant using its name

```
PERCENTAGE
```

and outside the class you can access the constant as

```
className.PERCENTAGE
```

respectively. The data kept in the memory location PERCENTAGE cannot be modified and its data type is char. Similarly, the second statement instructs the compiler to allocate 4 bytes in memory and to store the integer 4000. The memory location will be known as MAXIMUM_ALLOWED_WEIGHT at the class level. It cannot be modified and its data type is int.

The meaning of the remaining two statements is similar except that they are available only at the method level. Therefore, you can access the constant ERROR_ALLOWED only inside the method it is declared.

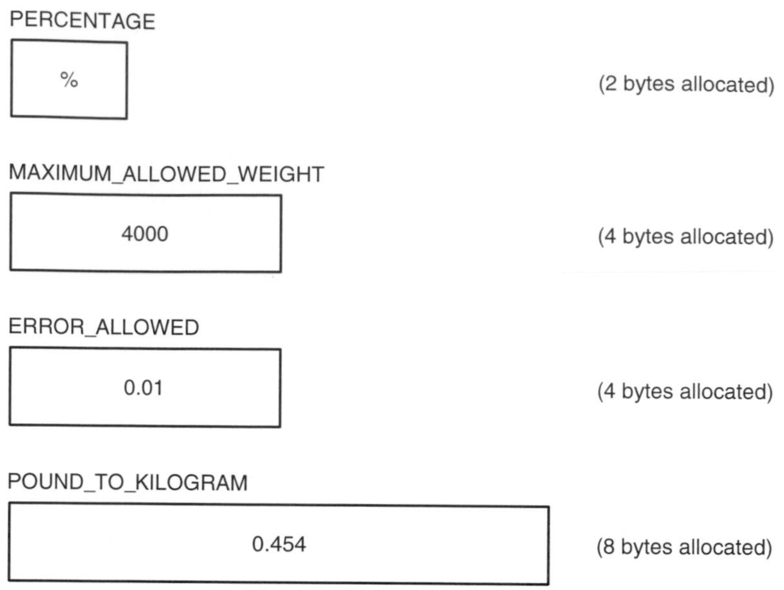

FIGURE 2.2 Named constants.

The effect of the four statements given in Example 2.20 is illustrated in Figure 2.2.

Note 2.13 Using a named constant in place of a literal has many advantages. First, if the value needs to be changed, you need to modify only one line of code. Second, a typographical error in a literal cannot be detected by a compiler. However, a typographical error in a named constant will be detected during the compilation process.

During the execution of a program, new values are computed. These values also need to be stored in memory. For example, in a program for computing the maximum and minimum temperatures of a day, after each new temperature reading either the minimum temperature or the maximum temperature may change. Similarly, consider a program that tracks the inventory of a convenience shop. Observe that after each sale the number of items in the inventory changes. Thus, there is a need for memory cells whose contents can be modified during program execution. In all programming languages, memory cells whose contents can be modified during the execution of a program are called *variables.*

In Java, there are four categories of variables. They are *instance variables* and *static variables* declared at class level and *parameter variables* and *local variables* declared at *block* level. A *block* is a sequence of statements enclosed within a pair of braces {and}. Every `main` method in this chapter has exactly one block. You will see many block statements, including a block statement inside another one, in Chapter 4. The syntax to declare instance variables is

```
[accessModifier] dataType identifierOne[[=LOne], ...,identifierN
                                                        [=LN]];
```

The syntax to declare class variables (also known as static variables) is

```
[accessModifier] static dataType identifierOne[[=LOne], ...,
                                   identifierN[=LN]];
```

The syntax to declare parameter variables is

```
dataType identifierOne[,..., dataType identifierN]
```

The syntax to declare local variables is

```
dataType identifierOne[[=LOne], identifierTwo[=LTwo],...,
                                        identifierN[=LN]];
```

Recall that an item inside a pair of square brackets in a syntax template is optional.

Good Programming Practice 2.12

> Use letters only as identifiers for variables and start the variable in lowercase. Use uppercase letters only at the beginning of each new word.

Declaring local variables is covered here. Instance variables and parameter variables are covered in Chapter 3 and the static variables in Chapter 6.

Consider the following variable declarations:

```
char nextCharacter;
int totalGamesPlayed;
double stateTaxRate = 6.85;
String nextLine = "Have a close look at me!";
```

The first statement instructs the compiler to allocate 2 bytes of memory and no value is stored. Further, the memory location will be known as nextCharacter in the program and can be modified so long as the modified value is of type `char`. Similarly, the second statement instructs the compiler to allocate 4 bytes of memory. The memory location will be known as `totalGamesPlayed` in the program and can be modified as long as the modified value is of data type `int`. The third statement allocates 8 bytes and stores 6.85 as the initial value. The memory allocated will be known in the program as `StateTaxRate`.

The behavior of the fourth statement is quite different from the other three statements. This is due to the fact that `String` is a class in Java and not a primitive data type. In the case of classes, the only variables you can declare are reference variables. Now a reference variable is one that can keep the reference of an object. Thus in the case of the fourth statement, a memory location is allocated and labeled as `nextLine`. However, memory location `nextLine` does not contain the `String` `"Have a close look at me!"`. Instead, the `String` is stored somewhere else in memory and the *reference* of the String is kept in the reference variable `nextLine`. For the sake of visualization, assume that the compiler uses memory address as reference. (Recall that in an object-oriented environment, the information is hidden from the user. Thus, we do not know how the compiler maintains the reference.) Assume that the compiler stored the `String` `"Have a close look at me!"` at memory location 12084. Then the variable `nextLine` will contain 12084. We shall use ? in our drawings to indicate that no value has been stored in a memory location (Figure 2.3).

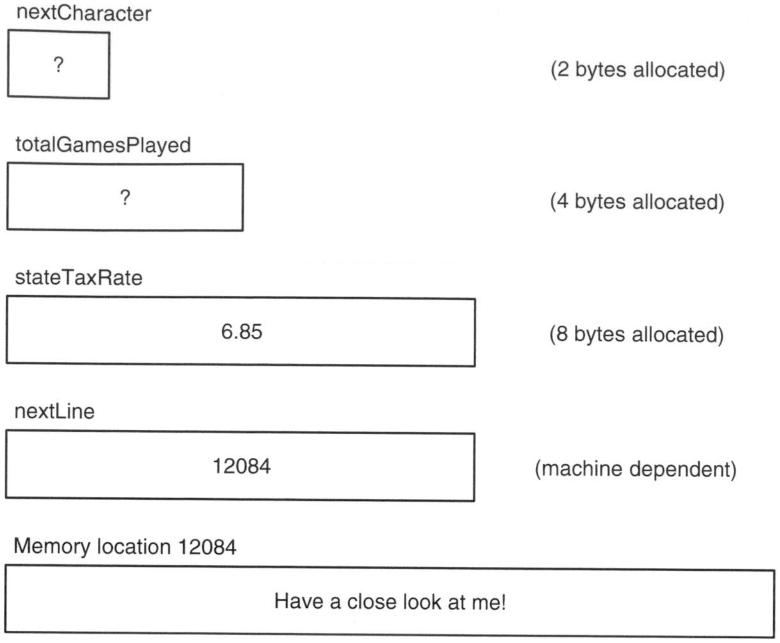

FIGURE 2.3 Variables.

The above discussion can be summarized as follows:

1. In the case of a primitive data type, as you declare a variable, the compiler allocates necessary memory depending on the data type and labels the memory location using the variable name. If you specify an initial value for the variable, then the compiler will store that initial value at the variable location.
2. In the case of reference variables, as you declare a variable, the compiler allocates necessary memory for storing a memory address and labels the memory location using the variable name. If an initial object is specified, the compiler will allocate a new memory location and store the object. Further, the address of the object is placed in the reference variable.

In the case of primitive data types, the value is kept at the variable location; however, in the case of objects, the reference variable does not keep the object, it keeps the reference of the object.

Fortunately, as a programmer, you need not deal with memory locations. All you need to know is the existence of this subtle difference between variables of primitive data types and reference variables.

Note 2.14 In Java, all named constants and variables must be declared before using them. Using a named constant or a variable without declaring is a syntax error, and as such Java compiler will generate an error message. In the case of primitive data types, you must also initialize the variables before using them.

Self-Check

29. Declare a variable to store product description.
30. Declare a variable to store number of dependents of an employee.

Changing Data Values of Variable

In Java, you can change the value of a variable either through an assignment statement or through increment or decrement operators.

Assignment Statement

The assignment statement has three parts: a variable, followed by the equal sign (known as the assignment operator), and then an expression. Thus, the syntax of an assignment statement is

```
variable = expression;
```

Example 2.21

Consider the following:

```
char     grade;
double currentScore, totalScore = 0.0;

grade = 'A';                              // (1)
currentScore = 30.0 ;
totalScore = 95.7;
```

The statement 1 changes the value stored at grade to `'A'`. Once the statement 1 is executed, any value previously kept in grade will be lost forever.

In mathematics, if you say y = 5, you are indicating the fact that the value of the variable y is 5. For example, you may start with an equation such as 3y − 15 = 0 and then arrive at the conclusion that y = 5. The semantics of an assignment statement is quite different. For example, statement 1 specifies that the character 'A' is stored at memory location grade.

In an assignment statement, the expression on the right side is evaluated and then the value computed is stored at the variable on the left side. Therefore, in an assignment statement, the value of the `expression` on the right-hand side and the `variable` on the left-hand side should match in data type. Just as in the evaluation of mixed expressions, if the variable on the left-hand side is a data type of wider range than the data type at right-hand side, the right-hand side value will be promoted to match the left-hand side data type. If the variable on the left-hand side is a data type of narrower range than the right-hand data type, then the compiler will issue an error message similar to the following:

```
possible loss of precision
found : right_hand_side_data_type
required: left_hand_side_data_type
```

A statement that places a value in a variable for the first time is called an *initialization* statement.

Example 2.22

Assignment Operator

Expression on the Right-Hand Side	Left-Hand Variable Type is int	Left-Hand Variable Type is double
37	37 is assigned. (data types match)	37.0 is assigned. (37 is promoted to 37.0 double)
12.5F	Error:possible loss of precision (int is narrower than float)	12.5 is assigned. (12.5 is promoted to double)
56.8	Error:possible loss of precision (int is narrower than double)	56.8 is assigned. (data types match)

Example 2.23

Consider the following program:

```
/**
    Demonstration of assignment operator
*/
public class Assignments
{

    public static void main(String[] args)
    {
        int numberOfStudents;
        float toalPoints;
        String heading;
        boolean smartStudents;

        numberOfStudents = 25.0;
        toalPoints = 2000;
        heading = 2004;
        smartStudents = "All students";

    }
}
```

If you compile the above program, the following error messages will be generated:

```
C:\dirName\Assignments.java:11: possible loss of precision
found : double
required: int
          numberOfStudents = 25.0;
 ^
```

```
C:\dirName\Assignments.java:13: incompatible types
found : int
required: java.lang.String
          heading = 2004;
 ^
C:\dirName\Assignments.java:14: incompatible types
found : java.lang.String
required: boolean
          smartStudents = "All students";
 ^
3 errors
```

The first error message is generated due to line 11 of the file `C:\dirName\Assignments.java`. The error is then summarized as follows:

```
possible loss of precision
found : double
required: int
```

In other words, the right-hand side of the assignment statement is found to be of data type `double`, whereas variable on the left-hand side is of the data type `int`. Note that in this case, the error is "possible loss of precision." Interpretation of second and third error messages is similar. However, the error is stated as "incompatible types." Thus, if you mismatch the data types within numerical data types, system generates a "possible loss of precision" error message. If you assign a `String` or `boolean` to a number (or assign a `String` or number to `boolean`; or assign a number or `boolean` to a `String`) the error message produced will be "incompatible types."

The "possible loss of precision" can be avoided through the use of the *cast operator*. The cast operator is explained later in this chapter.

Example 2.24

The following Java statements illustrate the assignment statements for different data types:

```
int hoursWorked;
double hourlyRate, weeklySalary;
char status;
String fullName;

hoursWorked = 5 * 8 - 3;
hourlyRate = 16.75;
weeklySalary = hourlyRate * hoursWorked;
status = 'S';
fullName = "James F. Kirk";
```

Consider the assignment statement:

```
hoursWorked = 5 * 8 - 3;
```

The computer first evaluates the expression on the right to an `int` data type of value 37. Now the variable on the left-hand side is of type `int`. Therefore, `int` value 37 is stored at memory location `hoursWorked`.

Semantics of other assignment statements are similar.

A Java statement such as

```
k = k + 1;
```

means take the current value of variable `k`, add 1 to it, and assign the new value to the memory location `k`. In the case of an assignment statement, the expression on the right side is evaluated first irrespective of the variable on the left-hand side. Once the right-hand side value is computed, the result is stored in the memory location specified by the variable on the left side. Thus, if value of k was 10 before the execution of the statement, k becomes 11 after the execution of the statement.

Semantics of Assignment Operator

Statement	i	j	k	p	q
`int i;`	0				
`int j = 10;`		10			
`int k = j +2;`			12		
`double p;`				0.0	
`double q = 1.234;`					1.234
`i = j * i + k;`	12				
`k = j % i;`			10		
`p = i / (k + 2) + q;`				2.234	
`k = i / (k + 2) + q;`			error		

Suppose that `a`, `b`, `c`, and `d` are variables of the same data type. The following statement is legal in Java:

```
a = b = c = d;
```

In this statement, first the value of `d` is assigned to `c`. Since = is an operator, `c = d` produces (or returns) a value and the value retuned is equal to that of `c`. Therefore, this statement is equivalent to the following two statements:

```
c = d;
a = b = c;
```

By the same argument, the statement `a = b = c;` is equivalent to the following two statements:

```
b = c;
a = b;
```

Thus, the statement `a = b = c = d;` is equivalent to the following three statements:

```
c = d;
b = c;
a = b;
```

Note 2.15 The *associativity* of the *assignment operator* is from right to left. Therefore, the assignment operator, =, is evaluated from right to left.

Self-Check

31. Assume that `x` and `y` are two variables of type `int` and let `x` be `10` and `y` be `20`. After executing the statement `x = y`, the value of `x` is ______ and the value of `y` is ______.
32. Write the assignment statement that will change the value of `x` to two times the current value of `x` plus three times the current value of `y`.

INPUT STATEMENT

You have learned how to output. You have learned how to use variables to store and manipulate data. In this section, you will learn how to get data from the standard input device and place it in variables using Java's input operations and related classes. Note that the standard input device is the keyboard and the standard output device is the monitor. You have already used `System.out` object, which for all practical purposes can be thought of as the standard output device. Similarly, `System.in` can be thought of as the standard input device.

With `System.in`, we can input either a single byte or a sequence of bytes. However to be useful to us, we want to break the input into meaningful units of data called *tokens*. For example, the following input

```
123-45-6789 James Watts 56 5432.78
```

contains five tokens. They are

```
6789-6789-6789   String   (social security number)
James            String   (first name)
Watts            String   (last name)
56               int      (age)
5432.78          double   (monthly salary)
```

Java provides a `Scanner` class with necessary operations to get individual tokens.

First, we need to create an object belonging to the `Scanner` class. This is achieved by the following Java statement:

```
Scanner scannedInfo = new Scanner(System.in);
```

You may observe that this statement is in fact a declaration and initialization statement very similar to the following:

```
int k = 10;
```

In the case of declaration and initialization of a primitive data type such as `int`, three actions take place in sequence. They are as follows:

1. Reserve enough space in memory (in this case int is of size 4 bytes)
2. Label the memory location with variable name (in this case variable name is k)
3. Place data in the memory location (in this case data is 10)

However, the effect of the following statement

```
Scanner scannedInfo = new Scanner(System.in);
```

can be summarized as follows:

1. A reference variable `scannedInfo` of type `Scanner` is created. Since a reference variable can hold a memory address, so in this case, `scannedInfo` can hold the address of an object belonging to the `Scanner` class.
2. A scanner object is created in memory using `System.in` as argument. Note that `new` operator creates a new object of the class `Scanner` and returns its memory address.
3. The assignment operator places the address of the scanner object in the reference variable `scannedInfo`.

Now, `Scanner` class has operations such as `nextInt()` that can get the next token and convert it as `int`, `nextDouble()` that can get the next token and convert it as `double`, and `next()` that can get the next token and convert it as a `String`. The `nextLine()` method can be used to read the entire line delimited by the current setting of the delimiter characters. By default, the delimiter characters are *whitespace* characters. In Java, there are five whitespace characters. They are the space character, the horizontal tab, the form feed, the carriage return, and the linefeed. Therefore, if the input is

```
123-45-6789 James Watts 56 5432.78
```

you need to first observe that there are five tokens. Of those five tokens, the first three are `Strings`. The next token needs to be treated as an `int` data and finally, the last token is to be interpreted as `double`. Therefore, you need to use operation `next()` three times to handle the first three `Strings`. Then use the operation `nextInt()` once to handle the integer data. Finally, you need to use operation `nextDouble()` to process the last token as `double`. Further, to store these data items, there must be three `String` variables, one `int` variable, and one `double` variable. Thus, the Java statements required to process the above input data are as follows:

```
String socialSecNum, firstName, lastName;
int age;
double monthlySalary;
Scanner scannedInfo = new Scanner(System.in);
socialSecNum = scannedInfo.next();
firstName = scannedInfo.next();
lastName = scannedInfo.next();
age = scannedInfo.nextInt();
monthlySalary = scannedInfo.nextDouble();
```

Self-Check

33. `NextInt` is a method of the _______ class.
34. List three whitespaces in Java.

PACKAGES AND `import` STATEMENT

In Java, classes are grouped into packages. Most commonly used classes are in the package `java.lang` and this package is available to all programs. The `Scanner` class is in the package `java.util`. To use the `Scanner` class, you need to import either the `Scanner` class or the package `java.util`. This is accomplished by placing an `import` statement in the beginning of the source program:

```
import java.util.*;            // imports java.util package
```

or

```
import java.util.Scanner;      // imports Scanner class
```

The documentation on all Java classes can be found at `http://java.sun.com/javase/6/docs/api/`. The reader is strongly encouraged to explore various Java classes.

In Chapter 6, you will learn how to create your own packages and use them in your programs.

Example 2.25

The following Java program illustrates the input and output of various data types:

```
import java.util.Scanner;
/**
    Demonstration of input statements
*/
public class ScannerOne
{
    public static void main(String[] args)
```

```
    {
        String socialSecNum;
        String firstName;
        String lastName;
        int age;
        double monthlySalary;

        Scanner scannedInfo = new Scanner(System.in);

        socialSecNum = scannedInfo.next();
        firstName = scannedInfo.next();
        lastName = scannedInfo.next();
        age = scannedInfo.nextInt();
        monthlySalary = scannedInfo.nextDouble();

        System.out.println(socialSecNum);
        System.out.println(firstName);
        System.out.println(lastName);
        System.out.println(age);
        System.out.println(monthlySalary);

    }
}
```

Output

```
123-45-6789
James
Watts
56
5432.78
```

Self-Check

35. In Java, classes are grouped into ________.
36. `Scanner` class belongs to the ________.

Single Character Input

To read a single character, you can use the operation `next` of the `Scanner` class. This in fact produces a `String` of length 1. For almost all applications, you could use a `String` of length 1 rather than a char data type. However, if you are really interested in keeping the character as a char data type, you could do that as well. Recall that the characters of a `String` are numbered 0, 1, 2, and so on. Therefore, from a `String` of length 1, to get the first character, you can use the `charAt` operation of the `String` class with argument 0.

Thus, to process an input such as

```
123-45-6789 James T Watts 56 5432.78
```

you need to introduce two more variables:

```
String middle;
char middleInitial;
```

to first store T as a `String` and then T as a char. You can now use `next` and `charAt` as

```
middle = scannedInfo.next();
middleInitial = middle.charAt(0);
```

to store the single character `T` in the `char` variable `middleInitial`. It is possible to combine the above two lines into one as follows:

```
middleInitial = scannedInfo.next().charAt(0);
```

In this case, there is no need for the `String` variable `middleInitial`. Thus, the modified program is as follows:

```
import java.util.Scanner;

/**
    Demonstration of input statements
    Modified version
*/
public class ScannerTwo
{

    public static void main(String[] args)
    {
        String socialSecNum, firstName, lastName;
        char middleInitial;
        int age;
        double monthlySalary;

        Scanner scannedInfo = new Scanner(System.in);

        socialSecNum = scannedInfo.next();
        firstName = scannedInfo.next();
        middleInitial = scannedInfo.next().charAt(0);
        lastName = scannedInfo.next();
        age = scannedInfo.nextInt();
        monthlySalary = scannedInfo.nextDouble();
```

```
        System.out.println(socialSecNum);
        System.out.println(firstName);
        System.out.println(middleInitial);
        System.out.println(lastName);
        System.out.println(age);
        System.out.println(monthlySalary);

    }
}
```

Note 2.16 It is totally irrelevant how many lines are used to input the data. You may type in all data items on one line or you may type in your data in many lines including blank lines. All that is important is that data must match the type and a token must be kept as one.

For example,

```
123-45-6789 James T Watts 56 5432.78
```

where all six tokens are in one line or in 10 lines as shown below are acceptable. Here ↵ stands for position where return is entered.

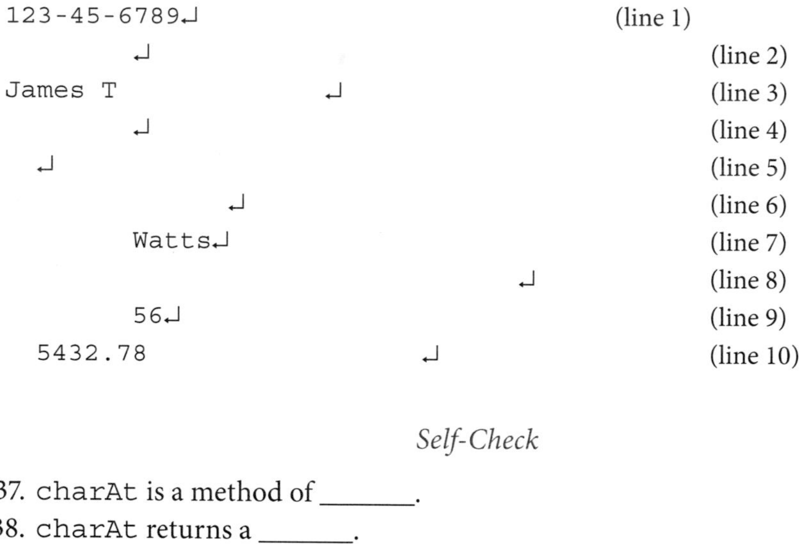

123-45-6789↵ (line 1)
↵ (line 2)
James T ↵ (line 3)
↵ (line 4)
↵ (line 5)
↵ (line 6)
Watts↵ (line 7)
↵ (line 8)
56↵ (line 9)
5432.78 ↵ (line 10)

Self-Check

37. `charAt` is a method of _______.
38. `charAt` returns a _______.

INTERACTIVE MODE AND PROMPT LINES

A computer program that waits for some input during the execution is said to be in *interactive mode*. In this section, it may be quite clear to you that the order in which you enter data is quite important. Therefore, it is important for the user to know the type of data the program is expecting during the execution, especially if the program is interactive. You could use operations `print` and `println` for providing helpful messages on the monitor. For example, if the application is expecting last name, you could use a statement similar to the one that follows:

```
System.out.print("Enter last name: ");
```

You may think that the above message in fact will appear on the monitor. Unfortunately, that may not happen. The reason being, all the output generated by the operation `print` is kept in a memory location called *output buffer*. Once output buffer is full, the entire output in the buffer is sent to the monitor. Recall that the attribute `out` of the `System` class is an object of the `PrintStream` class. The `PrintStream` class has an operation, `flush`, to empty the buffer. Note that `flush` operation empties the buffer and sends information to the monitor. The `flush` operation does not change the cursor position. You need the following two statements:

```
System.out.print("Enter last name: ");

System.out.flush();
```

to prompt the user for the last name. In this case, the prompt line will appear as follows with the cursor at the position shown:

```
Enter last name: |
```

Self-Check

39. The attribute `out` of the `System` class is an object of ________ class.
40. A computer program that waits for some input during execution is said to be in ________ mode.

EXPLICIT DATA–TYPE CONVERSION

Consider the following statements:

```
double totalArea = 2586.24;
int minimumRoomArea = 120;
int numberOfRoomsPossible;

numberOfRoomsPossible = totalArea / minimumRoomArea;
```

From our discussion on mixed expressions, you know that the result of dividing a `double` value by an `int` value is a `double` value. Therefore, the expression on the right hand of the last assignment statement above is of type `double`. However, the variable on the left-hand side is of data type `int`. Therefore, the last assignment statement will cause an error during the compilation.

To avoid such compiler errors, Java provides for explicit data–type conversion using the cast operator. The syntax of the *cast operator* is as follows:

```
(NewDataTypeName) expression
```

The semantics of the cast operator can be explained as follows. First, the `expression` is evaluated as explained in the section on mixed expressions. The value is converted to new data type as specified by NewDataTypeName. Note that converting a floating point number to an integer using the cast operator results in loosing the decimal part of the floating point number. In particular, observe that the floating point number is truncated and not rounded. For example, casting both numbers 12.0001 and 12.9999 to an integer type will produce 12. The following examples further illustrate how cast operators work.

Example 2.26

Cast Operator and Literals

Expression	Value	Explanation
(int)(1000.9999)	1000	Decimal part is dropped
(int)(1000.0000)	1000	Decimal part is dropped
(double)(267)	267.0	Decimal part is added
(double)(150-84)	66.0	= (double)(66) = 66.0
(double)(151) % 2	1.0	(double)(151) = 151.0/2 = 1.0
(double)(151 % 2)	1.0	(151 % 2) = 1 = (double)(1) = 1.0
(double)(151)/2	75.5	(double)(151) = 151.0 = 151.0/2.0 = 75.5
(double)(151/2)	75.0	(151/2)=75 = (double)(75) 75.0
(int)(2.5 * (double)(1)/2)	12	= (int)(2.5 * 0.5) = (int)(12.5) = 12
(int)(2.5 * (double)(1/2))	0	= (int)(2.5 * 0.0) = (int)(0) = 0

Example 2.27

For this example, consider the following statements:

```
int a = 7, b = 10;
double c = 4.25, d = 5.80;
```

Cast Operator and Expressions

Expression	Value	Explanation
(double)(b/a)	1.0	(10/7) = 1. So (double)(1) = 1.0
(int)(c) + (int)(d)	9	4 + 5 = 9
(int)(c + d)	10	(4.25 + 5.80) = 10.05 (int)(10.05) = 10
(double)(a/b) + c	4.25	(7/10) = 0. 0 + 4.25 = 4.25
(double)(a)/b + d	6.50	7.0/10 = 0.7 0.7 + 5.80 = 6.50

You can also use cast operators to explicitly convert between `char` and `int` data values. For example, in the Unicode character set, `(int)('J')` is 74 and `(int)('0')` is 48. Therefore, `(int)('J') + (int)('0') = 74 + 48 = 122`. Note that `(int)('J') + (int)('0') = 'J' + '0'`. In this case, the cast operators can be omitted. However, you need to use the cast operator to convert from `int` to `char`. Thus, `(char)(74)` is `'J'` and `(char)(48)` is `'0'`.

Self-Check

41. What is `(char)(65)`?
42. True or false: Cast operation can convert "123" to 123.

Advanced Topic 2.7: Increment and Decrement Operators

Suppose `numberOfItems` is an integer variable used to keep track of the number of items found so far. Every time a new item is encountered, `numberOfItems` has to be incremented by one. The Java statement to accomplish this can be written as follows:

```
numberOfItems = numberOfItems + 1;
```

Recall from our discussion on assignment statements, to execute the above statement, the expression on the right-hand side is evaluated first and the result is assigned to the variable. Therefore, if the current value of `numberOfItems` is 25, the right-hand side evaluates to 26, and thus the variable `numberOfItems` becomes 26. Similarly, if the variable `itemsOnStock` is used to keep track of items currently on stock, as soon as an item is sold, the value has to be decremented by 1. The Java statement to accomplish this can be written as follows:

```
itemsOnStock = itemsOnStock - 1;
```

If the current value of `itemsOnStock` is 157, the right-hand side evaluates to 156, and thus the variable `itemsOnStock` becomes 156.

Thus, statements that increment a variable by 1 and statements that decrement a variable by 1 occur quite often in a program. Recognizing this fact, Java (C, C++, C#) provides the *increment operator*, ++, which increments the value of a variable by 1, and the *decrement operator*, --, which decrements the value of a variable by 1. Increment and decrement operators each come in two forms: pre (before the variable) and post (after the variable). Thus, we have the following:

Increment and Decrement Operators

Operator Name	Operator	Example	Equivalent Statement
Preincrement	++	`++items`	`items = item + 1`
Postincrement	++	`items++`	`items = item + 1`
Predecrement	--	`--items`	`items = item - 1`
Postdecrement	--	`items--`	`items = item - 1`

As a stand-alone statement, there is no difference between the pre- and postincrement operators. Similarly, as a stand-alone statement, there is no difference between the pre- and postdecrement operators. However, they are different if used as part of another expression or an assignment statement.

Suppose that `item` is an integer-type variable. If `++item` is used in an expression or an assignment statement, `item` is incremented by `1`, and then the new value of `item` is used to evaluate the expression or the right-hand side value of an assignment statement. However, if `item++` is used in an expression or right-hand side value of an assignment statement, or current value of `item` is used in the expression or right-hand side value of an assignment statement, then the `item` value gets incremented by `1`.

Example 2.28

Consider the following Java statement:

```
int i = 137;
int j = 65;
```

Increment and Decrement Operators and Assignment

Statement	i	j	Equivalent Statements
`i = ++j;`	66	66	`j = j + 1;` `i = j;`
`i = j++;`	65	66	`i = j;` `j = j + 1;`
`i = --j;`	64	64	`j = j - 1;` `i = j;`
`i = j--;`	65	64	`i = j;` j = j – 1;

Example 2.29

Consider the following Java statement:

```
int i = 137;
int j = 65;
int k = 10;
```

Increment and Decrement Operators in an Expression

Statement	i	j	Equivalent Statements
`i = (++j) * k;`	660	66	`j = j + 1;` `i = j * k;`
`i = (j++) * k;`	650	66	`i = j * k;` `j = j + 1;`
`i = (--j) * k;`	640	64	`j = j - 1;` `i = j * k;`
`i = (j--) * k;`	650	64	`i = j * k;` `j = j - 1;`

Good Programming Practice 2.13

Use the increment and decrement operators exclusively in stand-alone statements. Avoid using increment or decrement operator as part of any expression or assignment statement.

Advanced Topic 2.8: Compound Assignment Operators

The compound assignment operators are of the form ⊗=, where ⊗ is any one of the binary operators. The syntax of an assignment statement involving compound assignment operator is

```
variable ⊗= (expression);
```

and the above statement is equivalent to the following assignment statement:

```
variable = variable ⊗ (expression);
```

Example 2.30

This example illustrates the use of compound statements.

Compound Assignment Operator

	Compound Assignment Statement	Equivalent Simple Assignment Statements
1	`total += 10;`	`total = total + 10;`
2	`total += newValue;`	`total = total + newValue;`
3	`available -= sold;`	`available = available - sold;`
4	`salary *= 1 + raise;`	`salary = salary *(1 + raise);`
5	`Cost/= 100.0;`	`cost = cost/100.0;`

6	`String s1 = "Hello ";`
	`String s2 = "there";`
	`s1 += s2;`
		`String s1 = "Hello ";`
		`String s2 = "there";`
		`s1 = s1 + s2;`

Good Programming Practice 2.14

Use the compound operators += and −= only. Use compound operators to increment or decrement a variable by a constant value or by another variable as in the entries 1–3 of table in Example 2.30.

Note 2.17 Java statements fall under two categories: *declaration statements* and *executable statements*. Declaration statements are employed to declare attributes in a class and to declare variables within an operation. Executable statements include all assignment statements and output statements.

REVIEW

1. A class is a collection of items with identical behavior.
2. Attributes keep necessary information for the class.
3. A comment is ignored by the compiler.
4. Every Java program is a collection of one or more classes.
5. A Java application program must have a main method.
6. Every character in an identifier is a letter (A to Z, a to z), or a digit (0 to 9), or the underscore character (_), or the dollar sign ($). The first character of an identifier cannot be a digit.
7. A reserved word cannot be used as an identifier.
8. Java is a case-sensitive language. Therefore, identifiers Cat and cat are not the same.
9. The concatenation operator + can also be used to join a `String` and a numeric value or a character.
10. Every reserved word in Java has lowercase alphabets only.
11. Data type is a set of values along with a set of meaningful basic operations on those data values.
12. Java has eight primitive data types: boolean, byte, char, short, int, long, float, and double.
13. Just as an identifier or a keyword is treated as one token, Java treats special symbols also as one token.
14. Java uses the Unicode character set containing 65536 characters.
15. There are five operations on numeric data types: addition (symbol used is +), subtraction (symbol used is −), multiplication (symbol used is *), division (symbol used is /), and modulus (symbol used is %).
16. The operator precedence rule determines the order in which each of the operators in an expression is being evaluated.
17. Java converts the data type of lesser range to the other data type for the purpose of this expression evaluation. That is, data values are converted as follows: `byte` ➔ `short` ➔ `int` ➔ `long` ➔ `float` ➔ `double` or any combination of these in the same direction.
18. Java allows you to allocate memory using a named constant and store a data value. The value stored remains fixed during the program execution.
19. A named constant is declared and initialized at the same time.
20. In the case of primitive data types, the value is kept at the variable location and in the case of objects, the reference variable do not keep the object, rather reference variable keeps the reference of the object.
21. In an assignment statement, the expression on the right side is evaluated and then the value computed is stored at the variable on the left side.

22. An assignment statement changes the value of the variable on the left-hand side.
23. Using a variable in an expression does not change its contents.
24. If a program receives some input from a human during its execution, then the program is said to be executing in interactive mode.
25. Java provides for explicit data–type conversion using a cast operator.
26. Java statements fall under two categories: declaration statements and executable statements.
27. Declaration statements are employed to declare attributes in a class and to declare variables within a method.
28. Executable statements include all assignment statements and output statements.
29. A class has attributes and operations.
30. A method is the implementation of an operation.
31. A Java statement ends with a semicolon.

EXERCISES

1. Mark the following statements as true or false:
   a. Dog is a class.
   b. My dog Mr. Boomer is an instance of the Dog class.
   c. An object is an instance of a class.
   d. Two objects of a class have identical behavior.
   e. In Java, identifiers Cat and cat are the same.
   f. In Java, a class name must start with an uppercase letter.
   g. An identifier can start with a $ character.
   h. In Java, there is no difference between literals `"a"` and `'a'`.
   i. The operands of + operator must be numbers.
   j. The operands of % can be any two numbers.
   k. In Java, `char` data type variable takes 8 bits of memory.
   l. If one of the operands of the modulus operator is a `double` value, the computation produces a `double` value.
   m. The number `75.` is a literal of type `float`.
   n. Each primitive data type has its own set of operations.
   o. Let `first` and `second` be two `double` variables with values `3.5` and `10.7`, respectively. Then after the statement `first = second` executes, value of `second` changes to `3.5`.
   p. Statement `m = k + 1 + j;` is equivalent to `m = ++k + j;`
   q. Statement `k = k + 1 + j;` is equivalent to `k += j + 1;`

r. An expression is evaluated using one operator at a time.

s. Every Java application program must have exactly one method `main`.

t. A Java application program can have any number of classes.

u. Let `x` and `y` be `3` and `5`, respectively. After the statements `z = x; x = y; y = z` are executed, value of `y` is `3`.

2. Mark the identifier as valid or invalid. If invalid, explain why it is invalid.

a. `James Bond`

b. `wireLess`

c. `Why?`

d. `mail.com`

e. `10DowningSt`

f. `throw`

g. `Println`

h. `Tiger'sPaw`

i. `%profit`

j. `Int`

k. `_ output`

l. `Byte`

m. `AAA`

n. `L235`

o. `-23`

p. `ON-OFF`

q. `Yes/No`

r. `_ $ _ $ _ $`

s. `out`

t. `Public`

3. Select the best answer.

a. The value of `27/2` is

(i) `13` (ii) `13.5` (iii) `13.4999` (iv) none of these

b. The value of `27/3` is

(i) `9.0` (ii) `9` (iii) 8.9999 (iv) none of these

c. The value of `27 % 10` is

(i) `2` (ii) `3` (iii) `7` (iv) none of these

d. The value of `10 % 27` is

(i) `0` (ii) `10` (iii) `2` (iv) none of these

e. The value of `27./2` is

   (i) `13` (ii) `13.5` (iii) `13.0` (iv) none of these

f. The value of `27.0 % 10` is

   (i) `2.0` (ii) `3.0` (iii) `7.0` (iv) none of these

g. The value of `27 - 1/2` is

   (i) `27` (ii) `13` (iii) `13.0` (iv) none of these

h. The value of `- 27 * 2 + 4.0 * 8/3` is

   (i) `−43.3333` (ii) `-46.0` (iii) `-100.0` (iv) none of these

4. Given

```
int a, b, c;
double x, y;
```

Determine the value assigned by each of the following assignment statements. If an assignment statement is not valid, state the reason.

a. `b = 5 ; a = 7 ; c = 9 ;`

b. `c = 2 + a = 2 + b = 5 ;`

c. `x = y = 10.7 ;`

d. `(a + b) = c ;`

e. `a = 2 ; b = 5 * a ; x = 7 * b ;`

f. `x = y = a = b = c = 7;`

g. `x = 2.5; b = x ;`

h. `c = 7; x = 5.0 / 7;`

i. `b = 9; y = b % 7; x = 7 % b;`

j. `x * y = 5.7;`

k. `x = 7.0; y = x + 1; x = y + 3;`

l. `b = 25; c = 35; a = b; b = c; c = a;`

m. `a = 0; b = 10; c = b % a;`

n. `x = 17 % 3;`

o. `y = 17.0 % 3;`

p. `x = 10; y = x % 3;`

q. `b = 10; c = 20; b = b + c; c = b - c; b = b - c;`

5. Which of the following variable declarations are correct? If a variable declaration is not correct, provide the correct variable declaration.

a. `short J,I,j = 12;`

b. `double x = y = z = 5;`

c. `int a, b = 5; c;`

d. `boolean ready, okay = 1;`

e. `char space = " ";`

f. `String welcome = "Hello!"`

6. Which of the following are valid assignment statements? Assume that `b, c, d` are variables of type `double` and `j, k, t` are variables of type `int`, respectively. Further assume that all these variables are already initialized with nonzero values.

   a. `j = j + k / t;`

   b. `d + b = c;`

   c. `j = (k < t) ? 10 : 5;`

   d. `b = ++c - j`

   e. `c = k % t;`

   f. `t = d % 5;`

   g. `k *= k;`

   h. `d = b = c;`

7. Write Java statements that accomplish the following tasks:

   a. Declare two variables `a` and `b` of type `double`.

   b. Declare and initialize a logical variable `fine` to `true`.

   c. Modify the value of a `double` variable `x` by subtracting `17.5`.

   d. Initialize a `double` variable `d` with three times the sum of two `double` variables `b` and `c`.

   e. Increment an `int` variable `k`.

   f. Initialize a `double` variable `d` through an input statement.

   g. Initialize a variable `k` of type `int` through an input statement.

   h. Output statement for variable `c` of type `double`;

   i. Output statement to print the `String` "How are you".

   j. Output statement to print the `String` "\n is the newline character".

8. Suppose `b, c, d,` and `e` are variables of type `double`. What are their values after the execution of the last statement?

```
b = 17.7;
c = 13;
d = c - b;
e = c + b;
b = d + e + 2 * b - c / 4;
c = b - 3 * e + d;
e = b / c;
d = -e + 27 / 2;
```

9. Suppose b, c, d, and e are variables of type int. What are their values after the execution of the last statement?

```
b = 17;
c = 13;
d = c - 2 * b / 3;
e = c + b * d % 4;
b = d / e + 2 * b - c / 4;
c = b % 3 / e * d + 1;
e = b / c + e * 5;
d = -e + 7 / 10 ;
```

10. Suppose b, c are variables of type int and d, e are variables of type double, respectively. What are their values after the execution of the last statement?

```
b = 17;
d = 13.3;
c = 10 - 2 * b / 3;
e = c + d * d / 4;
b = b - c / 3 + 5;
c = b % 3 - 8;
e = b / c - e * 5 - 2;
d = 6 - e + 7 / 2 + 5;
```

11. Suppose b, c are variables of type int such that b = 7 and c = 9. What is the output produced? If any one of the output statement has an error, explain the exact nature of the error.

a. `System.out.println("b = " + b + ", c = " + c");`

b. `System.out.println("b + c = " + b + c);`

c. `System.out.println("b - c = " + b - c);`

d. `System.out.println("b * c = " + b * c);`

e. `System.out.println("b / c = " + b / c);`

f. `System.out.println("b % c = " + b % c);`

g. `System.out.println(b + c + " = b + c");`

h. `System.out.println((b + c) + " = (b + c)");`

12. Repeat Exercise 11. However, assume that b and c are of data type double. Also assume that b and c are 7.0 and 9.0, respectively.

13. Write the output statement necessary for the following:

a. `Good Morning America!`

b. `Good`

```
Morning
America
```

c. `"Good Morning America"`

d. `\Good Morning America\`

e. `/Good Morning America\`

14. Correct the syntax errors.

a.

```
public void class SyntaxErrOne
{
final ratio = 1.8;
static public main(String args())
    {
        int b = c = 10;
        d = c + 5;
        c = C - 10;
        b = d / c;
        ratio = b / d;
        system.out.println("b = + ", b);
            System.out.println("c = ", b);
            System.out.println("Ratio = " + b / d);
    }
}
```

b.

```
public class syntaxErrorTwo{
    static final char new = \7777\;
    static final int TEN;

    public void main(String[] param){
        int a, b, c;

        TEN = a + 5;
        b = a + new;
        c = a + 5 * TEN;
        (b + c) = b;
        System.out.print("a = " + a);
        System.out.print("new = " + new);
        System.out.print("TEN = " + TEN);
    }
}
```

c.

```
void public class SyntaxErrorThree
{
    final int OFFSET 32;
    void public main(String args())
    {
        int b; c = 7 ; d;
        d = c + 5;
        c = d - b;
        b = d / c;
        c = b % OFFSET++;
            c++ = d + b;
        system.out.println("B = + ", b);
            System.out.println("C = ", c);
            System.out.println("D " + d);
    }
}
```

d.

```
public class syntaxErrorFour{
    static final char new = "A";
    static final int SIXTY_SIX = "B";
    public void main(String args){
        int a, b, c,

        10 = a;
        b = a + new;
        c = a + 5 * SIXTY_SEVEN;
        a = ++(b + c);
        System.out.print("a : " + a);
        System.out.print("b : " + b);
        System.out.print("(a+b) : " + (a+b));
    }
}
```

15. Write an equivalent compound statement. If no such statement is possible, write "No such statement." Assume that `u, v, w` are variables of data type `int`.

    a. `u--;`

    b. `u = v - w + u;`

    c. `u = v - w * u`

    d. `u = 3*(v + w) + u;`

    e. `u = u % v;`

16. Write an equivalent simple statement. If no such statement is possible, write "No such statement." Assume that `u, v, w` are variables of data type `int`.

    a. `u -= (v + w);`

    b. `u -= w % v - v;`

    c. `v *= u++;`

    d. `u /= --u;`

    e. `u += v % w;`

17. For each of the objects listed below, identify the class. Suggest possible attributes (if any) and operations (if any) for the class.

    a. This text book.

    b. Your car.

    c. The house at 1256 Farnam Circle.

    d. Your cellular phone.

    e. The Zoo in Great Inland.

18. For each of the sentences below, identify the objects and their classes.

    a. Let me throw a ball at you.

    b. Jill was reading "A passage to South America" by R.G. Wells.

    c. Jack took his neighbor Mark's dog for a walk.

    d. Switch on the TV to watch "This day is history."

    e. Joy just bought a new bike from "Cycle sport."

PROGRAMMING EXERCISES

1. Write a program that prompts the user to input temperature in Centigrade. The program should then output the temperature in Fahrenheit.
2. Write a program that prompts the user to input length, width, and height of a box. The program then outputs the surface area and volume.
3. Write a program that prompts the user to input days, hours, minutes, and seconds it took a mail to reach a destination. The output is total time in seconds for the mail to reach its destination.
4. Given delay time in seconds, determine the number of days, hours, minutes, and seconds it took a mail to reach its destination.
5. Write a program to evaluate the expression $ax^2 + bx + c$. User will be prompted to enter `a, b, c,` and `x`.
6. Write a program to solve the linear equation $ax + b = 0$.
7. According to the grading policy, final grade is determined by the average of four test scores. Create a program to read student's first name, last name, and four test scores. The program outputs the student's last name, four test scores, and the average.

8. Given the monthly salary of an employee, compute the bonus. The bonus is $1000 plus 2% of the amount above $7000 of the employee's annual salary. Assume that every employee has annual salary above $7000.
9. Find the average character code value of a five character word. Print the word, each of its characters along with their codes, and the average.
10. Given an amount between 1 and 100 in cents, determine the number of quarters, dimes, nickels, and cents to be returned. Also print out the total number of coins to be returned.
11. Write a program to compute the average temperature of the day from five readings in Fahrenheit. Program outputs the average temperature in Centigrade.
12. Write a program to estimate the cost of an upcoming vacation. There are four types of expenses: Gas , food, boarding, and entertainments. Gas expense is computed on the basis of cost per mile and estimated miles of travel. Food, boarding, and entertainments are based on cost per day and estimated days for each one of them.
13. From first name, middle initial, last name, and social security number, create a password as in the following example. Example: Meera S. Nair 123-45-6789 will have password M6S4N1.
14. From first name, middle initial, last name, and social security number, create a password as in the following example. Example: Meera S. Nair 123-45-6789 will have password m6s4n1.
15. Write a program to create a tip table similar to the one shown below.

The Tip Table

Amount	15%	20%
10	1.5	2.0
20	3.0	4.0
40	6.0	8.0
60	9.0	12.0
100	15.0	20.0

ANSWERS TO SELF-CHECK

1. Three attributes: First Name, Last Name, Phone Number; one operation: compute grade point average
2. Two attributes: length, width; two operations: compute area, compute perimeter
3. Attribute: radius; operations: compute area, compute perimeter
4. Two attributes: velocity, acceleration; two operations: change velocity, liftoff
5. Within a pair of square brackets
6. keyword
7. `Fine, FiNe, fiNE, fINE`

8. Conveys no purpose or meaning
9. no. class is a reserved word
10. True
11. False
12. True
13. False
14. `System.out.println("All is \"well\" in the eastern border!");`
15. `"NorthAmerica"`
16. `"South Africa"`
17. 19
18. 20
19. `String`
20. data type
21. `double`
22. `int`
23. 0
24. 17
25. 31
26. 5
27. 6.0
28. 6.454545454545455
29. `String productDescription;`
30. `int numOfDependents;`
31. 20, 20
32. `x = 2*x + 3*y;`
33. `Scanner`
34. space, horizontal tab, form feed
35. packages
36. `java.util`
37. `String` class
38. `char`
39. `PrintStream`
40. interactive
41. `char`
42. False

CHAPTER 3

Class Design

In this chapter you learn

- Object-oriented concepts
  - Encapsulation, information hiding, interface, service, and message passing
- Java concepts
  - Class, object, parameter passing, method invocation, method creation, categories of variables, and default constructor
- Programming skills
  - Design, implement, and test a simple Java program in a purely object-oriented way

In Chapter 2, you have learned how to create a simple application program. Every program in Chapter 2 has exactly one class and one method, the `main`, and every variable and executable statement is placed inside the method `main`. In this chapter, you will learn to create new classes. Recall that a class encapsulates both data and operations. You will also learn to create objects or instances of a class. Thus, in this chapter you learn to design a program in a purely object-oriented way.

In keeping with the pedagogical principle of introducing difficult concepts in an incremental fashion, this chapter is intentionally made short and simple. Although various aspects of object-oriented design are introduced in this chapter, many ideas will be explored in more detail in later chapters. For instance, the method invocation and the method creation discussed in this chapter are limited to methods with at most one formal parameter. Methods having more than one formal parameter are introduced in Chapter 6. Similarly, we introduce the concept of a constructor through default constructor. The general case is discussed in Chapter 6.

CLASS

Recall that keeping data along with operations is known as *encapsulation* and is one of the fundamental principles of object-oriented design. Another, different but closely related principle of object-oriented design is that of *information hiding* and it refers to the fact that the user need not and should not have access to the internal parts. The user is provided with a set of operations and all that a user needs to know is the specifications of each of those operations.

Attributes store the data and operations access and manipulate attributes. As you have already seen in Chapter 2, the basic syntax for a `class` definition is

```
[accessModifier] class ClassName
{
    [members]
}
```

where `ClassName` is an identifier you wish to give to the class and `members` consists of attributes and operations of the class. As a general rule, we keep classes `public`. Thus, we use the following simplified syntax template for user-defined classes:

```
public class ClassName
{
    [members]
}
```

Self-Check

1. Consider your television. To use the television, you need not know about its internal parts is an example of ________.
2. In Java, `class` is a ________.

Attributes

Attributes of a class are used to store data and they can be classified into two categories: *instance variables* and *class variables*. Certain attributes are such that each object or instance has its own variable and hence the name instance variable. However, certain attributes are shared by all objects or instances of a class. These attributes are known as class variables and will be introduced in Chapter 7.

The syntax template of an instance variable is

```
[accessModifier] dataType identifierOne[[=LOne], ...,
                                        identifierN[=LN]];
```

Have you ever thought how unwise it is to keep data as `public`? If you have access to your bank's data, you can change your account balance to any value. By the same token, if you are running a business, if you allow your customers complete access to their account, they can modify their account balance as they please. In short, we have complete chaos! Therefore, adhering to the principle of information hiding, unless there is a very compelling reason, we keep all instance variables as `private`. Being private means, only the object alone has access to its instance variables.

Thus, the simplified syntax template for instance variable is

```
private dataType variableName;
```

Example 3.1

In this example, we begin the creation of a new class `Stock` to store information about a stock. Assume that you want to keep information about the number of stocks owned, stock symbol, and dividend per year. The number of stocks currently owned by a person is an integer value. The stock symbol is a string. The dividend per year is a real number. The instance variables of the `Stock` class can be declared as follows:

```
/**
    Keeps ticker symbol, number of shares and dividend
*/
class Stock
{
    private int numberOfShares;
    private String tickerSymbol;
    private double dividend;
    // add code to implement operations
}
```

It is useful to visualize a `Stock` class as a unit of three slots, each slot having its own label (Figure 3.1).

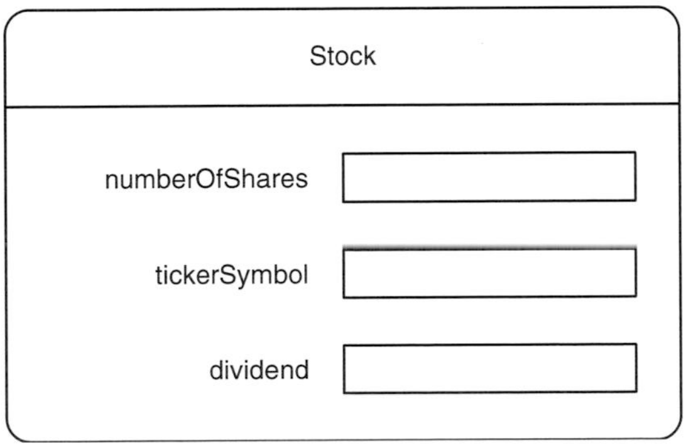

FIGURE 3.1 Visual representation of `Stock` class.

Self-Check

3. Declare an instance variable to keep track of the number of students enrolled in a course.
4. Declare an instance variable to keep track of the maximum grade point average among students enrolled in a course.

Operations

Data being kept as `private`, it is not *directly* accessible to other objects. The operations are `public` and operations provide the necessary *interface* to

- Initialize or modify an instance variable
- Retrieve the current value of an instance variable
- Compute a new value based on current value of an instance variable

Recall that the implementation of an operation is called a method. Before we indulge in creating methods, let us look into the usage or invocation of a method to get some insight.

Introspection

`displayTime` is an operation of all clocks. Can the `displayTime` method of an electronic clock be the same as the `displayTime` method of a mechanical clock?

Eat can be an operation of all mammals. Do humans and dogs eat the same way?

Self-Check

5. As a general rule, keep operations ________.
6. List an operation of your television other than switch on and switch off.

METHOD INVOCATION

A Java program is a collection of *collaborating* objects. By the term *collaborating* we mean whenever an object needs some task to be done, it *invokes* a method of a class that could provide the *service*. In fact, we have done this many times already! In all the programs you have encountered so far, whenever you need to output, you had invoked methods `print` or `println` of the `PrintStream` class. Similarly, in the section on input in Chapter 2, you have invoked methods such as `next`, `nextInt`, and `nextDouble` of the `Scanner` class.

Assume that `scannedInfo` is a reference variable of the type `Scanner`, and `age` is a variable of the type `int`. The following Java statement

```
age = scannedInfo.nextInt();
```

invokes the method nextInt of the Scanner class. The syntax template for invoking a method using a reference variable is

```
referenceVariableName.methodName([explicitParameters]);
```

The reference variable referenceVariableName is known as the *implicit parameter*. The method nextInt *returns* an int value. Therefore, we have declared an int variable age in our program and used the assignment operator = to store the value returned by nextInt in the variable age. If we do not use an assignment statement, the return value will be lost forever. Since method nextInt returns a value, we call nextInt a *value returning* method. Further, since nextInt returns a value of type int, we say method nextInt is of the type int. A value returning method can be used in any expression where we may use a value. To be specific, the nextInt method can be used in any expression where we may use an int.

A value returning method can be invoked in three different ways:

1. *Invoke it on the right-hand side of an assignment statement.* In this case, the value returned by the method is stored in the variable on the left-hand side of the assignment statement. Using this technique, the value returned by the method can be used in other expressions or output statements.
2. *Invoke it as part of an expression.* In this case, value returned by the method is used only once. Since the value is not stored in a variable, it cannot be used in any other calculations or any other output statements.
3. *Invoke it as part of an output statement.* The value is not stored in a variable, and therefore it cannot be used in any other calculations or any other output statements.

Not all methods return values. For example, println method of the PrintStream class does not return any value. Since method println does not return any value, we say println is of type void or a void *method*. You cannot use void methods as part of an expression. A value returning method is invoked as part of an expression or stand-alone statement, whereas a void method can only be invoked as a stand-alone Java statement.

Example 3.2

This example illustrates various ways of method invocations (Table 3.1).

Self-Check

7. What is the type of the method charAt?
8. What is the type of the method nextDouble?

TABLE 3.1 Illustration of Method Invocation

Method Invocation	Remark
`System.out.println("Hello");`	This is a valid invocation. The method `println` is a `void` method.
`String word;` `word = System.out.println("Hello");`	Invalid. The method `println` is a `void` method and cannot appear in an assignment statement.
`String word = "Try me!";` `word.charAt(1);`	Invalid. The method `charAt` is a value returning method. The returned value `'r'` is lost.

`String word = "Try me!";`
`boolean` `letter;`
`letter = word.charAt(1);`
	Invalid. The method `charAt` returns a `char` value. The returned value `'r'` must be assigned to variable of type `char`.

`String word = "Try me!";`
`char` `letter;`
`letter = word.charAt(1);`
	Valid. The variable `letter` contains the character `'r'`.

`Scanner scannedInfo =`
   `new` `Scanner(System.in);`
`scannedInfo.nextInt();`
	Invalid. The value returned by the method `nextInt` is lost.

`Scanner scannedInfo =`
   `new` `Scanner(System.in);`
`int` `salary;`
`salary = scannedInfo.nextDouble();`
	Invalid. The value returned by the method is of type `double` and hence assigning to an `int` will result in a syntax error.

`Scanner scannedInfo =`
   `new` `Scanner(System.in);`
`double` `salary;`
`salary = scannedInfo.nextDouble();`
	Valid. The variable `salary` contains the `double` value returned by `nextDouble` method.

METHOD DEFINITION

You are already familiar with the syntax template of the method `main` as shown below:

```
public static void main (String[] args)
{
    [statements]
}
```

Observe that the method consists of two parts: the header

```
public static void main (String[] args)
```

and the body

```
{
    [statements]
}
```

The header starts with the keyword `public` that makes the method accessible to other objects. If you use the keyword `private`, the method cannot be invoked by other objects.

The keyword `static` refers to the fact that method `main` is shared by all objects belonging to the class. This topic requires further explanation and therefore it is described in more detail in Chapter 7. For the present, note that we define all the methods other than the `main` as nonstatic methods and as such the keyword `static` will be omitted.

The next keyword is the return type of the method. The method `main` is a `void` method. In a method, the correct return type needs to be specified.

After the return type, you must provide an identifier as the name of the method. The naming convention for methods is identical to that of an instance variable.

The header ends with a list of *formal parameters* (or *arguments*) that are enclosed within a pair of parentheses. The formal parameter list in this case is `String[] args.` You need to know arrays for a full explanation. The list of formal parameters in a method header can be empty. However, the pair of parentheses enclosing the formal parameter cannot be omitted.

The syntax template of a method is as follows:

```
[accessModifier] [abstract|final] [static] returnType
                                    name(paramList)
{
    [statements]
}
```

The access modifier of all the methods in this chapter is `public`. The optional keywords `abstract`, `final`, and `static` will be explained in later chapters. The vertical bar between `abstract` and `final` indicates that both of them cannot appear simultaneously. All the methods you create in this chapter will have an empty parameter list or one formal parameter. Thus, the syntax template of all the methods you create in this chapter other than the method `main` can be given as follows:

```
public returnType methodName()
{
    [statements]
}
```

or

```
public returnType methodName(dataType variable)
{
    [statements]
}
```

All methods of a class have access to its instance variables. In the case of methods with one formal parameter, the formal parameter modifies the behavior of the method. For example, consider the following Java statements:

```
char seventhChar = eighthChar = ' ';
String currentLine = "It is quite sunny today."
seventhChar = currentLine.charAt(6);                    // (1)
eighthChar = currentLine.charAt(7);                     // (2)
```

It may be noted that in Statement 1 `charAt` returns `'q'` and in Statement 2 the same method returns the character `'u'`. Thus, the *actual parameter* values such as 6 and 7 are crucial in determining the value returned.

In the case of `void` methods, arguments play an important role as well. For instance, consider the following two `println` statements:

```
System.out.println("Happy Birthday to you");
```

outputs `Happy Birthday to you,` and

```
System.out.println("Good Night; Sweet Dreams!");
```

outputs `Good Night; Sweet Dreams!.` Thus, arguments provide the additional information necessary to carry out the service specified by a method.

The syntax for declaring a formal parameter is

```
dataType parameterName
```

where `dataType` is the data type of the formal parameter `parameterName`. When a method is invoked, the actual parameter value is copied into the formal parameter.

Inside the method, you can declare additional variables and they are known as local variables. A local variable is not accessible outside the block it is declared. Therefore, no access modifier is required for a local variable. Local variables are discussed in Chapter 2.

Self-Check

9. What is the type of the formal parameter of `charAt`?
10. True or false: A method may not have any formal parameter.

CATEGORIES OF VARIABLES

So far, you have seen three kinds of variables: instance variable, local variable, and formal parameter. The following discussion summarizes the similarities and dissimilarities among them. There is one more type of variables in Java called, class variables (also known as `static` variables). They are introduced in Chapter 7.

Syntax Template

Instance variable

```
[accessModifier]  dataType identifierOne[[=LOne], ...,
                                   identifierN[=LN]];
```

Local variable

```
dataType identifierOne[[=LOne], identifierTwo[=LTwo], ...,
                                      identifierN[=LN]];
```

Formal parameter

```
dataType identifierOne[, ..., dataType identifierN]
```

Self-Check

11. A local variable is declared inside a ________.
12. True or false: A local variable has no access modifier.

Initialization

An instance variable can be initialized through a constructor as well. The concept of a constructor is explained later in this chapter. Therefore, quite often an instance variable is not initialized during declaration.

It is the programmers' responsibility to initialize a local variable before using it. Therefore, programmers quite often initialize a local variable during declaration.

A formal parameter is initialized during the method invocation and the actual parameter value is copied to the formal parameter before the execution of the very first executable statement of the method.

Self-Check

13. Declare and initialize a local variable to store age.
14. Declare and initialize an instance variable to store interest rate.

Scope

An instance variable is available only within the object (if it is declared `private`).

A local variable is available only within the block from its point of declaration. Recall that a block is a sequence of statements enclosed within a pair of braces {and}. You will see block statements in Chapter 4.

A formal parameter is available within the method.

Self-Check

15. True or false: Scope of a local variable is less than that of a formal parameter.
16. True or false: All local variables have the same scope.

Existence

An instance variable exists as long as the associated object exists.

A local variable is created every time declaration statement is executed during the method execution and the variable ceases to exist upon the completion of the block it is declared.

A formal parameter is created at the beginning of the method invocation and it ceases to exist upon the completion of the method.

Self-Check

17. True or false: A local variable exists throughout the method.
18. True or false: A formal parameter exists throughout the method.

`return` STATEMENT

The last statement of all value returning methods in this chapter will have the following syntax:

```
return [returnValue];
```

where `returnValue` is a literal value or a variable or an expression that matches the return type of the method. This statement returns the `returnValue` and makes it available where the method was invoked. Further, the control goes back to the Java statement that invoked the method.

In the case of a `void` method, no `return` statement is necessary. However, you can have `return` statement in a `void` method so long as `returnValue` is omitted.

The word `return` is a keyword in Java.

Self-Check

19. True or false: Every value returning method must return a value.
20. True or false: A `void` method need not have any return statement.

JAVADOC CONVENTION

The comments for a method to be processed by the javadoc utility is placed immediately above the method between /** and */ similar to the comments for a class. The first line must describe the functionality of the method. Subsequent lines start with @param and provide brief description of each formal parameter of the method. In the case of a value returning method, the last line starts with @return. We will follow these conventions for all methods presented in this book.

ACCESSOR METHOD

A method that does not modify any instance variable of an object is called an accessor method. All other methods are known as mutator methods.

Let `dataMember` be an instance variable of a class of the type `dataType`. Since a `dataMember` is declared as `private`, it cannot be accessed outside the object. Therefore, if the value of `dataMember` is required outside the class, you must provide a method. An accessor method that returns the value of an instance variable needs no additional information. Thus, no formal parameter is required. Therefore, we start with the syntax template for a method with no formal parameters. That is,

```
public returnType methodName()
{
    [statements]
}
```

The method returns a value of data type `dataType`, the method is of type `dataType`. By convention, quite often followed by Java programmers, the name of the method is of the form `getDataMember`. In other words, the name of the method is created by concatenating the word `get` and the name of the instance variable with the first letter of the instance variable's name changed to uppercase. Finally, the only statement required in the body is a `return` statement to return the `dataMember`. The syntax template of an accessor method that returns the value of `dataMember` is

```
public dataType getDataMember()
{
    return dataMember;
}
```

Example 3.3

In this example, we provide the accessor methods that return the value of an instance variable of the `Stock` class introduced in Example 3.1.

```
/**
    Accessor method for the number of shares
    @return the number of shares
*/
public int getNumberOfShares()
{
    return numberOfShares;
}

/**
    Accessor method for the ticker symbol
    @return the ticker symbol
*/
```

```
public String getTickerSymbol()
{
    return tickerSymbol;
}

/**
    Accessor method for the dividend
    @return the dividend
*/
public double getDividend()
{
    return dividend;
}
```

Self-Check

21. True or false: An accessor method may modify some instance variables.
22. True or false: An accessor method returns a value.

MUTATOR METHOD

Accessor methods do not change the value of any instance variable. A method that changes the value of one or more instance variables is called a mutator method.

Let `dataMember` be an instance variable of the type `dataType` of a class. Then due to the principle of information hiding, `dataMember` can only be modified by the object. Therefore, a mutator method to modify the `dataMember` is often essential and such a mutator method requires one formal parameter of the type `dataType`. Therefore, we start with the syntax template for a method with one formal parameter. That is,

```
public returnType methodName(dataType variable)
{
    [statements]
}
```

A mutator method that modifies an instance variable has nothing to return and thus is a `void` method. Once again, by convention often followed by Java programmers, a mutator method that modifies `dataMember` has the name `setDataMember`. In other words, the name of the method is created by concatenating the word `set` and the name of an instance variable with the first letter of the instance variable name changed to uppercase. As observed above, such a mutator method requires one argument of the type `dataType`. There is no convention for naming formal parameters. Throughout this book we name the single argument appearing in a mutator method by concatenating the word `in` and the name of an instance variable with the first letter of the instance variable name changed to uppercase. The syntax template of all mutator methods that modify an instance variable of the class is

```
public void setDataMember (dataType inDataMember)
{
    dataMember = inDataMember;
}
```

Note that the only statement required in the body is an assignment statement, which assigns the incoming value to the instance variable.

Example 3.4

In this example, we develop the mutator methods of the `Stock` class introduced in Example 3.1.

```
/**
    Mutator method to set the number of shares
    @param inNumberOfShares new value for the number of
                                                 shares
*/
public void setNumberOfShares(int inNumberOfShares)
{
    numberOfShares = inNumberOfShares;
}

/**
    Mutator method to set the ticker symbol
    @param inTickerSymbol new value for the ticker symbol
*/
public void setTickerSymbol(String inTickerSymbol)
{
    tickerSymbol = inTickerSymbol;
}

/**
    Mutator method to set the dividend
    @param inDividend new value for the dividend
*/
public void setDividend(double inDividend)
{
    dividend = inDividend;
}
```

Self-Check

23. True or false: A method can be both an accessor and a mutator.
24. True or false: A mutator method may not have a `return` statement.

toString METHOD

It is a good programming practice to provide a toString method. The toString method returns a String that contains essential information. Observe that toString is an accessor method that returns a String. Thus, a toString method has the following form:

```
public String toString()
{
    String str;

    // create the String str to be returned.
    return str;
}
```

Example 3.5

In this example, we develop the toString method for the Stock class. For a stock, the most essential information is the number of shares currently owned and the stock symbol. Thus, we have the following toString method for the Stock class:

```
/**
    The toString method
    @return number of shares and ticker symbol
*/
public String toString()
{
    String str;
    str = numberOfShares + " " + tickerSymbol;
    return str;
}
```

Self-Check

25. The toString is a ________ method.
26. True or false: Every class must have a toString method.

APPLICATION-SPECIFIC METHODS

In general, every class needs accessor methods and mutator methods for every instance variable. In addition, it is a good idea to include the accessor method toString. Apart from these, you may need additional methods. Quite often, these additional methods are application specific. In this section, we illustrate application specific methods for the Stock class.

Example 3.6

A method that returns the total dividend for a stock may be a useful method. The return type of the method is double and the method is best named yearlyDividend. Thus, we have the following accessor method.

```
/**
    Computes and returns yearly dividend
    @return the yearly dividend
*/
public double yearlyDividend()
{
    double totalDividend;
    totalDividend = numberOfShares * dividend;
    return totalDividend;
}
```

CONSTRUCTOR

As mentioned before, a constructor is used to initialize instance variables. The syntax template of a constructor can be thought of as a mutator method with no return type and name the same as the name of the class. Thus, a constructor for the Stock class can be written as follows:

```
/**
    Constructs a Stock with zero shares
*/
public Stock()
{
    numberOfShares = 0;
    tickerSymbol = "[NA]";
    dividend = 0.0;
}
```

Constructors are covered in detail in Chapter 6.

Self-Check

27. True or false: A constructor is a method.
28. True or false: One of the purposes of a constructor is to initialize instance variables.

PUTTING ALL PIECES TOGETHER

We are now ready to put all the pieces of a class together. There is no specific order for methods or instance variables. However, throughout this book the following order is maintained:

1. Instance variables
2. Constructors
3. Application specific methods
4. Accessor methods returning the instance variable value
5. Mutator methods modifying the instance variable
6. `toString` method

Once again, we stress that this order is purely the author's choice and you need not follow this. The `Stock` class we developed is presented in Example 3.7.

Example 3.7

```
/**
    Keeps ticker symbol, number of shares and dividend
                                            information.
*/
class Stock
{
    private int numberOfShares;
    private String tickerSymbol;
    private double dividend;

    /**
        Constructs a Stock with zero shares
    */
    public Stock()
    {
        numberOfShares = 0;
        tickerSymbol = "[NA]";
        dividend = 0.0;
    }

    /**
        Computes and returns yearly dividend
        @return the yearly dividend
    */
```

```
public double yearlyDividend()
{
    double totalDividend;
    totalDividend = numberOfShares * dividend;
    return totalDividend;
}

/**
    Accessor method for the number of shares
    @return the number of shares
*/
public int getNumberOfShares()
{
    return numberOfShares;
}

/**
    Accessor method for the ticker symbol
    @return the ticker symbol
*/
public String getTickerSymbol()
{
    return tickerSymbol;
}

/**
    Accessor method for the dividend
    @return the dividend
*/
public double getDividend()
{
    return dividend;
}

/**
    Mutator method to set the number of shares
    @param inNumberOfShares new value for the number
                                            of shares
*/
public void setNumberOfShares(int inNumberOfShares)
{
    numberOfShares = inNumberOfShares;
}
```

```
    /**
        Mutator method to set the ticker symbol
        @param inTickerSymbol new value for the ticker
                                                symbol
    */
    public void setTickerSymbol(String inTickerSymbol)
    {
        tickerSymbol = inTickerSymbol;
    }


    /**
        Mutator method to set the dividend
        @param inDividend new value for the dividend
    */
    public void setDividend(double inDividend)
    {
        dividend = inDividend;
    }


    /**
        The toString method
        @return number of shares and ticker symbol
    */
    public String toString()
    {
        String  str;
        str = numberOfShares + " " + tickerSymbol;
        return str;
    }
}
```

Advanced Topic 3.1: Representing Class in UML 2

As mentioned in Chapter 1, the unified modeling language (UML) is a standard language for software specification. The UML uses mostly graphical notations to express the design of software projects. The unified modeling language version 2 (UML 2) is the latest version of UML. The UML 2 notation of the `Stock` class is shown in Figure 3.2.

There are three distinct areas: class name area, instance variable area, and operations area. The − sign and + sign indicate the access modifiers `private` and `public`, respectively. A simplified notation for the `Stock` class in UML 2 notation is shown in Figure 3.3 and it is quite useful in showing relationships among classes.

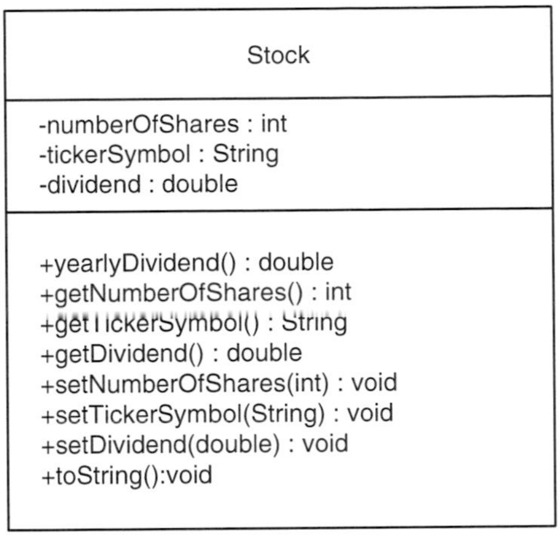

Stock
-numberOfShares : int -tickerSymbol : String -dividend : double
+yearlyDividend() : double +getNumberOfShares() : int +getTickerSymbol() : String +getDividend() : double +setNumberOfShares(int) : void +setTickerSymbol(String) : void +setDividend(double) : void +toString():void

FIGURE 3.2 Class diagram of `Stock` class.

Stock

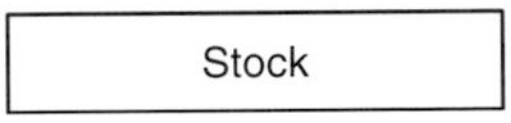

FIGURE 3.3 Simplified class diagram of `Stock` class.

TESTING

Every class you create must be tested thoroughly before it is used in any application. The most general approach to test a class is to create an application program and test every method of the class you have developed. This approach is followed in this book and is illustrated in the next example. Java development environments such as BlueJ provide the facility to test a class without an application program. The readers are encouraged to test their class by creating an application program.

Example 3.8

In this example we create an application to test the `Stock` class.

As you know by now, every application has at least one class and one of the classes of the application has a method `main`. Therefore, we create an application having exactly one class that contains the method `main` as shown below:

```
public class StockTesting
{
    public static void main (String[] args)
    {
        //Java statements to test Stock class
    }
}
```

To begin with, to test any method of the Stock class, you must create an instance of the Stock class. The required Java statement is

```
Stock testStock = new Stock();      //(1)
```

The left-hand side of the above assignment statement declares a reference variable testStock of the type Stock. The right-hand side creates a new object belonging to the Stock class and returns the reference. In fact the new on the right-hand side is an operator that creates an instance of the class on the basis of the constructor that follows. For example, Stock() is a constructor with no formal parameter. Such a constructor is known as *default constructor* and is present in every class by default so long as there is no user-defined constructor. A default constructor initializes every instance variable of the object with default values. In this example, we have already included a constructor with no formal parameter, and therefore the right-hand side of Statement 1 creates an instance of the Stock class, initializes numberOfShares to 0, tickerSymbol to "[NA]", and dividend to 0.0. The new operator returns the reference of the object created. Thus, the variable testStock has the reference of the new object created. Note that Statement 1 is equivalent to the following two statements:

```
Stock testStock;                 //(2)
testStock = new Stock();         //(3)
```

Statement 2 declares a local variable testStock. Statement 3 creates a new instance of Stock and initializes the reference variable testStock with the reference of the newly created instance returned by the new operator. In other words, Statement 3 instantiates the local variable testStock.

Now to test a pair of accessor and mutator methods corresponding to an instance variable, the best way is to get an input value, use the mutator method to store it, and then output it with the help of corresponding accessor method. Once all instance variables received valid data values, application specific methods and toString method can be tested. Thus, we have the following Java application program:

```
import java.util.Scanner;

/**
    An application class to test Stock class
*/
public class StockTesting
{
    public static void main (String[] args)
```

```
{
    //Create an object belonging to the class Stock
    Stock testStock = new Stock();

    //Declare local variables
    int inputNumberOfShares, outputNumberOfShares;
    String inputTickerSymbol, outputTickerSymbol;
    double inputDividend, outputDividend;
    double outputYearlyDividend;

    //Get input values
    Scanner scannedInfo = new Scanner(System.in);
    System.out.print("Enter numbers own, stock symbol"
                            + " and dividend : ");
    System.out.flush();
    inputNumberOfShares = scannedInfo.nextInt();
    inputTickerSymbol = scannedInfo.next();
    inputDividend = scannedInfo.nextDouble();

    System.out.println();

    //Test mutator and accessor methods
    testStock.setNumberOfShares(inputNumberOfShares);
    testStock.setTickerSymbol(inputTickerSymbol);
    testStock.setDividend(inputDividend);

    outputNumberOfShares
                = testStock.getNumberOfShares ();
    outputTickerSymbol = testStock.getTickerSymbol();
    outputDividend = testStock.getDividend();

    System.out.println("Number of shares : " +
                      outputNumberOfShares);
    System.out.println("Stock symbol is : " +
                       outputTickerSymbol);
    System.out.println("Dividend per share is : " +
                                 outputDividend);

    //Test yearlyDividend method
    outputYearlyDividend = testStock.yearlyDividend();
    System.out.println("Yearly Dividend is : " +
                        outputYearlyDividend);

    //Test toString method
```

```
        System.out.println(testStock.toString());
    }
}
```

Output

```
Enter numbers own, stock symbol and dividend: 500 XYZ 1.43

Number of shares : 500
Stock symbol : XYZ
Dividend per share : 1.43
Yearly Dividend : 715.0
500 XYZ
```

Self-Check

29. True or false: Every class needs to be tested.
30. Testing will help identify ________ errors.

Advanced Topic 3.2: Representing Relationship in UML 2

The UML 2 class diagram of `StockTesting` application is presented in Figure 3.4. The association is the most common relationship that exists among classes and it represents the relationship among instances of classes. The multiplicity of the association denotes

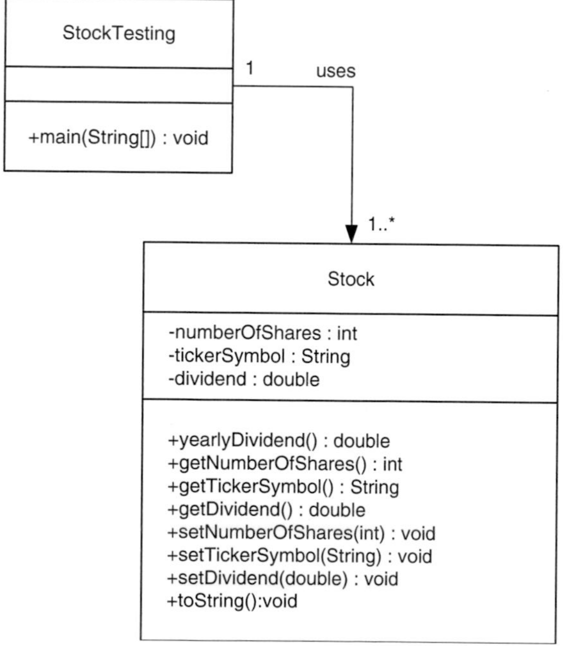

FIGURE 3.4 Class diagram of `StockTesting` application.

the number of objects participating in such a relationship. For example, an instance of the class `StockTesting` uses one or more instances of the class `Stock`. Thus, there is an association between class `StockTesting` and class `Stock`. The number 1 appearing near the class `StockTesting` indicates that corresponding to one `Stock` object there is 1 `StockTesting` object and the notation 1..* appearing near the class `Stock` indicates that corresponding to one `StockTesting` object there can be 1 to many `Stock` objects. Both 1 and 1..* are known as the *multiplicity*.

Advanced Topic 3.3: Class Design, Implementation, and Testing

In this section, we continue with the software development process outlined in Chapter 1. As mentioned in Chapter 1, design, implementation, and testing phases are covered in this section. We list them first for easy reference and then proceed to elaborate on each of those phases through an example.

Phase 2. Design

Step 1. Decide on attributes

Step 2. Decide on methods

Phase 3. Implementation (or create classes using Java constructs)

Phase 4. Testing

Design

The primary aim of this phase is to decide on various classes required and assign responsibilities to each one of them. Design of the software begins with a formal specification of the intended product. The following is an example of a formal specification of a circular counter.

Example 3.9

A circular counter counts 0, 1, 2, ..., limit − 1. Once the counter reaches limit − 1, the next value is not limit, rather it is 0. In other words, a circular counter can count up to limit − 1 and then it resets to 0. The value of the limit can be any integer greater than 1.

From the above specification, it is quite clear that the software we develop must provide a "get counter value" service to the user. Without such a service, the counter is of no use. Being a counter, there needs to be service to "increment the counter value." Another useful service is "set counter value." Being a circular counter, there is a limit value and there needs to be services to set and get the limit value. On the basis of this analysis, we can create the use case diagram given in Figure 3.5.

The design phase starts where the use case analysis left off. Thus in this example, from the use case diagram given in Figure 3.5, we tentatively decide to have one class named `CircularCounter`.

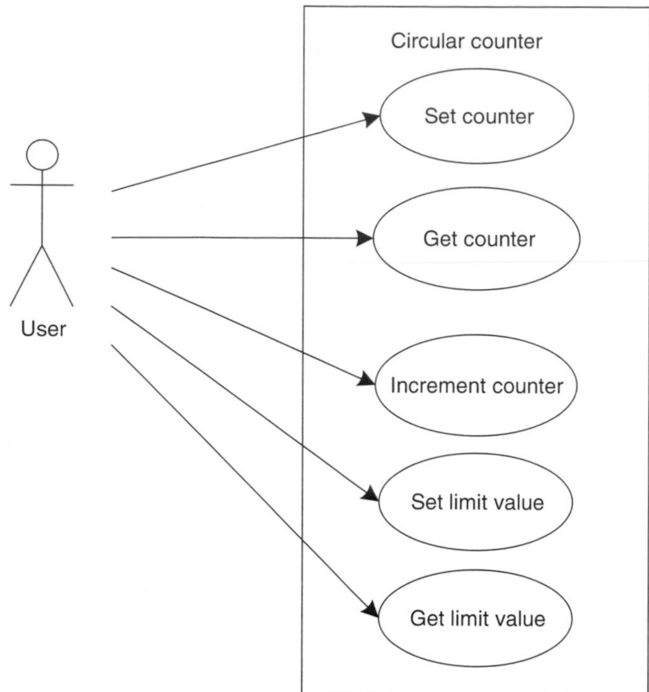

FIGURE 3.5 Use case diagram for the circular counter.

Decide on Attributes

From the use case diagram, it is quite clear that the circular counter must keep the current value of the counter and limit value. Clearly, both of these values can be of `int` data type. Therefore, the `CircularCounter` class needs the following two attributes:

```
private int counterValue;
private int limitValue;
```

Decide on Methods

In this step you need to decide on methods. As explained before, you may include accessor and mutator methods corresponding to each instance variable. You may also provide a `toString` method. So in this section the focus is on additional methods that are needed for this class to meet its specifications.

The "increment counter" is the only use case that needs to be addressed. Therefore, there must be a method to increment the counter value by one. In this case, you must clearly state preconditions and postconditions. *Preconditions* refer to the necessary conditions for the method to behave as specified. *Postconditions* refer to the conditions satisfied by instance variables at the completion of the method invocation. The set of values of all instance variables is called the *state* of an object. Thus, preconditions and postconditions specify the states of an object before and after a method have been invoked. In this case, the preconditions are limitValue > 0 and 0 ≤ counterValue < limitValue. The postconditions are limitValue > 0, 0 ≤ counterValue < limitValue and counterValue is incremented by 1 in the circular order 0, 1, 2, . . . , limitValue - 1, 0.

Thus, the class has the following six methods:

```
public void incrementCounterValue()
//     Preconditions :   limitValue > 0;
//                       0 ≤ counterValue < limitValue.
//     Postconditions :  limitValue > 0;
//                       0 ≤ counterValue < limitValue;
//                       counterValue is incremented by 1
//                       in the circular order
//                       0, 1, 2, ..., limitValue - 1, 0.
public int getCounterValue()
public int getLimitValue()
public void setCounterValue(int inCounterValue)
public void setLimitValue(int inLimitValue)
public void toString()
```

Consider the following expression:

```
(counterValue + 1) % limitValue
```

Assume that limitValue is 5. Now counterValue has to be one of the following: 0, 1, 2, 3, and 4. Let us evaluate the above expression for each of those values (see Table 3.2).

Thus, the above expression computes the next value based on counterValue as desired. Therefore, the Java statement

```
counterValue = (counterValue + 1) % limitValue
```

increments instance variable counterValue in circular order 0, 1, 2, . . . , limitValue - 1, 0.

The circular counter can be visualized as in Figure 3.6 and the class diagram is given in Figure 3.7.

TABLE 3.2 Illustration of Circular Increment

counterValue	**(counterValue +1) % limitValue**
0	(0 + 1) % 5 = 1 % 5 = 1
1	(1 + 1) % 5 = 2 % 5 = 2
2	(2 + 1) % 5 = 3 % 5 = 3
3	(3 + 1) % 5 = 4 % 5 = 4
4	(4 + 1) % 5 = 5 % 5 = 0

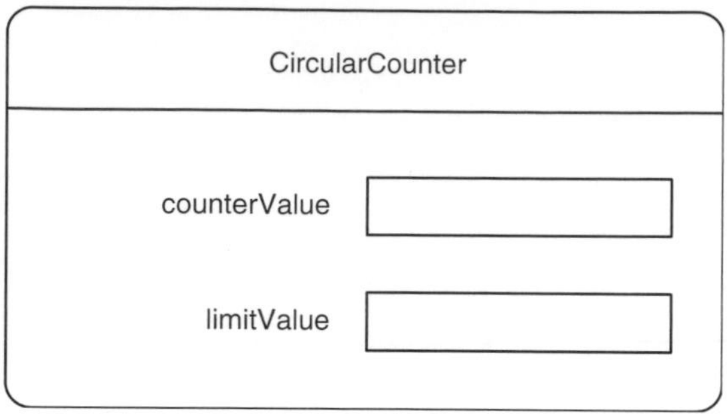

CircularCounter

counterValue

limitValue

FIGURE 3.6 Visualization of `CircularCounter`.

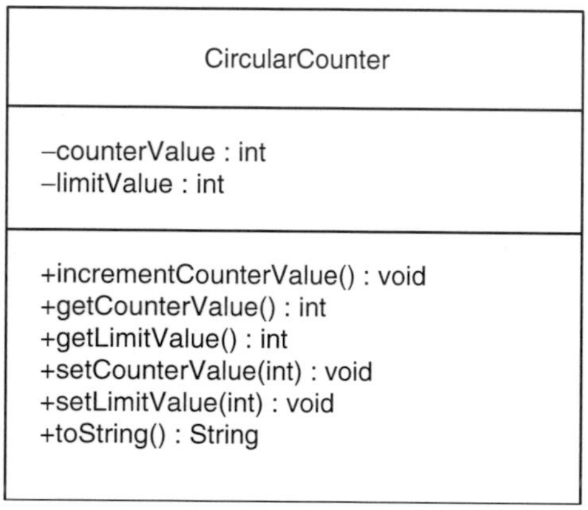

CircularCounter

–counterValue : int
–limitValue : int

+incrementCounterValue() : void
+getCounterValue() : int
+getLimitValue() : int
+setCounterValue(int) : void
+setLimitValue(int) : void
+toString() : String

FIGURE 3.7 Class diagram of `CircularCounter` class.

Implementation

In this phase, we use Java programming language to code the class(es) designed in the design phase.

```
/**
    Circular counter counts 0, ..., limit - 1, 0
*/
public class CircularCounter
{
    private int counterValue;
    private int limitValue;

    /**
        Constructs a circular counter limitValue 100;
                                     counterValue 0.
    */
    public CircularCounter()
    {
        counterValue = 0;
```

```
    limitValue = 100;
}

/**
    Increments the circular counter
*/
public void incrementCounterValue()
{
    counterValue = (counterValue + 1) % limitValue;
}

/**
    Accessor method for the counter value
    @return the counter value
*/
public int getCounterValue()
{
    return counterValue;
}

/**
    Accessor method for the limit value
    @return the limit value
*/
public int getLimitValue()
{
    return limitValue;
}

/**
    Mutator method to set the counter value
    @param inCounterValue new value for the counter value
*/
public void setCounterValue(int inCounterValue)
{
    counterValue = inCounterValue % limitValue;
}

/**
    Mutator method to set the limit value
    @param inLimitValue new value for the limit value
*/
public void setLimitValue(int inLimitValue)
```

```
    {
        limitValue = inLimitValue;
    }

    /**
        The toString method
        @return counter value and limit information
    */
    public String toString()
    {
        String  str;
        str = "Counter (0 to " + (limitValue - 1) + ") value : "
                                                  + counterValue;
        return str;
    }

}
```

Testing

In this phase we create an application program to test the class `CircularCounter`. The UML 2 diagram is shown in Figure 3.8 and the code is as follows:

```
import java.util.Scanner;

/**
    The application tester class for circular counter
*/
public class CircularCounterTesting
{
    public static void main (String[] args)
    {

        //Create an instance of CircularCounter
        CircularCounter testCircularCounter = new
                                  CircularCounter();

        //Declare variables to input and output counterValue
        //     and limitValue

        int inputCounterValue, outputCounterValue;
        int inputLimitValue, outputLimitValue;
        int outputIncrementCounterValue;

        //Get two input values
        Scanner scannedInfo = new Scanner(System.in);
```

```
        System.out.print("Enter counter value and
                                limit value : ");
        System.out.flush();
        inputCounterValue = scannedInfo.nextInt();
        inputLimitValue = scannedInfo.nextInt();
        System.out.println();

        //Test mutator and accessor corresponding
        //to instance variables.
        testCircularCounter.setCounterValue(inputCounterValue);
        testCircularCounter.setLimitValue(inputLimitValue);

        outputCounterValue =
            testCircularCounter.getCounterValue();
        outputLimitValue =
              testCircularCounter.getLimitValue();

        System.out.println("Counter value : "+
                          outputCounterValue);
        System.out.println("Limit value : "+
                          outputLimitValue);

        //Test incrementCounterValue method
        System.out.println("Counter is incremented
                                    five times ");
        System.out.println(testCircularCounter);
        testCircularCounter.incrementCounterValue();
        System.out.println(testCircularCounter);
        testCircularCounter.incrementCounterValue();
        System.out.println(testCircularCounter);
        testCircularCounter.incrementCounterValue();
        System.out.println(testCircularCounter);
        testCircularCounter.incrementCounterValue();
        System.out.println(testCircularCounter);
        testCircularCounter.incrementCounterValue();
        System.out.println(testCircularCounter);
    }
}
```

Output

```
Enter counter value and limit value : 3 5

Counter value : 3
Limit value : 5
```

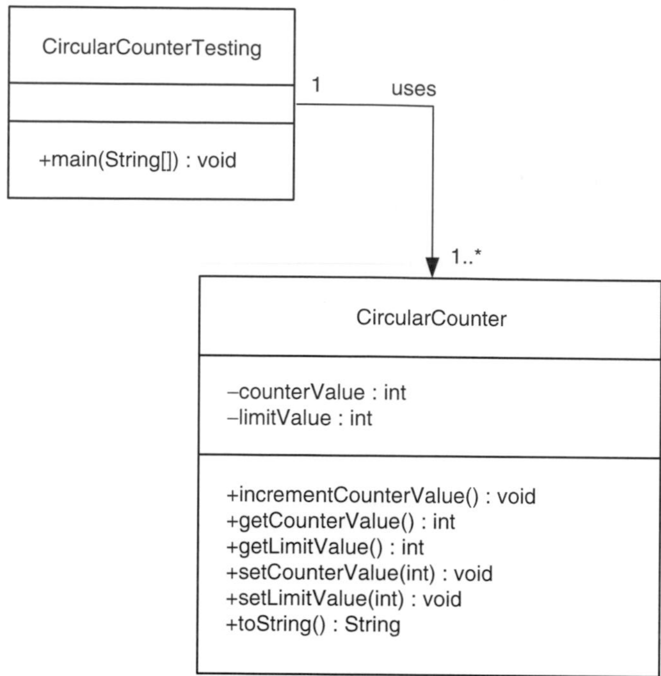

FIGURE 3.8 Class diagram of circular testing program.

```
Counter is incremented five times
Counter (0 to 4) value : 3
Counter (0 to 4) value : 4
Counter (0 to 4) value : 0
Counter (0 to 4) value : 1
Counter (0 to 4) value : 2
Counter (0 to 4) value : 3
```

Note 3.1 Compare Figures 3.4 and 3.8. Observe that the essential difference is that in Figure 3.4 we have the `Stock` class and in Figure 3.8 we have `CircularCounter` class. In other words, once the UML 2 class diagram for a class is created, the UML 2 class diagram for the corresponding testing program is quite obvious. Therefore, we omit such figures.

REVIEW

1. A class has attributes and operations: attributes are used to store the data, and operations access and manipulate the data values.
2. Keeping data along with operations is known as encapsulation.
3. Unless there is a very compelling reason, all attributes are declared as `private`.
4. A `private` attribute (or operation) is accessible only to the object.

5. A `public` attribute (or operation) is accessible to any object.
6. Operations of a class are generally declared as `public`.
7. Implementation of an operation is called a method.
8. Operations provide the necessary interface to initialize or modify an attribute, retrieve the current value of an attribute, and compute a new value based on current value of attributes.
9. A Java program is a collection of collaborating objects.
10. The syntax for invoking a method using a reference variable is

```
referenceVariableName.methodName();
```

    The reference variable is called an explicit parameter.
11. There are two types of methods: value returning methods and `void` methods.
12. A `void` method cannot be invoked in an expression. It is invoked as a stand-alone Java statement.
13. A value returning method can be invoked in an assignment statement or can be used in an expression or as part of an output statement. If the value returned is not stored using an assignment statement, it will be lost forever.
14. A method has two parts: the header

```
public static void main (String[] args)
```

    and the body

```
{
     [statements]
}
```

15. The header ends with a list of formal parameters (or arguments) enclosed within a pair of parentheses. The list of formal parameters in a method header can be empty. However, the pair of parentheses enclosing the formal parameter cannot be omitted.
16. As part of the method invocation, the actual parameter value is copied into the formal parameter.
17. Inside a method, you can declare additional variables. All variables declared inside a method, excluding the formal parameters in the header, are known as local variables.
18. A local variable is not accessible outside the block. Therefore, no access modifier is required.

19. An instance variable is available only within the object (if it is declared `private`). A local variable is available only within the block from its point of declaration. A formal parameter is available within the method.
20. An instance variable exists as long as the associated object exists. A local variable is created every time declaration statement is executed during the method execution and the variable ceases to exist upon the completion of the block. A formal parameter is created at the beginning of the method invocation and it ceases to exist upon the completion of the method.
21. At the beginning of the method invocation, the actual parameter value is copied to the formal parameter.
22. Once the return statement is executed, the control goes back to the Java statement that invoked the method.
23. A method that does not modify any instance variable of an object is called an accessor method. All other methods are known as mutator methods.

EXERCISES

1. Mark the following statements as true or false:
   a. An accessor method can access only one attribute.
   b. There are two types of methods: value returning and `void`.
   c. A mutator method modifies at least one attribute.
   d. A local variable can be made accessible outside the method by declaring it as `public`.
   e. It is okay for a `void` method to return 0.
   f. Every value returning method must have a `return` statement.
   g. Every class must have a `toString` method.
   h. The default constructor has no formal parameters.
   i. A `void` method cannot be used on the right-hand side of an assignment statement.
   j. The method `main` need not be a `static` method.
   k. The method `charAt` of the `String` class is of type `char`.
   l. Every method must have a `return` statement.
2. Mark the following method invocation as valid or invalid. If invalid, explain why it is invalid. Assume that `scannedInfo` is a reference variable of type `Scanner`, `word` is a reference variable of the type `String`, that references the `String` "Okay, Ready to go!" and `stk` is a reference variable of the type `Stock`.
   a. `int a = scannedInfo.next();`
   b. `int a = scannedInfo.nextInt();`
   c. `int = scannedInfo.nextInt(a);`

d. `int a; scannedInfo.nextInt(a);`

e. `char ch = charAt(word, 2);`

f. `char ch = Character.charAt(word, 2);`

g. `char ch = word.charAt(2);`

h. `char ch = word.charAt(0);`

i. `int n = word.charAt(1);`

j. `word.charAt(0) = 'w';`

k. `System.out.println;`

l. `System.out.write("Hello");`

m. `String str = System.out.println("Hi, There");`

n. `System.out.println("Hi") + System.out.println(", There");`

o. `String str = "Hello"; System.out.println(str.charAt(2));`

p. `String str = word.toString();`

q. `String str = toString(word);`

r. `int gain = Stock.getDividend();`

s. `double gain = stk.getDividend();`

t. `double gain = stk.getDividend(2008);`

u. `stk = setNumberOfShares(100);`

3. Select the best answer.

   a. As a general rule, an attribute is

      (i) `private` (ii) `static` (iii) `public` (iv) none of these

   b. As a general rule, an operation is

      (i) `private` (ii) `static` (iii) `public` (iv) none of these

   c. A `void` method returns

      (i) `null` (ii) nothing (iii) `0` (iv) none of these

   d. An accessor method may modify _____ attributes of a class.

      (i) some (ii) none (iii) all (iv) none of these

   e. The `yearlyDividend` method of the class `Stock` is an example of _____ method.

      (i) mutator (ii) accessor (iii) `static` (iv) none of these

   f. A default constructor has _____ explicit parameters.

      (i) `0` (ii) `0 or 1` (iii) many (iv) none of these

   g. An implicit parameter is a _________ variable.

      (i) primitive (ii) reference (iii) parameter (iv) none of these

   h. You need not know the internal parts of a radio to use it is an example of

      (i) encapsulation (ii) interface (iii) information hiding (iv) none of these

4. Given

```
int a, b, c;
double x, y;
Scanner scannedInfo = new Scanner(System.in);
```

Determine the validity of the assignment statements. If an assignment statement is invalid, state the reason.

a. `b = scannedInfo.nextInt();`
b. `c = scannedInfo.next();`
c. `x = scannedInfo.nextDouble();`
d. `a = scannedInfo.nextDouble();`
e. `a = System.out.println();`
f. `x = System.out.println(b + c);`
g. `x = System.out.println(y);`
h. `c = 7 + scannedInfo.nextInt();`
i. `b = scannedInfo.nextInt() + a;`

5. Which of the following method heading is incorrect? Explain.

a. `static void public trial()`
b. `static public sum(void)`
c. `boolean void public find()`
d. `public double next();`
e. `private String static try`
f. `public void test(10, 20)`
g. `public double back(int, int)`

6. Assume that you are implementing a class having three private attributes: `symbol` a `String`, `cost` a `double`, and `quantity` an `int`. Correct the errors, if any.

a. `public void getCost() { return cost; }`
b. `public String getAmount(){ return amount;}`
c. `public String getsymbol{ return symbol;}`
d. `public int setQuantity(int qty){quantity = qty;}`
e. `public void setAmount(double amt){double amount = amt;}`
f. `public void set symbol(String smbl){smbl = symbol}`
g. `public int returnQuantity(){int a = quantity; return a;}`
h. `public void updateCost(double b){cost = cost + b;}`
i. `public double find _ cost(){return; }`

7. Assume that you are implementing a class having five private attributes: `productName` a `String`, `price` a `double`, `onHand` an `int`, `isBackOrder` a `boolean`, `deptCode` a `char`. Write Java statements that accomplish the following tasks:
   a. Get methods for each of the instance variables.
   b. Set methods for each of the instance variables.
   c. `toString` method that returns a `String` with information about `productName` and `price`.
   d. A method `update` to change the value of `onHand` by a given value. If `onHand` was 2000, `update(−200)` will change `onHand` value to 1800.
   e. A method `priceChange` to change the price by a given percentage. If `price` is 100.00, `priceChange(.1)` will change `price` to 110.00.

PROGRAMMING EXERCISES

1. Create a class `Square` and test it. The class `Square` has exactly one data member `length`. Your class must provide `getLength`, `setLength`, and `toString` methods. There are two application specific methods: `getArea` and `getPerimeter`.
2. Create a `Circle` class and test it. The class `Circle` has exactly one data member `radius`. Your class must provide `getRadius`, `setRadius`, and `toString` methods. There are two application specific methods: `getArea` and `getCircumference`. Use `Math.PI`, a constant defined in `Math` class in your code for the value of pi.
3. Create a `PhoneNumber` class and test it. The instance variables are first name, last name, and phone numbers. The application specific method returns a `String` of the form last name, first name, and phone number.
4. Create an employee class and test it. The instance variables are first name, last name, and annual salary. There are two application specific methods. The first method returns the monthly salary. The second method returns bonus calculated as a percentage of the annual salary plus 1000. The percentage is an explicit parameter of this method.
5. According to the grading policy, final grade is determined by the average of four test scores. Design and test a class to compute the average of four test scores (do not forget the student name).
6. Write a program that prompts the user to input time in seconds. The program should then output in day hour minute second format. Design and use appropriate class(es).
7. Write a program that prompts the user to input distance in inches. The program should then output the distance in miles, furlong, yard, feet, and inches. Design and use appropriate class(es).
8. Write a program to create a shorter version of the name from fullname. A name such as Meera S. Nair will have a shorter format M.S. Nair. Design and use appropriate class(es).

9. Write a program to estimate the profit from a particular product for a month. Information such as product name, unit cost, sale price, and average number of items sold per month are available. Note that product name may consist of many words such as "Hunter Miller 56in Ceiling Fan." Design and use appropriate class(es).
10. Write a program to convert between Centigrade and Fahrenheit. Design and use appropriate class(es).
11. Write a program to convert between days, hours, minutes, seconds, and total seconds. Design and use appropriate class(es).
12. Write a program to evaluate the expression $ax^2 + bx + c$. Design and use appropriate class(es).
14. Create a cylinder class to compute area and volume. The attributes can be either radius and height or base of the type circle and height. The second option is more challenging. In the case of second option you need to create a circle class as specified in Programming Exercise 2.
15. Create a box class to compute area and volume. The attributes are length, width, and height or base of type rectangle and height. The second option is more challenging. It involves creating a rectangle class with two attributes length and width.

ANSWERS TO SELF-CHECK

1. information hiding
2. reserved word
3. `private int noOfStudents;`
4. `private double averageGpa;`
5. `public`
6. `changeChannel`
7. `char`
8. `double`
9. `int`
10. True
11. block
12. True
13. `int age = 18;`
14. `double interestRate = 5.0;`
15. True
16. False
17. False
18. True

19. True
20. True
21. False
22. True
23. False
24. True
25. accessor
26. False
27. False
28. True
29. True
30. logical

CHAPTER 4

Decision Making

In this chapter you learn

- Object-oriented concepts
  - Analysis and design of classes
- Java concepts
  - Boolean variables, logical operators and expressions, equality and relational operators, control structures `if`, `if ... else`, and `switch`, nesting of control structures, and enumerated data types
- Programming skills
  - Design, implement, and test Java programs capable of decision making

Programs you wrote so far did not make any decision. Given five test scores, you know how to write a program to add them to obtain a cumulative test score. If you want to translate the cumulative test score into a letter grade, then you need a control structure that can make decisions based on cumulative test score. Such a control structure is called a *selection structure*. In this chapter you will learn about various selection structures available in Java. However, the fundamental principles you learn in this chapter can be applied in many scripting languages and programming languages, including C, C++, and C#.

CONTROL STRUCTURES

Programs you have written so far are executed in *sequence*. Once a statement is executed, the immediately following statement is executed next. Therefore, if sequence is the only control structure available, every statement in a program or a method is always executed in the same order. Clearly, such programs cannot make decisions.

Introspection

Programs you have seen so far are like a one-way street with single entrance and single exit. Your car just rolls through the same predictable, predetermined path.

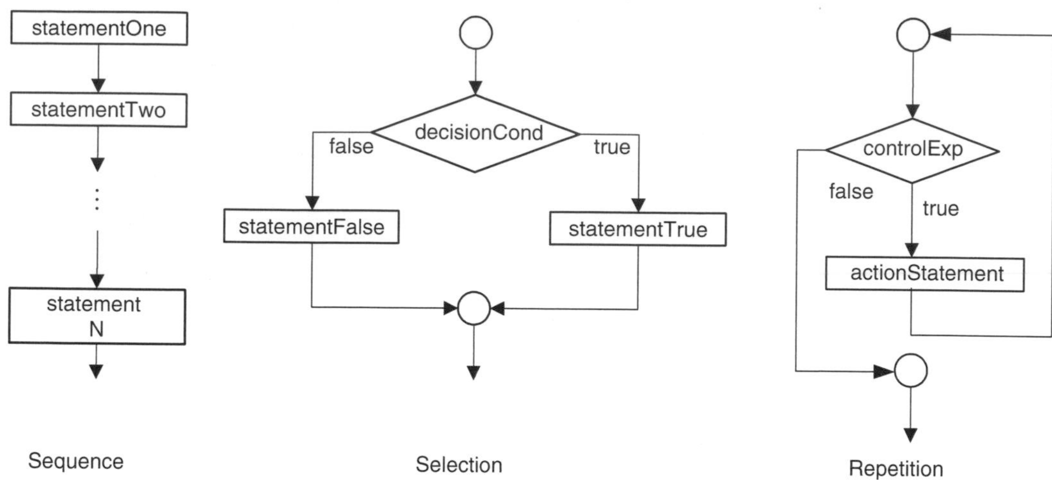

FIGURE 4.1 Control structures.

Every computer program can be constructed from three basic control structures, *sequence*, *selection*, and *repetition*, shown in Figure 4.1. This is the essence of the *structure theorem*. So far, all the programs you have encountered consist of only the sequence structure. Thus, program starts with executing statements one after another from the first executable statement of the method `main`. Selection structure enables you to selectively execute a certain part of a code while skipping some other parts. Repetition structure, however, allows you to repeat a certain part of the code again and again. Thus, both selection and repetition structures alter the order of execution of statements. The selection structure is discussed in this chapter and repetition structure is described in Chapter 5.

To better understand the issues involved, let us look at the following problem. Mr. James Jones is a college student. Lately, James has noticed that he ends up wasting lots of time doing mundane day-to-day chores. James decided to get organized and for that matter, to have a schedule for every day. He started creating a list of things to do as follows:

Get up in the morning

Eat breakfast

Go to the University

Attend lectures

Have lunch at noon

Visit library

Come back in the evening

Have dinner

Study

Watch TV for an hour

Go to bed

TABLE 4.1 James's Schedule

Weekday Schedule	Weekend Schedule
Get up in the morning	Get up in the morning
Eat breakfast	Eat breakfast
Go to the University	Finish all assignments
Attend lectures	Have lunch at noon
Have lunch at noon	Have some recreation
Visit the library	Have dinner
Come back in the evening	Go to bed
Have dinner	
Study	
Watch TV for an hour	
Go to bed	

James started in earnest and strictly followed the schedule. On weekends, James does not go to the University. He could not follow his weekday schedule on weekends. Therefore, James came up with the schedules given in Table 4.1.

Thus, James has two schedules. On weekdays, he *selects* one schedule and on weekends he *selects* the other schedule. James's algorithm can be stated as follows:

If today is a weekday then do the following:

Get up in the morning

Eat breakfast

Go to the University

Attend lectures

Have lunch at noon

Visit library

Come back in the evening

Have dinner

Study

Watch TV for an hour

Go to bed

If today is not a weekday then do the following:

Get up in the morning

Eat breakfast

Finish assignments

Have lunch at noon

Have recreation

Have dinner

Go to bed

Consider the statement:

Today is a weekday.

This statement is either true or false and you could at any time determine whether or not the above statement is true. In logic, such statements are known as propositions.

Introspection

If it is raining outside, then let us play an indoor game. If it is not raining, then let us play an outdoor game.

LOGICAL EXPRESSION AND OPERATORS

An expression is called logical if it evaluates to true or false. We evaluate many logical expressions in our day-to-day life. During shopping you may decide to buy an item based on whether or not it is on sale. You may decide to go for jogging if outside temperature is between 45 and 80°F. We evaluate so many logical expressions every day without being aware of the fact that we are dealing with logical expressions. The difference between everyday logical expression and the one you are going to see in connection with programming is very little. In the case of day-to-day logical expressions, we tend to answer yes or no rather than true or false.

Java has a primitive data type `boolean` introduced in Chapter 2. The `boolean` data type can have two values: `true` and `false`. Both `true` and `false` are reserved words in Java.

Example 4.1

The following program illustrates the declaring, the initializing, the inputting, the outputting, and the assigning of `boolean` data types. In particular, note that in the case of inputs, `boolean` value can be entered using upper or lowercase letters. For example, the `true` can be entered in any one of the following forms: true, TrUE, TRUE, or TruE. However, when used as a literal inside the program to initialize a `boolean` variable, `true` and `false` need to be entered exactly as it is.

```
import java.util.Scanner;

/**
    Illustrates declaration, initialization, input and
    output of logical variables
*/
public class LogicalData
{

    public static void main (String[] args)
    {
        boolean numberFound;                    //declaration
```

```
        boolean processFinished = false;
                                              //initialization
        boolean errorReport = true;           //initialization

        Scanner scannedInfo = new Scanner(System.in);

        System.out.println
           ("processFinished = " + processFinished);    //output
        System.out.println
                   ("errorReport = " + errorReport);    //output

        System.out.print
                     ("Enter two boolean values : ");
        System.out.flush();
        numberFound = scannedInfo.nextBoolean();    //input
        processFinished = errorReport;              //assignment
        errorReport = scannedInfo.nextBoolean();    //input
        System.out.println();

        System.out.println("New values are as follows:");

        System.out.println
                ("numberFound (first input) = " + numberFound);
                                                        //output
        System.out.println
        ("processFinished (previous value of errorReport) = " +
                                  processFinished); //output
        System.out.println
               ("errorReport (second input) = " + errorReport);
                                                        //output

    }
}
```

Output

```
processFinished = false
errorReport = true
Enter two boolean values : TRue FaLSe

New values are as follows:
numberFound (first input) = true
processFinished (previous value of errorReport) = true
errorReport (second input)  = false
```

Note 4.1 Literals appearing in the code are processed during the compile time. However, input values are processed at the execution time. Input value is treated as a `String` and converted to appropriate `boolean` value using the method, `nextBoolean`.

Self-Check

1. True or false: "Bill is taller than George" is a logical expression.
2. True or false: "How are you?" is a logical expression.

LOGICAL OPERATORS

Java has five binary operators and one unary operator to form logical expressions. These operators are shown in Table 4.2.

We will address the logical operators `&` and `|` later in this chapter in the subsection on short-circuit evaluation. It is customary to specify each of these operators through truth tables. Truth tables list the outcome for each possible input combination.

The `not` operator is a unary operator and it changes a `true` value to `false` and vice versa. For example, if a `boolean` variable `itemFound` is `true`, then the expression `(!itemFound)` is `false` and `(!(! itemFound))` is `true`.

The truth table of the `not` operator is shown in Table 4.3.

Example 4.2

Consider the following program segment:

```
boolean fileFound;
boolean fileMissing = false;                    //(1)
fileFound =   !fileMissing;                     //(2)
```

Observe that in Line 1, the `boolean` variable `fileMissing` is initialized to `false`. Therefore, you can initialize `fileFound` as shown in Line 2. Note that in Line 2 `fileFound` receives the value `true`. However, value of the variable `fileMissing` is still `false`.

The logical operator `and` is a binary operator. Thus, there are four different possible combinations and the truth table is shown in Table 4.4. An `and` expression is `true` only if both operands are `true`. For instance, consider the following two statements:

Today is Monday.

Chris is more than 6 ft tall.

Note that the statement

Today is Monday and Chris is more than 6 ft tall.

TABLE 4.2 Logical Operators in Java

Operator Symbol	Operator Name	Common Name
`!`	`Logical not`	`not`
`&&, &`	`Logical and`	`and`
`||, |`	`Logical or`	`or`
`^`	`Logical exclusive or`	`xor`

TABLE 4.3 The Truth Table of the Logical `not` Operator

operand	!(operand)
false	true
true	false

TABLE 4.4 The Truth Table of the Logical and Operator

operandOne	operandTwo	operandOne && operandTwo
false	false	false
false	true	false
true	false	false
true	true	true

TABLE 4.5 The Truth Table of the Logical or Operator

operandOne	operandTwo	operandOne || operandTwo
false	false	false
false	true	true
true	false	true
true	true	true

TABLE 4.6 The Truth Table of the Logical xor Operator

operandOne	operandTwo	operandOne ^ operandTwo
false	false	false
false	true	true
true	false	true
true	true	false

is `true` only if both the statements are `true`.

The logical operator `or` is a binary operator. An `or` expression is `false` only when both operands are `false`. For example, consider the following two statements:

Today is Monday.

Chris is more than 6 ft tall.

The statement

Today is Monday or Chris is more than 6 ft tall

is `false` only if both the statements are `false`. The logical operator `or` is also known as inclusive `or`. The truth table of the `or` operator is shown in Table 4.5.

The logical operator `xor` (exclusive or) is a binary operator. A `xor` expression is `true` only when exactly one of the operands is `true`. This operator is very rarely used in programming. The truth table of the `xor` operator is presented in Table 4.6.

Example 4.3

The following program prints the truth tables of all four logical operators discussed in this section:

```
/**
    A class to illustrate logical operations
*/
public class LogicalOperator
{
    public static void main (String[] args)
```

```
    {
        boolean trueValueOne = true;
        boolean trueValueTwo = true;
        boolean falseValueOne = false;
        boolean falseValueTwo = false;

        System.out.println("\n\nLogical not Operator");

        System.out.println("!false = " + (!falseValueOne));
        System.out.println("!true  = " + (!trueValueOne));

        System.out.println("\n\nLogical and Operator");

        System.out.println("false && false = " +
                           (falseValueOne && falseValueTwo));
        System.out.println("true  && false = " +
                           (trueValueOne && falseValueTwo));
        System.out.println("false && true  = " +
                           (falseValueOne && trueValueTwo));
        System.out.println("true  && true  = " +
                           (trueValueOne && trueValueTwo));

        System.out.println("\n\nLogical or Operator");

        System.out.println("false || false = " +
                           (falseValueOne || falseValueTwo));
        System.out.println("true  || false = " +
                           (trueValueOne || falseValueTwo));
        System.out.println("false || true  = " +
                           (falseValueOne || trueValueTwo));
        System.out.println("true  || true  = " +
                           (trueValueOne || trueValueTwo));

        System.out.println("\n\nLogical xor Operator");

        System.out.println("false ^ false = " +
                           (falseValueOne ^ falseValueTwo));
        System.out.println("true  ^ false = " +
                           (trueValueOne ^ falseValueTwo));
        System.out.println("false ^ true  = " +
                           (falseValueOne ^ trueValueTwo));
        System.out.println("true  ^ true  = " +
                           (trueValueOne ^ trueValueTwo));

    }
}
```

Output

```
Logical not Operator
!false = true
!true  = false

Logical and Operator
false && false = false
true  && false = false
false && true  = false
true  && true  = true

Logical or Operator
false || false = false
true  || false = true
false || true  = true
true  || true  = true

Logical xor Operator
false ^ false = false
true  ^ false = true
false ^ true  = true
true  ^ true  = false
```

Self-Check

3. What is `(false && false)||true`?
4. What is `false && (false||true)`?

RELATIONAL OPERATORS

In our day-to-day life, we make decisions by comparing values. For example, you may buy an item only if its price is below a certain amount that you have in mind. In this case, you are comparing the price of the item with the amount you have decided to spend on it. In other words, if the price of the item is less than the amount you have decided to spend on it, you will buy the product. Otherwise, you may go to another shop in search of a better price. Yet another example is as follows. The life expectancy of females is longer than the life expectancy of males. Therefore, one of the determining factors, in the case of a life insurance premium is the gender of the person. Thus, there are many situations where decisions are made after comparing the values. Computers can also perform similar tasks.

Java supports six relational operators. The result of applying relational operator is a logical value. The six relational operators are shown in Table 4.7.

TABLE 4.7 Relational Operators

Relational Operator	Semantics
<	Less than
<=	Less than or equal to
>	Greater than
>=	Greater than or equal to
==	Equal to
!=	Not equal to

Self-Check

5. True or false: The relational operator <= can also be written as =<.
6. Java supports ________ relational operators.

RELATIONAL OPERATORS AND NUMERICAL DATA TYPES

The relational operators can be used in connection with numerical values. Values can be literals or variables.

Example 4.4

Consider the following declarations:

```
int numberOne = 22;
int numberTwo = 7;
double fractionOne = 5.0 / 11.0;
double fractionTwo = 6.0 / 11.0;
```

Relational Expression	Comparison Done	Result
`numberOne < 22`	22 is less than 22	`false`
`numberOne <= 22`	22 is less than or equal to 22	`true`
`numberOne >= 22`	22 is greater than or equal to 22	`true`
`numberOne > 22`	22 is greater than 22	`false`
`numberOne == 22`	22 is equal to 22	`true`
`numberOne != 22`	22 is not equal to 22	`false`
`numberOne <= numberTwo`	22 is less than or equal to 7	`false`
`numberOne >= numberTwo`	22 is greater than or equal to 7	`true`
`22 >= numberTwo`	22 is greater than or equal to 7	`true`
`fractionOne < 1.0`	(5.0/11.0) is less than 1.0	`true`
`fractionOne <= fractionTwo`	(5.0/11.0) is less than or equal to (6.0/11.0)	`true`
`0.45 < fractionOne`	0.45 is less than (5.0/11.0)	`true`

Example 4.5

Consider the following Java program dealing with relational operator == on data type `double`. From the output it can be seen that `(1.0/11.0)` added `11` times is more than `1.0`! As a result, relational operators may produce results not consistent with basic arithmetic.

```
/**
    Illustrates inexact arithmetic of floating-point numbers
*/
public class RelationalOperator
{
    public static void main (String[] args)
    {
        double fractionOne = 1.0 / 11.0;
        double fractionTwo = fractionOne + fractionOne
        +fractionOne+ fractionOne + fractionOne +fractionOne;
        double fractionThree = fractionTwo + fractionOne
        +fractionOne+ fractionOne + fractionOne +fractionOne;

        System.out.println(" fractionOne = " + fractionOne );
        System.out.println(" fractionTwo = " + fractionTwo );
        System.out.println(" fractionThree = " + fractionThree );
        System.out.println(" (fractionThree == 1.0)   is " +
                                    (fractionThree == 1.0));
        System.out.println(" (fractionThree > 1.0)   is " +
                                    (fractionThree > 1.0));

    }
}
```

Output

```
fractionOne = 0.09090909090909091
fractionTwo = 0.5454545454545455
fractionThree = 1.0000000000000002
(fractionThree == 1.0)   is false
(fractionThree > 1.0)   is true
```

Common Programming Error 4.1

Comparing two floating-point numbers for equality may cause unexpected results.

Note 4.2 Instead of testing two floating-point numbers for equality, test whether or not the absolute value of their difference is closer to zero. Java has a `static` method `abs` in the `Math` class. `Math.abs(x - y)` computes the absolute value of the difference between two values `x` and `y`. Therefore, instead of the expression `(x == y)` use the expression `(Math.abs(x - y) <= ERROR_ALLOWED)`. Here, `ERROR_ALLOWED` is a named constant that can be declared at the method level as follows:

```
final double ERROR_ALLOWED = 1.0E-14;
```

Self-Check

7. What is (3 <= 3)?
8. What is (3 < 3)?

RELATIONAL OPERATORS AND CHARACTER DATA TYPES

You can compare two `char` data based on their character codes. For example, the character code of the space character is 32, and thus it is smaller than all letters and digits. Similarly, the character code of the character `'A'` is 65 and that of `'a'` is 97. Thus, `'A'` is smaller than `'a'`. In fact, `'a'` is larger than all uppercase letters. Note that `'8'` has a Unicode value 56. As a consequence, `'8'` < 8 evaluates to `false`.

Example 4.6

Consider the following Java program:

```
/**
    A class to illustrate relational operators and characters
*/
public class CharRelationalOperator
{
    public static void main (String[] args)
    {
        char space = ' ';
        char upperA = 'A';
        char lowerA = 'a';
        char upperZ = 'Z';
        char char8 = '8';

        System.out.println(" space < 'A' is " + (space < 'A'));
        System.out.println(" space < '8' is " + (space < '8'));
        System.out.println(" space == 32 is " + (space == 32));
        System.out.println
                           (" 'A' < 'a' is " + (upperA < lowerA));
        System.out.println
                           (" 'Z' < 'a' is " + (upperZ < lowerA));
        System.out.println (" '8' < 8 is " + (char8 < 8));

    }
}
```

Output

```
space < 'A' is true
space < '8' is true
space == 32 is true
```

```
'A'   < 'a' is true
'Z'   < 'a' is true
'8'   <  8 is false
```

Self-Check

9. What is (`'A'` <= 65)?
10. What is (`'a'` > `'Z'`)?

Advanced Topic 4.1: Relational Operators and Objects

In the case of objects, there is no predetermined ordering of items. You may compare two objects based on the application. For example, you may order two diamonds based on clarity, purity, brilliance, and so on. So in the case of objects, comparison of objects is made possible through methods. However, you could use equality operators == and `!=`. We illustrate this fact in detail in Chapter 6. Since you are familiar with the `String` class, methods in the `String` class for comparing strings is presented next. The use of equality operators in conjunction with `String` references is also presented.

LEXICOGRAPHICAL ORDERING OF STRINGS

Recall that a string is a sequence of zero or more characters. If you look for the words *like* and *lake* in dictionary, the word *lake* appears before the word *like*. Thus, you could say that the word *lake* is smaller than the word *like*. Thus, there is an ordering of words in a dictionary. Similarly, the word *like* is smaller than the word *live* and the word *live* is smaller than the word *liver*. This ordering is called lexicographical ordering.

In the lexicographical ordering, strings are compared character by character, from the beginning of the string. You have already seen that there is an ordering within the Unicode character set based on the collating sequence. If a mismatch occurs, as in the case of words *like* and *lake*, the character-by-character comparison stops and the order of the mismatched characters determines the order of words. Thus, in our example, the mismatch occurs at second character position. Now, the letter *a* is smaller than the letter *i*; therefore, the word *lake* is smaller than the word *like*. Similarly, considering words *like* and *live*, a mismatch occurs at third character. Again the letter *k* is smaller than the letter *v*; therefore, the word *like* is smaller than the word *live*. If no mismatch occurs, then eventually one or both strings may end. These possibilities are explored next.

Consider the case in which character-by-character comparison continues and one of the strings ends. In our discussion, this will be the situation for words *live* and *liver*. You could treat this as a mismatch at fifth character position. The fifth character of the word *live* is a null character. Recall that null character is the first character in the collating sequence. Thus, any other character is larger than null character. In particular, character *r* is larger than null character. Thus, the string *live* is smaller than *liver*. Thus, in general, if two strings match until one of them ends, the string that ended is smaller than the other.

The only case that remains is both strings end simultaneously. In this case, both strings are identical or both strings are equal.

The class String has two methods to compare strings: equals and compareTo. The syntax to use both methods is as follows:

```
strOne.equals(strTwo)
strOne.compareTo(strTwo)
```

In the above syntax, strOne must be a String reference variable, whereas strTwo can be either a String reference variable or a String literal. The equals method returns a logical value. If both the strings are equal, then equals method returns true; otherwise returns false. The compareTo method returns an integer value. If strOne is smaller than strTwo, then the value returned is a negative integer. If strOne is larger than strTwo, then the value returned is a positive integer. If both strings are equal, the method compareTo returns integer 0. You should not relay on the actual integer returned by compareTo method. Rather, you should make your decisions on the basis of the sign of the number.

Example 4.7

Consider the following statements:

```
String strOne = "America the beautiful";
String strTwo = "America the beautiful!";
String strThree = "Maple leaf";
String strFour = "Maple Leaf";
String strFive = "Maple Leaf";
```

Table 4.8 illustrates the behavior of equals and compareTo methods.
The following program verifies Table 4.8:

```
/**
    Illustration of methods equals and compareTo in String class
*/
public class EqualsCompareToStringMethods
{
    public static void main(String[] args)
    {
        String strOne = "America the beautiful";
        String strTwo = "America the beautiful!";
        String strThree = "Maple leaf";
        String strFour = "Maple Leaf";
        String strFive = "Maple Leaf";

        System.out.println("strOne.equals(strTwo) is "
                    +   strOne.equals(strTwo));
        System.out.println("strOne.compareTo(strTwo) is "
                    +   strOne.compareTo(strTwo));
```

```
        System.out.println("strTwo.equals(strOne) is "
                    +   strTwo.equals(strOne));
        System.out.println("strTwo.compareTo(strOne) is "
                    +   strTwo.compareTo(strOne));
        System.out.println("strThree.equals(strFour) is "
                    +   strThree.equals(strFour));
        System.out.println("strThree.compareTo(strFour) is "
                    +   strThree.compareTo(strFour));
        System.out.println("strThree.equals((\"Maple leaf\") is "
                    +   strThree.equals(("Maple leaf"));
        System.out.println("strFour.compareTo(strFive) is "
                    +   strFour.compareTo(strFive));

    }

}
```

Output

```
strOne.equals(strTwo) is false
strOne.compareTo(strTwo) is -1
strTwo.equals(strOne) is false
strTwo.compareTo(strOne) is 1
strThree.equals(strFour) is false
strThree.compareTo(strFour) is 32
strThree.equals("Maple leaf") is true
strFour.compareTo(strFive) is 44
```

Self-Check

11. Assume the assignment statement `strOne = "Bad";`. What is `strOne.compareTo("Good")`?
12. Assume the assignment statement `strOne = "Better";`. What is `strOne.compareTo("Best")`?

TABLE 4.8 `String` Class Methods `equals` and `compareTo`

Method Invocation	Value Retuned
`strOne.equals(strTwo)`	`false`
`strOne.compareTo(strTwo)`	`An integer < 0`
`strTwo.equals(strOne)`	`false`
`strTwo.compareTo(strOne)`	`An integer > 0`
`strThree.equals(strFour)`	`false`
`strThree.compareTo(strFour)`	`An integer > 0`
`strThree.equals("Maple leaf")`	`true`
`strFour.compareTo(strFive)`	`An integer > 0`

Advanced Topic 4.2: Equality Operators and String Class

Equality operators == and != can be applied to reference variables of String type. For example, if strOne and strTwo are two reference variables of the String type, both expressions

```
strOne == strTwo
strOne != strTwo
```

are legal in Java. The expression strOne == strTwo returns true if the reference kept in both variables are identical. In other words, strOne == strTwo returns true if both refer to the same String object. Similarly, strOne == strTwo returns false if both refer to different String objects.

In Java, String is a class. However, it is different from other classes in many respects. During compilation, a String literal is stored once only. Therefore, the following two statements

```
String strOne = "Have a pleasant day!"              //(1)
String strTwo = "Have a pleasant day!"              //(2)
```

create only one String object. To be more specific, as compiler encounters (1), the following steps are carried out:

a. Creates a reference variable strOne of String type
b. Creates a String object corresponding to the literal "Have a pleasant day!"
c. Assigns the reference of the object created in (b) to reference variable strOne

As compiler encounters (2), the following steps are carried out:

a. Creates a reference variable strTwo of String type.
b. Recognizes the fact that a String object corresponding to the literal "Have a pleas ant day!" already exists. Therefore, no new object is created.
c. Assigns the reference of the existing object to the reference variable strTwo.

Therefore, the expression strOne == strTwo evaluates to true. However, if you input two identical String literals and then compare them using equality operator, the Java system will return a false value. In this case, system in fact creates two String objects. Thus, creating only one String object for a String literal is done during compilation and not during execution. These ideas are illustrated in the following example:

```
/**
    Illustration of equality operator in String class
*/
public class EqualityOperatorsOnString
{
```

```
    public static void main(String[] args)
    {
        String strOne = "Have a nice day!";
        String strTwo = "Have a nice day!";
        String strThree = "Happy birthday to you";
        String strFour = strThree;
        String strFive = "America";

        Scanner scannedInfo = new Scanner(System.in);

        System.out.println("strOne == strTwo is "
                                 +   (strOne == strTwo));
        System.out.println("strThree == strFour is "
                                 +   (strThree == strFour));
        System.out.println("strOne == strThree is "
                                 +   (strOne == strThree));

        System.out.print("Input the word America twice : ");
        strOne = scannedInfo.next();
        strTwo = scannedInfo.next();
        System.out.flush();
        System.out.println("strOne is " + strOne);
        System.out.println("strTwo is " + strTwo);

        System.out.println("strOne == strTwo is "
                                 +   (strOne == strTwo));
        System.out.println("strOne.equals(strTwo) is "
                                 +   strOne.equals(strFive));

        System.out.println("strOne == strFive is "
                                 +   (strOne == strFive));
        System.out.println("strOne.equals(strFive) is "
                                 +   strOne.equals(strFive));
    }
}
```

Output

```
strOne == strTwo is true
strThree == strFour is true
strOne == strThree is false
Input the word America twice : America America
strOne is America
strTwo is America
strOne == strTwo is false
strOne.equals(strTwo) is true
```

```
strOne == strFive is false
strOne.equals(strFive) is true
```

Note that associated with each input of the word "America", Java system created a new String object. As a consequence, strOne == strTwo is false. Further, even though there is a String object with string value "America", during execution a new object is created. Thus, strOne == strFive is false.

Note 4.3 String literals appearing in the code are processed during the compile time. However, input values are processed at the execution time. As a consequence, storing a String literal once only rule applies during compile time and not at the execution time.

Self-Check

13. True or false: Let strOne and strTwo be two String references. If strOne == strTwo is true, then strOne.equal(strTwo) is also true.
14. True or false: If strOne.equal(strTwo) is false, then strOne == strTwo is false.

PRECEDENCE RULES

In Chapter 2, you had seen precedence rules for arithmetic operations. These rules determine the priority and associativity of operators. Since an expression may involve arithmetic, relational, and logical operators, we need a precedence rule for all those operators. As mentioned in Chapter 2, it is customary to state these rules in the form of a table (see Table 4.9). For a complete list see Appendix A.

Table 4.9 may be quite overwhelming for anyone, particularly for a beginner. With practice, you will get better at it. However, the following observations may be helpful.

1. Assignment operator has the lowest precedence. Therefore, there is no need to enclose the expression on the right-hand side with a pair of parentheses.

   In other words,

   ```
   variable = (expression);
   ```

   is equivalent to

   ```
   variable = expression;
   ```

2. Binary logical operator has lower precedence than all arithmetic and relational operators. Therefore, there is no need to put parentheses around the operands of a binary logical operator.

   Thus,

   ```
   (expressionOne) && (expressionTwo)
   (expressionOne) || (expressionTwo)
   ```

TABLE 4.9 Precedence Rules

Operator	Operand Types	Operation	Level	Group	Associativity
++ −−	Numeric variable	Postincrement Postdecrement	1	Post	LR
++ −−	Numeric variable	Preincrement Predecrement	2	Pre	RL
+ −	Number	Unary plus Unary minus	2	Unary	
!	Logical	Logical not			
* / %	Number, number	Multiplication Division Modulus	4	Arithmetic	LR
+ −	Number, number	Addition Subtraction	5		
< <= > >=	Number, number	Less than Less than or equal Greater than Greater than or equal	7	Relational	
== !=	Any type, the same type	Equality operators	8		
^	Boolean, boolean	Logical XOR	10	Logical	
&&	Boolean, boolean	Logical AND	12		
||	Boolean, boolean	Logical OR	13		
=	Variable, same type	Assignment	15	Assignment	RL
*=	Variable, number	Assignment with operator			
/= %=	Variable, integer				
+= −=	Variable, number				

are equivalent, respectively, to

```
(expressionOne && expressionTwo)
(expressionOne || expressionTwo)
```

3. Relational operators have lower precedence than all arithmetic operators. Therefore, there is no need to put parentheses around arithmetic expressions.
   In other words,

```
(exp1 / exp2) <= (exp3 - exp4)
```

are equivalent to

```
(exp1 / exp2 <= exp3 - exp4)
```

Example 4.8

Consider the following declarations:

```
boolean workCompleted = true;
boolean errorFound = false;
char letter = 'J';
int  numOne = 7, numTwo = 9, numThree = 20;
double valueOne = 2.25, valueTwo = 0.452;
```

Evaluation of Logical Expressions

Expression	Value	Explanation
`!errorFound`	`true`	`!errorFound = !false = true.`
`!workCompleted`	`false`	`!workCompleted = !true = false.`
`workCompleted && !errorFound`	`true`	Order of evaluation: `!, &&.` `true && true = true.`
`letter == 'P'`	`false`	`letter` is `'J'.`
`workCompleted || letter == 'P'`	`true`	Order of evaluation: `==, ||.` `workCompleted` is `true;` `letter == 'P'` is `false;` `true || false = true.`
`!workCompleted && letter != 'P'`	`false`	Order of evaluation: `!, !=, &&.` `!workCompleted` is `false;` `letter != 'P'` is `true;` `false && true = false.`
`numOne + numTwo < 15`	`false`	Order of evaluation: `+, <.` `numOne + numTwo` is `7 + 9 = 16.` Therefore, `16 > 15` evaluates to `false.`
`numOne >= 0 && numOne <= 9`	`true`	Order of evaluation: `>=, <=, &&.` `numOne = 7.` Note that `7 >= 0` and `7 >= 9` evaluate to `true. true && true = true.`
`letter >= 'a' && letter <= 'z'`	`false`	Order of evaluation: `>=, <=, &&` `letter is 'J'.` Note that `'J' >= 'a'` is `false` and `'J' <= 'z'` is `true. false && true = false.`
`letter >= 'A' && letter <= 'Z'`	`true`	Order of evaluation: `>=, <=, &&` `letter` is `'J'.` Thus, `'J' >= 'A'` is `true` and `'J' <= 'Z'` is `true.` `true && true = true.`
`letter >= 'a' && letter <= 'z' || letter >= 'A' && letter <= 'Z'`	`false`	Order of evaluation: `>=, <=, &&, ||` `false || true = true.`
`valueTwo > 0.25 && valueTwo < 0.45`	`false`	Order of evaluation: `>, <, &&` `valueTwo` is `0.452.` Thus, `0.452 > 0.25` is `true.` Also, `0.452 < 0.45` is `false. true && false = false.`

(continued)

Expression	Value	Explanation
valueOne < 7.75	true	valueOne is 2.25. Thus, 2.25 < 7.75 is true
valueOne < 7.75 || valueTwo > 2.5 && valueTwo < 10	false	true || false = true.
(valueOne < 7.75 || valueTwo > 2.5)&& valueTwo < 10	true	(true || true) && false = true && false = false.

The following example verifies our computations through a Java program.

Example 4.9

```
/**
    Illustrates the operator precedence involving logical
    operators
*/
public class OperatorPrecedence
{
    public static void main(String[] args)
    {
        boolean workCompleted = true;
        boolean errorFound = false;
        char letter = 'J';
        int   numOne = 7, numTwo = 9, numThree = 20;
        double valueOne = 2.25, valueTwo = 0.452;

        System.out.println("!errorFound is "
                          + (!errorFound));
        System.out.println("!workCompleted is "
                          + (!workCompleted));
        System.out.println("workCompleted && !errorFound is "
                          + (workCompleted && !errorFound));
        System.out.println("letter == 'P' is "
                          + (letter == 'P'));
        System.out.println("workCompleted || letter == 'P' is "
                          + (workCompleted || letter == 'P'));
        System.out.println("!workCompleted && letter != 'P' is "
                          + (!workCompleted && letter != 'P'));
        System.out.println("numOne + numTwo < 15 is "
                          + (numOne + numTwo < 15));
        System.out.println("numOne >= 0 && numOne <= 9 is "
                          + (numOne >= 0 && numOne <= 9));

        System.out.println("letter >= 'a' && letter <= 'z' is "
                          + (letter >= 'a' && letter <= 'z'));
```

```
        System.out.println("letter >= 'A' && letter <= 'Z' is "
                         + (letter >= 'A' && letter <= 'Z'));

        System.out.println("letter >= 'a' && letter <= 'z' || "
                   + "letter >= 'A' && letter <= 'Z' is "
                         + (letter >= 'a' && letter <= 'z' ||
                         letter >= 'A' && letter <= 'Z' ));
        System.out.println("valueTwo > 2.5 && valueTwo < 0.45
               is " + (valueTwo > 2.5 && valueTwo < 0.45));
        System.out.println("valueOne < 7.75 is "
                         + (valueOne < 7.75));
        System.out.println("valueOne < 7.75 || valueTwo >
                         0.25 " + "&& valueTwo < 0.45 is "
                       + (valueOne < 7.75 || valueTwo > 0.25
                         && valueTwo < 0.45));
        System.out.println("(valueOne < 7.75 || valueTwo >
                         0.25) "+ "&& valueTwo < 0.45) "

                   + ((valueOne < 7.75 || valueTwo > 0.25)
                         && valueTwo < 0.45));
    }
}
```

Output

```
!errorFound is true
!workCompleted is false
workCompleted && !errorFound is true
letter == 'P' is false
workCompleted || letter == 'P' is true
!workCompleted && letter != 'P' is false
numOne + numTwo < 15 is false
numOne >= 0 && numOne <= 9 is true
letter >= 'a' && letter <= 'z' is false
letter >= 'A' && letter <= 'Z' is true
letter >= 'a' && letter <= 'z' || letter >= 'A' && letter <=
                                                  'Z' is true
valueTwo > 0.25 && valueTwo < 0.45 is false
valueOne < 7.75 is true
valueOne < 7.75 || valueTwo > 0.25 && valueTwo < 0.45 is true
(valueOne < 7.75 || valueTwo > 0.25) && valueTwo < 0.45) false
```

Note 4.4 Parentheses can be used to change the order of execution. In particular, consider the last two expressions in Example 4.9.

```
valueOne < 7.75 || valueTwo > 0.25 && valueTwo < 0.45 is true   (1)
(valueOne < 7.75 || valueTwo > 0.25) && valueTwo < 0.45) false (2)
```

In expression 1 above, no parentheses were used. Thus, logical and operation is carried out before the logical `or`. In expression 2, due to the parenthesis, logical `or` is carried out before the logical and. Thus, in the case 1 the expression is evaluated to `true`, whereas in case 2 it is evaluated to `false`.

Self-Check

15. True or false: Relational operators have higher precedence than assignment operators.
16. True or false: Logical operators have higher precedence than arithmetic operators.

Advanced Topic 4.3: Syntax Error Explained

The logical expressions `(letter >= 'a' && letter <= 'z')` and `('a' <= letter && letter <= 'z')` are equivalent. However, it cannot be written as `('a' <= letter <= 'z')`. Many beginning programmers wonder why one could write `(valueOne + valueTwo + valueThree)` without any syntax error; while `('a' <= letter <= 'z')` is not legal in Java. The reason can be explained as follows: The expression `(valueOne + valueTwo + valueThree)` is evaluated in two steps. First, `valueOne + valueTwo` is computed. The result of this computation is a numeric value and is added to `valueThree`. However, `'a' <= letter` is a logical expression. Therefore, `'a' <= letter` is either `true` or `false`. Therefore, `('a' <= letter <= 'z')` is either `(true <= 'z')` or `(false <= 'z')`. Since a relational operator cannot be used to compare a logical value with a character, `('a' <= letter <= 'z')` is an illegal expression.

Advanced Topic 4.4: Short-Circuit Evaluation

In the case of logical operation `or`, you know that

`true || false` is `true`

and

`true || true` is `true`

In other words, if the first operand evaluates to `true`, the value of the second operand has no bearing on the final result. Thus, in a logical expression of the form

```
(logicalExpressionOne || logicalExpressionTwo)
```

if `logicalExpressionOne` evaluates to `true`, there is no need to evaluate `logicalExpressionTwo`.

Similarly, in the case of logical operation and

```
false && false = false
```

and

```
false && true = false
```

Thus, if the first operand evaluates to `false`, the value of the second operand has no bearing on the final result. Thus, in a logical expression of the form

```
(logicalExpressionOne && logicalExpressionTwo)
```

if `logicalExpressionOne` evaluates to `false`, there is no need to evaluate `logicalExpressionTwo`.

Java compiler makes use of these facts and skips the evaluation of operands accordingly. This method of evaluating a logical expression is called the *short-circuit evaluation*.

Advanced Topic 4.5: Additional Logical Operators

Java provides two other operators `&` and `|`. You can use these operators instead of `&&` and `||`, respectively, to avoid short-circuit evaluation. Some programmers use them to achieve certain *side effects* as shown in the following example. Author does not recommend their approach. The following example is given only to illustrate operators `&` and `|`.

Example 4.10

Consider the following declarations:

```
int numOne = 7, numTwo = 9, numThree = 20;
```

In the case of the following expression,

```
(numOne <= numTwo || numTwo == numThree++)
```

the first operand of the `or` operator `numOne <= numTwo` evaluates to `true`. Due to short-circuit evaluation, second operand `(numTwo == numThree++)` is never evaluated. Therefore, the variable `numThree` is not incremented by 1. Thus, `numThree` is `20`. However, in the following expression,

```
(numOne <= numTwo | numTwo == numThree++)
```

no short-circuit evaluation is performed and the second operand `(numTwo == numThree++)` is evaluated. The variable `numThree` is incremented by 1. Thus, `numThree becomes 21`.

For a slightly more complex example, consider the following declaration and the logical expression:

```
int numFour = 107, numFive = 109, numSix = 120;
```

```
(numFour >= numFive && 200 <= numFive++ || 300 == numSix++)
```

In this case, since first operand numFour >= numFive of the logical and operation is false, the second operand 200 <= numFive++ of the logical and operations is never performed. Thus, we have the intermediate result

```
false || 300 == numSix++
```

Now, the first operand of a logical or operation is false. Therefore, no short-circuit evaluation is possible. Recall that in the case of a logical or operation, compiler will employ short-circuit evaluation only if the first operand is true. Thus, in this case, compiler evaluates the operand 300 == numSix++. Therefore, after evaluating the statement

```
(numFour >= numFive && 200 <= numFive++ || 300 == numSix++)
```

the variable numFive is not incremented by 1 while the variable numSix is incremented by 1.

However, the following expression evaluates every operand, and thus as a side effect increments both variables numFive and numSix:

```
(numFour >= numFive & 200 <= numFive++ | 300 == numSix++)
```

In this book we will consistently use expressions with no side effects. Therefore, we will not be using logical operators & and |.

Example 4.11

This example provides a Java program to verify Example 4.10.

```
/**
    Operator precedence involving logical operators
*/
public class ShortCircuitEvaluation
{
    public static void main(String[] args)
    {

        int   numOne = 7, numTwo = 9, numThree = 20;
        int   numFour = 107, numFive = 109, numSix = 120;

        System.out.println("(numOne <= numTwo ||"
               + " numTwo == numThree++) is "
               + (numOne <= numTwo ||   numTwo == numThree++));
        System.out.println("numThree is " + numThree);
        System.out.println("(numOne <= numTwo |"
               + " numTwo == numThree++) is "
               + (numOne <= numTwo |   numTwo == numThree++));
```

```
            System.out.println("numThree is " + numThree);

            System.out.println("(numFour >= numFive &&"
                   + " 200 <= numFive++ ||  300 == numSix++) is "
                   + (numFour >= numFive && 200 <= numFive++
                   || 300 == numSix++));
            System.out.println("numFive is " + numFive);
            System.out.println("numSix is " + numSix);

            System.out.println("(numFour >= numFive &"
                   + " 200 <= numFive++ |  300 == numSix++) is "
                   + (numFour >= numFive & 200 <= numFive++
                   | 300 == numSix++));
            System.out.println("numFive is " + numFive);
            System.out.println("numSix is " + numSix);
        }
    }
```

Output

```
(numOne <= numTwo ||  numTwo == numThree++) is true
numThree is 20
(numOne <= numTwo |  numTwo == numThree++) is true
numThree is 21
(numFour >= numFive && 200 <= numFive++ ||  300 == numSix++)
is false
numFive is 109
numSix is 121
(numFour >= numFive & 200 <= numFive++ |  300 == numSix++) is
false
numFive is 110
numSix is 122
```

Advanced Topic 4.6: Positive Logic

Research studies have shown that it is easy to understand positive logic compared to negative logic. In this section, we discuss De Morgan's rule and other techniques for converting a negative logic expression into a positive one.

Consider the following expression:

`!(numOne > 7).`

This is an example of negative logic. To convert this logical expression into a positive logic, replace the relational operator > with its complement relational operator <=. Thus, the logical expression `!(numOne > 7)` can be replaced by the following equivalent logical expression:

`(numOne <= 7)`

TABLE 4.10 The Complement of a Relational Operator

Operator	Complement	Negative Expression	Positive Expression
`<`	`>=`	`!(numOne < 7)`	`(numOne >= 7)`
`<=`	`>`	`!(numOne <= 7)`	`(numOne > 7)`
`>`	`<=`	`!(numOne > 7)`	`(numOne <= 7)`
`>=`	`<`	`!(numOne >= 7)`	`(numOne < 7)`
`==`	`!=`	`!(numOne == 7)`	`(numOne != 7)`
`!=`	`==`	`!(numOne != 7)`	`(numOne == 7)`

Table 4.10 shows each relational operator and its complement operator.

Now let us consider a more complex expression involving logical operators `&&` and `||`. In this case, you can apply De Morgan's rules as follows:

`!(expOne && expTwo)` is equivalent to `!(expOne) || !(expTwo)`

and

`!(expOne || expTwo)` is equivalent to `!(expOne) && !(expTwo)`

Example 4.12

Consider the following declarations:

```
boolean workCompleted = true;
char letter = 'J';
int    numOne = 7, numTwo = 9, numThree = 20;
```

Positive Logic

<table>
<tr><th>Expression</th><th>Modified Expression and Explanation</th></tr>
<tr><td>!(numOne < 17 || NumTwo > 9)</td><td>= !(numOne < 17) && !(NumTwo > 9)
= (numOne >= 17) && (NumTwo <= 9)
= (numOne >= 17 && NumTwo <= 9).</td></tr>
<tr><td>!(numOne <= 17 && NumTwo > 9)</td><td>= !(numOne <= 17) || !(NumTwo > 9)
= (numOne > 17) || (NumTwo <= 9)
= (numOne > 17 || NumTwo <= 9).</td></tr>
<tr><td>!(numOne <= 17 && NumTwo > 9
|| letter == 'P')</td><td>= !((numOne <= 17 && NumTwo > 9) ||
(letter == 'P'))
= !(numOne <= 17 && NumTwo > 9) &&
!(letter == 'P')
= (!(numOne <= 17) || !(NumTwo > 9))
&& !(letter == 'P')
= ((numOne > 17) || (NumTwo <= 9))
&&(letter != 'P')
= (numOne > 17 || NumTwo <= 9) &&
(letter != 'P').</td></tr>
<tr><td>!(numOne >= 17 || NumTwo < 9
&& letter != 'P')</td><td>= !((numOne >= 17) || (NumTwo < 9 &&
letter != 'P'))
= !(numOne >= 17) && !(NumTwo < 9 &&
letter != 'P')</td></tr>
</table>

(continued)

<table>
<tr><th>Expression</th><th>Modified Expression and Explanation</th></tr>
<tr><td></td><td>= !(numOne >= 17) && (!(NumTwo < 9)
||!(letter != 'P'))
= (numOne < 17) && ((NumTwo >= 9) ||
(letter == 'P'))
= (numOne < 17 && (NumTwo >= 9 ||
letter == 'P').</td></tr>
<tr><td>!(numOne > 17 || NumTwo <= 9
|| !workCompleted)</td><td>= !((numOne > 17 || NumTwo <= 9) ||
(!workCompleted))
= !(numOne > 17 || NumTwo <= 9) &&
!(!workCompleted)
= !(numOne > 17) && !(NumTwo <= 9) &&
!(!workCompleted)
= (numOne <= 17) && (NumTwo > 9) &&
(workCompleted)
= (numOne <= 17 && NumTwo > 9 &&
workCompleted).</td></tr>
<tr><td>!(numOne > 17 && NumTwo != 9
&& !workCompleted)</td><td>= !((numOne > 17 && NumTwo != 9) &&
(!workCompleted))
= !(numOne > 17 && NumTwo != 9) ||
!(!workCompleted)
= !(numOne > 17) || !(NumTwo != 9) ||
!(!workCompleted)
= (numOne <= 17) || (NumTwo == 9) ||
(workCompleted)
= (numOne <= 17 || NumTwo == 9 ||
workCompleted)</td></tr>
</table>

SELECTION STRUCTURES

Java provides three structures for selection and decision making. In the next section, we discuss the one-way selection structure `if`. The two-way selection structure `if ... else` and the multiway selection structure `switch` are introduced later in this chapter. Although Java provides three different structures, any decision making can in fact be performed using a sequence of `if` structures. In other words, a two-way decision structure can be replaced by two one-way decision structures. Similarly, a `n`-way decision-making structure can in fact be replaced by `n` one-way decision-making structures. These issues will be further explored later in this chapter.

ONE-WAY SELECTION STRUCTURE

A university may place a student in its prestigious dean's list if the student has at least 3.75 grade point average (gpa) out of a possible 4.0. Thus, if the gpa of a student is greater than or equal to 3.75, student's name is entered into dean's list. However, if the gpa of a student is less than 3.75, no action needs to be carried out. This is a situation where a one-way selection structure is appropriate. Similarly, an automobile insurance firm may apply a

20% discount to drivers with no accident claim in past 3 years. Again, the discount applies to drivers with no accident claim in past 3 years and no action needs to be taken if a driver had an accident in past 3 years.

The syntax template of one-way selection structure `if` is

```
if (decisionCondition)
    actionStatement
```

Thus, the one-way selection structure `if` has three parts. It begins with the reserved word `if`. The second part, known as *decision condition*, is a logical expression enclosed within a pair of left and right parentheses. The third part is any executable Java statement, known as the *action statement*. Java treats all three parts together as one Java statement and for the sake of discussion, we call it the `if` statement.

The semantics of the `if` statement is as follows. If the decision condition evaluates to `true`, then the action statement is executed. However, if the decision condition evaluates to `false`, then the action statement is not executed. The statement immediately following the `if` statement is always executed.

Consider the following:

```
statementBefore
if (decisionCondition)
    actionStatement
statementAfter
```

The semantics of the `if` statement can be illustrated through the diagram shown in Figure 4.2. Here the `statementBefore` and `statementAfter` stands for Java statements immediately before and after the `if` statement.

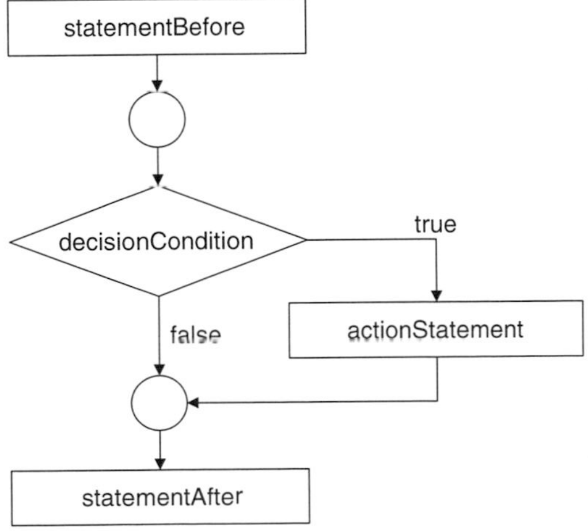

FIGURE 4.2 Control structure `if`.

Example 4.13

Consider the following segment of code:

```
char characterFound;
characterFound = 'Q';
if (characterFound >= 'A' && characterFound <= 'Z')
    System.out.println("The character found is uppercase. ");
System.out.println("Good Bye!");
```

In the above code, the decision condition `(characterFound >= 'A' && characterFound <= 'Z')` evaluates to `true`. Therefore,

```
The character found is uppercase.
```

is printed. Then as a next line,

```
Good Bye!
```

is also printed. However, if `characterFound = 'Q';` is replaced by `characterFound = 'q';` statement, then the decision condition `(characterFound >= 'A' && characterFound <= 'Z')` evaluates to `false`. Therefore,

```
The character found is uppercase.
```

will not be printed. However,

```
Good Bye!
```

will be printed.

Let us write a Java program to test our reasoning. The following program reads a character and outputs

```
The character found is uppercase.
Good Bye!
```

if the character is uppercase, and outputs

```
Good Bye!
```

in all other cases.

```
/**
    Example: One-way selection - Uppercase letter
*/
import java.util.Scanner;

public class OnewayUppercase
{

    public static void main (String[] args)
    {

        char characterFound;
```

```
        Scanner scannedInfo = new Scanner(System.in);
        System.out.print("Input a character : ");
        System.out.flush();
        characterFound = scannedInfo.next().charAt(0);
        System.out.println();
        if (characterFound >= 'A' && characterFound <= 'Z')
            System.out.println
                        ("The character found is uppercase.");
        System.out.println("Good Bye!");
    }
}
```

Output

Case 1. Input is an uppercase letter

```
Input a character: Q
The character found is uppercase.
Good Bye!
```

Case 2. Input is not an uppercase letter

```
Input a character : q
Good Bye!
```

Example 4.14

In this example, the preceding program is modified so that if the input is an uppercase letter, the following messages are printed:

```
The character found is uppercase.
Good Bye!
```

If the input happens to be a lowercase letter, the following messages are printed:

```
The character found is lowercase.
Good Bye!
```

In all other cases, program outputs

```
Good Bye!
```

Recall that in the case of uppercase letters, the statement that was instrumental in producing the message is the following `if` statement:

```
if (characterFound >= 'A' && characterFound <= 'Z')
    System.out.println("The character found is uppercase.");
```

Therefore, to produce a similar message corresponding to lowercase letters, all you need is another `if` statement with appropriate decision condition and action statement. Thus, we have the following Java statement:

```
if (characterFound >= 'a' && characterFound <= 'z')
    System.out.println("The character found is lowercase.");
```

The modified program is given as follows:

```
/**
    Example: One-way selection - Uppercase and lowercase
*/
import java.util.Scanner;

public class OnewayUpperLowercase
{

    public static void main (String[] args)
    {

        char characterFound;

        Scanner scannedInfo = new Scanner(System.in);

        System.out.print("Input a character : ");
        System.out.flush();
        characterFound = scannedInfo.next().charAt(0);
        System.out.println();

        if (characterFound >= 'A' && characterFound <= 'Z')
            System.out.println
                          ("The character found is uppercase.");

        if (characterFound >= 'a' && characterFound <= 'z')
            System.out.println
                          ("The character found is lowercase.");

        System.out.println("Good Bye!");

    }
}
```

Output

Case 1. Input is an uppercase letter

```
Input a character : Q

The character found is uppercase.
Good Bye!
```

Case 2. Input is a lowercase letter

```
Input a character : q

The character found is lowercase.
Good Bye!
```

Case 3. Input is not a letter

```
Input a character : %

Good Bye!
```

The one-way selection structure is quite useful in validating data. In Chapter 3, you have encountered `setNumberOfShares` of the class `Stock`.

```
public void setNumberOfShares(int inNumberOfShares)
{
    numberOfShares = inNumberOfShares;
}
```

Note that `inNumberOfShares` can have any `int` value. However, number of shares owned by a person must always be a nonnegative value. Therefore, if `inNumberOfShares` is a negative value such as −200, you may want to assign 0. Thus, you need the following one-way decision statement:

```
if (inNumberOfShares < 0)
    inNumberOfShares = 0;
```

Therefore, the `setNumberOfShares` method can be written as follows:

```
public void setNumberOfShares(int inNumberOfShares)
{
    if (inNumberOfShares < 0)
        inNumberOfShares = 0;
    numberOfShares = inNumberOfShares;
}
```

Example 4.15

In this example, we revisit the `setCounterValue` method of the class `CircularCounter` introduced in Chapter 3. For the sake of convenience, the method is shown below:

```
public void setCounterValue(int inCounterValue)
{
    counterValue = inCounterValue;
}
```

Note that the `inCounterValue` is assigned to `counterValue`. However, recall that the `counterValue` is supposed to be an integer between `0` and `limitValue - 1`. Therefore, you need to have a strategy to deal with situation where `inCounterValue` is outside the range. That is, `inConterValue` is less than `0` or `inConterValue` is greater than or equal to `limitValue`. A possible approach is to assign `counterValue` with `0` whenever `inCounterValue` is outside the allowed range. The required Java statement is

```
if (inCounterValue < 0 || inCounterValue >= limitValue)
    inCounterValue = 0;
```

Thus, the modified version of the `setCounterValue` method is as follows:

```
public void setCounterValue(int inCounterValue)
{
    if (inCounterValue < 0 || inCounterValue >=
                                        limitValue)
        inCounterValue = 0;
    counterValue = inCounterValue;
}
```

Self-Check

17. Rewrite the `setNumberOfShares` method such that if the number of shares is less than 1, then the instance variable is set to 0.
18. Rewrite the `setCounterValue` method such that the relational operator `>=` is replaced by another relational operator without changing the logic.

BLOCK STATEMENT

The `if` structure has one action statement. However, there are many situations where you may want to execute more than one action statement if the decision condition evaluates to `true`. To accommodate this need, Java provides *block statement*. A block statement is a sequence of Java statements enclosed within a pair of braces. The syntax of a block statement is as follows:

```
{
    actionStatementOne
    actionStatementTwo
            .
            .
            .
    actionStatementN
}
```

Example 4.16

Consider the following situation. Mr. Jones has two accounts with ABC bank. One is a checking account and the other is a savings account. Since checking accounts do not pay any interest, Mr. Jones keeps essentially all his savings in his savings account. However, he has bills to pay. So, first of every month, Mr. Jones checks his current balance in his checking account. If his current checking account balance happens to be less than or equal to $500.00, he transfers $1000.00 from his savings account to his checking account.

Assume the following declarations:

```
double savingsBalance;      // keeps track of savings acct.
                                                  balance
double checkingBalance;     // keeps track of checking acct.
                                                  balance

final double MINIMUM_BALANCE = 500.00;
final double TRANSFER_AMT    = 1000.00;
```

Now, the segment of code relevant to our discussion can be written as follows. First, check whether or not the checking balance is below `MINIMUM_BALANCE`. If the decision condition evaluates to `true`, then you need to withdraw `TRANSFER_AMT` from savings account and deposit `TRANSFER_AMT` to checking account. Thus, we have the following:

```
if (checkingBalance <= MINIMUM_BALANCE)
{
    savingsBalance  = savingsBalance - TRANSFER_AMT;
    checkingBalance = checkingBalance + TRANSFER_AMT;
}
```

Common Programming Error 4.2

In Java, = is an assignment operator. Therefore, using = in a logical expression is a syntax error.

For example, the following code is incorrect:

```
if (checkingBalance = MINIMUM_BALANCE) //ERROR
{
    savingsBalance = savingsBalance - TRANSFER_AMT;
    checkingBalance = checkingBalance + TRANSFER_AMT;
}
```

In this context, using == is a logical error. The logic dictates that you use <= and not ==.

Common Programming Error 4.3

In the case of an `if` statement, the decision condition must always be within a pair of parentheses. Thus, omitting the parenthesis enclosing a decision condition is an error.

The following code illustrates the error:

```
if checkingBalance <= MINIMUM_BALANCE  //ERROR
{
    savingsBalance  = savingsBalance - TRANSFER_AMT;
    checkingBalance = checkingBalance + TRANSFER_AMT;
}
```

Common Programming Error 4.4

In the case of an `if` statement, there is no semicolon immediately following the right parenthesis enclosing the decision condition. Thus, placing a semicolon immediately following the right parenthesis after decision condition is an error.

The following code has the Common Programming Error 4.3. Statements 1 and 2 will always be executed.

```
if (checkingBalance <= MINIMUM_BALANCE);           //ERROR
{
    savingsBalance  = savingsBalance - TRANSFER_AMT; //(1)
    checkingBalance = checkingBalance + TRANSFER_AMT; //(2)
}
```

Common Programming Error 4.5

In Java, a block statement is always enclosed within a pair of braces. Thus, omitting braces in a block statement is an error.

For example, in the following code, (1) alone is the action statement. Thus, Statement 2 is always executed.

```
if (checkingBalance <= MINIMUM_BALANCE) //ERROR
    savingsBalance  = savingsBalance - TRANSFER_AMT; //(1)
    checkingBalance = checkingBalance + TRANSFER_AMT; //(2)
```

Self-Check

19. Can a block statement contain exactly one statement?
20. Modify the code so that amount is transferred only if the checking account balance falls below $300.00.

TWO-WAY SELECTION STRUCTURE

You learned how to handle multiple actions in a one-way selection structure. Just as there are many situations in which you must perform multiple actions when a decision condition evaluates to `true`, there are many situations in which you must perform two different sets of actions when a decision condition evaluates to `true` and `false`, respectively. For example, a university may place a student in its prestigious dean's list if the student has 3.75 gpa out of a possible 4.0. If the gpa of a student is greater than or equal to 3.75, the student's name is entered into dean's list. However, if the gpa of a student is less than 3.75, university may want to send an encouraging message to the student. In this case, you need a two-way selection structure. Similarly, an automobile insurance firm may apply a 20% discount to drivers with no accident claim in past 3 years. This discount applies to drivers with no accident claim in past 3 years. However, as a public relations matter, every customer is given 5% discount. This is another situation where a two-way selection is required.

Java provides a control structure `if ... else` to address this issue. The syntax of the control structure `if ... else` is as follows:

```
if (decisionCondition)
    actionStatementTrue
else
    actionStatementFalse
```

Thus, the two-way selection structure `if ... else` has five parts. First, it begins with the reserved word `if`. The second part is a decision condition enclosed within a pair of left and right parentheses. The third part is an executable Java statement. The fourth is the reserved word `else`. The fifth is an executable Java statement. Java treats all five parts together as one Java statement and for the sake of discussion, we call it the `if ... else` statement.

The semantics of the `if ... else` statement is as follows. If decision condition evaluates to `true`, the action statement `actionStatementTrue` is executed. However, if the decision condition evaluates to `false`, the action statement `actionStatementFalse` is executed. The statement immediately following the `if ... else` statement is always executed. Consider the following:

```
statementBefore
if (decisionCondition)
    actionStatementTrue
else
    actionStatementFalse
statementAfter
```

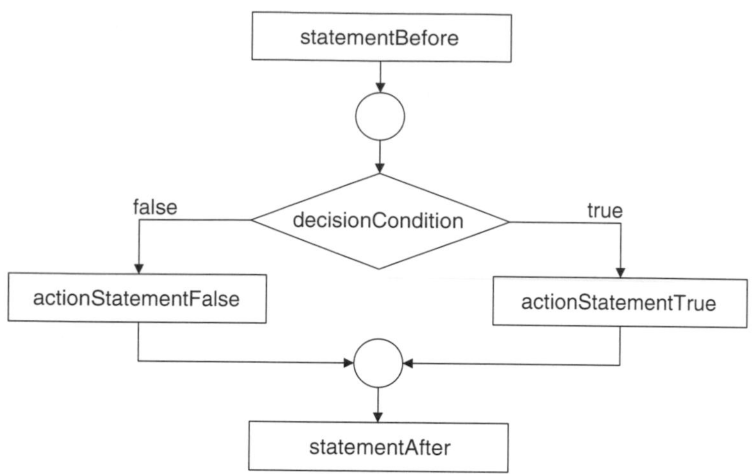

FIGURE 4.3 Control structure `if ... else`.

The semantics of the `if ... else` structure can be illustrated through the diagram shown in Figure 4.3. Here, the `statementBefore` and `statementAfter` stand for Java statements immediately before and after the `if ... else` statement.

As in the case of the `if` statement, any one or both the action statements can be a block statement.

Example 4.17

Consider the following segment of code:

```
int ageDifference;
.
.
.
    if (ageDifference <= 20)
        System.out.println("are of the same generation");
    else
        System.out.println("are of different generations");
```

In the above code, let `ageDifference` contain the absolute difference between ages of two individuals. Then, if the decision condition `(ageDifference <= 20)` evaluates to `true`, then the following message is printed:

```
They are of the same generation
```

However, if the expression `(ageDifference <= 20)` evaluates to `false`, then the following message is printed:

```
They are of different generations
```

Let us write a Java program to test our reasoning.

The following program reads first names, last names, and ages of two individuals and determines whether or not they are of the same generation. Assume that ageOne and ageTwo are the variables containing the ages of two individuals. Then (ageTwo - ageOne) is the difference between two age values. However, the above difference can be positive or negative. Therefore, we need to find the absolute value (ageTwo - ageOne). Recall that Java has a static method abs in the Math class. Math.abs(ageTwo - ageOne) computes the absolute difference between ageTwo and ageOne.

```
/**
Example: Two-way selection - same generation
*/
import java.util.Scanner;

public class TwowayAge
{
    public static void main (String[] args)
    {
        String firstNameOne, lastNameOne;
        String firstNameTwo, lastNameTwo;
        int ageOne, ageTwo, ageDifference;

        Scanner scannedInfo = new Scanner(System.in);

        System.out.print
                    ("Enter first name, last name and age : ");
        System.out.flush();
        firstNameOne = scannedInfo.next();
        lastNameOne = scannedInfo.next();
        ageOne = scannedInfo.nextInt();
        System.out.println();

        System.out.print
                    ("Enter first name, last name and age : ");
        System.out.flush();
        firstNameTwo = scannedInfo.next();
        lastNameTwo = scannedInfo.next();
        ageTwo = scannedInfo.nextInt();
        System.out.println();

        ageDifference = Math.abs(ageTwo - ageOne);

        System.out.print(firstNameOne + " " + lastNameOne
         + " and " + firstNameTwo + " " + lastNameTwo + " " );
```

```
        if (ageDifference <= 20)
            System.out.println("are of the same generation.");
        else
            System.out.println("are of different generations.");

    }
}
```

Output

Case 1: `ageOne < ageTwo`, same generation

```
Enter first name, last name and age : Joy Mathew 37
Enter first name, last name and age : Chris Cox 57
Joy Mathew and Chris Cox are of the same generation.
```

Case 2: `ageOne > ageTwo`, different generation

```
Enter first name, last name and age : Adam Smith 67
Enter first name, last name and age : Mike McCoy 46
Adam Smith and Mike McCoy are of different generations
```

Example 4.18

In this example, we use the `String` method `compareTo`. This example also illustrates the use of block statements in an `if ... else` statement.

Let us call the name of a person ascending if the last name is lexicographically larger than the first name. This example verifies whether or not a person's name is ascending. Recall that to compare two strings, you use `compareTo` method of the `String` class and if `strOne` and `strTwo` are two strings, `strOne.compareTo(strTwo)` returns a negative value if `strOne` is smaller than `strTwo`. Therefore, if `firstName` and `lastName` are `String` variables for first name and last name of a person, `firstName.compareTo(lastName)` is less than `0` if the name of the person is an ascending name. In other words, `(firstName.compareTo(lastName) < 0)` is `true` for ascending names. The complete program is as follows:

```
/**
Example: Two-way selection - Ascending Name
*/
import java.util.Scanner;

public class AscendingName
{
    public static void main (String[] args)
    {
        String firstName, lastName;
```

```
        Scanner scannedInfo = new Scanner(System.in);

        System.out.print
                          ("Enter first name, last name: ");
        System.out.flush();
        firstName = scannedInfo.next();
        lastName = scannedInfo.next();
        System.out.println();

        if (firstName.compareTo(lastName) < 0)
        {
            System.out.println("Hi " + firstName + " " +
                                              lastName);
            System.out.println
                             ("You got an ascending Name!");
        }
        else
        {
            System.out.println("Sorry! " + firstName + " " +
                                                  lastName);
            System.out.println
                      ("Your name is not an ascending one!");
        }
    }
}
```

Output

Case 1: Ascending name

```
Enter first name, last name: James Jones

Hi James Jones
You got an ascending Name!
```

Case 2: Not an ascending name

```
Enter first name, last name: Elaine Cox

Sorry! Elaine Cox
Your name is not an ascending one!
```

Self-Check

21. Write the code to print “Set for life!” if the net worth is more than $10 million dollars and print “Not there yet!” otherwise.
22. Write the code to print “Expensive item” if the cost is more than $25.00 and print “Reasonable Item” otherwise.

PRIMITIVE DATA TYPE boolean

In Chapter 2, you have seen the primitive data type boolean. You can declare and initialize boolean variables similar to other primitive data types.

```
boolean errorFound;
boolean jobCompleted = true;
boolean isLowercaseLetter;

errorFound = false;
```

You can also use a logical expression on the right-hand side of an assignment statement as shown below:

```
isLowercaseLetter = ('a' <= charOne && charOne <= 'z');
```

Note that above statement assigns true to isLowercaseLetter if charOne is a character between 'a' and 'z' and assigns false otherwise.

Note 4.5 There is no need to use an if structure to assign a boolean variable true or false based on a boolean expression. For example, instead of the following if statement,

```
if ('a' <= charOne && charOne <= 'z')
    isLowercaseLetter = true;
else
    isLowercaseLetter = false;
```

use

```
isLowercaseLetter = ('a' <= charOne && charOne <= 'z');
```

Note 4.6 Observe that true and false have no numerical value and hence you cannot use any of the relational operators. However, equality operators can be used to test whether or not two boolean variables have the same logical value.

Note 4.7 The expression isLowercaseLetter == true is true if isLowercaseLetter is true and isLowercaseLetter == true is false if isLowercaseLetter is false. Therefore, isLowercaseLetter == true is equivalent to isLowercaseLetter. Similarly, the expression isLowercaseLetter == false is true if !isLowercaseLetter is true, and isLowercaseLetter == false is false if !isLowercaseLetter is false. Thus, isLowercaseLetter == false is equivalent to !isLowercaseLetter. Therefore, there is no need to compare a boolean variable and a boolean literal. If isLowercaseLetter is a boolean variable, use isLowercaseLetter instead of isLowercaseLetter == true and use !isLowercaseLetter instead of isLowercaseLetter == false.

Self-Check

23. Write a statement similar to

    ```
    isLowercaseLetter = ('a' <= charOne && charOne <= 'z');
    ```

    for the boolean variable isUppercaseLetter.

24. Can the expression `('a' <= charOne && charOne <= 'z')` be replaced by `('a' <= charOne <= 'z')`?

NESTED STRUCTURES

James was following his schedule and things seemed to work well. He thought it would be a great idea to organize a party on a weekend. Definitely, he cannot follow his regular weekend schedule, if he is throwing a party. In other words, his weekend schedule has two possible options and it depends on the fact that whether or not he is throwing a party. Thus, his schedule has the following outline:

```
if (weekday)
  //follow weekday schedule
else
    if (throwing a party)
        //follow party schedule
    else
        //follow regular weekend schedule
```

Observe the presence of a decision structure inside another decision structure. Placing one control structure within another control structure is called *nesting* and the control structure obtained through nesting is known as a *nested* control structure.

Consider the syntax of `if` and `if ... else` structure you have already learned. For the sake of convenience, those structures are reproduced here.

```
if (logicalExpression)
    ActionStatement

if (logicalExpression)
    ActionStatement
else
    ActionStatement
```

In the above structures, `ActionStatement` stands for any executable Java statement, including a block statement. In particular, `ActionStatement` can be another `if` or `if ... else` statement.

Example 4.19

In this example, we graphically illustrate some of the possible nested structures using `if` and `if ... else`.

The structure shown in Figure 4.4 is an `if ... else` structure nested inside an `if` structure. The two structures shown in Figure 4.5 are obtained by nesting one `if` structure inside an `if ... else`.

The structure shown in Figure 4.6 is created by nesting two `if` structures inside an `if ... else` structure.

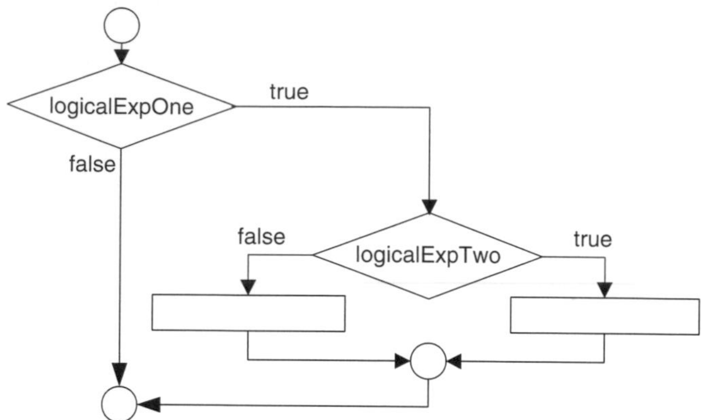

FIGURE 4.4 Nesting control structure 1.

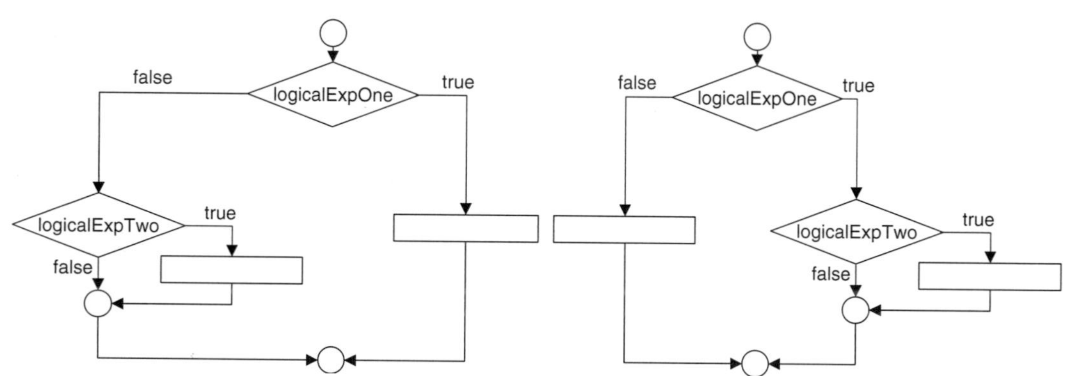

FIGURE 4.5 Nesting control structure 2.

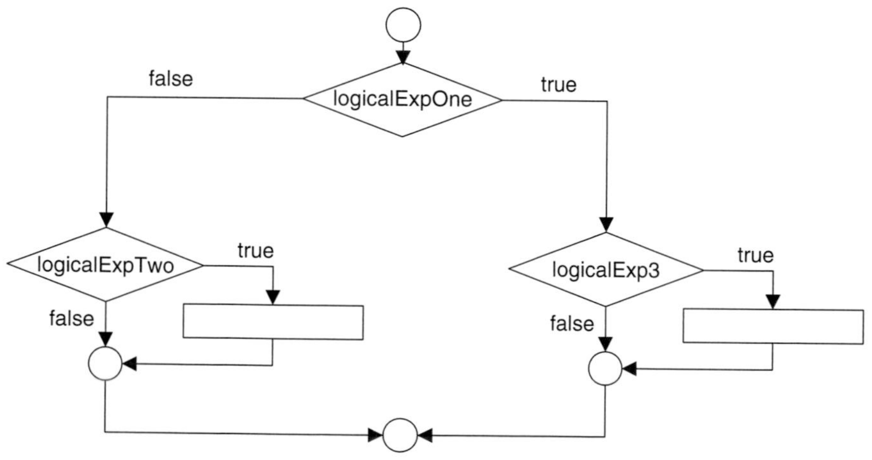

FIGURE 4.6 Nesting control structure 3.

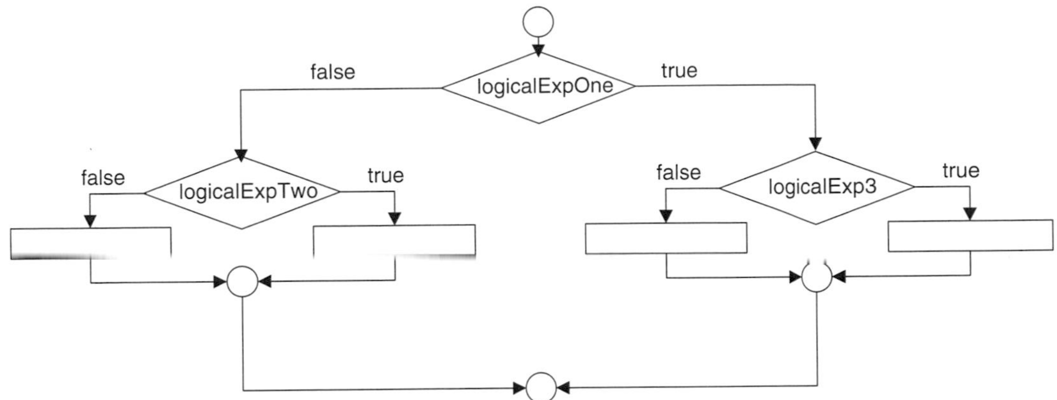

FIGURE 4.7 Nesting control structure 4.

The structure shown in Figure 4.7 shows the nesting of two `if ... else` structures inside an `if ... else` structure. Observe that it can make a four-way decision. The four different selections are

- `logicalExpOne is false and logicalExpTwo is false`
- `logicalExpOne is false and logicalExpTwo is true`
- `logicalExpOne is true and logicalExp3 is false`
- `logicalExpOne is true and logicalExp3 is true`

Example 4.19 shows that a wide variety of structures can be created through nesting of `if` and `if ... else` structures.

Example 4.20

Consider the income tax rules of a certain country. There is no income tax for the first $25,000.00. The next $75,000.00 is taxed at the rate of 15% and the amount above 100,000.00 is taxed at the rate of 25%.

Note that there are two groups of people. One group has the income less than or equal to 25,000.00 and as such do not pay any income tax at all. The other group has income greater than 25,000.00 and has to pay income tax. Thus, we can start with the following `if` statement:

```
if (totalIncome > 25000.00)
    "compute income tax"
else
    incomeTax = 0.0;
```

Now consider the group of people who pay income tax. This group can be classified into two subgroups: those with income less than or equal to 100,000.00 and those with income greater than 100,000.00. Income tax for those with income less than or equal to 100,000 can be computed using the formula

$$\text{incomeTax} = (\text{totalIncome} - 25000.00) * .15$$

and for people with income greater than 100,000.00 it can be computed using the formula

$$\begin{aligned}\text{incomeTax} &= 75000.00 * .15 + (\text{totalIncome} - 100000.00) * .25 \\ &= 11250.00 + (\text{totalIncome} - 100000.00) * .25\end{aligned}$$

Thus, for the group with income more than 25,000.00, we have the following:

```
if (totalIncome > 100000.00)
    incomeTax = 11250.00 + (totalIncome - 100000.00) * .25
else
    incomeTax = (totalIncome - 25000.00) * .15
```

Now, the above code applies to people with total income more than 25,000.00. Thus, we have the following:

```
if (totalIncome > 25000.00)                                          // (1)
    if (totalIncome > 100000.00)                                     // (2)
        incomeTax = 11250.00 + (totalIncome - 100000.00) * .25;
    else                                                             // (3)
        incomeTax = (totalIncome - 25000.00) * .15;
else                                                                 // (4)
    incomeTax = 0.0;
```

Later in this chapter you will see a more elegant solution using a multiway selection structure. Another improvement to the code can be achieved by using named constants for 25,000.00 and 100,1000.00.

In Java, there is no `else` structure. Every `else` must be part of an `if ... else` structure. As you start creating nested `if` and `if ... else` structure, you may be wondering which `else` corresponds to which `if`? The rule is quite simple. An `else` is always paired with last `if` statement without `else`. Thus, `else` in line (3) pairs with `if` in line (2) and `else` in line (4) pairs with the `if` in line (1).

Self-Check

25. Could you write the code for income tax computation without nesting?
26. True or false: Nesting improves understanding of the logic involved.

Advanced Topic 4.7: Better Coding Options

Note that above computation can be accomplished by the following sequence of `if` statements:

```
if (totalIncome <= 25000.00)
    incomeTax = 0.0;
if (totalIncome > 25000.00 && totalIncome <= 100000.00)
    incomeTax = (totalIncome - 25000.00) * .15;
if (totalIncome > 100000.00)
    incomeTax = 11250.00 + (totalIncome - 100000.00) * .25;
```

There is a natural tendency among many students to code as shown above. You are strongly encouraged to refrain from that practice. Compared to our first solution, the one above is less efficient since all three logical expressions will be evaluated irrespective of a person's total income. However, in our first solution, if total income is less than or equal to 25,000.00, only one logical expression will be evaluated and in all other cases two logical expressions will be evaluated.

Example 4.21

A certain university has a grading policy shown in Table 4.11.

Let wats denote the weighted average of all tests rounded to the nearest integer. Since grade assigned can be more than one character, you need a `String` variable to store the grade assigned. Thus, we have the following declarations:

```
int wats;
String gradeAssigned;
```

The highest grade a student can get is A. Therefore, you need to test whether the student is eligible for A. So, you can start the code as follows:

```
if (wats >= 90)
    gradeAssigned = "A"
else
    // a grade other than A is assigned
```

TABLE 4.11 The Grading Policy

Weighted Average of Test Scores (wats)	Grade Assigned
wats >= 90	A
85 <= wats < 90	A–
80 <= wats < 85	B
75 <= wats < 80	B–
70 <= wats < 75	C
60 <= wats < 70	D
wats < 60	F

For a student who has not received A, the next highest possible grade is A–. Hence, you need to test whether the student is eligible for A–. Thus, the code can be written as follows:

```
if (wats >= 90)
    gradeAssigned = "A"
else
    if (wats >= 85)
        gradeAssigned = "A-"
    else
        // a grade other than A or A- is assigned
```

Repeating the above analysis, we arrive at the following segment of code:

```
if (wats >= 90)
    gradeAssigned = "A"     ;
else
    if (wats >= 85)
        gradeAssigned = "A-" ;
    else
        if (wats >= 80)
            gradeAssigned = "B" ;
        else
            if (wats >= 75)
                gradeAssigned = "B-" ;
            else
            if (wats >= 70)
                gradeAssigned = "C"     ;
                else
                    if (wats >= 60)
                        gradeAssigned = "D" ;
                        else
                            gradeAssigned = "F" ;
```

Similar situations arise quite often in programming. Therefore, to reduce the indentation, the above code is quite often written as follows:

```
if (wats >= 90)
    gradeAssigned = "A";
else if (wats >= 85)
    gradeAssigned = "A-";
else if (wats >= 80)
    gradeAssigned = "B";
else if (wats >= 75)
    gradeAssigned = "B-";
```

```
else if (wats >= 70)
    gradeAssigned = "C";
else if (wats >= 60)
    gradeAssigned = "D";
else
    gradeAssigned = "F";
```

The above code is in fact a seven-way selection statement.

In general, a multiway selection structure has the following syntax:

```
if (logicalExpOne)
    actionStatementOne
else if (logicalExpTwo)
    actionStatementTwo
else if (logicalExp3)
    actionStatement3
.
.
.
else if (logicalExpN)
    actionStatementN
else
    actionStatement(N+1)
```

The following code we developed in Example 4.20

```
if (totalIncome > 25000.00)
    if (totalIncome > 100000.00)
        incomeTax = 11250.00 + (totalIncome - 100000.00) * .25;
    else
        incomeTax = (totalIncome - 25000.00) * .15;

else
    incomeTax = 0.0;
```

is better programmed as a three-way selection since there are three disjoint groups of people: the total income greater than 100,000.00, the total income less than or equal to 100,000.00 but greater than 25,000.00, and the total income less than or equal to 25,000.00.

```
if (totalIncome > 100000.00)
    incomeTax = 11250.00 + (totalIncome - 100000.00) * .25;
else if (totalIncome > 25000.00)
    incomeTax = (totalIncome - 25000.00) * .15;
else
    incomeTax = 0.0;
```

Advanced Topic 4.8: Order of Logical Expressions

In the case of a multiway selection structure, the order in which the logical expressions are evaluated is quite crucial. Unless great care is taken, your program may not behave the way you have expected. For example, consider the grade assignment code presented in Example 4.21. If you change the order of logical expressions, the code will not work as desired. For instance, the following code is not correct:

```
//Following code is not correct
if (wats >= 70)
    gradeAssigned = "C" ;
else if (wats >= 85)
    gradeAssigned = "A-" ;
else if (wats >= 80)
    gradeAssigned = "B" ;
else if (wats >= 75)
    gradeAssigned = "B-" ;
else if (wats >= 90)
    gradeAssigned = "A" ;
else if (wats >= 60)
    gradeAssigned = "D" ;
else
    gradeAssigned = "F" ;
```

Note that the above code is equivalent to the following segment of code:

```
if (wats >= 70)
    gradeAssigned = "C" ;
else if (wats >= 60)
    gradeAssigned = "D" ;
else
    gradeAssigned = "F" ;
```

Thus, anyone with wats greater than or equal to 70 will receive C grade and no one will receive A, A–, B, or B–.

However, you can avoid the above pitfall through the use of better logical expressions. For instance, the following code is correct:

```
if (wats >= 70 && wats < 75)
    gradeAssigned = "C" ;
else if (wats >= 85 && wats < 90)
    gradeAssigned = "A-" ;
else if (wats >= 80 && wats < 85)
    gradeAssigned = "B" ;
```

```
else if (wats >= 75 && wats < 80)
    gradeAssigned = "B-" ;
else if (wats >= 90)
    gradeAssigned = "A" ;
else if (wats >= 60 && wats < 70)
    gradeAssigned = "D" ;
else
    gradeAssigned = "F" ;
```

Advanced Topic 4.9: Overriding `if ... else` Pairing Rule

The following example illustrates the need and technique involved in overriding the `if ... else` pairing rule.

Example 4.22

A certain automobile insurance company assigns risk factor based on the following facts. Every driver has a risk factor 1.0. Anyone in the age group 16–25 is considered of high risk and has a risk factor 5.0. Anyone not in the high-risk category but has children in high-risk category is considered of modest risk and has a risk factor of 3.0.

You may be tempted to write the following code:

```
int driverAge;
double riskFactor;
boolean hasHighRiskChildren;
.
.
.
riskFactor = 1.0;
if (driverAge >= 26)                               // (1)
    if (hasHighRiskChildren)                       // (2)
        riskFactor = 3.0;
else                                               // (3)
    riskFactor = 5.0;
```

However, the above code is not correct. Recall that an `else` is always paired with the immediate `if` with no `else` part. Therefore, `else` in Line 3 is paired with `if` in Line 2 and not with `if` in Line 1. Note that indentation has no significance on the way `if` and `else` are paired. To pair an `else` with an `if` of your choice, you must create a block statement as shown below:

```
riskFactor = 1.0;
if (driverAge >= 26)                               // (1)
```

```
{
    if (hasHighRiskChildren)                     // (2)
        riskFactor = 3.0;
}
else                                             // (3)
    riskFactor = 5.0;
```

Note that the above code may in fact be written as multiway selection statement as shown below:

```
if (driverAge <= 25)
    riskFactor = 5.0;
else if (hasHighRiskChildren)
    riskFactor = 3.0;
else
    riskFactor = 1.0;
```

Advanced Topic 4.10: Ternary Operator

Java provides a ternary operator `?:` that can be used in place of an `if` ... `else` structure in certain cases. Consider an `if` ... `else` structure of the following form:

```
if (logicalExpression)
    variable = valueTrueCase ;
else
    variable = valueFalseCase ;
```

Note that in the above statement, `if (logicalExpression)` evaluates to `true`, `variable = valueTrueCase` is executed. Otherwise, `variable = valueFalse Case` is executed. In other words, the purpose of the above `if` structure is to assign one value or another to the same variable based on the truth or falsehood of a logical expression. In such cases, you can replace the above `if` ... `else` structure by the following semantically equivalent Java statement:

```
variable = (logicalExpression) ? valueTrueCase : valueFalseCase ;
```

The expression appearing on the right-hand side of the above assignment statement is called a *conditional expression.* Consider the following `if` ... `else` structure:

```
if (valueOne <= valueTwo)
    minValue = valueOne;
else
    minValue = valueTwo;
```

Using the conditional operator, the above statement can be written as follows:

```
minValue = (valueOne <= valueTwo)? valueOne : valueTwo;
```

MULTIWAY STRUCTURE switch

You have already seen a multiway structure created through the nesting of `if ... else` structure. In this section, a new structure called **switch** is presented. The general syntax of the `switch` statement is as follows:

```
switch(expression)
{
case valueOne:
    statementsOne
    [break;]
case valueTwo:
    statementsTwo
    [break;]
.
.
.
case valueN:
    statementsN
    [break;]
[default:
    statementsD
    [break;]]
}
```

Note that `break`, `case`, `default`, and `switch` are reserved words and `statementsOne`, `statementsTwo`, and so on can be one or more executable statements. The semantics of the `switch` statement can be explained as follows. First, the `expression` is evaluated. If the value of the `expression` is equal to `valueOne`, then `statementsOne` are executed. If the `break` statement is present, then no other statements within the structure are executed. However, if there is no `break` statement, `statementsTwo` are executed. If the `break` statement is present, then no other statements within the structure are executed. However, if there is no `break` statement, `statementsThree` are executed, and so on.

Similarly, if the value of the `expression` is equal to `valueTwo`, then `statementsTwo` are executed. If the `break` statement is present, then no other statements within the structure are executed. However, if there is no `break` statement, `statementsThree` are executed, and so on. If none of the values listed in case statements match the expression, then `statementsD` are executed.

The value of expression can only be an integral or enumerated data type; and is called the *controlling expression*. The enumerated data types will be introduced in the next section. The literal value that appears after a `case` in a `case` statement is called a *label*. A specific label can appear only in one `case` statement and a `case` statement can list only one label.

However, you can list case statements one after another without any action statements for all of them except the last one. Thus, all these case statements will have the same action statements. Since syntax allows us to have multiple statements corresponding to a case, there is no need to use braces. All break statements are optional. In particular, the break statement associated with default value has no significance at all and can be omitted.

The semantics of a switch structure can be summarized as follows (see Figure 4.8):

1. If the expression evaluates to a case label, then the statements are executed starting from that matching label until either a break statement or end of the switch structure is encountered. Once a break statement is encountered, no other statement in the switch structure is executed.
2. If the expression evaluates to a value that does not exist as a case label, then statements are executed starting from the label default.
3. If the expression evaluates to a value that does not exist and the switch structure has no default label, then no statement in the switch structure is executed.

Example 4.23

A university's admission criteria include points of the athletic participation. The points are awarded as follows: 25 points for international level participation, 18 points for national level participation, 12 points for state level participation, 6 points for district level participation, and 3 points for school level participation. This situation can be coded using a switch statement as follows:

```
switch(participationLevel)
{
case 1:                                // school level
    athleticPoints = 3;
    break;
case 2:                                // district level
    athleticPoints = 6;
    break;
case 3:                                // state level
    athleticPoints = 12;
    break;
case 4:                                // national level
    athleticPoints = 18;
    break;
case 5:                                // international level
    athleticPoints = 25;
    break;
default:                               // no participation
    athleticPoints = 0;
    break;
}
```

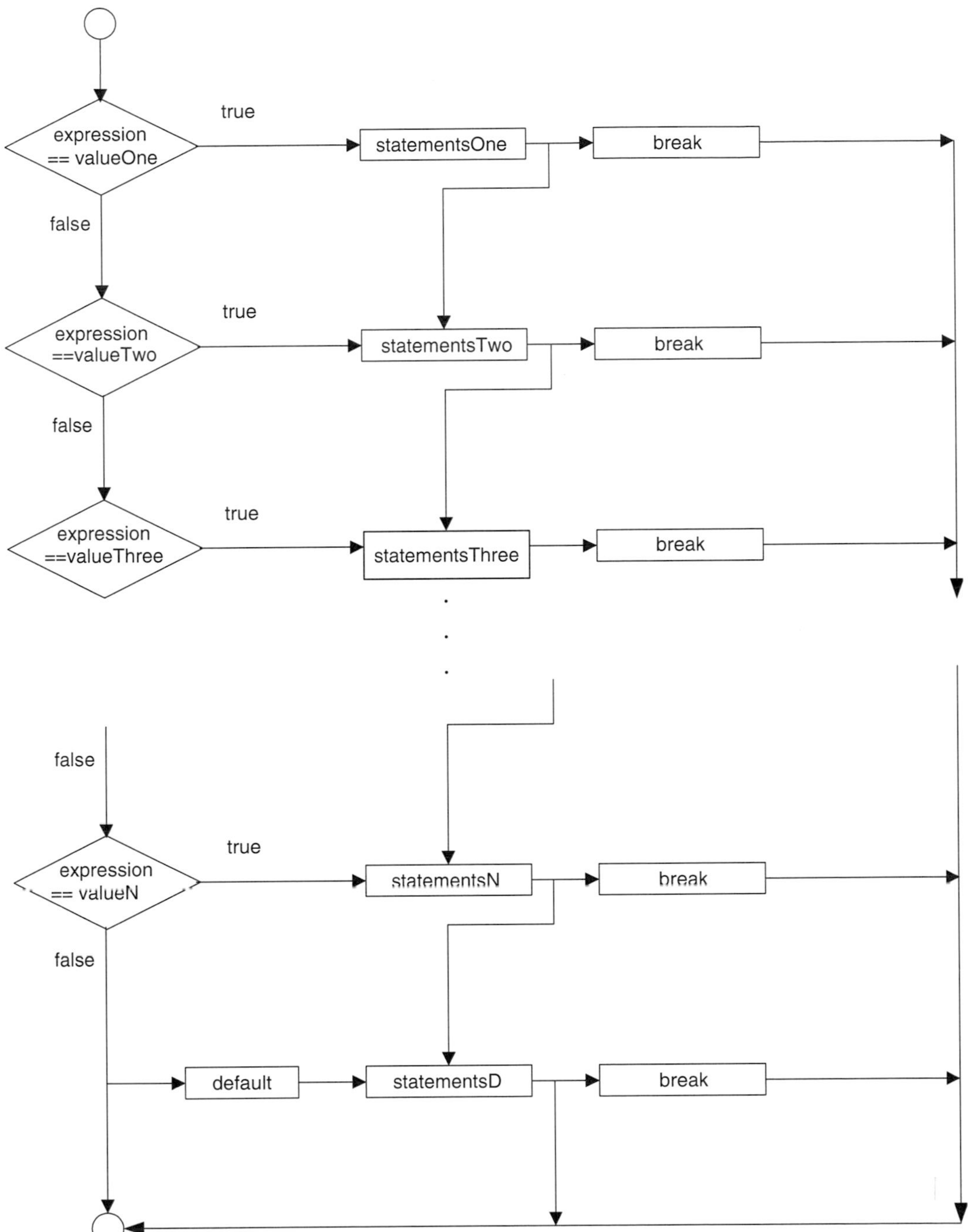

FIGURE 4.8 The control structure `switch`.

In this example, the control expression participationLevel is an int variable. The labels are int literals 1 through 5.

The above switch statement can be replaced by the following semantically equivalent multiway if ... else statement:

```
if (participationLevel == 1)
    athleticPoints = 3;
else if (participationLevel == 2)
    athleticPoints = 6;
else if (participationLevel == 3)
    athleticPoints = 12;
else if (participationLevel == 4)
    athleticPoints = 18;
else if (participationLevel == 5)
    athleticPoints = 25;
else
    athleticPoints = 0;
```

Self-Check

27. True or false: In the switch statement to compute the athletic points, the order of case statements is immaterial.
28. Is above statement true in general?

Advanced Topic 4.11: Sharing Code in a switch Statement

Example 4.24

The following switch statement illustrates the sharing of common code for more than one label. Assume that month and numberOfDays are int variables and leapYear is a boolean variable.

```
switch(month)
{
case 1   :
case 3   :
case 5   :
case 7   :
case 8   :
case 10  :
case 12  :
    numberOfDays = 31;
    break;
case 2   :
    if (leapYear)
        numberOfDays = 29;
```

```
        else
            numberOfDays = 28;
        break;
    case 4   :
    case 6   :
    case 9   :
    case 11  :
        numberOfDays = 30;
        break;
    default:
        numberOfDays = 0;
        break;
}
```

The above switch statement can be replaced by the following if … else statement:

```
if (month == 1 || month == 3 || month == 5 || month == 7 ||
                       month == 8 || month == 10 || month == 12 )
    numberOfDays = 31;
else if (month == 4 || month == 6 || month == 9 || month == 11)
    numberOfDays = 30;
else if (month == 2)
    if (leapYear)
        numberOfDays = 29;
    else
        numberOfDays = 28;
else
    numberOfDays = 0;
```

Example 4.25

A credit card company issues four different types of credit cards: basic, silver, gold, and platinum. As one moves up the level, one has all the benefits of the level below and some additional benefits. A switch statement is quite useful in this context.

```
System.out.println("You are entitled to the following:");
System.out.println();
switch(cardType)
{
case 1   :                          //platinum
    System.out.println("\t$100,000 in travel insurance");
case 2   :                          //gold
    System.out.println("\tThree month price protection");
case 3   :                          //silver
    System.out.println("\tFree rental car insurance");
```

```
case 4    :                          //basic
    System.out.println("\tFree credit protection against loss");
    System.out.println("\t24/7 toll free customer service");
    System.out.println();
    break;
default :

    // no action required.
}
```

The above code is equivalent to the following segment of code:

```
System.out.println("You are entitled to the following:");
System.out.println();
if (cardType == 1)                   //platinum
{
    System.out.println("\t$100,000 in travel insurance");
    System.out.println("\tThree month price protection");
    System.out.println("\tFree rental car insurance");
    System.out.println("\tFree credit protection against loss");
    System.out.println("\t24/7 toll free customer service");
    System.out.println();
}
else if (cardType == 2)                   //gold
{
    System.out.println("\tThree month price protection");
    System.out.println("\tFree rental car insurance");
    System.out.println("\tFree credit protection against loss");
    System.out.println("\t24/7 toll free customer service");
    System.out.println();
}
else if (cardType == 3)                   //silver
{
    System.out.println("\tFree rental car insurance");
    System.out.println("\tFree credit protection against loss");
    System.out.println("\t24/7 toll free customer service");
    System.out.println();
}
else                                      //basic
{
    System.out.println("\tFree credit protection against loss");
    System.out.println("\t24/7 toll free customer service");
    System.out.println();
}
```

Advanced Topic 4.12: Limitations of a switch Statement

As you can observe from the preceding examples, a switch statement can always be implemented as a multiway if ... else structure. However, since switch statement relies on equality operator and the controlling expression cannot be a floating-point value, not every if ... else structure can be replaced by an elegant switch statement. Even though there are no broadly accepted rules to determine whether or not to use a switch structure to implement multiway selections, readability and maintainability can be one of the criteria in your decision making. When an equality comparison is involved with three or more alternatives, a switch statement may be more readable.

Example 4.26

In this example, we make use of the code developed in Example 4.25 to generate personalized messages for every customer. We begin by creating a CreditCardCustomer class. This class has four data members: salutation, firstName, lastName, and cardType. The only application-specific method is createWelcomeMessage that returns a personalized message as a string.

```
/**
    Credit card message generator class
*/
public class CreditCardCustomer
{
    private String salutation;
    private String firstName;
    private String lastName;
    private int cardType;

    /**
        Creates and returns the customized message
        @return message as a string
    */
    public String createWelcomeMessage()
    {
        String outStr;
        outStr = "\n\n\n\t\t\t" +
            "Hi " + salutation + " " +
            firstName + " " + lastName +
            "\n\n\n" +
            "\n\tYou are entitled to the following:\n";

    switch(cardType)
    {
    case 1   :
        outStr = outStr +"\t$100,000 in travel insurance\n";
```

```
        case 2   :
            outStr = outStr +"\tThree month price protection\n";
        case 3   :
            outStr = outStr +"\tFree rental car insurance\n";
        case 4   :
            outStr = outStr +"\tFree credit protection against
                                                    loss\n";
            outStr = outStr + "\t24/7 toll free customer service";
        }
        outStr = outStr + "\n\n\n\n";
        return outStr;
    }

    /**
        Accessor method for salutation
        @return salutation
    */
    public String getSalutation()
    {
       return salutation;
    }

    /**
        Accessor method for first name
        @return first name
    */
    public String getFirstName()
    {
        return firstName;
    }

    /**
        Accessor method for last name
        @return last name
    */
    public String getLastName()
    {
        return lastName;
    }

    /**
        Accessor method for card type
        @return card type
    */
```

```
public int getCardType()
{
    return cardType;
}

/**
    mutator method for salutation
    @param inSalutation new value for salutation
*/
public void setSalutation(String inSalutation)
{
    salutation = inSalutation;
}

/**
    mutator method for first name
    @param inSalutation new value for first name
*/
public void setFirstName(String inFirstName)
{
    firstName = inFirstName;
}

/**
    mutator method for last name
    @param inSalutation new value for last name
*/
public void setLastName(String inLastName)
{
    lastName = inLastName;
}

/**
    mutator method for card type
    @param inSalutation new value for card type
*/
public void setCardType(int inCardType)
{
cardType = inCardType;
}

/**
    toString method returns name as a string
    @return name of the customer
*/
```

```
    public String toString()
    {
        String   str;
        str = firstName + " " + lastName;
        return str;
    }
}
```

The application program creates an object of the class `CreditCardCustomer`. Then it reads salutation, first name, and last name, followed by credit card type code (4 for the basic, 3 for the silver, 2 for the gold, and 1 for the platinum) and sets those values. A personalized welcome message is produced using the `createWelcomeMessage` method.

```
import java.util.Scanner;

/**
    Tester class for credit card customer
*/
public class CreditCardCustomerTesting
{

    public static void main (String[] args)
    {
        String custSalutation, custFirstName, custLastName;
        int custCardType;

        CreditCardCustomer customer = new CreditCardCustomer();
        Scanner scannedInfo = new Scanner(System.in);

        System.out.print("Enter salutation, first name, last
                              name and credit card type: ");
        System.out.flush();
        custSalutation = scannedInfo.next();
        custFirstName = scannedInfo.next();
        custLastName = scannedInfo.next();
        custCardType = scannedInfo.nextInt();

        customer.setSalutation(custSalutation);
        customer.setFirstName(custFirstName);
        customer.setLastName(custLastName);
        customer.setCardType(custCardType);
        System.out.print(customer.createWelcomeMessage());
    }
}
```

Advanced Topic 4.13: Enumerated Types

Consider Example 4.25. As mentioned, a credit card can have only four different types: basic, silver, gold, and platinum. In Example 4.26 we just assigned values 1 through 4 for those four different types. A better option is to use enumerated types. We can create a new data type called `CardType` as follows:

```
public enum CardType {PLATINUM, GOLD, SILVER, BASIC};
```

You can declare a variable `customerCardType` of `CardType` as follows:

```
CardType customerCardType;            // (1)
```

Now the following assignment statement is legal:

```
customerCardType =   CardType.GOLD; // (2)
```

Just as in the case of any other data type, you can combine Lines 1 and 2 as follows:

```
CardType customerCardType = CardType.GOLD;
```

Trying to assign any value other than `CardType.BASIC`, `CardType.SILVER`, `CardType.GOLD`, and `CardType.PLATINUM` is an error. For example, values such as 7 and `CardType.DIAMOND` will result in compilation error.

You can use equality operators == and `!=`. Quite often, we create an enumerated type within a class. In that case, enumerated data values can be accessed using the following syntax template:

```
ClassName.EnumeratedDataTypeName.DataValue
```

For example, if `CardType` is created within a class `CreditCardCustomer`, you can use the enumerated data value GOLD outside the class as `CreditCardCustomer.CardType.GOLD`.

Example 4.27

This example illustrates the use of enumerated type. We redesign the `CreditCardCustomer` class presented in Example 4.26. Apart from changing the data type of the data member `cardType`, the only change required is in the `setCardType` method. You can still use the same application program without any modifications.

```
/**
    Illustration class for enumerated data type
*/
public class CreditCardCustomerEnumVersion
{
    public enum CardType {PLATINUM, GOLD, SILVER, BASIC};
    private String salutation;
    private String firstName;
    private String lastName;
    private CardType cardType;
```

```
/**
    Creates and returns the customized message
    @return message as a string
*/
public String createWelcomeMessage()
{
    String outStr;
    outStr = "\n\n\n\t\t\t" +
        "Hi " + salutation + " " +
        firstName + " " + lastName +
        "\n\n\n" +
        "\n\tYou are entitled to the following:\n";

    switch(cardType)
    {
    case PLATINUM   :
        outStr = outStr + "\t$100,000 in travel insurance\n";
    case GOLD   :
        outStr = outStr + "\tThree month price protection\n";
    case SILVER   :
        outStr = outStr + "\tFree rental car insurance\n";
    case BASIC :
        outStr = outStr +
                "\tFree credit protection against loss\n";
        outStr = outStr + "\t24/7 toll free customer service";
    }

    outStr = outStr + "\n\n\n\n";

    return outStr;
}

/**
    Accessor method for salutation
    @return salutation
*/
public String getSalutation()
{
    return salutation;
}

/**
    Accessor method for first name
    @return first name
```

```
*/
public String getFirstName()
{
    return firstName;
}
/**
    Accessor method for last name
    @return last name
*/
public String getLastName()
{
    return lastName;
}
/**
    Accessor method for card type
    @return card type
*/
public CardType getCardType()
{
    return cardType;
}
/**
    Mutator method for salutation
    @param inSalutation new value for salutation
*/
public void setSalutation(String inSalutation)
{
    salutation = inSalutation;
}
/**
    Mutator method for first name
    @param inFirstName new value for first name
*/
public void setFirstName(String inFirstName)
{
    firstName = inFirstName;
}
/**
    Mutator method for last name
    @param inLastName new value for last name
*/
```

```
    public void setLastName(String inLastName)
    {
        lastName = inLastName;
    }

    /**
        Mutator method for card type
        @param inCardType new value for card type
    */
    public void setCardType(int inCardType)
    {
        switch(inCardType)
        {
        case 1   :
            cardType = CardType.PLATINUM;
        case 2   :
            cardType = CardType.GOLD;
        case 3   :
            cardType = CardType.SILVER;
        case 4 :
            cardType = CardType.BASIC;
        }
    }

    /**
        toString method returns name as a string
        @return name of the customer
    */
    public String toString()
    {
        String  str;
        str = firstName + " " + lastName;
        return str;
    }
}
```

CASE STUDY 4.1: PAYROLL FOR A SMALL BUSINESS

Specification

Mr. Jones is the proud owner of Heartland Cars of America (HCA), a car dealership. Mr. Jones has three types of employees. In the case of full-time employees, Mr. Jones pays a base salary for first 80 h of work and 150% of hourly rate for each additional hour.

A part-time employee is paid by hour. Employees in sales are compensated by a base salary plus 1% of the sales amount for the period.

Mr. Jones pays his employees on alternate Fridays. One of the full-time employees happened to be Ms. Smart, a Java programmer. Her first task is to write Java program to help Mr. Jones print a payroll stub on each payday.

Input

Input varies by employee type. For full-time employees, the input values are character F (indicating the employee is full-time), first name, last name, base salary, and hours worked. In the case of a part-time employee, the input values are character P (indicating the employee is part-time), first name, last name, hourly compensation, and hours worked. For employees in sales, the input values are character S (indicating the employee is in sales), first name, last name, base salary, and sales amount.

Output

Pay stub for every employee.

We begin by performing a use case analysis. Clearly, there needs to be a use case to prepare the pay stub. To prepare the pay stub, you may need to compute the compensation for each employee. Finally, there are three different types of employees: full-time, part-time, and sales. Thus, we arrive at the use case diagram shown in Figure 4.9.

Decide on Classes

From the use case diagram, it is quite clear that there are three different types of employees. Therefore, let us start with the following three classes: `FullTimeEmp`, `PartTimeEmp`, and `SalesEmp`.

Decide on Attributes

From the above specification, the following design decision is quite clear:

Class: **FullTimeEmp**

Data members:

```
private String firstName;
private String lastName;
private double baseSalary;
private int hoursWorked;
```

Class: **PartTimeEmp**

Data members:

```
private String firstName;
private String lastName;
private double payPerHour;
private int hoursWorked;
```

Class: **SalesEmp**

Data members:

```
private String firstName;
private String lastName;
private double baseSalary;
private double salesVolume;
```

Decide on Methods

In this step you need to decide on methods. As explained before, you may include accessor and mutator methods corresponding to each data member. Further, it is nice to have `toString` in each class. So in this section, the focus is on additional methods that are needed in each of the above three classes. Once again, from the use case diagram, each of the three classes needs at least two operations: `computeCompensation` and `createPayStub`.

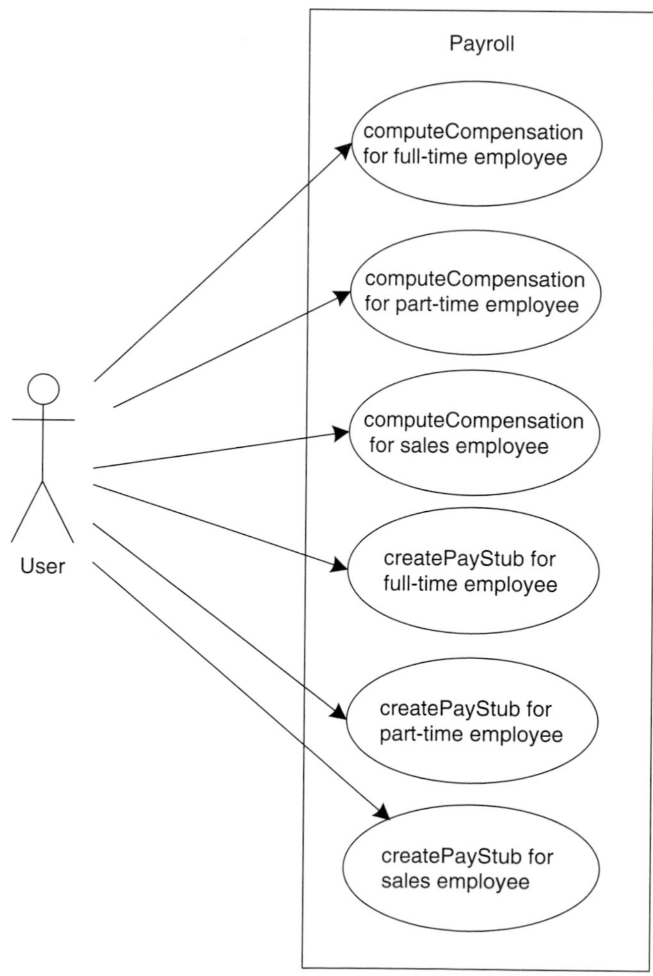

FIGURE 4.9 Use case diagram for the payroll.

Class: **FullTimeEmp**

Methods:

```
public double computeCompensation()
public void createPayStub()
```

From the problem specification, it follows that the compensation can be computed using the formula, If the `hoursWorked` > 80.

```
compensation = baseSalary + (hoursWorked - 80) * 1.5 * payPerHour.
```

where `payPerHour` can be computed using the formula, `payPerHour = baseSalary / 80`.

If `hoursWorked <= 80`, then the following formula applies: `compensation = baseSalary`.

The `createPayStub()` method creates a string containing all the data in the employee's pay stub. It creates an object `currencyFormat` of the class `DecimalFormat` as follows:

```
DecimalFormat currencyFormat = new DecimalFormat("0.00");
```

The `String` parameter `"0.00"` in the constructor specifies a decimal format with two decimal positions. To format a double variable value such as the `salary`, the required expression is `currencyFormat.format(salary)`. Note that the expression invokes the `format` method of the `DecimalFormat` class.

Class: **PartTimeEmp**

Methods:

```
public double computeCompensation()
public void createPayStub()
```

From the problem specification the compensation is computed using the following formula:

```
compensation = payPerHour * hoursWorked
```

The `createPayStub()` method creates a string containing all the data in the employee's pay stub.

Class: **SalesEmp**

Methods:

```
public double computeCompensation()
public void createPayStub()
```

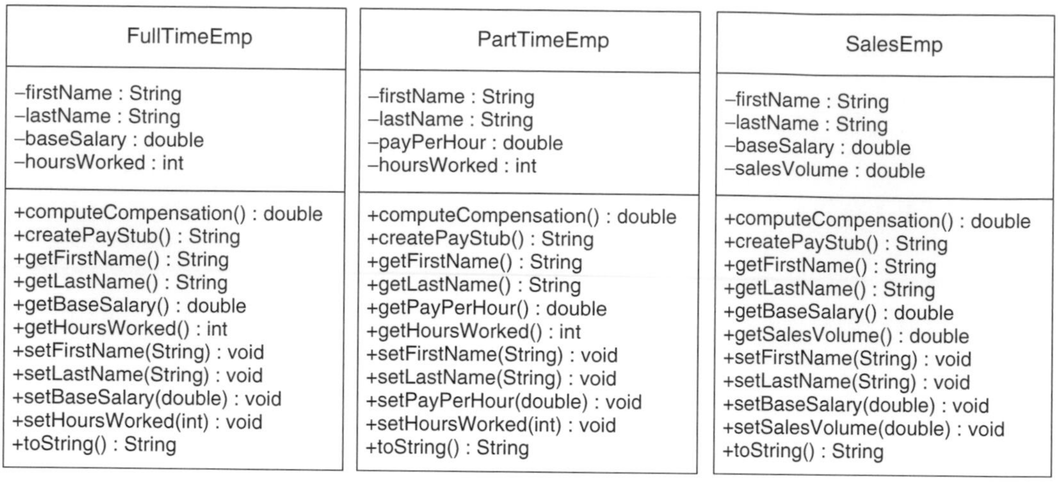

FullTimeEmp	PartTimeEmp	SalesEmp
–firstName : String –lastName : String –baseSalary : double –hoursWorked : int	–firstName : String –lastName : String –payPerHour : double –hoursWorked : int	–firstName : String –lastName : String –baseSalary : double –salesVolume : double
+computeCompensation() : double +createPayStub() : String +getFirstName() : String +getLastName() : String +getBaseSalary() : double +getHoursWorked() : int +setFirstName(String) : void +setLastName(String) : void +setBaseSalary(double) : void +setHoursWorked(int) : void +toString() : String	+computeCompensation() : double +createPayStub() : String +getFirstName() : String +getLastName() : String +getPayPerHour() : double +getHoursWorked() : int +setFirstName(String) : void +setLastName(String) : void +setPayPerHour(double) : void +setHoursWorked(int) : void +toString() : String	+computeCompensation() : double +createPayStub() : String +getFirstName() : String +getLastName() : String +getBaseSalary() : double +getSalesVolume() : double +setFirstName(String) : void +setLastName(String) : void +setBaseSalary(double) : void +setSalesVolume(double) : void +toString() : String

FIGURE 4.10 Class diagram of employees.

From the problem specification, it follows that the compensation can be computed using the following formula if the `hoursWorked > 80`:

```
compensation = baseSalary + 0.02 * salesVolume
```

The `createPayStub()` method creates a string containing all the data in the employee's pay stub (Figure 4.10) shows the UML 2 class diagram of three types of employees.

Implementation

```
/**
    Full-time employee class
*/
public class FullTimeEmp
{

    private String firstName;
    private String lastName;
    private double baseSalary;
    private int hoursWorked;

    /**
        Computes and returns the compensation
        @return compensation
    */
    public double computeCompensation()
    {
        double compensation, payPerHour;

        payPerHour = baseSalary / 80;
```

```
    if (hoursWorked > 80)
    {
        compensation = baseSalary +
            (hoursWorked - 80) * 1.5 * payPerHour;
    }
    else
    {
        compensation = baseSalary;
    }

    return compensation;
}

/**
    Creates and returns a String for Paystub
    @return paystub information
*/
public String createPayStub()
{
    DecimalFormat currencyFormat = new DecimalFormat("0.00");
    double salary;

    salary = computeCompensation();
    String outStr;

    outStr = "\n\n\n\t\t\t" +
            "HEARTLAND CARS OF AMERICA" +
            "\n\n\n\t" +
            firstName + " " + lastName +
            "\n\n\n" +
            "\n\tBasic Salary \t$" +
            currencyFormat.format(baseSalary) +
            "\n\tHours Worked \t " + hoursWorked +
            "\n\tPay              \t$" +
            currencyFormat.format(salary) +
            "\n\n\n\n";
    return outStr;
}

/**
    Accessor method for first name
    @return first name
*/
```

```
public String getFirstName()
{
    return firstName;
}

/**
    Accessor method for last name
    @return last name
*/
public String getLastName()
{
    return lastName;
}

/**
    Accessor method for base salary
    @return base salary
*/
public double getBaseSalary()
{
    return baseSalary;
}

/**
    Accessor method for hours worked
    @return hours worked
*/
public int getHoursWorked()
{
    return hoursWorked;
}

/**
    Mutator method for first name
    @param inFirstName new value for first name
*/
public void setFirstName(String inFirstName)
{
    firstName = inFirstName;
}

/**
    Mutator method for last name
    @param inLastName new value for last name
*/
```

```
    public void setLastName(String inLastName)
    {
        lastName = inLastName;
    }

    /**
        Mutator method for base salary
        @param inBaseSalary new value for base salary
    */
    public void setBaseSalary(int inBaseSalary)
    {
        baseSalary = inBaseSalary;
    }

    /**
        Mutator method for hours worked
        @param inHoursWorked new value for hours worked
    */
    public void setHoursWorked (int inHoursWorked)
    {
        hoursWorked = inHoursWorked;
    }

    /**
        toString method returns name as a string
        @return name of the customer
    */
    public String toString()
    {
        String   str;
        str = firstName + " " + lastName;
        return str;
    }
}

/**
    Part-time employee class
*/
public class PartTimeEmp
{
    private String firstName;
    private String lastName;
    private double payPerHour;
    private int hoursWorked;
```

```
/**
    Computes and returns the compensation
    @return compensation
*/
public double computeCompensation()
{
    double compensation;
    compensation = payPerHour * hoursWorked;
    return compensation;
}

/**
    Creates and returns a String for Paystub
    @return paystub information
*/
public String createPayStub()
{
    DecimalFormat currencyFormat = new
        DecimalFormat("0.00");
    double salary;

    salary = computeCompensation();
    String outStr;

    outStr = "\n\n\n\t\t\t" +
                "HEARTLAND CARS OF AMERICA" +
                "\n\n\n\t" +
                firstName + " " + lastName +
                "\n\n\n" +
                "\n\tSalary/Hour \t$" +
                currencyFormat.format(payPerHour) +
                "\n\tHours Worked \t " + hoursWorked +
                "\n\tPay               \t$" +
                currencyFormat.format(salary) +
                "\n\n\n\n";

    return outStr;
}

/**
    Accessor method for first name
    @return first name
*/
```

```
public String getFirstName()
{
    return firstName;
}

/**
    Accessor method for last name
    @return last name
*/
public String getLastName()
{
    return lastName;
}

/**
    Accessor method for pay per hour
    @return pay per hour
*/
public double getPayPerHour()
{
    return payPerHour;
}

/**
    Accessor method for hours worked
    @return hours worked
*/
public int getHoursWorked()
{
    return hoursWorked;
}

/**
    Mutator method for first name
    @param inFirstName new value for first name
*/
public void setFirstName(String inFirstName)
{
    firstName = inFirstName;
}

/**
    Mutator method for last name
    @param inLastName new value for last name
*/
```

```
    public void setLastName(String inLastName)
    {
        lastName = inLastName;
    }

    /**
        Mutator method for pay per hour
        @param inPayPerHour new value for pay per hour
    */
    public void setPayPerHour (double inPayPerHour)
    {
        payPerHour = inPayPerHour;
    }

    /**
        Mutator method for hours worked
        @param inHoursWorked new value for hours worked
    */
    public void setHoursWorked (int inHoursWorked)
    {
        hoursWorked = inHoursWorked;
    }

    /**
        toString method returns name as a string
        @return name of the customer
    */
    public String toString()
    {
        String  str;
        str = firstName + " " + lastName;
        return str;
    }
}

/**
    Sales employee class
*/
public class SalesEmp
{
    private String firstName;
    private String lastName;
    private double baseSalary;
    private double salesVolume;
```

```
/**
    Computes and returns the compensation
    @return compensation
*/
public double computeCompensation()
{
    double compensation;

    compensation = baseSalary + 0.02 * salesVolume;

    return compensation;
}

/**
    Creates and returns a String for Paystub
    @return paystub information
*/
public String createPayStub()
{
    DecimalFormat currencyFormat = new DecimalFormat("0.00");
    double salary;

    salary = computeCompensation();
    String outStr;

    outStr = "\n\n\n\t\t\t" +
                "HEARTLAND CARS OF AMERICA" +
                "\n\n\n\t" +
                firstName + " " + lastName +
                "\n\n\n" +
                "\n\tBasic Salary \t$" +
                currencyFormat.format(baseSalary) +
                "\n\tSales Volume \t$" +
                currencyFormat.format(salesVolume) +
                "\n\tPay              \t$" +
                currencyFormat.format(salary) +
                "\n\n\n\n";

    return outStr;
}

/**
    Accessor method for first name
    @return first name
*/
```

```
public String getFirstName()
{
    return firstName;
}

/**
    Accessor method for last name
    @return last name
*/
public String getLastName()
{
    return lastName;
}

/**
    Accessor method for base salary
    @return base salary
*/
public double getBaseSalary()
{
    return baseSalary;
}

/**
    Accessor method for sales volume
    @return sales volume
*/
public double getSalesVolume()
{
    return salesVolume;
}

/**
    Mutator method for first name
    @param inFirstName new value for first name
*/
public void setFirstName(String inFirstName)
{
    firstName = inFirstName;
}

/**
    Mutator method for last name
    @param inLastName new value for last name
*/
```

```
    public void setLastName(String inLastName)
    {
        lastName = inLastName;
    }

    /**
        Mutator method for base salary
        @param inBaseSalary new value for base salary
    */
    public void setBaseSalary(int inBaseSalary)
    {
        baseSalary = inBaseSalary;
    }

    /**
        Mutator method for sales volume
        @param inSalesVolume new value for sales volume
    */
    public void setSalesVolume (double inSalesVolume)
    {
        salesVolume = inSalesVolume;
    }

    /**
        toString method returns name as a string
        @return name of the customer
    */
    public String toString()
    {
        String   str;
        str = firstName + " " + lastName;
        return str;
    }
}
```

Application Program

The class diagram of the application program is shown in Figure 4.11 and the application program is as follows:

```
import java.util.Scanner;
/**
    Heartland Cars of America pay roll
*/
```

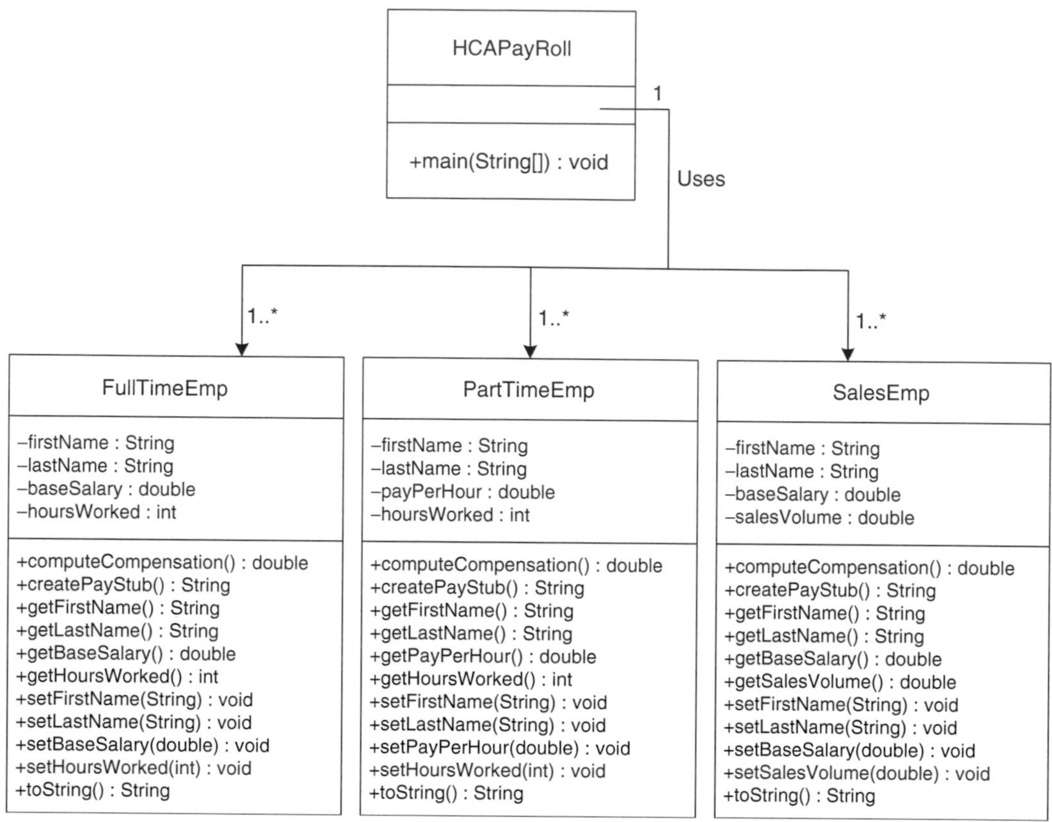

FIGURE 4.11 Class diagram of HCA payroll.

```
public class HeartlandCarsOfAmericaPayRoll
{
    public static void main (String[] args)
    {

        //Create reference variable of all three employee types
        FullTimeEmp fullTimeEmployee;
        PartTimeEmp partTimeEmployee;
        SalesEmp salesEmployee;

        //Declare variables to input

        char inputEmployeeType;
        String inputFirstName;
        String inputLastName;
        double inputBaseSalary;
        double inputPayPerHour;
        int inputSalesVolume;
        int inputHoursWorked;
```

```
//Get two input values
Scanner scannedInfo = new Scanner(System.in);
System.out.print("Enter Employee Type : ");
System.out.flush();
inputEmployeeType = scannedInfo.next().charAt(0);
System.out.println();

switch (inputEmployeeType)
{
case 'F' :
case 'f' :

    //get necessary values as input
    System.out.print("Enter First Name, " +
                    "Last Name, Base Salary, Hours : ");
    System.out.flush();
    inputFirstName = scannedInfo.next();
    inputLastName = scannedInfo.next();
    inputBaseSalary = scannedInfo.nextDouble();
    inputHoursWorked = scannedInfo.nextInt();
    System.out.println();

    //create an object and initialize data members
    fullTimeEmployee = new FullTimeEmp();
    fullTimeEmployee.setFirstName(inputFirstName);
    fullTimeEmployee.setLastName(inputLastName);
    fullTimeEmployee.setBaseSalary(inputBaseSalary);
    fullTimeEmployee.setHoursWorked(inputHoursWorked);

    //invoke the printPayStub method
    System.out.print(fullTimeEmployee.createPayStub());

    break;

case 'P' :
case 'p' :
    //get necessary values as input
    System.out.print("Enter First Name, Last Name, "+
                    "Pay per hour, Hours : ");
    System.out.flush();
    inputFirstName = scannedInfo.next();
    inputLastName = scannedInfo.next();
    inputPayPerHour = scannedInfo.nextDouble();
    inputHoursWorked = scannedInfo.nextInt();
    System.out.println();
```

```
        //create an object and initialize data members
        partTimeEmployee = new PartTimeEmp();
        partTimeEmployee.setFirstName(inputFirstName);
        partTimeEmployee.setLastName(inputLastName);
        partTimeEmployee.setPayPerHour(inputPayPerHour);
        partTimeEmployee.setHoursWorked(inputHoursWorked);

        //invoke the printPayStub method
        System.out.print(partTimeEmployee.createPayStub());

        break;

    case 'S' :
    case 's' :
        //get necessary values as input
        System.out.print("Enter First Name, Last Name, "+
                        "Base Salary, Sales Volume : ");
        System.out.flush();
        inputFirstName = scannedInfo.next();
        inputLastName = scannedInfo.next();
        inputBaseSalary = scannedInfo.nextDouble();
        inputSalesVolume = scannedInfo.nextInt();
        System.out.println();

        //create an object and initialize data members
        salesEmployee = new SalesEmp();
        salesEmployee.setFirstName(inputFirstName);
        salesEmployee.setLastName(inputLastName);
        salesEmployee.setBaseSalary(inputBaseSalary);
        salesEmployee.setSalesVolume(inputSalesVolume);

        //invoke the printPayStub method
        System.out.print(salesEmployee.createPayStub());

        break;
    }
}
}
```

Testing

You need to test three totally different sets of data: full-time employee, part-time employee, and sales employees. Among full-time employees, there are two categories: one who worked 80 h or less and others. In particular, you must test full-time employees who worked exactly 80 h.

Thus, test data can be as follows:

```
F    Adam Smith       2450.00    87
F    Joyce Witt       3425.67    80
F    Mike Morse       1423.56    75
P    Chris Olsen      34.56      34
S    Patrick McCoy    1040.57    856985
```

Output

```
Enter Employee Type : F

Enter First Name, Last Name, Base Salary, Hours : Adam Smith
      2450.00 87

                                HEARTLAND CARS OF AMERICA

            Adam Smith

            Basic Salary        $2450.00
            Hours Worked        87
            Pay                 $2771.56

Enter Employee Type : P

Enter First Name, Last Name, Pay per hour, Hours : Chris Watt
      18.75 94
```

REVIEW

1. Control structures determine the flow of control.
2. Every computer program can be constructed from the following three control structures: sequence, selection, and repetition.
3. Selection structure enables you to selectively execute a certain part of a code while skipping some other parts.
4. Repetition structure, however, allows you to repeat a certain part of the code again and again.
5. A logical expression always evaluates to either `true` or `false`.
6. There are five binary logical operators: && (and), & (and), || (or), | (or), and ^ (xor) in Java.
7. The only logical unary operator in Java is ! (not).
8. There are six relational operators in Java. They are == (equal to), != (not equal to), < (less than), <= (less than or equal to), > (greater than), and > = (greater than or equal to).

9. It is a syntax error to have a space character in the middle of any one of the following relational operators: ==, <=, >=, and !=.
10. The assignment operator has lower precedence than relational operators.
11. The relational operator has lower precedence than arithmetic operators.
12. Java provides three structures for selection and decision making. They are the one-way selection structure `if`, the two-way selection structure `if ... else`, and the multiway selection structure `switch`.
13. The semantics of the one-way selection structure

    ```
    if(decisionCondition)
        actionStatement
    ```

    is as follows: if the `decisionCondition` is `true`, then the `actionStatement` is executed; otherwise, the `actionStatement` will be skipped.
14. A block statement is a sequence of Java statements enclosed within a pair of braces.
15. In Java, = is an assignment operator. Therefore, using = instead of == in a logical expression is a syntax error.
16. In the case of an `if` statement, the decision condition must always be within a pair of parentheses.
17. In the case of an `if` statement, there is no semicolon immediately following the right parenthesis enclosing the decision condition.
18. The semantics of the two-way selection structure

    ```
    if (decisionCondition)
        actionStatementTrue
    else
        actionStatementFalse
    ```

    is as follows: if decision condition evaluates to `true`, the action statement `actionStatementTrue` is executed. However, if the decision condition evaluates to `false`, the action statement `actionStatementFalse` is executed.
19. In Java, there is no selection structure `else`.
20. An `else` is always paired with the most recent `if` that has not been paired with any other `else`.
21. Placing one control structure inside another control structure is called nesting.
22. A conditional operator can replace certain `if ... else` structures.
23. The `switch` control structure can replace certain multiple selections.
24. Use of enumerated types makes a program more readable.
25. You can use equality operators with objects and enumerated types.

EXERCISES

1. Mark the following statements as true or false:
   a. A logical expression can be assigned to a `boolean` variable.
   b. Logical expression `true + true` evaluates to `true`.
   c. Logical literal `true` has numeric value 1.
   d. Any program can be rewritten without any two-way or multiway selection structures.
   e. Every `else` must have a corresponding `if`.
   f. Every `if` must have a corresponding `else`.
   g. A `switch` statement need not have any `default` case.
   h. The logical operators `||` and `&&` have the same precedence.
   i. Addition has lower precedence than the logical operator `!`
2. What is the output produced by the following codes? If there is no output, then write "No output". Indicate any syntax errors.

   a.
```
int x = 2;
if(x <= 2);
    System.out.print("Good");
    System.out.print(" Morning");
```

   b.
```
int x = 2;
int y = 7;

if(x > y || y < 4)
    System.out.print("Good");
    System.out.print(" Afternoon");
```

   c.
```
int x = 2;
int y = 7;
int z = 20;

if(z > y * x && x < y - 2)
    System.out.print("Good");
    System.out.print(" Night");
```

   d.
```
int x = 10;
int y = 8;

if(!(x < y))
    System.out.print("Good");
    System.out.print(" Night");
```

   e.
```
int x = 12;
if(x < 12);
    System.out.print("Good");
```

```
else
    System.out.print(" Morning");
```

f.
```
int x = 8;
int y = 17;
if(x >= y || y <= 14)
    System.out.print("Good");
    else
        System.out.print(" Afternoon");
```

g.
```
int x = 2;
int y = 7;
int z = 20;

if(z > y * x && x < y - 2)
    System.out.print("Good");
else
    System.out.print(" Night");
```

h.
```
int x = 11;
int y = 5;

if(!(x > y))
    System.out.print("Good");
else
    System.out.print(" Night");
```

i.
```
int x = 8;
int y = 17;

if(!(x > y || y <= 14))
    System.out.print("Good");
else
    System.out.print(" Day");
```

j.
```
int x = 2;
int y = 7;
int z = 20;

if(!(z > y * x && x < y - 2))
    System.out.print("Good");
else
    System.out.print(" Night");
```

k.
```
int x = 2;
int y = 7;
int z = 20;
```

```
    if(z > y )
        System.out.print("Excellent");
    if(x > y )
        System.out.print("Good");
    else
        System.out.print("Acceptable");
```

l.
```
    int x = 2;
    int y = 7;
    int z = 20;

    if(z < y - x )
        System.out.print("Accept");
    if(x > y )
        System.out.print("Reject");
    else
        System.out.print("undecided");
    System.out.print("?");
```

m.
```
    int x = 2;
    int y = 7;
    int z = 20;

    if(z > y )
        System.out.print("Excellent");
    if(x > y )
        System.out.print("Good");
    else
        System.out.print("Acceptable");
    else
        System.out.print("Unacceptable");
```

n.
```
    int x = 2;
    int y = 7;
    int z = 20;

    if(z < y - x )
        System.out.print("Red");
    else
        System.out.print("Blue");
    if(x > y )
        System.out.print("Green");
    else
        System.out.print("Yellow");
```

3. What is the output produced by the following codes? Check the `switch` statements for syntax errors. Explain each error identified.

a.
```
double value = 1000;
switch(value)
{
case 1000: System.out.println("High");
    break;
case 100: System.out.println("Medium");
    break;
case 1: System.out.println("Low");
    break;
case default: System.out.println("Perfect");
}
```

b.
```
int value = 100;
switch(value < 10000)
{
case 1: System.out.println("Value is reasonable");
    break;
case 10: System.out.println("Value out of range");
case 1000: System.out.println("Value computed is wrong");
    break;
}
```

c.
```
int num = 1000;
switch (num % 4 + 1)
{
case 1:
case 2: System.out.println("Too Small");
    break;
case 2:
case 3: System.out.println("Perfect");
    break;
case 3:
case 4: System.out.println("Too Big");
}
```

d.
```
int n = 7;
int num = 0;
switch(num = n%2)
    {
    case 0: System.out.println("Even number");
    break;
    case 1: System.out.println("Odd number");
    break;
    }
```

e.
```
int num = 10;
switch (num == 10 )
{
case false  : System.out.println("False case");
case true   : System.out.println("True case");
case all    : System.out.println("All cases");
}
```

f.
```
int num = 10;
switch (num % 2 )
{
case 0      : num = num / 2;
System.out.println("even case");
case 1      : num = 3*num + 1;
System.out.println("odd case");
}
```

4. (a) What are the outputs produced by the following segments of code?
   (b) Rewrite the code without `switch` statement.
   (c) What is the output, if the control variable has the value 3?

```
int k = 4;
int n = 0;
int m = 0;
switch(k)
{
case 1: k = k - 2;
    break;
case 2: m = 2 * k;
    break;
case 3: n = 3 * k; k = k + 3;
case 4:
case 5: m = 4 * k; k = k + 4;
case 6: k = 8;
    break;
    default: m = k;
}
System.out.println("k = " + k);
System.out.println("n = " + n);
System.out.println("m = " + m);
```

5. (a) What are the outputs produced by the following segments of code?
   (b) Rewrite the code without `switch` statement.
   (c) What is the output, if the control variable has the value 5?

```
int k = 3;
```

```
int n = 0;
int m = 0;
    switch(k)
    {
    case 1:
    case 2: m = 2 * k; k = k + 2;
        break;
    case 3: n = 3 * k; k = k + 3;
    case 5: m = 4 * k; k = k + 4;
        break;
    case 6: k = 8;
        break;
    default: m = k;
        break;
    }
    System.out.println("k = " + k);
    System.out.println("n = " + n);
    System.out.println("m = " + m);
```

6. (a) What are the outputs produced by the following segments of code?
   (b) Rewrite the code without `switch` statement.
   (c) What is the output, if the control variable has the value 3?

```
int k = 1;
int n = 0;
int m = 0;
switch(k)
{
case 1:
case 2:
case 3: k = k + 2;
case 4: k = k + 3;
case 5: m = 4 * k; k = k + 4;
    case 6: k = 8;
default: m = k;
}
System.out.println("k = " + k);
System.out.println("n = " + n);
System.out.println("m = " + m);
```

PROGRAMMING EXERCISES

1. Design and test the following class:
   a. Create a class `BankAccount` with data members `firstName`, `lastName`, `accountNumber`, and `balance`.

b. Provide a method that returns true if the balance is greater than the minimum balance, which currently is 500.00.

c. Implement a method `withdraw` that will deduct the amount specified from the balance, provided balance is greater than or equal to the amount to be withdrawn.

2. Ms. Erin Cook just bought a little puppy. Each Sunday, she will weigh her puppy to make sure it is neither overfed nor underfed. Design a class to help Ms. Cook to keep the weight of her puppy within the range. (*Hint*: There are three attributes: lowerLimit, upperLimit, and current weight. The method `checkPuppyWeight` that will output a message depending on the puppy's weight.)
3. Design a class `student` with the following six data members to store first name, last name, and four test scores. Provide a method `getLetterGrade()` that returns the letter grade based on the policy outlined in Example 4.21.
4. Design a class `Student` with the following three data members to store first name, last name, and levelofparticipation (school, district, state, national, international). Provide a method that returns athletic points as specified in the chapter. (*Hint*: Level of participation may be maintained as an integer value as in the text.)
5. Redo Programming Exercise 4 using enumerated types and switch structure.
6. Redo Programming Exercise 5 so that user can enter "school," "district," "state," "national," "international" as data rather than integer values. Make your program more user-friendly by allowing user to mix uppercase and lowercase letters as they input participation level. (*Hint*: Use `equalsIgnoreCase` method of the String class.)
7. Enhance your solution to Programming Exercise 6 by allowing user to make spelling mistakes while they enter "school," "district," "state," "national," "international" as data. If the first two letters are s and c, it will be treated as school. If the first letter is d (s, n, i), then it will be treated as district (state, national, international).
8. Mr. Jones insurance coverage is as follows: He has $500 deductable and as such he gets nothing for first 500.00. Next 5000, he gets 90% and for the next 5000 he gets 80% and he gets 70% for anything over 10,500 from the insurance company. Design a class for Mr. Jones to determine the amount he can expect from the insurance company.
9. The income tax of a certain country is based on two factors: income and whether or not the taxfiler is a resident of the country. The tax structure can be summarized as follows: A resident pays no tax for the first 37,500.00. Anything above 37,500.00 and less than 120,000.00 is taxed at the rate of 17%. Anything above 120,000.00 is taxed at the rate of 21%. A nonresident pays 10% for any amount up to 20,000.00. Next 100,000.00 is taxed at the rate of 20% and any amount over 120,000.00 is taxed at the rate of 35%. Design a class to perform the tax computation.
10. Design, implement, and test a class to solve quadratic equations. The roots of the quadratic equation $ax^2 + bx + c = 0, a \neq 0$ are given by the following formula:

$$\frac{-b \pm \sqrt{b^2 - 4ac}}{2a}$$

The term $b^2 - 4ac$ is called the discriminant. If discriminant is nonnegative, then the equation has two real solutions. Otherwise, the equation has two complex solutions. (*Hint*: Have three data members: a, b, and c. Provide methods `hasRealRoots` that will return `true` if discriminant is nonnegative and `hasEqualRoots` that will return `true` if discriminant is zero. Also, provide following four methods: `SolutionOne`, `SolutionTwo`, `realPart`, and `imaginaryPart`. The method `sqrt` in the `class Math` can be used to compute the square root of the discriminant, if the discriminant is nonnegative.)

11. Create a class with three data members and three methods: maxValue, middleValue, and minValue returning maximum, middle, and minimum values, respectively.
12. The Great Eastern Bank (GEB) offers savings accounts. In the case of savings account, there is no charge if the minimum balance is at least 300.00. Otherwise, each customer must pay $15.00 toward maintenance fee. A savings account also allows 10 free withdrawals per month. For each additional withdrawal, the customer is charged $5.00. Create the Java program to compute the monthly charges for a customer.
13. Design a class `FortuneCookie` with a method `getFortune` that returns a fortune cookie message. Your class must produce at least eight different messages (use `Math.Random()` to generate a random number). Select the fortune cookie message based on the random number generated.
14. Create a class `Classifier` as follows. A `Classifier` object keeps the x–y coordinates of four points in Cartesian plane. Two of them are "normal points" and the other two are "abnormal points." If the new point is closer to two normal points, then the new point is classified as normal and if the new point is closer to two abnormal points, then the new point is classified as abnormal. Otherwise, the point cannot be classified. To make it more challenging, create a class `point` and use it in your program.
15. Create a class `Triangle`. A `Triangle` object keeps the x–y coordinates of three points in Cartesian plane. Provide three methods, `isTriangle`, `isIsosceles`, and `isEquilateral`. (*Hint*: Three points form a triangle, if sum of the lengths of any two sides is greater than the third side. A triangle is isosceles if at least two of its sides are equal. A triangle is equilateral if all three of its sides are equal.) To make it more challenging, create a class `point` and use it in your program.

ANSWERS TO SELF-CHECK

1. True
2. False
3. `true`
4. `false`
5. False
6. six
7. `true`

8. `false`
9. `true`
10. `true`
11. Negative integer
12. Positive integer
13. True
14. True
15. True
16. False
17. Replace `(inNumberOfShares < 0)` with `(inNumberOfShares < 1)`
18. Replace `inCounterValue >= limitValue` with `limitValue <= inCounterValue`
19. Yes
20. Make the following changes: `MINIMUM_BALANCE = 300.0;` and `(checkingBalance < MINIMUM_BALANCE)`
21.
```
if (cost > 25.0)
    System.out.println("Set for life!");
else
    System.out.println("Not there yet!");
```
22.
```
if (networth > 10000000.0)
    System.out.println("Expensive Item");
else
    System.out.println("Reasonable Item");
```
23. `isUppercaseLetter = ('A' <= charOne && charOne <='Z');`
24. No
25. Yes
26. True
27. True
28. No

CHAPTER 5

The Power of Repetition

In this chapter you learn

- Java concepts
  - Repetition structures `while`, `for`, and `do ... while`
  - Control statements `break` and `continue`
  - Exceptions
- Programming skills
  - Process multiple sets of data
  - Appropriate use of repetition structures
  - Nesting of control structures
  - Using text files for input and output

In Chapter 2, you have studied the control structure sequence. In Chapter 4, you have seen the control structure selection. In this chapter, you will study the control structure *repetition*. A program with sequence structure alone is capable of executing one statement after another in sequence. Such a program has no ability to make decisions. With the introduction of selection structure, programs became capable of making decisions based on data values. For example, using sequence structure alone it is possible to write a program to compute the cumulative test score by adding five test scores. If you want to translate the cumulative test score into letter grades, you need a control structure that can examine and make decisions based on the test score. Suppose you want to find the class average of a test. If the class consists of only a handful of students, you can definitely use the sequence structure to find the average. However, such a technique is not practical if there are 75 students. What you need is a structure capable of repeating an action. To find the sum of the test scores of all students in a class, you need a control structure that can read a number and add the number read to a partial sum repeatedly. Such a structure is called a *repetition structure*. In this chapter, you will learn about various repetition structures available in Java. However, the fundamental principles you learn in this chapter can be applied in many scripting languages and programming languages including C, C++, and C#.

CONTROL STRUCTURES

Recall that every computer program can be constructed from three basic control structures, *sequence*, *selection*, and *repetition*, shown in Figure 5.1. Repetition structure allows you to repeat a certain code. Note that both selection and repetition structures alter the order of execution of statements. We study the repetition structure in this chapter.

Let us revisit the pay stub–printing program presented in Chapter 4. Heartland Cars of America turned out to be a very successful business. Now it has 40 employees. However, this introduced two major issues for Ms. Smart. Every 2 weeks, Ms. Smart has to enter lots of data items. However, a big majority of the data values remain the same from one pay period to the next. For example, consider a full-time employee like Adam Smith. Ms. Smart has to enter four values: first name, last name, base salary, and hours worked. First name and last name are not going to change. Base salary may change once a year or so. Therefore, the only data item that may change in a 2 week period is the hours worked. Ms. Smart need not type in all these values every time. Instead, Ms. Smart can keep these values in a file, and all she needs to do is to modify the hours worked. In Chapter 4, Ms. Smart printed the pay stub for each employee as she entered the data. A better alternative is to store the pay stub information for all employees in a file and then print that file. This approach has two advantages. First, it makes the printing more efficient. Second, it permits the preservation of the pay stub data for future reference. Ms. Smart decided to modify the program using the following algorithm:

1. Read the data for an employee from a file
2. Process the information for an employee
3. Write the pay stub information of an employee to a file
4. Repeat steps 1 through 3 for every employee

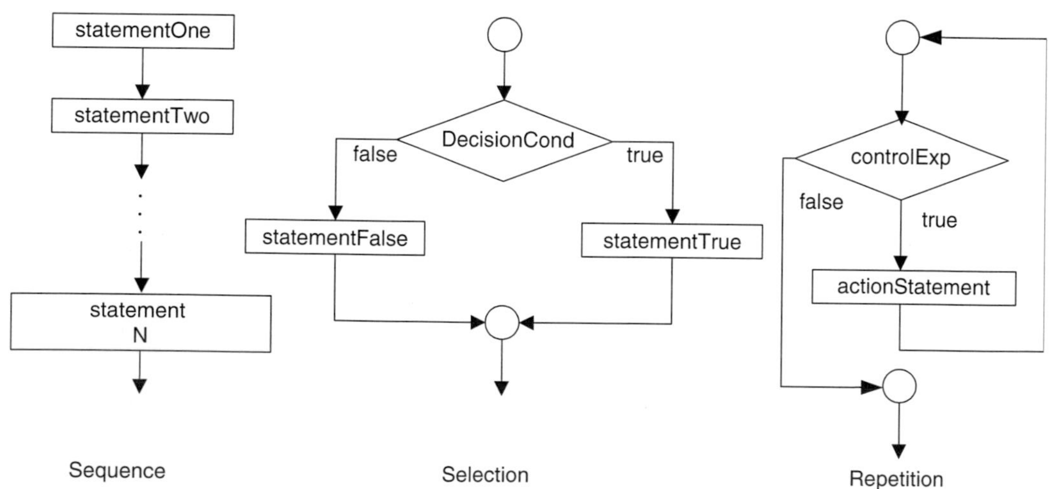

FIGURE 5.1 Control structures.

Self-Check

1. What are the three control structures?
2. To implement step 4, Ms. Smart needs a ________ control structure.

USING TEXT FILE FOR INPUT

In Chapter 2, you have seen how to input data from standard input device. Recall that with `System.in`, you can input either a single byte or a sequence of bytes. To be useful, you want to break the input into meaningful units of data called *tokens*. Java provides a `Scanner` class with necessary methods to get individual tokens. You need to create an object belonging to the `Scanner` class. Recall from Chapter 2 that this is achieved by the following Java statement:

```
Scanner scannedInfo = new Scanner(System.in);
```

You can replace `System.in` with a file object and input can be read from a file object. If the data is in a file `C:\myfile.dat`, you can modify the above statement as follows:

```
Scanner scannedInfo = new Scanner(new File("C:\\myfile.dat"));
```

Observe that you need two backslash characters inside a string to produce one backslash character. Further, `new File("C:\\myfile.dat")` creates a file object using the `new` operator. The program in Example 5.1 reads data from a file instead of a standard input. This program is obtained by replacing

```
Scanner scannedInfo = new Scanner(System.in);
```

by

```
Scanner scannedInfo = new Scanner(new File("C:\\myfile.dat"));
```

in Example 2.25. The `File` class belongs to `java.io` package and therefore `java.io` package needs to be imported.

Self-Check

3. Write the Java statement to import the package containing the class `File`.
4. Write the necessary Java statements to create a scanner object that can be used to read data from a file `testData.txt` in the `C` drive.

Declaring Exceptions

Unlike `System.in`, there is no guarantee that the file named in the program is available in the location specified. If the file cannot be found, then it is an error and such errors are classified as exceptions in Java. Exceptions are discussed in detail in Chapter 11. For the present, just note that certain types of exceptions must be declared in a program. File not found is one such exception, and corresponding to that exception there is a `FileNotFoundException` class in package `java.io`. To declare a file not found exception, you add "`throws FileNotFoundException`" as follows:

```
public static void main (String[] args) throws
                         FileNotFoundException
```

Another exception that needs to be declared is the `IOException`. An `IOException` occurs, whenever a mismatch in data type happens during an I/O operation. For example, if an integer is expected and instead a string is encountered as input, an I/O exception occurs.

Example 5.1

```
import java.io.*;
import java.util.Scanner;

/**
    Illustration of reading from an input file
*/
public class ScannerInputFile
{
    public static void main (String[] args) throws
                FileNotFoundException, IOException
    {
        String socialSecNum;
        String firstName;
        String lastName;
        int age;
        double monthlySalary;

        Scanner scannedInfo = new Scanner(new
                    File("C:\\myfile.dat"));

        socialSecNum = scannedInfo.next();
        firstName = scannedInfo.next();
        lastName = scannedInfo.next();
        age = scannedInfo.nextInt();
        monthlySalary = scannedInfo.nextDouble();

        System.out.println(socialSecNum);
        System.out.println(firstName);
        System.out.println(lastName);
```

```
            System.out.println(age);
            System.out.println(monthlySalary);
        }
    }
```

Output

```
123-45-6789
James
Watts
56
5432.78
```

Self-Check

5. True or false: If the social security number appeared as `123X45X6789`, an exception will occur during the program execution.
6. If the social security number appeared as `123   45   6789`, the value of `socialSecNum` will be ________.

USING FILE FOR OUTPUT

To create an output file `C:\myfile.out` you need a statement similar to

```
PrintWriter output = new PrintWriter
  (new FileWriter("C:\\myfile.out"));
```

Recall that the new operator creates an object. The above statement first creates a `FileWriter` object and then using this object a `PrintWriter` object is created. Note that both `FileWriter` and `PrintWriter` classes are in the package `java.io` and there is no need to import any other packages. This newly created `PrintWriter` object can be referenced using the reference variable `output`. Now, the `PrintWriter` class has methods such as `print` and `println` to print various data types. The program in Example 5.1 can be modified so that the output is written in `C:\myfile.out`.

Self-Check

7. Write the Java statement to import the package containing the class `PrintWriter`.
8. Write the necessary Java statements to create a `PrintWriter` object that can be used to write data to a file `info.txt` in the `C` drive.

Method close

It is a good programming practice to close `PrintWriter` objects before program terminates. The `close` method, in particular, will flush any data remaining in the output buffer to the output file. To close the `PrintWriter` object `output`, you need the following statement:

```
output.close();
```

The program in Example 5.1 can be modified so that the output is written in a file C:\myfile.out.

Example 5.2

```
import java.io.*;
import java.util.Scanner;

/**
    Illustration of reading/writing from/to an input/output file
*/
public class ScannerInputOutputFile
{
    public static void main (String[] args) throws
                FileNotFoundException, IOException
    {
        String socialSecNum;
        String firstName;
        String lastName;
        int age;
        double monthlySalary;

        Scanner scannedInfo =
                    new Scanner(new File("C:\\myfile.dat"));
        PrintWriter output =
             new PrintWriter(new FileWriter("C:\\myfile.out"));

        socialSecNum = scannedInfo.next();
        firstName = scannedInfo.next();
        lastName = scannedInfo.next();
        age = scannedInfo.nextInt();
        monthlySalary = scannedInfo.nextDouble();

        output.println(socialSecNum);
        output.println(firstName);
        output.println(lastName);
        output.println(age);
        output.println(monthlySalary);

        scannedInfo.close();
        output.close();
    }
}
```

Having studied the basic file processing techniques, you are now ready to learn the repetition structure. The first repetition structure you will be introduced to is the `while` statement, which is the subject of our next section.

Self-Check

9. True or false: In Java, you must always close all files explicitly.
10. True or false: The `close` method will flush any data remaining in the output buffer to the output file.

REPETITION STRUCTURE: while

Consider a program to compute the average of three floating-point numbers. The algorithm can be outlined as follows:

- Get the values in three variables, say, `valueOne`, `valueTwo`, and `valueThree`.
- Add the three values together and divide by 3.

```
averageValue = (valueOne + valueTwo + valueThree)/3.0
```

The following example presents a program to compute the average of three floating-point numbers.

Example 5.3

```
import java.util.Scanner;

/**
    Computes the average of three double values
*/
public class AverageOfThreeValues
{
    public static void main (String[] args)
    {
        double valueOne;
        double valueTwo;
        double valueThree;
        double averageValue;

        Scanner scannedInfo = new Scanner(System.in);

        System.out.print("Enter three double values : ");
        System.out.flush();
        valueOne = scannedInfo.nextDouble();
        valueTwo = scannedInfo.nextDouble();
        valueThree = scannedInfo.nextDouble();
```

```
            System.out.println();
            averageValue = (valueOne + valueTwo + valueThree)/3.0;

            System.out.println("The average value is " +
                                              averageValue);
        }
}
```

Output

```
Enter three double values : 34.67 56.98 45.69

The average value is 45.78
```

Suppose you want to find the average of 100 numbers. The above approach is quite cumbersome. Further, if you would like to find the average of 10,000 data items, the above approach is practically impossible. Let us approach the problem in a different way. To begin with, notice that once the sum of all the numbers is known, you could compute the average by dividing the sum by the number of items. Therefore, the real issue is how to compute the sum of a large set of data.

Have a look at the following numbers:

```
4.1        2.2        3.3
```

One way to add them is to just add one number at a time to a partial sum. To begin with, partial sum is zero. The next number is 4.1. We add the next number to the partial sum. The partial sum becomes $0 + 4.1 = 4.1$. Now the next number is 2.2. Therefore, after adding next number the partial sum becomes 6.3. The next number is 3.3. After adding the next number the partial sum becomes 9.6. At this point all three data values have been added to the partial sum and there is no more number to be added to the partial sum. All that is remaining is to divide 9.6, the partial sum, by 3. This logic can be explained as follows:

1. Initialize the partial sum to zero
2. Get the next number and add the next number to the partial sum
3. Get the next number and add the next number to the partial sum
4. Get the next number and add the next number to the partial sum
5. Divide the partial sum by 3

In the above description, note that steps 2 through 4 are essentially identical. Therefore, another way of expressing the above logic is

1. Initialize the partial sum to zero
2. Repeat three times the following: get the next number and add the next number to the partial sum
3. Divide the partial sum by 3

Introspection

What exactly we mean by repetition? Clearly you are not adding the same number. However, you are "reading a number and adding to the partial sum."

You know how to code step 1. You need a variable, say sum and you need to initialize it to zero. The step 1 can be implemented as follows:

```
double sum = 0.0;
```

Similarly, step 3 can be coded as shown below assuming average is a variable of type `double`.

```
average = sum / 3;
```

Now, let us look into step 2. Getting a number can be implemented as follows:

```
nextValue = scannedInfo.nextDouble();
```

and `nextValue` can be added to the variable `sum` by

```
sum = sum + nextValue;
```

Here, `nextValue` is a variable of type `double` and `scannedInfo` is a reference variable as in Example 5.1. Now all that is left to implement is the logic "repeat three times the following." You need a repetition structure to implement it. The basic syntax of a `while` statement is as follows:

```
while (controlExp)
    actionStatement
```

Note that `while` is a reserved word and the pair of parentheses enclosing the `controlExp` is part of the syntax. The `control Exp` is a logical expression and therefore it evaluates to either `true` or `false`. The semantics of the `while` statement can be explained as follows: First, the `controlExp` is evaluated, and if the control expression evaluates to `true`, then the `actionStatement` is executed once. The `actionStatement` can be a single statement or a block statement. After executing the `actionStatement` once, the `controlExp` is evaluated again and if the control expression evaluates to `true`, then the `actionStatement` is executed once more and so on. In other words, so long as the `controlExp` evaluates to `true`, the `actionStatement` is executed and the `controlExp` is evaluated again. If the `controlExp` evaluates to `false`, then the `actionStatement` is not executed. Figure 5.2 illustrates the semantics of a `while` statement.

Generally speaking there are three types of `while` statements: counter-controlled, event-controlled, and data-controlled.

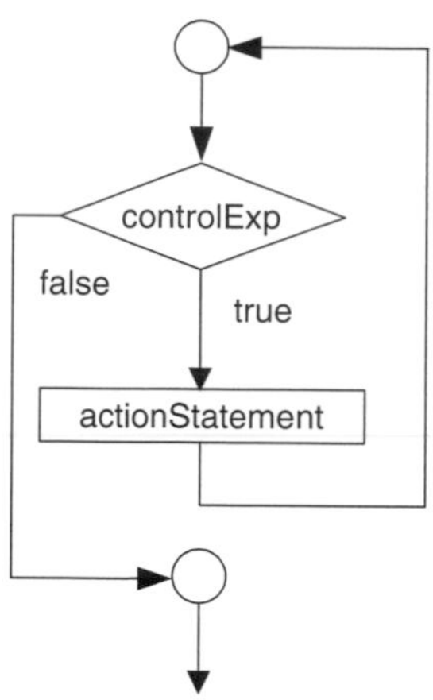

FIGURE 5.2 `while` statement.

Self-Check

11. True or false: It is possible to write a `while` statement such that the action statement gets executed an infinite number of times.
12. True or false: It is possible to write a `while` statement such that the action statement is never executed.

Counter-Controlled `while` Statement

Consider the following code:

```
double nextValue;
double sum;
double average;
int counter;

sum = 0;                                               // (1)
counter = 1;                                           // (2)
while (counter < 4)                                    // (3)
{
    nextValue = scannedInfo.nextDouble();              // (4)
    sum = sum + nextValue;                             // (5)

    counter = counter + 1;                             // (6)
}
average = sum / 3;                                     // (7)
```

Line 1 initializes the variable `sum` to zero and Line 2 initializes the `counter` to 1. At Line 3, `counter` value is 1 and the control expression `counter` < `4` is `true`. Therefore, Lines 4 through 6 are executed. The first value gets stored in the variable `nextValue`. Further, `sum` becomes the same as `nextValue` and `counter` becomes 2. Now the control goes back to Line 3. The control expression `counter < 4` is evaluated again. Note that `counter < 4` is `true`. Hence Lines 4 through 6 are executed. The second value gets stored in the variable `nextValue`. Further, `sum` becomes the sum of first and second values. The `counter` is incremented and becomes 3. The control goes back to Line 3. The control expression `counter < 4` evaluates to `true`, and therefore `nextValue` becomes the third input value and `sum` becomes the sum of all three input values. The `counter` is incremented and the `counter` value becomes 4. Therefore, as control goes back to Line 3, the control expression `counter < 4` evaluates to `false`. Consequently, Lines 4 through 6 are skipped and Line 7 gets executed. The above segment inputs three data values and computes their average.

Example 5.4

In this example `while` statement is used to compute the average of three numbers.

```
import java.util.Scanner;

/**
    Computes average of three values using while loop
*/
public class AverageThreeValuesUsingWhile
{
    public static void main (String[] args)
    {
        double nextValue, sum, averageValue;
        int counter;
        Scanner scannedInfo = new Scanner(System.in);

        sum = 0;
        counter = 1;
        while (counter < 4)
        {
              System.out.print("Enter a double value : ");
              System.out.flush();
              nextValue = scannedInfo.nextDouble();
              System.out.println();

              sum = sum + nextValue;
              counter = counter + 1;
        }
         averageValue = sum / 3;
         System.out.println("The average value is " +
                                           averageValue);

    }
}
```

Output

```
Enter a double value : 34.67

Enter a double value : 56.98

Enter a double value : 45.69

The average value is 45.78
```

The following points are worth mentioning:

1. In a counter-controlled `while` statement, the counter variable need not be named `counter`. However, to indicate the purpose of the variable names such as `counter`, `count`, and `cnt` are quite often used.
2. The counter variable must be initialized before executing the control expression of the `while` statement.
3. If the initialization of the counter variable is such that the control expression evaluates to `true`, within the action statement of the `while` statement the counter variable needs to be modified in such a way that the control expression eventually become `false`.

Segment of Code	Comment

`countValue = 1;`
`while (countValue < 11)`
`{`
`    ...`
`    countValue++`
`}`
	No error.
	The action statement will be executed 11 − 1 = 10 times

`CountValue = 1;`
`while (countValue < 11)`
`{`
`    ...`
`}`
	Error.
	`countValue` needs to be modified inside the action statement. As it stands, `while` loop will never terminate. It is an infinite loop.

`CountValue = 10;`
`while (countValue > 0)`
`{`
`    ...`
`   countValue--`
`}`
	No error.
	The action statement will be executed 10 − 0 = 10 times.

`countValue = 1;`
`while (countValue < 11)`
`{`
`    ...`
`    countValue--`
`}`
	Error.
	The `countValue` is always less than 11 and therefore, `while` will never terminate. It is an infinite loop.

(*Continued*)

Segment of Code	Comment

`counter = 1;`
`while (counter < 21)`
`{`
`    ...`
`    counter = counter + 3;`
`}`
	No error.
	The action statement will be executed `Math.ceil ((21 - 1)/3) = 1 + 20/3 = 7` times for the values 1, 4, 7, 10, 13, 16, and 19. Note that `Math.ceil` returns equal or next higher integer.

`counter = 1;`
`while (counter == 21)`
`{`
`    ...`
`    counter = counter + 3;`
`}`
	Error.
	`counter` value is initially 1. Thus, `counter == 21` evaluates to `false`. Therefore, `while` statement is never executed.

`counter = 1;`
`while (counter != 21)`
`{`
`    ...`
`    counter = counter + 3;`
`}`
	Error.
	`counter` value is initially 1. Thus, `counter != 21` evaluates to `true`. The `counter` value changes as 4, 7, 10, 13, 16, 19, 22, and so on. The `counter` never becomes 21. This is an infinite loop.

`countValue = 75;`
`while (countValue < 85)`
`{`
`    ...`
`    countValue++`
`}`
	No error.
	The action statement will be executed 85 – 75 = 10 times for the following `countValues`: 75, 76, 77, 78, 79, 80, 81, 82, 83, and 84.

The following template can be used to implement a counter-controlled `while` statement:

```
countValue = 1;                                     // (1)
while (countValue < numberOfRepetitionsPlusOne)     // (2)
{
    ...
    countValue++;                                   // (3)
}
```

In the above segment of code, Line 1 initializes the counter variable `countValue` to 1. In Line 2 `numberOfRepetitionsPlusOne` can be either a variable or a literal value equal to 1 more than the number of repetitions required. For instance, if number of repetitions required is 100, you can replace `numberOfRepetitionsPlusOne` by 101.

Self-Check

13. Write a `while` statement in which the counter assumes all even number values between 1 and 99 in the ascending order.

14. Write a `while` statement in which the counter assumes the values 99, 88, 77, 66, 55, and 44 in descending order.

Advanced Topic 5.1: Use of Counter inside Counter-Controlled `while` Statement

You can use the counter inside the `while` statement just like any other variable. However, counter can play very important role inside a counter-controlled `while` statement.

Consider the following problem: you want to create a table that prints first 10 powers of 2 for your easy reference. Clearly you want to print 10 items. You can start with the following counter-controlled `while` statement:

```
counter = 1;
while (counter < 11)
{

    ...

    counter++;
}
```

Now, if you include a statement

```
System.out.println("\t" + counter);
```

as shown below,

```
counter = 1;
while (counter < 11)
{

    System.out.println("\t" + counter);
    counter++;

}
```

the following output will be produced:

```
        1
        2
        3
        4
        5
        6
        7
        8
        9
        10
```

Therefore, all that is left to do can be summarized as follows:

Compute and print 2^1 when counter is 1
Compute and print 2^2 when counter is 2
Compute and print 2^3 when counter is 3
Compute and print 2^4 when counter is 4
Compute and print 2^5 when counter is 5
Compute and print 2^6 when counter is 6
Compute and print 2^7 when counter is 7
Compute and print 2^8 when counter is 8
Compute and print 2^9 when counter is 9
Compute and print 2^{10} when counter is 10

In other words, compute and print $2^{counter}$ for `counter = 1, 2, …, 10`. Now Java has a built-in `static` method `pow(x, y)` in the class `Math`. Here both `x` and `y` can be of type `int` or `double`. Being a `static` method, you can invoke the method pow as follows:

```
Math.pow(x,y)
```

and the method returns the `double` value x^y. Therefore, to compute $2^{counter}$ you can invoke the method pow as follows:

```
Math.pow(2, counter)
```

The segment of code necessary to print first 10 powers of 2 can be written as follows:

```
counter = 1;
while (counter < 11)
{
    System.out.println("\t" + counter + "\t\t\t" +
                          (int) Math.pow(2, counter));
    counter++;
}
```

The complete program is as follows:

```
public class FirstTenPowersOfTwo
{
    public static void main (String[] args)
    {
        int counter;
```

```
            counter = 1;
            while (counter < 11)
            {
                System.out.println("\t" + counter + "\t\t\t" +
                                    (int) Math.pow(2, counter));
                counter++;
            }
        }
}
```

Output

```
1       2
2       4
3       8
4       16
5       32
6       64
7       128
8       256
9       512
10      1024
```

Advanced Topic 5.2: Event-Controlled while Statement

In the previous section, you have seen the counter-controlled `while` statement. However, there are many situations where one may not know the number of repetitions required. For example, data may be in a file and you may want to process all data items in the file. In this case, you cannot use a counter-controlled `while` statement. However, you can check for the end of tokens in a file. The `Scanner` class has a method `hasNext()` that returns `true` if there are more tokens and returns `false` when there are no more tokens. Therefore, assuming the following declaration,

```
Scanner scannedInfo
        = new Scanner(new File("C:\\Studentdata.txt"));
```

an event-controlled `while` statement can be created as follows:

```
while (scannedInfo.hasNext())
{
    // read next data item and process it

}
//no more tokens left.
```

To illustrate these concepts, let us consider the following situation. Mr. Grace has just finished grading his first test. He decided to create a file in which each line has three entries: first name, last name, and test score of the student. It is always good to know the class average. Therefore, Mr. Grace decided to write a simple Java program to compute the class average. Mr. Grace started with the following segment of code:

```
Scanner scannedInfo
          = new Scanner(new File("C:\\Studentdata.txt"));

while (scannedInfo.hasNext())
{
    // read next data item and process it

}
//no more tokens left.
```

Mr. Grace noted that the first two tokens on each line are first name and last name, respectively, and as such there is no need to store those values. However, the third value is a `double` value and all those `double` values need to be added to get the sum of test scores for all students. Mr. Grace modified the above code as follows:

```
Scanner scannedInfo
          = new Scanner(new File("C:\\Studentdata.txt"));

sum = 0;                      // initialize sum
while (scannedInfo.hasNext())
{
    scannedInfo.next(); // skip first name
    scannedInfo.next(); // skip last name
    nextValue = scannedInfo.nextDouble();
    sum = sum + nextValue;    // add test score
}
```

To find the average, Mr. Grace needs to count the number of test scores. Therefore, he introduced one more variable `count`. Each time a new value is added to the `sum`, the `count` is incremented by 1. Mr. Grace created the program shown in Example 5.5.

Example 5.5

Mr. Grace's class average program:

```
import java.util.*;
import java.io.*;
import java.text.DecimalFormat;

/**
```

```
    Illustration of event-controlled while loop
*/
public class EventControlledWhile
{
    public static void main (String[] args) throws
                FileNotFoundException, IOException
    {
        double nextValue;
        double sum;
        double averageValue;
        int count;

        DecimalFormat centFormat = new DecimalFormat("0.00");
        Scanner scannedInfo =
               new Scanner(new File("C:\\Studentdata.txt"));

        sum = 0;
        count = 0;
        while (scannedInfo.hasNext())
        {
             scannedInfo.next(); // skip first name
             scannedInfo.next(); // skip last name
             nextValue = scannedInfo.nextDouble();
             sum = sum + nextValue;      // add test scores
             count++;        // count the number of test scores
        }
        averageValue = sum / count;

        System.out.println("The class average is " +
                  centFormat.format(averageValue));

        scannedInfo.close();

    }

}
```

Input File Content

```
Kimberly Clarke      98.5
Chris Jones          78.5
Brian Wills          85.0
Bruce Mathew         60.5
Mike Daub            56.6
```

Output

```
The class average is 75.82
```

Advanced Topic 5.3: Data-Controlled while Statement

Data Validation

There are many situations where you may wish to control the `while` statement through input data directly. In this section you will see two such cases.

Let us assume your high school is in the process of updating alumni records. To make sure that you were a student, they want you to enter your high-school graduation year. However, if you fail to enter a valid value, program would like to prompt you again and again until a correct value is entered. In this context, a `while` statement can be used as follows:

```
highSchoolYear = 0;
while (highSchoolYear < 1940 || highSchoolYear > 1990)
{
    System.out.print("Enter year of high school graduation ");
    System.out.flush();
    highSchoolYear = scannedInfo.nextInt();
    System.out.println();
}
```

Sentinel Data

Sometimes, programmers use out of range data values to indicate the end of data. Such a data value is called *sentinel data*. For example, if you are dealing with positive values any negative value can be used as a sentinel data.

To illustrate these concepts, let us consider the following situation. Mr. Grace has just finished grading his final test. His grading policy is that the final grade depends on the average of all tests attempted by a student. A student need not take all tests. Therefore, Mr. Grace decides to write a program, using sentinel data, to compute the final score for a student. Since all the test scores are nonnegative, Mr. Grace decides to use −1.0 as sentinel value. A typical input to the program is as follows:

Kate Currin   67.8   89.9   78.0   95.4   −1.0

Mr. Grace notes that first two tokens on each line are first name and last name, respectively. All other values in a line are of type `double`. Once a negative value is encountered, no more data needs to be processed. Mr. Grace starts with the following segment of code:

```
Scanner scannedInfo = new Scanner(System.in);
firstName = scannedInfo.next();
lastName = scannedInfo.next();
nextValue = scannedInfo.nextDouble();
```

```
while (nextValue > -1.0)
{

    // process nextValue
    // increment count

    // read next data item into nextValue
}
```

In this context, process data involves adding all the grades. Mr. Grace modified the code as follows:

```
Scanner scannedInfo = new Scanner(System.in);

sum = 0;
count = 0;                    // initialize sum
firstName = scannedInfo.next();
lastName = scannedInfo.next();
nextValue = scannedInfo.nextDouble();

while (nextValue > -1.0)
{
    sum = sum + nextValue;
    count++;
    nextValue = scannedInfo.nextDouble();
}
```

Example 5.6

Mr. Grace's average score program:

```
import java.util.*;
import java.text.DecimalFormat;

/**
    Illustration of data-controlled while loop
*/
public class DataControlledWhile
{
    public static void main (String[] args)
    {
        double nextValue;
        double sum;
```

```
        double averageScore;
        int count;
        String firstName;
        String lastName;

        DecimalFormat centFormat = new DecimalFormat("0.00");
        Scanner scannedInfo = new Scanner(System.in);

        System.out.println("Enter first name, last name, " +
                                   "test scores and -1.0");

        sum = 0;
        count = 0;                      // initialize sum
        firstName = scannedInfo.next();
        lastName = scannedInfo.next();
        nextValue = scannedInfo.nextDouble();

        while (nextValue > -1.0)
        {
            sum = sum + nextValue;
            count++;
            nextValue = scannedInfo.nextDouble();
        }
        averageScore = sum / count;
        System.out.println(firstName +" "+ lastName +" "+
                        centFormat.format(averageScore));
        scannedInfo.close();
    }

}
```

Output

```
Enter first name, last name, test scores and -1.0
Kate Currin 67.8 89.9 78.0 95.4 -1.0
Kate Currin 82.78
```

REPETITION STRUCTURE: for

You have seen three types of `while` statements: counter-controlled, event-controlled, and data-controlled. In fact, `while` statement is general enough to implement any repetition. The `for` statement introduced in this section can replace any `while` statement.

The general syntax of the `for` statement is

```
for (initialStmt; controlExp; updateStmt)
    actionStatement
```

Note that `for` is a reserved word in Java. The `initialStmt` can be any Java statement or a sequence of Java statements separated by comma. However, its intended purpose is to initialize variables involved in the `controlExp`. The `controlExp` can be any logical expression in Java. The `updateStmt` can be any Java statement or a sequence of Java statements separated by comma. However, its purpose is to modify variables involved in the `controlExp`. The `actionStatement` can be either a single Java statement or a block statement. The semantics of a `for` statement in Java can be explained using Figure 5.3.

A better understanding of the `while` statement and the `for` statement can be obtained from Figure 5.4.

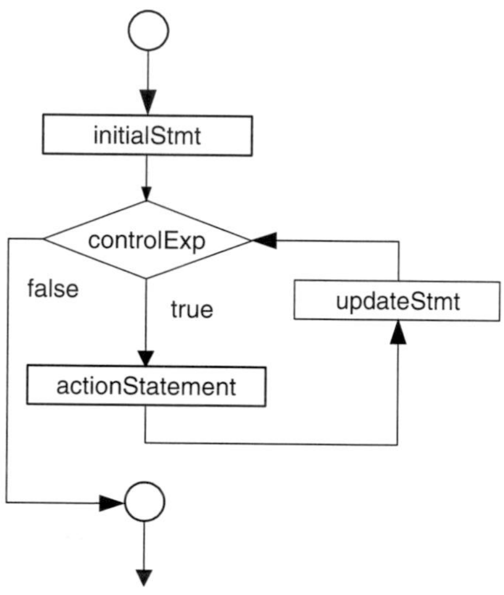

FIGURE 5.3 `for` statement.

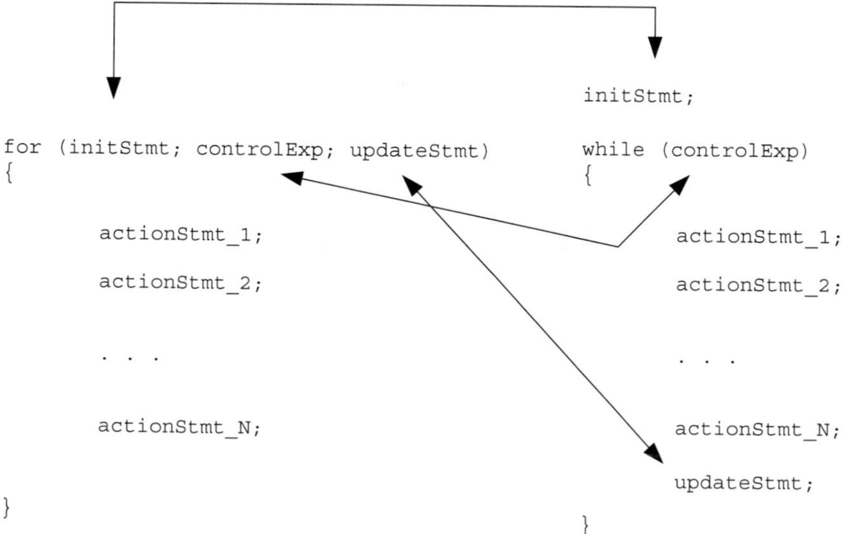

FIGURE 5.4 `while` and `for` statements.

Example 5.7

Throughout this example, assume the following declarations:

```
double nextValue;
double sum;
double average;
int counter;
```

Now the following segment of code

```
sum = 0;
counter = 1;
while (counter < 4)
{
    nextValue = scannedInfo.nextDouble();
    sum = sum + nextValue;
    counter++;
}
average = sum / 3;
```

can be written in any one of the following forms using the `for` statement:

a.
```
sum = 0;
for (counter = 1; counter < 4; counter++ )
{
    nextValue = scannedInfo.nextDouble();
    sum = sum + nextValue;
}
average = sum / 3;
```

b.
```
for (sum = 0,counter = 1; counter < 4; counter++ )
{
    nextValue = scannedInfo.nextDouble();
    sum = sum + nextValue;
}
average = sum / 3;
```

c.
```
for (sum = 0,counter = 1; counter < 4;
          counter++, sum += nextValue)
{
    nextValue = scannedInfo.nextDouble();
}
average = sum / 3;
```

d.
```
for (sum = 0,counter = 1; counter < 4;
        counter++, sum += scannedInfo.nextDouble());
average = sum / 3;
```

Even though all four `for` statements are semantically equivalent, you are encouraged to use the option a, which is easily understandable and hence easily maintainable.

Example 5.8

Consider the following segment of code you have seen in Example 5.4:

```
sum = 0;
counter = 1;
while (counter < 4)
{
    System.out.print("Enter a double value : ");
    System.out.flush();
    nextValue = scannedInfo.nextDouble();
    System.out.println();
    sum = sum + nextValue;
    counter = counter + 1;
}
```

The above segment of code can be replaced by the following:

```
sum = 0;
counter = 1;
for (counter = 1; counter < 4; counter++)
{
    System.out.print("Enter a double value : ");
    System.out.flush();
    nextValue = scannedInfo.nextDouble();
    System.out.println();
    sum = sum + nextValue;
}
```

The following points are worth mentioning:

1. The `initialStmt`, `controlExp`, `updateStmt`, and `actionStatement` are optional in a `for` statement. The smallest `for` statement that can be written is

   ```
   for (; ; );
   ```

2. The pair of left and right parentheses along with two semicolons appearing inside them is part of the syntax.
3. A missing `controlExp` evaluates to `true`.

4. If `controlExp` is `false` in the beginning, the `for` statement is never executed.
5. A semicolon following the right parenthesis is not a syntax error. It simply amounts to an empty `actionStatement`.

Segment of Code	Comment

`for (count = 1;count < 11; count++)`
`{`
`    ...`
`}`
	No error.
	The `actionStatement` will be executed 11 − 1 = 10 times.

`for (int cnt = 1;cnt < 11; cnt++)`
`{`
`    ...`
`}`
	No error.
	The `actionStatement` will be executed 11 − 1 = 10 times. The variable `cnt` does not exist outside the `for` statement.

`for (count = 10;count > 0;`
` count--)`
`{`
`    ...`
`}`
	No error
	The `actionStatement` will be executed 10 – 0 = 10 times.

`for (count = 10;count > 0; count++)`
`{`
`    ...`
`}`
	Logical error.
	The control expression `count > 0` is always `true`.

`for (count = 1;count < 11;`
` count--)`
`{`
`    ...`
`}`
	Logical error.
	The control expression `count < 11` is always `true`.

`for (count = 1;count < 11;`
` count++);`
`{`
`    ...`
`}`
	Logical error.
	The `actionStatement` is empty. The block statement that follows will be executed only once!

`for (count = 1;count < 11; )`
`{`
`    ...`
`}`
	Logical error.
	The `count` needs to be modified. In other words, this is an infinite loop.

`for (cnt = 1;cnt < 21;`
` cnt += 3)`
`{`
`    ...`
`}`
	No error.
	The `actionStatement` will be executed `Math.ceil((21 - 1)/3) = 7` times for the values 1, 4, 7, 10, 13, 16, and 19.

`for (count = 1; count == 21;`
` count++)`
`{`
`    ...`
`}`
	Logical error.
	The `count` value is initially 1. Thus, `count == 21` evaluates to `false`. The `for` statement is never executed.

(*Continued*)

Segment of Code	Comment

`for (cnt = 1;cnt != 21; cnt += 3)`
`{`
`    ...`
`}`
	Logical error.
	The `cnt` value is initially 1. Thus, `cnt != 21` evaluates to `true`. However, `cnt` value changes as 4, 7, 10, 13, 16, 19, 22, and so on. The `cnt` value never becomes 21 and hence this is an infinite loop.

`for (count = 75; count < 85;`
` count++)`
`{`
`    ...`
`}`
	No error.
	The `actionStatement` will be executed 85 − 75 =
	10 times for the following values of `count`: 75, 76, 77, 78, 79, 80, 81, 82, 83, and 84.

The following syntactic template can be used to implement a `for` statement:

```
for (cnt = 1; cnt < numberOfRepetitionsPlusOse; cnt++)
{
    ...
}
```

In the above segment of code, `numberOfRepetitionsPlusOne` can be either a variable or a literal value equal to 1 more than the number of repetitions required. For instance, if number of repetitions required is 1000, you can replace `numberOfRepetitionsPlusOne` by 1001.

Self-Check

15. Write a `for` statement in which the `count` assumes all even number values between 1 and 99 in ascending order.
16. Write a `for` statement in which the `count` assumes the values 99, 88, 77, 66, 55, and 44 in descending order.

Advanced Topic 5.4: Use of Counter inside `for` Statement

Consider the following problem: you want to create a table that has square and cube of integers 2 through 9. To print the eight lines, you can start with the following `for` statement:

```
for (cnt = 2; cnt < 10; cnt++)
{
    // you need a println statement
}
```

Now, the following statement

```
System.out.println("\t" + cnt +"\t" + cnt*cnt + "\t" + cnt*cnt*cnt );
```

prints a number, its square, and its cube. Thus, we have the following:

```
for (cnt = 2; cnt < 10; cnt++)
{
    System.out.println("\t" + cnt +"\t" + cnt*cnt +
                                   "\t" + cnt*cnt*cnt );

}
```

The complete program is as follows:

```
public class SquareCube
{
    public static void main (String[] args)
    {
        for (cnt = 2; cnt < 10; cnt++)
        {

            System.out.println("\t" + cnt +"\t" + cnt*cnt +
                                   "\t" + cnt*cnt*cnt );

        }
    }
}
```

Output

```
2       4       8
3       9       27
4       16      64
5       25      125
6       36      216
7       49      343
8       64      512
9       81      729
```

Advanced Topic 5.5: Repetition Statement : do … while

The general syntax of the do … while statement is

```
do
    actionStatement
while (controlExp);
```

Note that do and while are reserved words in Java.The `controlExp` can be any logical expression in Java. The `actionStatement` can be either a single Java statement or a block statement. The semantics of a `do … while` statement can be explained as follows. The `actionStatement` is executed once and then the control expression `controlExp` is evaluated. If the control expression evaluates to `true`, the `actionStatement` is executed once more and the control expression is evaluated again. However, if the control expression evaluates to `false`, the statement following the `while` statement is executed (see Figure 5.5).

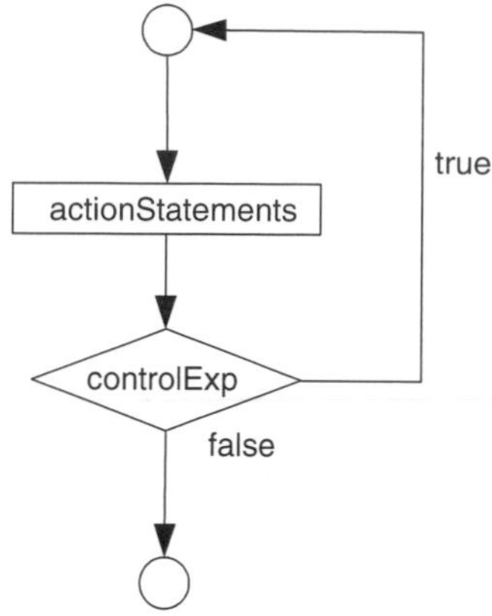

FIGURE 5.5 do … while statement.

The usefulness of a do … while statement in Java can be explained through the following example.

Example 5.9

Consider the following segment of code you have seen previously in the subsection on data validation:

```
highSchoolYear = 0;
while (highSchoolYear < 1940 || highSchoolYear > 1990 )
{
    System.out.print("Enter year of high school graduation ");
    System.out.flush();
    highSchoolYear = scannedInfo.nextInt();
    System.out.println();
}
```

In the above segment of code, the variable highSchoolYear is initialized to zero so that the control expression highSchoolYear < 1940 || highSchoolYear > 1990 evaluates to false. The associated action statement is executed at least once. Since do … while is a posttest structure, control expression is evaluated only after the action statement has been executed. Therefore, there is no need to initialize the variable highSchoolYear. The above segment of code is equivalent to the following do … while statement:

```
do
{
    System.out.print("Enter year of high school graduation ");
    System.out.flush();
    highSchoolYear = scannedInfo.nextInt();
```

```
    System.out.println();
} while (highSchoolYear < 1940 || highSchoolYear > 1990);
```

Example 5.10

Consider the following segment of code you have seen previously in the subsection on sentinel data:

```
nextValue = scannedInfo.nextDouble();

while (nextValue > -1.0)
{
    sum = sum + nextValue;
    count++;
    nextValue = scannedInfo.nextDouble();
}
```

The above code is equivalent to the following:

```
nextValue = scannedInfo.nextDouble();
if (nextValue > 1.0)
{
    do
    {
        sum = sum + nextValue;
        count++;
        nextValue = scannedInfo.nextDouble();
    } while (nextValue > -1.0);
}
```

In this case, use `while` statement for better readability.

Example 5.11

Consider the following segment of code you have seen in Example 5.4:

```
sum = 0;
counter = 1;
while (counter < 4)
{
    System.out.print("Enter a double value : ");
    System.out.flush();
    nextValue = scannedInfo.nextDouble();
    System.out.println();
```

```
        sum = sum + nextValue;
        counter = counter + 1;
    }
```

The above code is equivalent to the following:

```
sum = 0;
counter = 1;
do
{
    System.out.print("Enter a double value : ");
    System.out.flush();
    nextValue = scannedInfo.nextDouble();
    System.out.println();
    sum = sum + nextValue;
    counter = counter + 1;
} while (counter < 4)
```

The following points are worth mentioning:

1. The `controlExp` and `actionStatement` are optional in a `do … while` statement. The smallest `do … while` statement that can be written is

```
do while();
```

2. The pair of left and right parentheses along with the semicolons appearing after the right parenthesis is part of the syntax.
3. A missing `controlExp` evaluates to `true`.
4. The `actionStatement` is executed at least once, even if the `controlExp` is `false` in the beginning.

Advanced Topic 5.6: Guidelines for Choosing Repetition Structure

You have learned three ways of implementing a repetition structure. So a natural question you may have is which one of these structures is the best.

- Use the `for` statement in the case of counter-controlled structure. The `for` statement has the distinct advantage of keeping all three components of repetition structure, initialization, control expression, and update statements in one place.
- Use the `while` statement in all other cases and avoid using `do … while` if possible. The major drawback of a `do … while` statement is its low visibility. That is, it is hard to locate the beginning and the end of a `do … while` statement compared to other two repetition structures.

NESTING OF CONTROL STRUCTURES

In Chapter 4, you have learned that it is possible to nest various selection structures. In fact you can always nest various control structures as required by the programming logic. This section will illustrate this fact through examples.

Example 5.12

Consider the segment of code for grade computation presented in Chapter 4. For the sake of easy reference, the segment of code is listed below:

```
double wats;                    // wats: weighted average test score
String gradeAssigned;

if (wats >= 90)
    gradeAssigned = "A" ;
else if (wats >= 85)
    gradeAssigned = "A-" ;
else if (wats >= 80)
    gradeAssigned = "B" ;
else if (wats >= 75)
    gradeAssigned = "B-" ;
else if (wats >= 70)
    gradeAssigned = "C" ;
else if (wats >= 60)
    gradeAssigned = "D" ;
else
    gradeAssigned = "F" ;
```

The data for each student consists of first name, last name, and four test scores. The `wats` (weighted average test score) is computed by taking the average of all four test scores. The segment of code necessary for `wats` computation can be adopted from Example 5.4 as follows:

```
double nextValue;
double sum;
double wats;
int counter;
String firstName;
String lastName;
Scanner scannedInfo = new Scanner(System.in);

System.out.print("Enter first and last names : ");
System.out.flush();
```

```
firstName = scannedInfo.next();
lastName = scannedInfo.next();
System.out.println();

sum = 0;
counter = 1;
while (counter < 5)
{
    System.out.print("Enter grade for test "+ counter +":");
    System.out.flush();
    nextValue = scannedInfo.nextDouble();
    System.out.println();
    sum = sum + nextValue;
    counter = counter + 1;
}
wats = sum / 4;
```

Putting these two pieces together along with necessary output statements, we have the following segment of code:

```
double nextValue;
double sum;
double wats;
int counter;
String firstName;
String lastName;
String gradeAssigned;
Scanner scannedInfo = new Scanner(System.in);

// get data and compute wats

System.out.print("Enter first name and last name : ");
System.out.flush();
firstName = scannedInfo.next();
lastName = scannedInfo.next();
System.out.println();

sum = 0;
counter = 1;
while (counter < 5)
{
    System.out.print("Enter grade for test "+ counter + " : ");
    System.out.flush();
    nextValue = scannedInfo.nextDouble();
```

```
        System.out.println();
        sum = sum + nextValue;
        counter = counter + 1;
}
wats = sum / 4;

// use wats to assign proper grades

if (wats >= 90)
        gradeAssigned = "A" ;
else if (wats >= 85)
        gradeAssigned = "A-" ;
else if (wats >= 80)
        gradeAssigned = "B" ;
else if (wats >= 75)
        gradeAssigned = "B-" ;
else if (wats >= 70)
        gradeAssigned = "C" ;
else if (wats >= 60)
        gradeAssigned = "D" ;
else
        gradeAssigned = "F";

// produce output

System.out.println(firstName + " " + lastName + " \t" +
                                                gradeAssigned);
```

The above segment of code gets the data of one student and assigns the grade according to the grading policy. Suppose the class has 12 students. Then all you need to do is repeat the above statements 12 times. Thus, we have the following:

```
double nextValue;
double sum;
double wats;
int counter;
int StudentCnt;
String firstName;
String lastName;
String gradeAssigned;
Scanner scannedInfo = new Scanner(System.in);
```

```
for (StudentCnt = 1; StudentCnt < 13; StudentCnt++)
{
    // get data and compute wats
    System.out.print("Enter first and last names : ");
    System.out.flush();
    firstName = scannedInfo.next();
    lastName = scannedInfo.next();
    System.out.println();

    sum = 0;
    counter = 1;
    while (counter < 5)
    {
        System.out.print("Enter grade for test "+
                                    counter + " : ");
        System.out.flush();
        nextValue = scannedInfo.nextDouble();
        System.out.println();
        sum = sum + nextValue;
        counter = counter + 1;
    }
    wats = sum / 4;

    // use wats to assign proper grades
    if (wats >= 90)
        gradeAssigned = "A" ;
    else if (wats >= 85)
        gradeAssigned = "A-" ;
    else if (wats >= 80)
        gradeAssigned = "B" ;
    else if (wats >= 75)
        gradeAssigned = "B-" ;
    else if (wats >= 70)
         gradeAssigned = "C" ;
    else if (wats >= 60)
         gradeAssigned = "D" ;
    else
        gradeAssigned = "F" ;

    // produce output

    System.out.println(firstName + " " + lastName +
                             " \t" + gradeAssigned);
}
```

The complete program listing along with sample test runs as follows. Note that number of students and number of tests are also part of the input.

```
import java.util.Scanner;
public class StudentGrades
{
    /**
        Illustration of nesting of control structures
    */
    public static void main (String[] args)
    {
        double nextValue;
        double sum;
        double wats;
        int counter;
        int StudentCnt;
        int numStudents;
        int numTests;
        String firstName;
        String lastName;
        String gradeAssigned;
        Scanner scannedInfo = new Scanner(System.in);

        System.out.print("Enter no. of students &
                                                tests : ");
        System.out.flush();
        numStudents = scannedInfo.nextInt();
        numTests = scannedInfo.nextInt();
        System.out.println();

        for (StudentCnt = 1; StudentCnt < numStudents + 1;
                                                StudentCnt++)
        {
            // get data and compute wats

            System.out.print("Enter first and last names : ");
            System.out.flush();
            firstName = scannedInfo.next();
            lastName = scannedInfo.next();
            System.out.println();

            sum = 0;
            counter = 1;
```

```
            while (counter < numTests + 1)
            {
                System.out.print("Enter grade for test "+
                                        counter + " : ");
                System.out.flush();
                nextValue = scannedInfo.nextDouble();
                System.out.println();

                sum = sum + nextValue;
                counter = counter + 1;
            }
            wats = sum / numTests;

            // use wats to assign proper grades

            if (wats >= 90)
                gradeAssigned = "A";
            else if (wats >= 85)
                gradeAssigned = "A-" ;
            else if (wats >= 80)
                gradeAssigned = "B" ;
            else if (wats >= 75)
                gradeAssigned = "B-" ;
            else if (wats >= 70)
                gradeAssigned = "C";
            else if (wats >= 60)
                gradeAssigned = "D" ;
            else
                gradeAssigned = "F" ;

            // produce output

            System.out.println(firstName + " " + lastName +
                                  " \t" + gradeAssigned);
        }
    }
}
```

Output

```
Enter no. of students & tests : 1 4

Enter first and last names : Kelly Pederson

Enter grade for test 1 : 85.7
```

```
Enter grade for test 2 : 79.3
Enter grade for test 3 : 93.6
Enter grade for test 4 : 94.8
Kelly Pederson A-
```

Self Check

17. If there are 20 students and 4 tests, how many times the `flush` method will be invoked?
18. If there are 20 students and 4 tests, how many times the statement `counter = counter + 1;` will be executed?

Advanced Topic 5.7: Statements `break` and `continue`

You can use two keywords `break` and `continue` to alter the behavior of a repetition structure. The syntax is as follows:

```
break [label];
continue [label];
```

Observe that label is optional and most common use of `break` and `continue` is without label. Therefore, use of `break` and `continue` without any label is explained first.

- A `break` statement is used to terminate execution of the innermost repetition structure that contains the `break` statement.
- A `continue` statement is used to skip execution of the remaining statements in the innermost repetition structure that contains the `continue` statement for the current iteration.

Recall that a `break` statement can be used inside a `switch` statement and the semantics of the `break` statement is to stop executing any other statement in the `switch` statement and start executing the first statement following the `switch` statement. The semantics of a `break` statement inside any repetition structure is identical. That is, `break` statement transfers the control to the very first statement following the repetition structure as shown in Figure 5.6.

Note that `break` causes an immediate exit from the repetition structure. The semantics of a `continue` statement can be explained through Figure 5.7.

The `continue` statement affects only the current iteration of the repetition structure. More specifically, `continue` statement causes `actionStmtsAfter` being skipped for the current iteration.

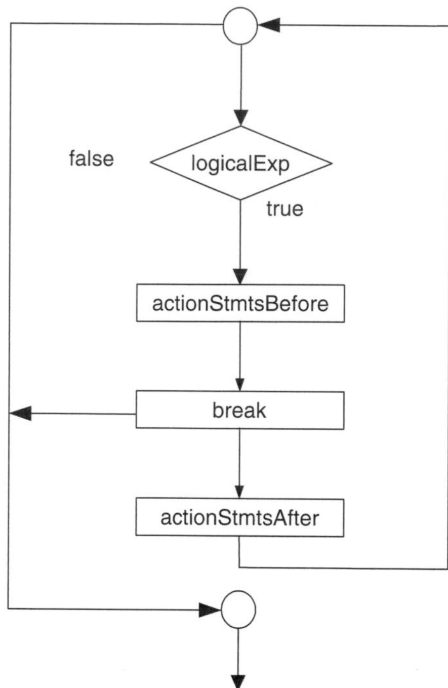

FIGURE 5.6 `break` statement.

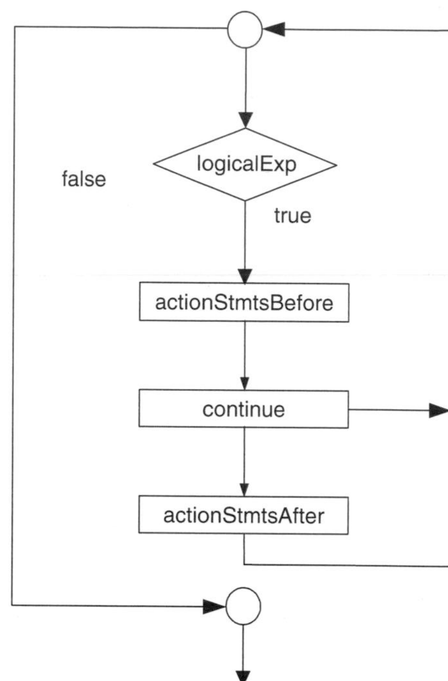

FIGURE 5.7 `continue` statement.

Good Programming Practice 5.1

The `continue` statement appearing inside a `while` statement or a `do ... while` statement causes control expression being evaluated next. However, in the case of a `for` statement, a `continue` statement causes the update statement being executed next, followed by the control expression. Therefore, if you use `continue` statement, a `for` statement is more preferable than other repetition structures. However, if you use `continue` inside a `while` statement or a `do ... while` statement, you must make sure that the necessary updates are done. Otherwise you may unintentionally create an infinite repetition structure.

Example 5.13

```
import java.util.Scanner;

/**
    Illustration of break and continue
*/
public class BreakAndContinue
{

    public static void main (String[] args)
```

```
    {
        int valueOne;
        int valueTwo;
        int cntOne;
        int cntTwo;

        Scanner scannedInfo = new Scanner(System.in);

        System.out.print("Enter two positive integers : ");
        System.out.flush();
        valueOne = scannedInfo.nextInt();
        valueTwo = scannedInfo.nextInt();
        System.out.println();
        for (cntOne = valueOne, cntTwo = valueTwo;
            cntOne + cntTwo > 0; cntOne--, cntTwo--)
        {
            System.out.println("\nFirst value is " + cntOne +
                        " and second value is " + cntTwo);
            System.out.println("Statement before Cont. /
                                            Break" );
            if (cntOne == 0 && cntTwo > 0)
            {
                System.out.println("First value is zero.");
                System.out.println("\tB R E A K!");
                break;
            }
            else if (cntOne > 0 && cntTwo == 0)
            {
                System.out.println("Second value is zero.");
                System.out.println("\tC O N T I N U E!");
                continue;
            }

                System.out.println("Statement after Cont./
                                            Break" );
        }
        System.out.println("\n\nExit Loop. Bye now." );
    }
}
```

Output

Case 1: First value > second value

```
Enter two positive integers : 5 2
```

```
First value is 5 and second value is 2
Statement before Cont./Break
Statement after Cont./Break

First value is 4 and second value is 1
Statement before Cont./Break
Statement after Cont./Break

First value is 3 and second value is 0
Statement before Cont./Break
Second value is zero.
    C O N T I N U E!

First value is 2 and second value is -1
Statement before Cont./Break
Statement after Cont./Break

Exit Loop. Bye now.
```

Case 2: First value < second value

```
Enter two positive integers : 2 5

First value is 2 and second value is 5
Statement before Cont./Break
Statement after Cont./Break

First value is 1 and second value is 4
Statement before Cont./Break
Statement after Cont./Break

First value is 0 and second value is 3
Statement before Cont./Break
First value is zero.
    B R E A K!

Exit Loop. Bye now.
```

Observe the behavior of `continue` in case 1 and that of `break` in case 2, respectively.

In case 1, continue statement is executed once. In that particular iteration, all statements following the `continue` statement are skipped. Note that both `cntOne` and `cntTwo` get decremented. Eventually, the control expression `cntOne + cntTwo > 0` becomes `false` and the repetition structure is exited. In case 2, once the `break` statement is executed, the repetition structure is exited.

Statements `break` and `continue` with Optional Label

Technically speaking every Java statement can have an optional label. However, the most common use of a label is in connection with `break` and `continue` statements. The `break` and `continue` statements without any label affect the behavior of the innermost repetition structure. Quite often, in a nested structure, you may want to `break` or `continue` to apply to the some outer repetition structure. Such situations can be handled by identifying a repetition structure by a label and including the label in the `break` or `continue` statement.

Example 5.14

```
import java.util.Scanner;

/**
    Illustration of break and continue with label
*/
public class BreakAndContinueWithLabel
{

    public static void main (String[] args)
    {
        int valueOne;
        int valueTwo;
        int cntOne;
        int cntTwo;

        Scanner scannedInfo = new Scanner(System.in);

        System.out.print("Enter two positive integers : ");
        System.out.flush();
        valueOne = scannedInfo.nextInt();
        valueTwo = scannedInfo.nextInt();
        System.out.println();
    outerLoop :
        for (cntOne = valueOne; cntOne > 0; cntOne--)
        {
            System.out.println("Outer loop: Before");
```

```
            for (cntTwo = valueTwo; cntTwo > 0; cntTwo--)
            {
                System.out.println("First value is " +
                cntOne +" and second value is " + cntTwo);
                System.out.println("Before Cont./Break" );
                if (cntOne % cntTwo == 0)
                {
                    System.out.println("\tB R E A K!\n");
                    break outerLoop;
                }
                else if (cntTwo % cntOne == 0)
                {
                    System.out.println("\tCONTINUE!\n");
                    continue outerLoop;
                }
                System.out.println("After Cont./Break" );
            }
            System.out.println("Outer loop: After");
        }
        System.out.println("\n\nOutside all loops. Bye now." );
    }
}
```

Output

First run:

```
Enter two positive integers : 7 3

Outer loop: Before
First value is 7 and second value is 3
Before Cont./Break
After Cont./Break
First value is 7 and second value is 2
Before Cont./Break
After Cont./Break
First value is 7 and second value is 1
Before Cont./Break
    B R E A K!

Outside all loops. Bye now.
```

Second run:

```
Enter two positive integers : 5 10

Outer loop: Before
First value is 5 and second value is 10
Before Cont./Break
    C O N T I N U E!

Outer loop: Before
First value is 4 and second value is 10
Before Cont./Break
After Cont./Break
First value is 4 and second value is 9
Before Cont./Break
After Cont./Break
First value is 4 and second value is 8
Before Cont./Break
    C O N T I N U E!

Outer loop: Before
First value is 3 and second value is 10
Before Cont./Break
After Cont./Break
First value is 3 and second value is 9
Before Cont./Break
    C O N T I N U E!

Outer loop: Before
First value is 2 and second value is 10
Before Cont./Break
    C O N T I N U E!

Outer loop: Before
First value is 1 and second value is 10
Before Cont./Break
    C O N T I N U E!

Outside all loops. Bye now.
```

Note the behavior of `break` in case 1 and that of `continue` in case 2, respectively. In case 1, `break` statement is executed once and the control is transferred to the first statement outside the outer repetition structure. In case 2, as soon as `continue` statement is executed, the control is transferred to the beginning of the outer repetition structure.

CASE STUDY 5.1: PAYROLL FOR SMALL BUSINESS: REVISITED

Specification

Mr. Jones currently has 30 employees. Ms. Smart decided to keep employee data in a file and print payroll information first into a file. Ms. Smart decided to rewrite the payroll program.

Input

Data is in a file `C:\employee.dat`. Other details are as before.

Output

Print the pay stub for every employee into a file `C:\payroll.dat`.

There is no change in classes. The only change is in the application program.

Application Program

The application program needs to be modified to reflect the changes. The changes to be made can be listed as follows:

- Read information from a file and not from standard input. Therefore, you need to declare exceptions in the method `main`'s heading and change the way `Scanner` object is created. Thus,

  ```
  public static void main (String[] args)
  ```

  is replaced by

  ```
  public static void main (String[] args) throws
              FileNotFoundException, IOException
  ```

  and

  ```
  Scanner scannedInfo = new Scanner(System.in);
  ```

  is replaced by

  ```
  Scanner scannedInfo = new Scanner
                    (new File("C:\\employee.dat"));
  ```

- Output is written into a file. Therefore, a `PrintWriter` object is created.

  ```
  PrintWriter outFile = new PrintWriter
                    (new FileWriter("C:\\payroll.dat"));
  ```

  All output statements of the form

  ```
  System.out.print(...);
  System.out.println(...);
  ```

are, respectively, replaced by output statements of the form

```
outFile.print(...);
outFile.println(...);
```

- Remove all prompt statements. The program is now executing in *batch mode* and not in an interactive mode. Therefore, there is no need for any prompt statements.
- Introduce a repetition structure. So long as there are more employees, the program must read the data from the input file and output the payroll information into the output file.

```
import java.util.*;
import java.io.*;

/**
    Modified program for Heartland Cars of America
    Each modified Java statement is kept as a comment
    To observe the difference.
*/
public class HeartlandCarsOfAmericaPayRollFileVersion
{
    //public static void main (String[] args)
    public static void main (String[] args) throws
                FileNotFoundException, IOException
    {
        //Create reference variable of all three employee
                                                    types
        FullTimeEmp fullTimeEmployee;
        PartTimeEmp partTimeEmployee;
        SalesEmp salesEmployee;

        //Declare variables to input
        char inputEmplyeeType;
        String inputFirstName;
        String inputLastName;
        double inputBaseSalary;
        double inputPayPerHour;
        int inputSalesVolume;
        int inputHoursWorked;

        //Get two input values
        //    Scanner scannedInfo = new Scanner(System.in);
        Scanner scannedInfo = new Scanner
                            (new File("C:\\employee.dat"));
```

```
            PrintWriter outFile = new PrintWriter
                        (new FileWriter("C:\\payroll.dat"));

        //   System.out.print("Enter Employee Type : ");
        //   System.out.flush();
        inputEmplyeeType = scannedInfo.next().charAt(0);
        //   System.out.println();

        while (scannedInfo.hasNext())
        {
            switch (inputEmplyeeType)
            {
                case 'F' :
                case 'f' :

                    //get necessary values as input
                    //System.out.print("Enter First Name, " +
                    //"Last Name, Base Salary, Hours : ");
                    //System.out.flush();
                    inputFirstName = scannedInfo.next();
                    inputLastName = scannedInfo.next();
                    inputBaseSalary = scannedInfo.nextDouble();
                    inputHoursWorked = scannedInfo nextInt();
                    //System.out.println();

                    //create an object and initialize data
                                                 members
                    fullTimeEmployee = new FullTimeEmp();
                    fullTimeEmployee.setFirstName(inputFirstName);
                    fullTimeEmployee.setLastName(inputLastName);
                    fullTimeEmployee.setBaseSalary
                                            (inputBaseSalary);
                    fullTimeEmployee.setHoursWorked
                                (inputHoursWorked);

                    //invoke the printPayStub method
                    outFile.print (fullTimeEmployee.
                                            createPayStub());

                break;

                case 'P' :
                case 'p' :
                    //get necessary values as input
                    //System.out.print("Enter First Name, Last
                                                  Name, "+
```

```
        //"Pay per hour, Hours : ");
        //System.out.flush();
        inputFirstName = scannedInfo.next();
        inputLastName = scannedInfo.next();
        inputPayPerHour = scannedInfo.nextDouble();
        inputHoursWorked = scannedInfo.nextInt();
        //System.out.println();

        //create an object and initialize data
                                          members
        partTimeEmployee = new PartTimeEmp();
        partTimeEmployee.
               setFirstName(inputFirstName);
        partTimeEmployee.setLastName(inputLastName);
        partTimeEmployee.setPayPerHour
                                   (inputPayPerHour);
        partTimeEmployee.setHoursWorked
                                   (inputHoursWorked);
        //invoke the printPayStub method
        outFile.print(partTimeEmployee
                                    createPayStub());

    break;

    case 'S' :
    case 's' :
        //get necessary values as input
        //System.out.print("Enter First Name, Last
                                          Name, "+
        //"Base Salary, Sales Volume : ");
        //System.out.flush();
        inputFirstName = scannedInfo.next();
        inputLastName = scannedInfo.next();
        inputBaseSalary = scannedInfo.nextDouble();
        inputSalesVolume = scannedInfo.nextInt();
        //System.out.println();

        //create an object and initialize data
                                          members
        salesEmployee = new SalesEmp();
        salesEmployee.setFirstName(inputFirstName);
        salesEmployee.setLastName(inputLastName);
        salesEmployee.setBaseSalary(inputBaseSalary);
        salesEmployee.setSalesVolume
                                   (inputSalesVolume);
```

```
                //invoke the printPayStub method
                outFile.print(salesEmployee. createPayStub());

                break;
                }          // End of switch
            }                  // End of while
        }                  // End of main
    }                      // End of class
```

Testing

Left as an exercise to the reader.

Output

Quite similar to the output shown at the end of Chapter 3.

REVIEW

1. Repetition structure allows you to repeat a certain code again and again.
2. If the data is in a file `C:\myfile.dat`, you can create a scanner object `scannedInfo` through the following statement:

```
Scanner scannedInfo = new
            Scanner(new File("C:\\myfile.dat"));
```

3. In Java, to create an output file `C:\myfile.out`, you need a statement similar to

```
PrintWriter output = new
      PrintWriter(new FileWriter("C:\\myfile.out"));
```

4. An `IOException` occurs whenever a mismatch in data type happens during an I/O operation.
5. The `close` method, in particular, will flush any data remaining in the output buffer to the output file.
6. Generally speaking, there are three types of `while` statements: counter-controlled, event-controlled, and data-controlled.
7. The counter variable must be initialized before executing the control expression of the `while` statement.
8. In the case of a `while` statement, the pair of left and right parentheses is part of the syntax.
9. In the case of a `for` statement, the pair of left and right parentheses along with two semicolons appearing inside them is part of the syntax.
10. In the following `while` statement, the `actionStatement` will be executed 11 − 1 = 10 times:

```
cnt = 1;
while (cnt < 11)
```

```
{
  ...
  cnt++
}
```

11. In the following for statement, the actionStatement will be executed 11 − 1 = 10 times:

```
for (cnt = 1; cnt < 11; cnt++)
{
  ...
}
```

12. In a repetition structure, an empty control expression evaluates to true.
13. In a repetition structure, the control expression must evaluate to false eventually to avoid infinite looping.
14. A do … while statement is executed at least once.
15. A break statement is used to terminate execution of the innermost repetition structure that contains the break statement.
16. A continue statement is used to skip execution of the remaining statements in the innermost repetition structure that contains the continue statement for the current iteration.

EXERCISES

1. Mark the following statements as true or false:
   a. In a while statement the control expression is initially false.
   b. In a do … while statement the control expression is always false.
   c. To have a finite while loop, control expression should evaluate to true during execution.
   d. It is possible to write a repetition structure such that it terminates for certain input values only.
   e. In the case of a counter-controlled while statement, the counter value cannot be changed.
   f. An empty control expression in a while statement always evaluates to false.
   g. Both while and do … while statements are executed at least once irrespective of the initial value of the control expression.
   h. In a counter-controlled while statement, the counter needs to be incremented by 1, and not by any other value.
   i. Every for statement can be replaced by a while statement.
   j. Within a repetition structure, you cannot have more than one break statement.
   k. It is legal to use both break and continue within a repetition structure.

2. What is the output produced by the following segment of code. In the case of infinite loop, write "infinite loop" and give first 10 iteration outputs. If there is a syntax error, write syntax error and explain the reason.

a.
```
int counter = 2;
int x = 10;
while (counter < x)
{
    System.out.println("x = " + x + " ,counter = " + counter);
    x++;
    counter = counter + 2;
};
```

b.
```
int counter = 2;
int x = 10;
while (counter < x)
    System.out.println("x = " + x + " ,counter = " + counter);
```

c.
```
int counter = 2;
int x = 10;
while (counter++ < --x)
    System.out.println("x = " + x + " ,counter = " + counter);
```

d.
```
int counter = 2;
int x = 10;
while ()
{
    System.out.println("x = " + x + " ,counter = " + counter);
    x++;
    counter = counter + 2;
    if ( x == 20) break;
};
```

e.
```
int counter = 2;
int x = 10;
while (true)
{
    System.out.println("x = " + x + " ,counter = " + counter);
    x++;
    counter = counter + 2;
    if ( x == 20) continue;
};
```

f.
```
int counter = 2;
int x = 10;
while (x + counter < 20)
    System.out.println("x = " + x + " ,counter = " + counter);
    x++;
```

```
        counter = counter + 2;
        if ( x == 20) continue;
```

g.
```
int counter = 2;
int x = 10;
while (x - counter < 20);
    System.out.println("x = " + x + " ,counter = " + counter);
    x++;
    counter = counter + 2;
    if ( x == 20) break;
```

3. What is the output produced by the following segment of code. In the case of infinite loop, write "infinite loop" and give first 10 iteration outputs. If there is a syntax error, write syntax error and explain the reason.

a.
```
int counter;
int x = 10;
for (counter = 0; counter < x; counter++)
{
    System.out.println("x = " + x + " ,counter = " + counter);
    x--;
};
```

b.
```
int counter;
int x = 10;
for (counter = 0; counter < x; )
{
    System.out.println("x = " + x + " ,counter = " + counter);
    x--;
};
```

c.
```
int counter;
int x = 10;
for (counter = 0; counter < x; counter++)
    System.out.println("x = " + x + " ,counter = " + counter);
```

d.
```
int counter = 2;
int x = 10;
for (;;)
{
    System.out.println("x = " + x + " ,counter = " + counter);
    x++;
    counter = counter + 2;
};
```

e.
```
int counter;
int x = 10;
for(counter = 2; true ;)
```

```
{
    System.out.println("x = " + x + " ,counter = " + counter);
    x++;
    counter = counter + 2;
    if ( x == 20) continue;
};
```

f.
```
int counter;
int x = 10;
for (counter = 0; counter < x; counter++)
{
    System.out.println("x = " + x + " ,counter = " + counter);
    x++;
    counter = counter + 2;
    if ( x == 20) break;
}
```

g.
```
int counter = 2;
int x = 30;
for (counter = 0; counter < x; counter++)
{
    System.out.println("x = " + x + " ,counter = " + counter);
    X = x - 3;
    counter = counter + 2;
    if ( x == 21) break;
}
```

4. If the data values are student grade points, suggest a sentinel value.
5. If the data values are student names, suggest a sentinel value.
6. Write a segment of code to sum 10 `int` values read from the standard input.
   a. Use `while` statement
   b. Use `for` statement
   c. Use `do ... while` statement
7. Redo Exercise 6 so that all positive values are added together and all negative values are added together.
8. What is the output produced by the following segment of code. In the case of infinite loop, write “infinite loop” and give first 10 outputs. If there is a syntax error, write syntax error and explain the reason.

a.
```
int counter = 1;
int x = 1;
while (counter < 10)
{
    System.out.println("x = " + counter+ " , counter = " +
                                                    counter);
```

```
    while (counter < 5)
    {
        System.out.println("x = " + x + " , counter = " +
                                                counter);

        x++;
    }
    counter++;
};
```

b.
```
int counter = 1;
int x = 8;
while (counter < 8)
{
    System.out.println("x = " + x + " , counter = " + counter);
    for (x = 0; x < 7; x = x + 2)
    {
        System.out.println("x = " + x + " , counter = " +
                                                counter);
    }
    counter++;
};
```

c.
```
int x = 1;
int counter;
for (counter = -5, counter < 6; counter = counter + 2)
{
    System.out.println("x = " + x + " , counter = " + counter);
    while (x < 9)
    {
        System.out.println("x = " + x + " , counter = " +
                                                counter);

        x++;
    }
};
```

d.
```
int x;
for (counter = -5; counter < 6; counter = counter + 2)
{
    System.out.println("x = " + x + " , counter = " +
                                            counter);

    x = 1;
    while (x < 9)
    {
        System.out.println("x = " + x + " , counter = " +
                                                counter);
```

```
            x++;
        }
    };
```

e.
```
int x = 1;
int counter = 8;
for (counter = -5; counter < 6; counter = counter + 2)
{
    System.out.println("x = " + x + " , counter = " +
    counter);
    for (x = 2; x > -5; x--)
    {
        System.out.println("x = " + x + " , counter = " +
                                                  counter);
    }
};
```

f.
```
int x = 1;
int counter = 8;
for (counter = -5; counter < 6; counter++)
{
    System.out.println("x = " + x + " , counter = " +
                                              counter);
    for (x = 2; x > -5; x--)
    {
        if (counter == x) continue;
        System.out.println("x = " + x + " , counter = " +
                                                  counter);
    }
};
```

g.
```
int x = 1;
int counter = 8;
for (counter = -5; counter < 6; counter++)
{
    System.out.println("x = " + x + " , counter = " +
                                              counter);
    for (x = 2; x > -5; x--)
    {
        if (counter == x) break;
            System.out.println("x = " + x + " , counter =
                                              " + counter);
    }
};
```

9. Calculate the number of iterations involved in each of the following repetition structures:

   a.
   ```
   for (int counter = 0; counter < 100; counter++)
   {
       ...
   };
   ```

   b.
   ```
   for (int counter = -50; counter < 50; counter++)
   {
       ...
   };
   ```

   c.
   ```
   for (int counter = 0; counter < 100; counter = counter + 3)
   {
       ...
   };
   ```

   d.
   ```
   for (int counter = 100; counter > -100; counter = counter - 7)
   {
       ...
   };
   ```

   e.
   ```
   for (int counter = 1; counter < 100; counter = counter * 2)
   {
       ...
   };
   ```

   f.
   ```
   for (int counter = 40; counter < 10; counter = counter + 3)
   {
       ...
   };
   ```

   g.
   ```
   for (int counter = 10; counter < 40; counter = counter--)
   {
       ...
   };
   ```

PROGRAMMING EXERCISES

1. Write a program that prompts the user to input a digit. The program should then output a square of that size using the digit. For example, if input is 5, then the output is as follows:

```
55555
55555
55555
55555
55555
```

2. Write a program that prompts the user to input an odd digit. The program should then output a rhombus of that size using the digit. For example, if input is 5, then the output is as follows:

```
  5
 555
55555
 555
  5
```

3. Write a program that prompts the user to input an odd digit. The program should then output a hollow rhombus of that size using the digit. For example, if input is 5, then the output is as follows:

```
  5
 5 5
5   5
 5 5
  5
```

4. Write a program to check the divisibility of an integer by 3. Your program must make use of the fact that an integer is divisible by 3 if and only if the sum of its digits is divisible by 3. You must use this fact repeatedly, till the sum reduces to a single digit. For example, 123456789 is divisible by 3 if and only if $1 + 2 + 3 + 4 + 5 + 6 + 7 + 8 + 9 = 45$ is divisible by 3. Now, 45 is divisible by 3 if and only if $4 + 5 = 9$ is divisible by 3. Observe that 9 is a single digit and is divisible by 3. Therefore, your program concludes that 123456789 is divisible by 3.
5. Write a program to check the divisibility of an integer by 11. Your program must make use of the fact that an integer is divisible by 11 if and only if the difference of the sum of odd digits and the sum of even digits is divisible by 11. You must use this fact repeatedly, till the sum reduces to a single digit. For example, 123456789 is divisible by 11 if and only if $(1 + 3 + 5 + 7 + 9) - (2 + 4 + 6 + 8) = 25 - 20 = 5$ is divisible by 11. Now 5 is a single digit other than 0 and therefore the program concludes that 123456789 is not divisible by 11.
6. Write a program to create a tip table. The table has five columns. First column has dollar values 5, 10, 15, 20, 25, 30, and so on and the largest multiple of 5 is determined through the user input. Next four columns contain tips at 10, 15, 20, and 25%, respectively.
7. Heartland Cars of America keeps the record of each sold car in a file. Once a car is sold, the model of the car, year of the car, cost basis, and sale price are entered into a file. At the end of each month, the file is processed to obtain the monthly sales volume and the profit made. Write a program to accomplish these tasks.
8. Heartland Cars of America sells three categories of cars: new, pre-owned, and used. A car is considered pre-owned if it is relatively new and in fairly good condition. Otherwise, it is classified as used. The cost basis does not include the sales commission. The sales commission is 10% of profit for the new, 15% for the pre-owned, and

20% for the used cars. Once a car is sold, category, the model of the car, year of the car, cost basis, and sale price are entered into a file. At the end of each month, the file is processed to obtain the monthly sales volume and the net profit made (after the sales commission) in each of the three categories. Design four classes, NewCar, PreOwnedCar, UsedCar, and ReportGenerator, to accomplish the task.

9. Physician's Clinic Inc. tracks its earnings on a weekly basis. There are four different revenue streams: patient consultation, patient procedure, and patient lab. The net earning in each of these categories is estimated to be 90, 80, and 60%, respectively. Physician's Clinic Inc. keeps an entry for each patient service as follows: The first letter C, P, or L indicates the type of service performed. Next are the last four digits of social security number of the patient followed by the amount charged by Physician's Clinic Inc. Write a program to generate a weekly report that indicates the revenue from each of the three streams as well as the total revenue.

10. Create a digital dice with six values 1, 2, 3, 4, 5, and 6. A dice is said to be fair if no face turns up more than 10% of any other face. Write a program to check whether or not your digital dice is fair. As an optional challenge, design your program so that there is a class `Dice` that has a method `roll` and a class `Experiment` that has a method `isFair`. There is a third class that allows you to input the number of rolls you would like to perform before you test the fairness.

11. Write a program to compute e^x, using the following power series up to a given precision. The power series expansion of $e^x = 1 + x/1! + x^2/2! + x^3/3! + x^4/4! + \ldots$.

12. Write a program to compute sin(x), using the following power series up to a given precision. The power series expansion of $\sin(x) = -x/1! + x^3/3! - x^5/5! + x^7/7! - \ldots$.

13. Design and implement an object-oriented Java program that can convert a string to corresponding telephone number. If it is an uppercase letter or a lower case letter, the program will substitute it with the corresponding digit. If it is already a digit, no substitution is done. Thus, "GOODCAR", "gooDCar", and "go6DC2r" will be translated to 4663227.

14. Write a program that prompts the user to input a digit. The program should then output a square of that size using the digit as shown in the examples. Example 1: If the input is 5, then the output is

```
55555
53335
53135
53335
55555
```

Example 2: If the input is 6, then the output is

```
666666
644446
642246
642246
666666
```

ANSWERS TO SELF-CHECK

1. sequence, selection, repetition
2. repetition
3. `import java.io.*;`
4. ```
   Scanner scannedInfo
     = new Scanner(new File("C:\\testData.txt"));
   ```
5. False
6. 123
7. `import java.io.*;`
8. ```
   PrintWriter output
     = new PrintWriter(new FileWriter("C:\\info.txt"));
   ```
9. False
10. True
11. True
12. True
13. ```
    count = 2;
    while (count < 100)
    {
        //statements
        count = count + 2;
    }
    ```
14. ```
    count = 99;
    while (count > 40)
    {
        //statements
        count = count - 11;
    }
    ```
15. ```
    for (count = 2; count < 100; count = count + 2)
    {
        //statements

    }
    ```
16. ```
    for (count = 99; count > 40; count = count - 11)
    {
        //statements
    }
    ```
17. 100
18. 80

CHAPTER 6

Methods and Constructors

In this chapter you learn

- Object-oriented concepts
  - Service, message passing, static variables and methods, method overloading, and role of constructors.
- Java concepts
  - Constructors, signature, method overloading, static and nonstatic methods, class variable, self-reference, and parameter passing.
- Programming skills
  - Use predefined methods, and create and use constructors and user-defined methods with any number of parameters.

Methods play a very important role in Java. In Chapter 3, you created new methods. This chapter provides an in-depth look into various aspects of method invocation and method creation.

CLASSIFICATION OF METHODS

Consider a DVD player. Every DVD player has play button, pause button, and a stop button. As you press any one of these buttons, you are in fact *passing a message* to the DVD player. For instance, as you press the play button, you are sending the message "play" to the DVD player. The DVD player receives the message and it responds by providing the "play" service. Similarly, as you press the button pause or stop, you are sending a message to the DVD player requesting a "pause" or "stop" service from the DVD player. Thus, your interaction with your DVD player can be viewed as you make a request for a certain *service* from the DVD player by sending the appropriate message.

Each of these buttons "play," "pause," and "stop" is a `public` method of the class DVD player. A `public` method can be invoked as long as you have access to the object involved. For instance, you can play a DVD on your DVD player. However, you cannot play a DVD on Mr. Jones DVD player. Thus, to "play" a movie on a DVD player two conditions need to be satisfied:

1. The DVD player must have a button play that is accessible.
2. The DVD player must be accessible.

As a general rule, in order for an object `X` to invoke a method `M` (or to request a service) of another object `Y`, two conditions need to be met:

1. The method `M` must be accessible to the requesting object `X`.
2. The object capable of providing the service (i.e., `Y`) must be accessible to the requesting object `X`.

In the case of a DVD player, you may be aware of the fact that there is an internal motor that rotates the DVD at a certain speed during the play. Methods associated with the motor are hidden (have `private` access) from the user of a DVD player. Therefore, condition 1 is not met and consequently, even though you own the DVD player and as such have complete access to your DVD player, you cannot control the speed of the motor. All `private` methods of a DVD player are hidden from the user of the DVD player and all `public` methods of a DVD player are available to the user of the DVD player. Methods you have encountered so far are `public` methods. As a general rule, methods are `public`. There are two more options of access control: `protected` and default. These concepts are discussed in Chapter 7.

From Chapter 3, you know that methods can be classified into two categories: methods *with no parameter* and methods *with parameters*. For example, methods `next`, `nextInt`, `nextDouble`, and `nextLine` of the class `Scanner` has no parameter and method `pow` of the class `Math` has two parameters. Again from Chapter 3, you know that methods can also be classified into two categories: *void methods* and *value returning methods*. Methods such as `next`, `nextInt`, and `nextDouble` of the class `Scanner` are value returning methods. Method `println` you have invoked many times is a good example of a `void` method.

Methods can be marked `static`. Technically speaking, a `static` method does not depend on the state of an object. In other words, a `static` method does not require an implicit parameter. A method, as you have seen quite often, manipulates or retrieves instance variables of an object. However, there are methods that perform some useful service in such a way that there is no need for the existence of an object itself. For instance, consider the method `pow` of the class `Math`. The `pow` method has two parameters of type `double` and returns a `double` value. Thus,

```
Math.pow(2.0, 5.0);
```

returns $2.0^{5.0}$ = 32.0. Note that to compute $2.0^{5.0}$ there is no need for an object. Thus, no implicit argument is required. Therefore, `pow` method is created as a `static` method. In contrast, the method `charAt` returns a character at a specified position in the implicit parameter. For example,

```
str.charAt(4);
```

returns the fifth character of the implicit parameter `str`. In this case, `charAt` method requires a `String` object. Consequently, `charAt` needs an implicit parameter and as such cannot be a `static` method. Throughout this book, a `static` method is always invoked using the syntax

```
ClassName.methodName([actualParameters])
```

and methods that are not marked as `static` are invoked using the syntax

```
objectReference.methodName([actualParameters])
```

Note 6.1 It is a syntax error to invoke a method using the syntax `ClassName.methodName([actualParameters])` unless it is `static`. Although Java lets you invoke a `static` method using the syntax `objectReference.methodName([actualParameters])`, throughout this book we use the syntax template `ClassName.methodName([actualParameters])`.

To completely understand the concept of `static` methods, you need to know about class variables. Therefore, these ideas are further explored later in this chapter.

Methods can also be classified as *predefined* or *user-defined*. Java language has a large collection of predefined classes to help application program development. All these classes in turn have predefined methods. As an application program developer you can use these classes and methods in your program. The collection of all predefined classes is called the Java application program interface (Java API). Java API is grouped under different units called packages. If you need to use a predefined method, you need to `import` the `class` containing the method. Recall that, you have the option of importing just a `class` or the `package` containing the class. Recall from Chapter 2 that there is a `package` in Java called `java.lang`. This `package` contains classes such as `System` and `String`. Almost any Java program may require this `package`. Therefore, the `package java.lang` is always imported by the Java compiler. Thus, you need not include a statement such as

```
import java.lang.*;
```

in your programs. You can also create your own packages and is discussed later in this chapter. Next, you will be introduced to some of the methods in classes `Math`, `Character`, and `String` of the `package java.lang`.

Self-Check

1. In `str.charAt(4)`, `str` is the ________, `charAt` is the ________, and 4 is the actual parameter.
2. A ________ method is invoked using the syntax

```
ClassName.methodName([actualParameters])
```

Math Class

In Java, all mathematical functions are `static` methods of the `Math` class in the package `java.lang`. Table 6.1 summarizes some of the mathematical methods of the class `Math` of the package `java.lang`.

TABLE 6.1 Selected Methods of the `Math` Class

java.lang.Math				
Method Invocation	**Argument Type(s)**	**Return Value**	**Return Type**	**Example**

`Math.abs(x)`	`int`
	`long`
	`float`
	`double`
		Absolute value of x	`int`
			`long`
			`float`
			`double`
				Math.abs(45) returns `45`, Math.abs(−45) returns `45`, `Math.abs(-2.1)` returns `2.1`.

`Math.acos(x)`	`double`	Arc cosine of angle in the range `0.0` to π	`double`	`Math.acos(0.0)` returns `pi/2`, `Math.acos(1.0)` returns `0.0`.
`Math.asin(x)`	`double`	Arc sine of angle in the range −π/2 to π/2	`double`	`Math.asin(0.0)` returns `0.0`, `Math.asin(1.0)` returns `pi/2`.
`Math.atan(x)`	`double`	Arc tangent of angle in the range −π/2 to π/2	`double`	`Math.atan(-1.0)` returns `-pi/4`, `Math.atan(1.0)` returns `pi/4`.
`Math.ceil(x)`	`double`	Numerically equivalent to next higher integer value	`double`	`Math.ceil(61.3)` returns `62.0`, `Math.ceil(-61.3)` returns `-61.0`.
`Math.cos(x)`	`double`	Cosine value of x	`double`	`Math.cos(pi/2)` returns `0.0`, `Math.cos(0.0)` returns `1.0`.
`Math.exp(x)`	`double`	e^x; where e is the Euler's constant, approx. `2.7183`	`double`	`Math.exp(2.0)` returns `7.38905609893065`, `Math.exp(-1.5)` returns `.22313016014842982`.

TABLE 6.1 Continued

java.lang.Math				
Method Invocation	**Argument Type(s)**	**Return Value**	**Return Type**	**Example**
`Math.floor(x)`	**`double`**	Numerically equivalent to next lower integer value	**`double`**	`Math.floor(61.3)` returns `61.0`, `Math.floor(-61.3)` returns `-62.0`.
`Math.log(x)`	**`double`**	Natural logarithm of x	**`double`**	`Math.log(2.0)` returns `0.6931471805599453`, `Math.log(-1.5)` returns NaN (Not a Number).
`Math.max(x,y)`	**`int,int`** **`long,long`** **`float,float`** **`double, double`**	Maximum of x and y	**`int`** **`long`** **`float`** **`double`**	`Math.max(2, 7)` returns `7`, `Math.max(1.5,-0.7)` returns `1.5`.
`Math.min(x,y)`	**`int,int`** **`long,long`** **`float,float`** **`double,double`**	Minimum of x and y	**`int`** **`long`** **`float`** **`double`**	`Math.min(2, 7)` returns `2`, `Math.min(1.5,-0.7)` returns `-0.7`.
`Math.pow (x, y)`	**`double, double`**	x^y	**`double`**	`Math.pow(2.0, 5.0)` returns `32.0` , `Math.pow(16.0, 0.5)` returns `4.0`.
`Math.random()`		Random value between `0.0` and `1.0`	**`double`**	`Math.random()` returns `0.9786309615836947`, `Math.random()` returns `0.6752079313199223`.
`Math.round(x)`	**`float`** **`double`**	Closest value of return type	**`int`** **`long`**	`Math.round(2.499)` returns `2.0`, `Math.round(2.50)` returns `3.0`.
`Math. toDegrees(x)`	**`double`**	Degree equivalent of x in radians	**`double`**	`Math.toDegrees(1.5)` returns `85.94366926962348`.
`Math. toRadians(x)`	**`double`**	Radian equivalent of x in degrees	**`double`**	`Math.toRadians(60)` returns `pi/3 = 1.0471975511965976`
`Math.sin(x)`	**`double`**	Sine value of x	**`double`**	`Math.sin(pi/2)` returns `1.0`, `Math.sin(0.0)` returns `0.0`.
`Math.tan(x)`	**`double`**	Tangent value of x	**`double`**	`Math.tan(pi/2)` returns `1.0`, `Math.tan(-pi/2)` returns `-1.0`.

Character Class

Table 6.2 lists some of the methods of the Character class.

String Class

Table 6.3 summarizes the methods of the class String.

TABLE 6.2 Selected Methods of the Character Class

java.lang.Character				
Method Invocation	**Argument Type**	**Return Value**	**Return Type**	**Example**
Character.isDigit(ch)	**char**	**true** if ch is a digit **false** otherwise	**boolean**	Character.isDigit('6') returns **true** Character.isDigit('<') returns **false**
Character.isLetter(ch)	**char**	**true** if ch is a letter **false** otherwise	**boolean**	Character.isLetter('J') returns **true** Character.isLetter('<') returns **false**
Character.isLetterOr-Digit (ch)	**char**	**true** if ch is a letter or digit **false** otherwise	**boolean**	Character.isLetterOrDigit ('J') returns **true** Character.isLetterOrDigit ('<') returns **false**
Character.isLower-Case(ch)	**char**	**true** if ch is a lowercase letter. **false** otherwise	**boolean**	Character.isLowerCase ('j') returns **true** Character.isLowerCase ('J') returns **false**
Character.isSpace-Char(ch)	**char**	**true** if ch is the space character. **false** otherwise	**boolean**	Character.isSpaceChar (' ') returns **true** Character.isSpaceChar ('J') returns **false**
Character.isUpper-Case(ch)	**char**	**true** if ch is an uppercase letter. **false** otherwise	**boolean**	Character.isUpperCase ('J') returns **true** Character.isUpperCase ('<') returns **false**
Character.isWhite-space(ch)	**char**	**true** if ch is a whitespace. That is, space, new line, tab or return character **false** otherwise	**boolean**	Character.isUpperCase ('\t') returns **true** Character.isUpperCase ('<') returns **false**
Character.toLower-Case(ch)	**char**	The corresponding lowercase letter if ch is a letter. ch otherwise	**char**	Character.toLowerCase ('J') returns 'j', Character.isLowerCase ('<') returns '<'
Character.toUpper-Case(ch)	**char**	The corresponding uppercase letter if ch is a letter. ch otherwise	**char**	Character.toUpperCase ('j') returns 'J' Character.toUpperCase ('<') returns '<'

TABLE 6.3 Selected Methods of the `String` Class

Method Invocation	Argument Type(s)	Return Value	Return Type	Example
		java.lang.String		
`str1.charAt(index)`	**int**	Character at index position First character is at index position 0	**char**	`strOne.charAt(3)` returns `'t'`, `strOne.charAt(4)` returns `' '`, `strOne.charAt(5)` returns `'a'`.
`str1.compareTo(str2)`	`String`	A negative integer if str1 is less than str2 A positive integer if str1 is greater than str2 Zero if str1 and str2 are equal	**int**	`strOne.compareTo("What's")` returns negative integer, `strOne.compareTo("WHat")` returns positive integer.
`str1.equals(str2)`	`String`	**true** if str1 and str2 are equal **false** otherwise	**boolean**	`strOne.equals(strTwo)` returns **false**.
`str1.indexOf(chs)`	**char** `String`	Index of the first occurrence of chs in the `String` Str1 –1 if chs not in `String` str1	**int** **int**	`strOne.indexOf('a')` returns 2, `strOne.indexOf('b')` returns –1. `strOne.indexOf("wor")` returns 17,
`str1.indexOf(chs, st)`	**char, int** `String,` **int**	Index of the first occurrence of chs starting from index st –1 if chs not in `String` str1 from index st	**int** **int**	`strOne.indexOf('a',5)` returns 5, `strOne.indexOf('a',6)` returns –1. `strOne.indexOf("wo", 10)` returns 17,
`str1.length()`		Length or number of characters	**int**	`strOne.length()` returns 23.
`str1.replace(ch, newCh)`	**char, char**	A new `String` in which every occurrence of ch in str1 is replaced by newCh	`String`	`strTwo.replace('R', 'T')` returns "Tunner", `strTwo.replace('n', 'd')` returns "Rudder".
`str1.substring(st, end)`	**int, int**	A new string `String` at st and ending at end-1 of str1	`String`	`strOne.substring(0,4)` returns "What". `strOne.substring(1,4)` returns "hat".
`str1.toLowerCase()`		A new `String` in which all uppercase letters of str1 are changed to corresponding lowercase letter	`String`	`strTwo.toLowerCase()` returns "runner".
`str1.toUpperCase()`		A new `String` in which all lowercase letters of str1 are changed to corresponding uppercase letter	`String`	`strTwo.toUpperCase()` returns "RUNNER".
`str1.trim()`		A new `String` in which all leading and trailing whitespace characters are removed from str1	`String`	`strThree.trim()` returns "J \ta \tv \ta".

Assume the following declarations and assignments:

```
String str1, str2;
String strOne = "What a wonderful world!";
String strTwo = "Runner", strThree = "\t J \ta \tv \ta\r";
```

METHOD INVOCATION

All the methods introduced in this chapter are public, value returning, and predefined. Therefore, possible classifications are based on whether methods have parameters and are static. Thus, the methods can be classified as follows:

1. static methods that have no parameters.
2. static methods that have parameters.
3. Methods that are not marked as static and have no parameters.
4. Methods that are not marked as static and have parameters.

Observe that all these methods are value returning methods. Therefore, method can be invoked as part of an expression on the right-hand side of an assignment statement or can be invoked inside an output statement such as System.out.print or System.out.println. Recall that if the value returned by a method is not stored in a variable, it will be lost.

Examples 6.1 through 6.4 will in turn illustrate the invocation of methods belonging to the above four categories. These examples are presented in this context to serve two different purposes:

1. To illustrate methods belonging to the above four categories
2. To learn some of the predefined methods

Example 6.1

This example illustrates the invocation of a static method with no parameter. The method random fits the bill. The method random returns a double value between 0 and 1. It is possible to use random method to generate a random number in any given range of values. For example, if you want to generate random values between 1 and 10, all you need to do is use the following expression:

```
1 + (int) (10 * Math.random())
```

Recall that random being a static method with no parameters, the method is invoked as Math.random(). The syntax template for invoking a static method without parameter is

```
ClassName.methodName()
```

The expression 10 * Math.random() is a double value between 0 and 10. Due to truncation, (int) (10 * Math.random()) is an int value between 0 and 9. Thus 1 + (int) (10 * Math.random()) is an int value between 1 and 10.

```
/**
    Illustration of static methods, no parameters
*/
public class StaticNoArguments
{
    public static void main (String[] args)
    {
        double nextValue;
        int number;
        int count;

        //method is part of an expression
        number = 1 + (int) (10 * Math.random());

        System.out.println(number +
                       "random numbers are as follows\n");
        for (count = 1; count < number + 1; count++)
        {
            //method in an assignment statement
            nextValue = Math.random();
            System.out.println(nextValue);
        }

            //method in output statement
            System.out.println("\nNext random number is"
                                    + Math.random());
    }
}
```

Output

```
10 random numbers are as follows:

0.9289086794637926
0.3623358184620312
0.18000213009794308
0.9325904849148384
0.009517094960630912
0.3019646502636888
0.4163229469179339
0.22850462123014104
0.6847559715976199
0.7281384727994286

Next random number is 0.5995440858527961
```

Example 6.2

Note that the actual parameters in a method invocation can be a literal, variable, or an expression. This example illustrates the invocation of `static` methods having parameters. Methods used for illustration are `ceil`, `round`, `floor`, and `pow`.

```
/**
    Illustration of static methods having parameters
*/
public class StaticArguments
{
    public static void main (String[] args)
    {
        double valueFive = 5;

        double value = 17.5;
        double valueThree = 3.0;
        int number;

        //method is part of an expression

        number = 10 * (int) Math.ceil(17.4999);
        System.out.println("Math.ceil(17.4999) times 10 is " +
                                                    number);
        number = 10 * (int) Math.round(17.4999);
        System.out.println("Math.round(17.4999) times 10 is " +
                                                    number);
        number = 10 * (int) Math.floor(17.4999);
        System.out.println("Math.floor(17.4999) times 10 is " +
                                                    number);

        System.out.println();

        //parameter can be a variable

        number = 10 * (int) Math.ceil(value);
        System.out.println("Math.ceil(value) times 10 is " +
                                                    number);
        number = 10 * (int) Math.round(value);
        System.out.println("Math.round(value) times 10 is " +
                                                    number);
        number = 10 * (int) Math.floor(value);
        System.out.println("Math.floor(value) times 10 is " +
                                                    number);
```

```
        System.out.println();

        //method in output statement

        System.out.println("Math.ceil(-17.5001) is " +
                                Math.ceil(-17.5001));
        System.out.println("Math.round(-17.5001) is " +
                                Math.round(-17.5001));
        System.out.println("Math.floor(-17.5001) is " +
                                Math.floor(-17.5001));

        System.out.println();

        System.out.println("Math.ceil(-17.5) is " +
                                Math.ceil(-17.5));
        System.out.println("Math.round(-17.5) is " +
                                Math.round(-17.5));
        System.out.println("Math.floor(-17.5) is " +
                                Math.floor(-17.5));

        System.out.println();
        //arguments can be literals, expression or variables

        System.out.println("Math.pow(5, valueThree) is " +
                                Math.pow(5, valueThree));
        System.out.println("Math.pow(2*2+1, 3) is " +
                                Math.pow(2*2+1, 3));
        System.out.println("Math.pow(valueFive, valueThree) is " +
                          Math.pow(valueFive, valueThree));

    }
}
```

Output

```
Math.ceil(17.4999) times 10 is 180
Math.round(17.4999) times 10 is 170
Math.floor(17.4999) times 10 is 170

Math.ceil(value) times 10 is 180
Math.round(value) times 10 is 180
Math.floor(value) times 10 is 170

Math.ceil(-17.5001) is -17.0
Math.round(-17.5001) is -18
Math.floor(-17.5001) is -18.0
```

```
Math.ceil(-17.5) is -17.0
Math.round(-17.5) is -17
Math.floor(-17.5) is -18.0

Math.pow(5, valueThree) is 125.0
Math.pow(2*2+1, 3) is 125.0
Math.pow(valueFive, valueThree) is 125.0
```

Note 6.2

If x is such that 17 $\leq$ x $<$ 17.5 then

`Math.round(x)` = `17` and `Math.floor(x)` = `17.0`

If x is such that 17.5 $\leq$ x $\leq$ 18 then

`Math.round(x)` = `18` and `Math.ceil(x)` = `18.0`

If x is such that -18 $\leq$ x $<$ -17.5 then

`Math.round(x)` = `-18` and `Math.floor(x)` = `-18.0`

If x is such that -17.5 $\leq$ x $\leq$ -17 then

`Math.round(x)` = `-17` and `Math.ceil(x)` = `-17.0`

Example 6.3

This example illustrates the invocation of methods that are not marked as `static` and have no parameter. The methods chosen for the illustration are from the class `String`. Recall that the syntax for the invocation of a method that is not marked as `static` and without parameters is

```
    objectReference.methodName()
/**
    Illustration of methods not marked as static having no
                                                 parameters
*/
public class NonStaticNoArguments
{
    public static void main (String[] args)
    {
        int number;
        String strOne = "What a wonderful world!";
        String strThree = "\r\t\t\tJ \ta \tv \ta\n\n\t";
        String strFour = "What a wonderful world";
        String str;

        //method is part of an expression
        number = 10 * strOne.length();
        System.out.println("strOne.length() times 10 is " +
                                                     number);
```

```
        number = 10 * strFour.length();
        System.out.println("strFour.length() times 10 is " +
                                                     number);

        System.out.println();
        str = strOne.toLowerCase();
        System.out.println("strOne.toLowerCase() returns " +
                                                        str);
        str = strOne.toUpperCase();
        System.out.println("strOne.toUpperCase() returns " +
                                                        str);

        //method in output statement
        System.out.println("strThree :" + strThree + ":");
        System.out.println("strThree.trim() returns :" +
                                strThree.trim()+ ":");

    }
}
```

Output

```
strOne.length() times 10 is 230
strFour.length() times 10 is 220

strOne.toLowerCase() returns what a wonderful world!
strOne.toUpperCase() returns WHAT A WONDERFUL WORLD!
                        J       a       v       a


        :
strThree.trim() returns :J     a      v      a:
```

Note 6.3 The method `trim` removes all whitespace characters appearing at the beginning and the end of a `String`. Thus in the above program, `strThree.trim()` returns the `String` `"J \ta \tv \ta"`.

Example 6.4

This example illustrates the invocation of methods that are not marked as `static` and have parameters. The methods chosen for illustration are from the class `String` of the `package` `java.lang`. Recall that the syntax for the invocation of a method that is not marked as `static` and with parameters is

```
    objectReference.methodName(actualParameters)
```

```
/**
    Methods that are not marked static and having
                                        parameters.
```

```
*/
public class NonStaticArguments
{
    public static void main (String[] args)
    {
        char ch;
        int number;
        boolean theSame;

        String strOne = "What a wonderful world!";
        String strFour = "What a wonderful world";

        //method is part of an expression
        ch = strOne.charAt(3);
        System.out.println("strOne.charAt(3) is " + ch);
        number = 10 * strOne.indexOf('a');
        System.out.println("strOne.indexOf('a') times 10 is " +
                                                      number);
        number = 10 * strOne.indexOf('a',6);
        System.out.println("strOne.indexOf('a',6) times 10 is " +
                                                      number);
        System.out.println();

        //parameter can be a variable
        number = strOne.compareTo(strFour);
        System.out.println("strOne.compareTo(strFour) is " +
                                                    number);
        number = strFour.compareTo(strOne);
        System.out.println("strFour.compareTo(strOne) is " +
                                                      number);
        theSame = strOne.equals(strFour);
        System.out.println("strOne.equals(strFour) is " +
                                                   theSame);
        theSame = strOne.equals(strOne);
        System.out.println("strOne.equals(strOne) is " +
                                                  theSame);
        System.out.println();

        //method in output statement
        System.out.println("strOne.substring(0,4) returns " +
                                        strOne.substring(0,4));
        System.out.println("strOne.substring(5,16) returns " +
                                        strOne.substring(5,16));
    }
}
```

Output

```
strOne.charAt(3) is t
strOne.indexOf('a') times 10 is 20
strOne.indexOf('a',6) times 10 is -10

strOne.compareTo(strFour) is 1
strFour.compareTo(strOne) is  1
strOne.equals(strFour) is false
strOne.equals(strOne) is true

strOne.substring(0,4) returns What
strOne.substring(5,16) returns a wonderful
```

USER-DEFINED METHODS

The user-defined methods were first introduced in Chapter 3. In this section, we have a closer look at user-defined `public` methods. User-defined `static` methods are discussed later in this chapter. Therefore, we classify user-defined methods based on whether they have parameters and are value returning. Thus, there are four different cases to consider. They are as follows:

1. Value returning methods having no parameters
2. Value returning methods having parameters
3. `void` methods having no parameters
4. `void` methods having parameters

Chapter 3 explains in detail cases 1 and 3. Further, Chapter 3 also covered cases 2 and 4 with single parameter. Therefore, in this chapter we discuss the general cases of 2 and 4. The syntax template of a method is as follows:

```
[accessModifier] [abstract|final][static] returnType methodName
                                                ([formalParam])
{
    [statements]
}
```

where the access modifier is one of the following: `public`, `private`, or `protected`. You are already familiar with access modifiers `public` and `private`. The term `protected` will be explained in Chapter 7. Keywords `abstract` and `final` are also covered in Chapter 7. The vertical bar between `abstract` and `final` in the syntax template indicates that both `abstract` and `final` cannot appear simultaneously. After the method name, all the formal parameters of the method are listed inside a pair of left and right parentheses.

Self-Check

3. Every _________ method must have a return type.
4. If a method does not return a value, it is a _________ method.

Formal Parameter List

The syntax of the formal parameter list is as follows:

```
dataType1 arg1[, dataType2 arg2, ..., dataTypeN argN]
```

In the given syntax template, `dataType1, ..., dataTypeN` can be any primitive data type or a class name and `arg1, ..., argN` are identifiers. Note that in a list of items, comma is used to separate individual members. Thus in the case of a formal parameter list, each item consists of a data type followed by an identifier separated by at least one white-space character.

Self-Check

5. In Java, items in a list are separated by _________.
6. Every formal parameter must be preceded by its _________.

Signature of a Method

The signature of a method consists of the name of the method along with the list of all data types in the order they appear in the formal parameter list of the method. Thus, the signature of a method with the heading

```
public returnType methodName(dataType1 arg1, ..., dataTypeN argN)
```

is

```
methodName(dataType1, ..., dataTypeN).
```

Example 6.5

Consider the method `compareTo` of the `String` class. The heading of the method is

```
public int compareTo(String str)
```

The formal parameter list is `String str` and the signature of the method is `compareTo(String)`.

Example 6.6

Consider the method `indexOf` of the `String` class. Note that there are four different methods in the `String` class, all having the name `indexOf`. The headings, formal parameter list, and signatures of all four `indexOf` methods are shown in Table 6.4.

TABLE 6.4 Method Overloading and `indexOf`

Method Heading	Formal Parameter List	Signature
`public int indexOf(char ch)`	`char ch`	`indexOf(char)`
`public int indexOf(String str)`	`String str`	`indexOf(String)`
`public int indexOf(char ch, int s)`	`char ch, int s`	`indexOf(char,int)`
`public int indexOf(String str, int s)`	`String str, int s`	`indexOf(String, int)`

TABLE 6.5 Method Overloading and `abs`

Method Heading	Formal Parameter List	Signature
`public int abs(int x)`	`int x`	`abs(int)`
`public long abs(long x)`	`long x`	`abs(long)`
`public float abs(float x)`	`float x`	`abs(float)`
`public double abs(double x)`	`double x`	`abs(double)`

In Java, method name need not be unique. There can be many methods all having the same name. You may be wondering if there are many methods all having the same name how would the compiler decide which method to execute? The answer is quite simple. It is not the name that distinguishes one method from the other. Rather, it is the signature of the method that distinguishes one method from the other. Therefore, it is possible to have many methods with identical names so long as no two of them have identical signatures. Observe that there are four different methods in the class `String` with the name `indexOf`. However, no two of them have identical signature. The programming language feature that allows the programmer to create more than one method within a class having the same name is known as *method overloading*. Java allows method overloading.

Method overloading. The programming language feature that allows the programmer to create more than one method within a class having the same name.

The following example further illustrates method overloading.

Example 6.7

Consider the method `abs` of `Math` class. There are four different methods, all having the same name `abs`. The headings, formal parameter list, and signatures of all four `abs` methods are shown in Table 6.5. Observe that the signatures are all different.

Note 6.4 Let `classOne` and `classTwo` be two classes. Let `xyz` be a method in both classes. This is not an example of method overloading.

Note 6.5 It is a compilation error to have two methods within a class having identical signatures.

Note 6.6 Return type is not part of the signature.

During the method invocation, the data type of each of the actual parameters must agree with corresponding formal parameter. For example, if the actual parameter is `int` and the corresponding formal parameter is `double`, the compiler will not issue any error message. In this case, `int` is implicitly promoted to `double`; and thus the actual parameter matches with the formal parameter. As an example, consider the method `pow` of the `Math` class. The method `pow` has two parameters, both `double`. However, `Math.pow(5, 6)` will not result in a compilation error. Due to implicit conversion, `Math.pow(5, 6)` is equivalent to `Math.pow(5.0, 6.0)`.

In contrast, consider the method invocation `strOne.charAt(3)` shown in Example 6.4. The method `charAt` has one formal parameter of type `int`. Therefore, `strOne.charAt(3.0)` results in type mismatch. Note that in this case, the actual parameter is of the type `double` and the formal parameter is of the type `int`. Further, a `double` value cannot be implicitly promoted to an `int` value. Therefore, `strOne.charAt(3.0)` results in a compilation error.

Self-Check

7. True or false: Return type is part of the signature of a method.
8. True or false: Signature of two overloaded methods cannot be identical.

Parameter Passing

Parameter passing mechanism in Java is known as *call by value*. That is, at the beginning of a method invocation, actual parameters are copied to formal parameters. A change in the formal parameter value during method execution has no impact on the value of the actual parameter. In fact, one could think that at the beginning of a method invocation, each actual parameter is assigned to the corresponding formal parameter. For this reason, let us review the concepts involved in an assignment statement.

Consider an assignment statement of the form

```
leftHandSide = RightHandSide;
```

where `leftHandSide` and `RightHandSide` are variables. Then we have the following:

- Assignment is a onetime operation. During its execution, the value at `rightHandSide` is copied to `leftHandSide`.
- Any change in the value of `leftHandSide` has no impact on `rightHandSide` and vice versa.

Assume the following declarations:

```
int number, index;
double valueOne, valueTwo;
boolean allDone, found;
```

Now, consider the following assignment statement:

```
number = index; // Right hand side is a variable of the same type.
```

In this case, number and index are of the same data type. Therefore, the current value of index is copied to the memory location labeled as number. The behavior of the following assignment statements is similar:

```
number = 71; // Right hand side is a literal of the same type.
number = 2*index + 1     // Right hand side is an expression
                         // of the same type.
```

Now consider the following assignment statement:

```
valueOne = number; // Right hand side is a variable of different
                                                          type.
                  // However, there exists implicit conversion
                                                       rules.
```

Observe that data type of number is int and that of valueOne is double. Since an int can be implicitly promoted to double, the above statement will not result in a compilation error. Similar comments apply to the following statements:

```
valueOne = 71;               // Right hand side is a literal.
valueOne = 2 * index + 1;  // Right hand side is an expression.
```

However, the following assignment statement will result in a compilation error:

```
number = valueOne; // Right hand side is a variable of different
                                                          type.
                   // There exists no implicit conversion rules.
```

In this case, double value on right-hand side cannot be implicitly converted into int, the data type of the variable on the left-hand side. Similar comments apply to the following assignment statements:

```
number = 71.234;              // Right hand side is a literal.
index = 2 * valueOne + 1.3; // Right hand side is an expression.
```

Further, recall that the compilation error can be avoided by explicit conversion using cast operators. Thus, the following statements are legal in Java:

```
number = (int) valueOne;
number = (int) 71.234;
number = (int) (2 * valueOne + 1.3);
```

Note that there exists no implicit or explicit rule to convert a `boolean` value to an `int` value. As a consequence, the following three assignment statements cannot be modified to make them legal:

```
number = allDone;              // Right hand side is a variable
number = true;                 // Right hand side is a literal
number = allDone && found;     // Right hand side is an expression
```

In the case of parameter passing, actual parameters play the role similar to the expression on the right-hand side of an assignment statement, and formal parameters play the role of the variable on the left-hand side of an assignment statement. In particular, actual parameter can be a variable, a literal, or an expression. The formal parameter must be a variable.

For instance, consider the method `indexOf`. The heading of the method is

```
public int indexOf(char ch, int start)
```

Note that the first actual parameter has to be a `char` and the second actual parameter has to be an `int`. However, you can also use other data types as long as there is an implicit conversion rule. For example, you can invoke the method with first actual parameter a `char` and second actual parameter a `short` data type. As the method is invoked, the actual parameters are "assigned" to corresponding formal parameters. Thus, if `indexOf` is invoked as

```
strOne.indexOf('a', 7);
```

where `strOne` is `String` reference variable. The first formal parameter `ch` gets the value `'a'` and the second formal parameter `start` gets the value 7.

Example 6.8

This example illustrates call by value of primitive data types. There are two classes in this example. The class `ParameterPassing` has one method `setAllData` with two formal parameters of type `int` and `double`, respectively. The class `ParameterPassingIllustration` invokes `setAllData` method with actual parameters `year` and `amount`, respectively.

```
/**
    Method illustrating parameter passing; primitive data types
*/
public class ParameterPassing
{
    private int number;
    private double value;
```

```
    public double setAllData(int num, double val)
    {
        System.out.println("\t\t\tInside setAllData method");
        System.out.println("("\t\t\tFirst formal parameter is
                                                        " + num);
        System.out.println("("\t\t\tSecond formal parameter
                                                   is  " + val);
        number = num;
        value = val;
        num = 2 * num + 1;
        System.out.println("\n");
        System.out.println("("\t\t\tFirst formal parameter is
                                                        " + num);
        System.out.println("("\t\t\tSecond formal parameter
                                                   is " + val);
        System.out.println("("\t\t\tExit : setAllData
                                                 method");
        return num * val;
    }
}


import java.util.Scanner;

/**
    Application illustrating parameter passing; primitive
                                                 data types
*/
public class ParameterPassingIllustration
{

    public static void main (String[] args)
    {
        double amount = 12.34;
        int years = 10;
        double valueReturned;

        ParameterPassing ppRef = new ParameterPassing();

        System.out.println("Just before entering setAllData
                                                    method");
```

```
        System.out.println("First actual parameter is " +
                                              amount);
        System.out.println("Second actual parameter is " +
                                               years);
        System.out.println("\n");
        valueReturned = ppRef.setAllData(years, amount);
        System.out.println("\n");
        System.out.println("Just after exiting setAllData
                                             method");
        System.out.println("First actual parameter is " +
                                              amount);
        System.out.println("Second actual parameter is " +
                                               years);
        System.out.println("The value returned is " +
                                     valueReturned);
    }
}
```

Output

```
Just before entering setAllData method
First actual parameter is 12.34
Second actual parameter is 10

                        Inside setAllData method
                        First formal parameter is 10
                        Second formal parameter is 12.34

                        First formal parameter is 21
                        Second formal parameter is 12.34
                        Exit : setAllData method

Just after exiting setAllData method
First actual parameter is 12.34
Second actual parameter is 10
The value returned is 259.14
```

The actual parameters just before the method invocation can be visualized as follows:

years	10	amount	12.34

During the method invocation, the actual parameter values are copied to formal parameters.

years	10	amount	12.34
num	10	val	12.34

From the output statements observe that value of `years` and `amount` are copied to formal parameters `num` and `val`, respectively. Further, `num` changes to 21 inside the method. This situation can be visualized as follows:

years	10	amount	12.34
num	21	val	12.34

However, the variable `years` is unaffected. Upon completion of the method, variables `num` and `val` do not exist any more; however, instance variables `years` and `amount` do exist. Thus, we have the following:

years	10	amount	12.34

The call by value of a reference variable is exactly the same. The value of an object reference in the actual parameter is copied to formal parameter during the parameter passing. Note that the object reference alone is copied and the object is not copied. Therefore during the method execution, both the actual parameter and the formal parameter refer to the same object. Consequently, any changes made in the object by the method are persistent. These ideas are illustrated in the next example.

Example 6.9

```
/**
    Method illustrating parameter passing; object reference
*/
public class ParameterPassingObjectRef
{
    private int number;
    private double value;

    public int replace(ParameterPassingObjectRef param, int
                                                          factor)
```

```
    {
        System.out.println("\t\t\tInside replace method");
        System.out.println("\t\t\tObject information: " +
                                                     param);
        System.out.println("\t\t\tThe factor value is " +
                                                     factor);
        param.number = factor;
        factor = 4;
        System.out.println("\n");
        System.out.println("\t\t\tObject information: " +
                                                     param);
        System.out.println("\t\t\tThe factor value is " +
                                                     factor);
        System.out.println("\t\t\tExit : replace method");

        return factor;
    }

    public void setData(int inNumber, double inValue)
    {

        number = inNumber;
        value = inValue;
    }

    public String toString()
    {

        String str;
        str = "(number = " + number + ", value = " +
                                        value + ")";

        return str;
    }
}


import java.util.Scanner;

/**
    Application illustrating parameter passing; object reference
*/
public class ParameterPassingObjectRefIllustration
```

```
{
    public static void main (String[] args)
    {
        int    multiplier = 10;
        double valueReturned;
        ParameterPassingObjectRef objectOne = new
                                   ParameterPassingObjectRef();
        ParameterPassingObjectRef objectTwo = new
                                   ParameterPassingObjectRef();
        objectOne.setData(10, 25.5);
        objectTwo.setData(30, 45.8);

        System.out.println("Just before entering replace
                                              method");
        System.out.println("Object information: " +
                                        objectTwo);
        System.out.println("The factor value is " + multiplier);
        System.out.println("\n");
        valueReturned = objectOne.replace(objectTwo,
                                        multiplier);
        System.out.println("\n");
        System.out.println("Just after exiting replace method");
        System.out.println("Object information: " + objectTwo);
        System.out.println("The factor value is " +
                                       multiplier);
    }
}
```

Output

```
Just before entering replace method
Object information: (number = 30, value = 45.8)
The factor value is 10

              Inside replace method
              Object information: (number = 30, value = 45.8)
              The factor value is 10

              Object information: (number = 10, value = 45.8)
              The factor value is 4
              Exit : replace method

              Just after exiting replace method
              Object information: (number = 10, value = 45.8)
              The factor value is 10
```

In the above example, ParameterPassingObjectRef class has two attributes: number and value. Thus, after executing the following two statements

```
ParameterPassingObjectRef objectTwo = new
                              ParameterPassingObjectRef();
objectTwo.setData(30, 45.8);
```

we have the following:

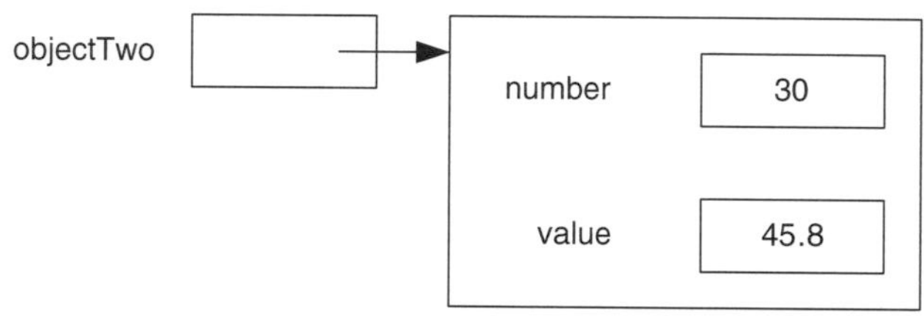

The actual parameters before the replace method invocation can be visualized as follows:

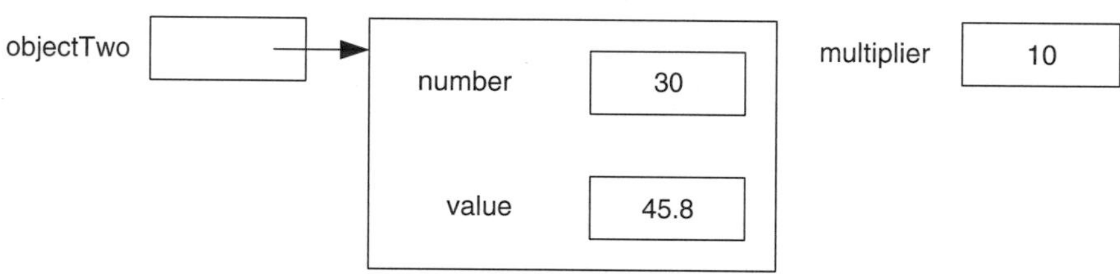

Passing objectTwo as an actual parameter results in copying the object reference in the variable objectTwo into the actual parameter param. Thus, both variables objectTwo and param contain the same object reference. The value contained in the primitive data type multiplier is copied to factor. Thus during method invocation, we have the following situation:

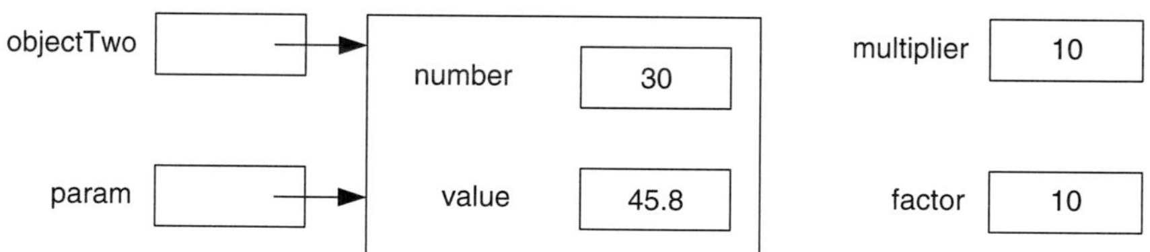

The execution of the statement

```
param.number = factor;
```

in the method replace results in the following change:

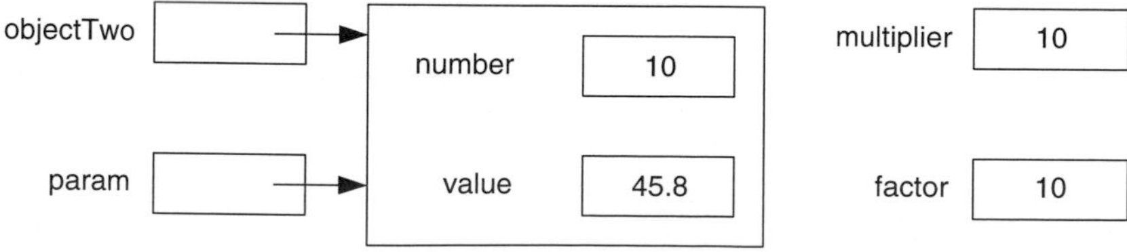

The statement

```
factor = 4;
```

changes `factor` to 4. Note that `multiplier` still contains 10.

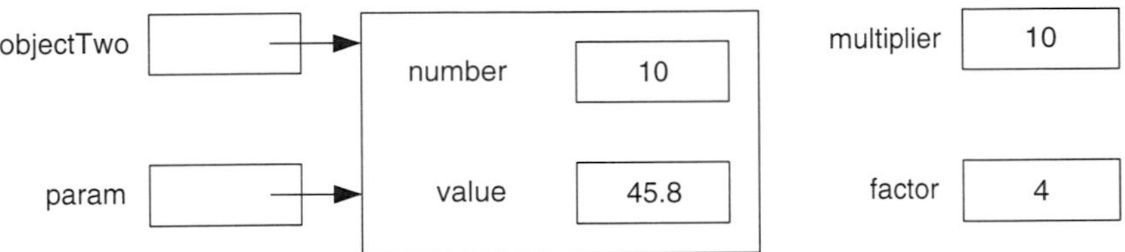

Once the method execution is completed, both variables `param` and `factor` do not exist and thus we have the following:

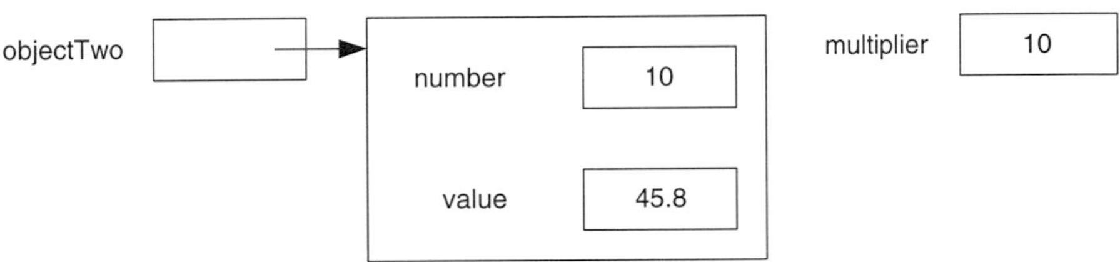

This shows that changes made to the instance variables of an object is persistent after the method invocation.

The following example illustrates the fact that since in Java, parameter passing is call by value, while it is possible to make persistent changes to the instance variables of an object during method execution, it is impossible to make persistent changes to the reference variable itself. In particular, observe that in the Example 6.8, no attempt was made to change the value of the actual parameter `objectTwo`.

Example 6.10

```
/**
    Method; parameter passing; object reference don't change
*/
public class ParameterPassingRef
{
    private int number;
    private double value;

    public int reAssign(ParameterPassingRef pRef)
    {
        System.out.println("\t\t\tInside reAssign method");
        System.out.println("\t\t\tObject information: " + pRef);
        pRef = new ParameterPassingRef();
        pRef.setData(100, 124.8);
```

```
        System.out.println("\n");
        System.out.println("\t\t\tObject information: " + pRef);
        System.out.println("\t\t\tExit : reAssign method");

        return 1;
    }

    public void setData(int inNumber, double inValue)
    {
        number = inNumber;
        value = inValue;
    }

    public String toString()
    {
        String str;
        str = "(number = " + number + ", value = " +
                                            value + ")";
        return str;
    }
}

/**

    Application; parameter passing; object reference don't
                                                    change
*/
public class ParameterPassingRefIllustration
{
    public static void main (String[] args)
    {
        int multiplier = 10;
        double valueReturned;

        ParameterPassingRef objectOne = new
                                    ParameterPassingRef();
        ParameterPassingRef objectTwo = new
                                    ParameterPassingRef();
        objectOne.setData(8, 12.2);
```

```
        objectTwo.setData(15, 37.8);
        System.out.println("Just before entering reAssign
                                                method");
        System.out.println("Object information: " + objectTwo);
        System.out.println("\n");
        valueReturned = objectOne.reAssign(objectTwo);
        System.out.println("\n");
        System.out.println("Just after exiting reAssign
                                               method");
        System.out.println("Object information: " + objectTwo);
    }
}
```

Output

```
Just before entering reAssign method
Object information: (number = 15, value = 37.8)


            Inside reAssign method
            Object information: (number = 15, value = 37.8)


            Object information: (number = 100, value = 124.8)
            Exit : reAssign method


Just after exiting reAssign method
Object information: (number = 15, value = 37.8)
```

The actual parameter `objectTwo` before the `reAssign` method invocation can be visualized as follows:

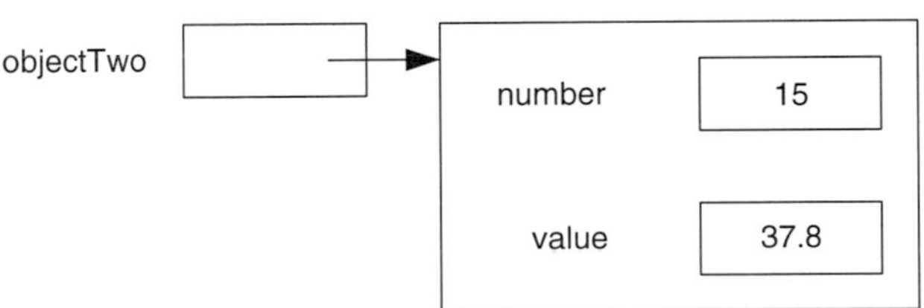

Now, passing `objectTwo` as an actual parameter results in copying the object reference in the variable `objectTwo` into the actual parameter `pRef`. Thus, both objectTwo and pRef variables contain the same object reference. Thus at the beginning of the method invocation, we have the following situation:

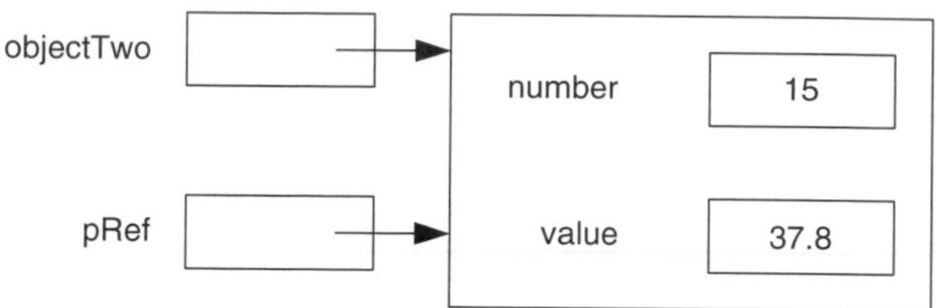

Now, the statement

```
pRef = new ParameterPassingRef();
```

creates a new object and its address is placed in the variable pRef.

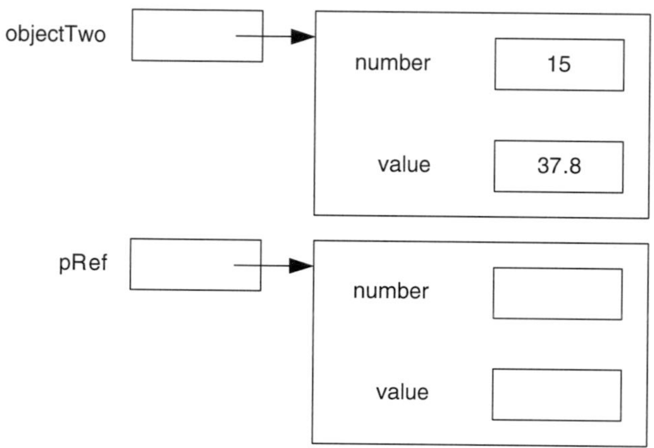

Further, the statement

```
pRef.setData(100, 124.8);
```

has the following effect. Note that after method invocation, a change in the reference variable pRef has no impact on the reference variable objectTwo.

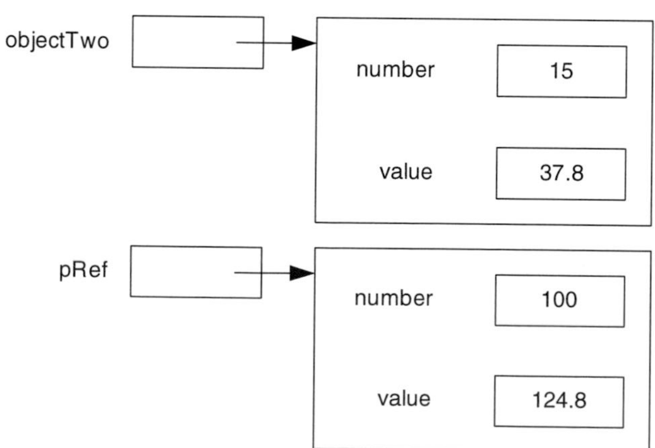

Upon completion of the method, the reference variable `pRef` does not exist and the reference variable `objectTwo` remains unaffected by the method invocation.

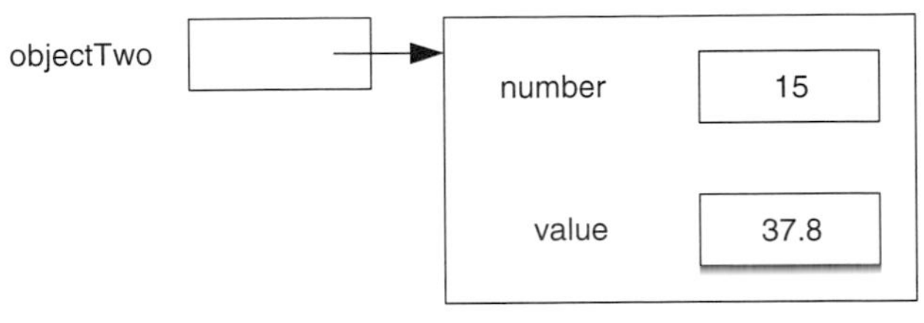

Self-Check

9. The parameter passing in Java is known as ________.
10. If the formal parameter is a primitive data type, during method invocation, the ________ of the actual parameter is copied to the formal parameter.

CONSTRUCTORS

Constructors you have seen so far are *default constructors*. Java automatically provides a default constructor if no constructor is included as part of the class definition; hence the name, default constructor. Note that the default constructor has no formal parameters.

In Java, constructor overloading is allowed. Thus, a class may have many constructors with different signatures. Further, once at least one constructor is included as part of the class definition, the compiler does not provide the default constructor. A constructor with no parameters is quite useful in various situations as you will see in subsequent chapters. Therefore, it is quite important that once a constructor is included as part of the class definition, there must be a constructor with no parameters as well.

The following facts on constructors are worth mentioning:

- The name of the constructor is the same as the class name.
- A constructor must not have a return type.
- A constructor can have an optional `return` statement.
- It is legal to overload a constructor. The compiler selects a constructor based on the signature (as is the case of a method).
- The constructors are invoked using the `new` operator.
- The Java compiler, by default, will provide a constructor if no constructor is included in the class definition. Such a constructor is known as default constructor. The default constructor has no parameters.
- Once a constructor is included as part of the class definition, Java compiler does not provide the default constructor. Therefore, it is quite important that once a constructor is included as part of the class definition, there must be a constructor with no parameter as well.
- One of the intended purposes of a constructor is the proper initialization of the attributes of a class.

The syntax template of a constructor is as follows:

```
accessModifier ClassName([formalParameterList])
{
    [statements]
}
```

The next example illustrates the overloading of constructors.

Example 6.11

In this example, we revisit the class `Stock` introduced in Chapter 3. The class has three attributes:

```
private int numberOfShares;
private String tickerSymbol;
private double dividend;
```

It is a good programming practice to start with a constructor with no parameters. In this case, you have to decide on possible default values for each of the attributes. In this example, number of shares can be initialized to `0`. The ticker symbol may be initialized to the string "[UA]" to indicate that the ticker symbol is unassigned. The attribute dividend can be initialized to `0.0`. Thus, we have the following constructor:

```
/**
    Constructor with no parameters
*/
public Stock()
{
    numberOfShares = 0;
    tickerSymbol = "[UA]";
    dividend = 0.0;
}
```

Next we create three more constructors as follows:.

```
/**
    Constructor that initializes all attributes
*/
public Stock(int inNumberOfShares,String inTickerSymbol,
                                        double inDividend)
{
    numberOfShares = inNumberOfShares;
    tickerSymbol = inTickerSymbol;
```

```
    dividend = inDividend;
}

/**
    Constructor; initializes no. of shares and ticker
                                                symbol
*/
public Stock(int inNumberOfShares,String inTickerSymbol)
{
    numberOfShares = inNumberOfShares;
    tickerSymbol = inTickerSymbol;
    dividend = 0.0;
}

/**
    Constructor that initializes ticker symbol, no. of
                                                shares
*/
public Stock(String inTickerSymbol, int inNumberOfShares)
{
    numberOfShares = inNumberOfShares;
    tickerSymbol = inTickerSymbol;
    dividend = 0.0;
}
```

Thus, there are four constructors in the class `Stock` with the following signatures:

```
Stock()
Stock(int, String, double)
Stock(int, String)
Stock(String, int)
```

The complete program listing along with the output follows. Additional `println` statements are included in constructors that appear in the complete listing for the purpose of identifying the constructor invoked in each of the cases.

Consider the following statement:

```
stockFour = new Stock("JKL", 400);
```

Note that in this case, the constructor with signature `Stock(String, int)` matches `Stock("JKL", 400)` and the corresponding code is being executed. The other constructor invocations can be understood in a similar manner.

```
class Stock
{
    private int numberOfShares;
```

```
    private String tickerSymbol;
    private double dividend;

    /**
        Constructor with no parameters
    */
    public Stock()
    {
        System.out.println("signature of the constructor
                                              invoked is");
        System.out.println("\t\tStock()");
        numberOfShares = 0;
        tickerSymbol = "[UA]";
        dividend = 0.0;
    }



    /**
        Constructor that initializes all attributes
    */
    public Stock(int inNumberOfShares,String
                        inTickerSymbol,double inDividend)
    {
        System.out.println("signature of the constructor
                                              invoked is");
        System.out.println("\t\tStock(int, String,
                                          double)");
        numberOfShares = inNumberOfShares;
        tickerSymbol = inTickerSymbol;
        dividend = inDividend;
    }

    /**
        Constructor; initializes no. of shares and ticker
                                                    symbol
    */
    public Stock(int inNumberOfShares,String inTickerSymbol)
    {
        System.out.println("signature of the constructor
                                              invoked is");
        System.out.println("\t\tStock(int, String)");
        numberOfShares = inNumberOfShares;
```

```
        tickerSymbol = inTickerSymbol;
        dividend = 0.0;
    }

    /**
        Constructor that initializes ticker symbol, no. of
                                                   shares
    */
    public Stock(String inTickerSymbol, int inNumberOfShares)
    {
        System.out.println("signature of the constructor
                                            invoked is");
        System.out.println("\t\tStock(String, int)");
        numberOfShares = inNumberOfShares;
        tickerSymbol = inTickerSymbol;
        dividend = 0.0;
    }


    /**
        Computes and returns yearly dividend
        @return the yearly dividend
    */
    public double yearlyDividend()
    {
        double totalDividend;
        totalDividend = numberOfShares * dividend;
        return totalDividend;
    }

    /**
        Accessor method for the number of shares
        @return the number of shares
    */
    public int getNumberOfShares()
    {
        return numberOfShares;
    }

    /**
        Accessor method for the ticker symbol
        @return the ticker symbol
    */
    public String getTickerSymbol()
```

```
{
    return tickerSymbol;
}

/**
    Accessor method for the dividend
    @return the dividend
*/
public double getDividend()
{
    return dividend;
}

/**
    Mutator method to set the number of shares
    @param inNumberOfShares the number of shares
*/
public void setNumberOfShares(int inNumberOfShares)
{
    numberOfShares = inNumberOfShares;
}



/**
    Mutator method to set the ticker symbol
    @param inTickerSymbol the ticker symbol
*/
public void setTickerSymbol(String inTickerSymbol)
{
    tickerSymbol = inTickerSymbol;
}

/**
    Mutator method to set the dividend
    @param inDividend the dividend
*/
public void setDividend(double inDividend)
{
    dividend = inDividend;
}

/**
    The toString method
```

```
        @return number of shares and ticker symbol
    */
    public String toString()
    {
        String str;
        str = numberOfShares + " " + tickerSymbol;
        return str;
    }

}

/**
    Application program to test constructors of Stock class
*/
public class StockTesting
{
    public static void main (String[] args)
    {
        Stock stockOne;
        Stock stockTwo;
        Stock stockThree;
        Stock stockFour;
        //Invoke the constructor with no parameters
        System.out.println("Constructor : Stock()");
        stockOne = new Stock();
        System.out.println(stockOne);
        System.out.println("\n");


        //Invoke the constructor with int, String, double
        System.out.println
        ("Constructor : Stock(200,\"ABC\", 1.60)");
        stockTwo = new Stock(200, "ABC", 1.60);
        System.out.println(stockTwo);
        System.out.println("\n");

        //Invoke the constructor with int, String
        System.out.println
        ("Constructor : Stock(300, \"XYZ\", 1.60)");
        stockThree = new Stock(300, "XYZ");
        System.out.println(stockThree);
        System.out.println("\n");
```

```
        //Invoke the constructor with int, String, double
        System.out.println("Constructor : Stock(\"JKL\", 400)");
        stockFour = new Stock("JKL", 400);
        System.out.println(stockFour);
        System.out.println("\n");
    }
}
```

Output

```
Constructor : Stock()
signature of the constructor invoked is
                Stock()
0 [UA] 0.0


Constructor : Stock(200, "ABC", 1.60)
signature of the constructor invoked is
                Stock(int, String, double)
200 ABC 1.6


Constructor : Stock(300, "XYZ", 1.60)
signature of the constructor invoked is
                Stock(int, String)
300 XYZ 0.0


Constructor : Stock("JKL", 400)
signature of the constructor invoked is
                Stock(String, int)
400 JKL 0.0
```

Example 6.12

Recall that a constructor can have `return` statement optionally. Thus, the following two constructors of the class `Stock` are semantically equivalent:

```
public Stock(int inNumberOfShares,String inTickerSymbol,
                                        double inDividend)
{
    numberOfShares = inNumberOfShares;
    tickerSymbol = inTickerSymbol;
```

```
        dividend = inDividend;
    }

    //constructor with explicit return statement
    public Stock(int inNumberOfShares,String inTickerSymbol,
                                              double inDividend)
    {
        numberOfShares = inNumberOfShares;
        tickerSymbol = inTickerSymbol;
        dividend = inDividend;
        return;
    }
```

Self-Check

11. True or false: A class can have only one constructor with no formal parameters.
12. The name of the constructor is the same as the ________.

Copy Constructor

There is one special type of constructor that is worth discussing in this context. A copy constructor is a constructor that creates a copy of an existing object. Thus, a copy constructor has one formal parameter that is a reference variable of the class type. As in the case of other constructors, a copy constructor also creates a new object. Further, the newly created object is an exact replica of the object referenced by the formal parameter.

Consider the following:

```
ClassName inObject, copyObject;
```

where `ClassName` is the name of a class. Assume that `inObject` has been instantiated. That is, `inObject` references an object of the class `ClassName`. Now a Java statement of the form

```
copyObject = new ClassName(inObject);
```

creates a new object of the class `ClassName` and instantiates `copyObject`. Further, objects referenced by variables `inObject`, `copyObject` are different, whereas their corresponding instance variables have identical values.

The syntax template of a copy constructor is

```
public ClassName(ClassName obj)
{
    [statements]
}
```

Note 6.7 Java has a `clone` method that is quite difficult to implement for a beginner. Therefore, this book has intentionally avoided using the `clone` method.

Example 6.13

In this example, we continue with the class Stock. The copy constructor of the Stock class can be written as follows:

```
public Stock(Stock obj)
{
    numberOfShares = obj.numberOfShares;
                    //primitive data type; copy value
    tickerSymbol = new String(obj.tickerSymbol);
                  //object reference; use copy constructor
    dividend = obj.dividend;
              //primitive data type; copy value
}
```

Note that numberOfShares and dividend are primitive data types. Therefore, assignment statements can copy data from the formal parameter to the new object. However, if the attribute is an object reference as in the case of tickerSymbol, a copy constructor of the appropriate class (in this example, String) must be invoked.

Assuming the following statements

```
Stock stockOne, stockTwo;
stockOne = new Stock(200, "ABC", 1.60);
```

the situation at this point can be shown as follows:

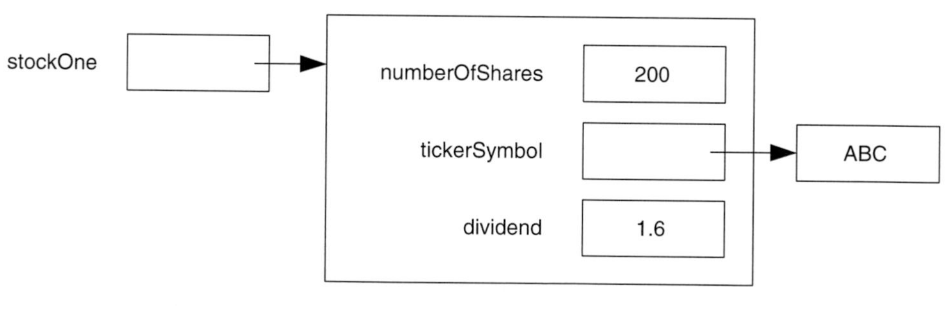

The invocation of the copy constructor in the statement

```
stockTwo = new Stock(stockOne);
```

results in creating a new stock object as shown in the following figure.

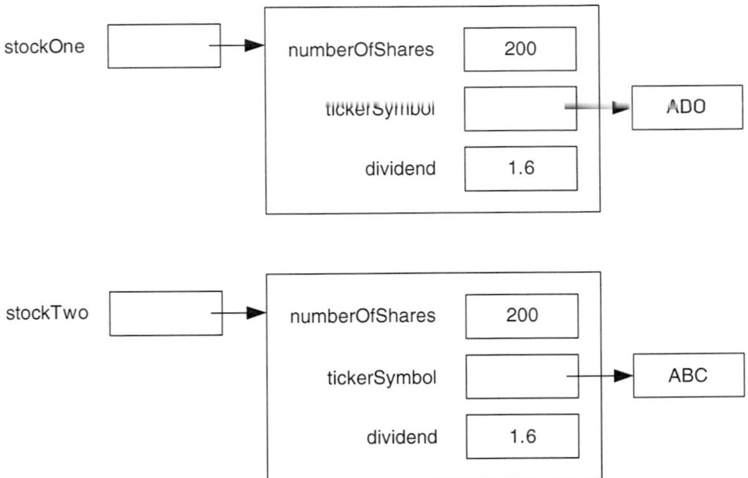

However, if the copy constructor was defined as follows:

```
public Stock(Stock obj)
{
    numberOfShares = obj.numberOfShares;
    tickerSymbol = obj.tickerSymbol;
    dividend = obj.dividend;
}
```

The result would have been quite different. Note that both `stackOne.ticker Symbol` and `stackTwo.tickerSymbol` reference the same `String` object.

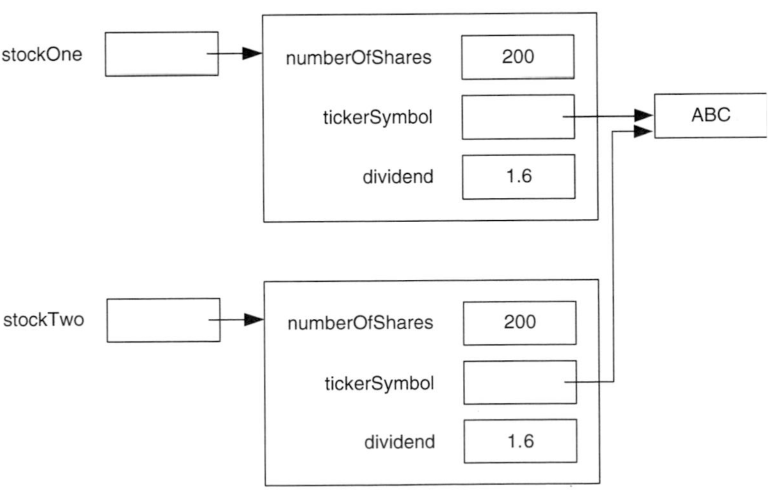

Self-Check

13. True or false: A class can have many copy constructors.
14. The copy constructor has ________ formal parameter(s).

Self-Reference

Consider the copy constructor presented in the previous subsection. A copy constructor creates a new object and initializes the newly created object using another object of the same class. For instance,

```
numberOfShares = obj.numberOfShares;
```

assigns attribute numberOfShares of the parameter obj to the attribute numberOf Shares of the newly created object. The above statement is equivalent to the following:

```
this.numberOfShares = obj.numberOfShares;
```

During the program execution, if the copy constructor is invoked as

```
stockOne = new stock(stockTwo);
```

then this contains the reference of the implicit parameter stockOne and the formal parameter obj contains the reference of explicit parameter stockTwo. Thus, this is a reference variable maintained by the compiler. As a programmer, you can access it; however, you cannot change it. The reference variable this is quite commonly known as the *self-reference*.

Example 6.14

The copy constructor of the previous subsection can be written as follows:

```
public Stock(Stock obj)
{
    this.numberOfShares = obj.numberOfShares;
    this.tickerSymbol = new String(obj.tickerSymbol);
    this.dividend = obj.dividend;
}
```

In the case of a method that is not marked as static, the self-reference this contains the reference of the implicit parameter. Thus, for example, the following method of the Stock class

```
public void setTickerSymbol(String inTickerSymbol)
{
    tickerSymbol = inTickerSymbol;
}
```

can also be written as follows:

```
public void setTickerSymbol(String inTickerSymbol)
{
    this.tickerSymbol = inTickerSymbol;
}
```

Self-Check

15. The self-reference `this` contains the reference of the ________ parameter.
16. During the method invocation `myStock.setTickerSymbol("ABC")`, the self-reference `this` contains the reference of ________.

Advanced Topic 6.1: Common Methods

In this section we present two quite useful common methods: `copy` and `equals`.

`copy` Method

An assignment operator copies the reference variable only. For example, if `StockOne` and `StockTwo` are two reference variables of the type `Stock`, the assignment statement

```
StockTwo = StockOne;
```

does not create another copy of the object referenced by `StockOne`. Rather, the reference variable `StockTwo` references the object referenced by `StockOne`. This type of copying is known as *shallow copying*. Quite often you may want to copy each attribute of `StockOne` to corresponding attribute of `StockTwo`. Such a copying is known as *deep copying*. The purpose of the `copy` method is to provide deep copying of the explicit argument to the implicit argument. For example, if `stockOne` and `StockTwo` are as shown

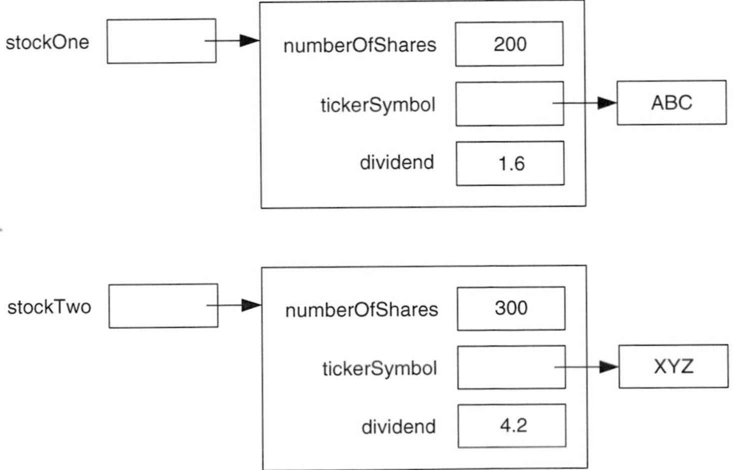

then the assignment statement

```
StockTwo = StockOne;
```

or shallow copying has the following effect

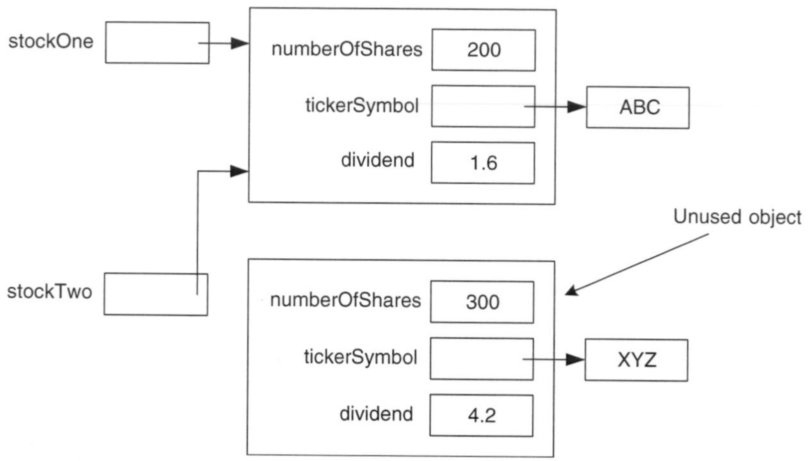

whereas deep copying has the following effect

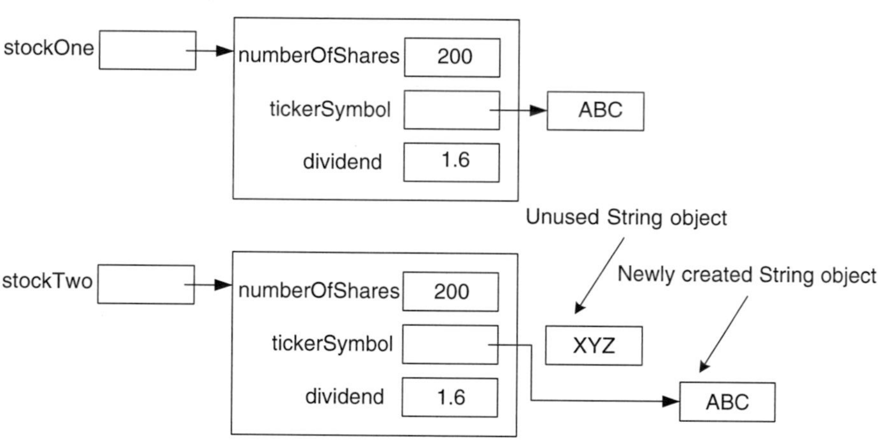

Ideally, we would even like deep copy `String` `"ABC"` onto `String` `"XYZ"`. However, Java does not have a deep `copy` method in the `String` class. Therefore, we are forced to use the copy constructor.

Table 6.6 summarizes the differences between a copy constructor and the `copy` method.

The general syntax of the `copy` method is

```
public void copy(ClassName obj)
{
    [statements]
}
```

TABLE 6.6 Copy Constructor versus Copy Method

Copy Constructor	Copy Method
A new object is created	No new object is created
The name, as in the case of any other constructor, is the same as the name of the class	The name, as in the case of any other method, is not the same as the name of the class. Throughout this book we use the name `copy`
There is no return type specification	Return type is `void`
Invoked using the `new` operator	Invoked as any other `void` method
The heading has the following syntax:	The heading has the following syntax:
`public ClassName (ClassName obj)`	`public void methodName(ClassName obj)` In this book, we use `copy` as the method name

and the `copy` method is invoked similar to other `void` methods. Thus if `objectOne` and `objectTwo` are two references of the type `ClassName`, the following statement

```
objectOne.copy(objectTwo);
```

copies `objectTwo` to `objectOne`. Note that there is no need to copy `objectTwo` to `objectOne` if both of them are already referencing the same object. Therefore, first you need to compare the references themselves. Observe that inside the method `copy`, the object reference `objectTwo` is available in the reference variable `obj` and the object reference of the implicit parameter `objectOne` is available in the self-reference `this`. Therefore, to compare the references `objectOne` to `objectTwo`, we need to compare `this` and `obj` inside the `copy` method. Thus we have the following:

```
if (this != obj)
{
    //perform copy.
}
```

The following example provides the `copy` method for the `Stock` class.

Example 6.15

The `copy` method of the `Stock` class can be written as follows:

```
/**
    Copy method
*/
public void copy(Stock obj)
{
    if (this != obj)
    {
        numberOfShares = obj.numberOfShares;
                                    // primitive data type, copy value
```

```
        tickerSymbol = new String(obj.tickerSymbol);
                       //object reference,
                       //use copy method if available
                       //otherwise use copy constructor


         dividend = obj.dividend;
                    //primitive data type, copy value
    }
}
```

Note that numberOfShares and dividend are primitive data types. Therefore, assignment statements are sufficient. However, if the attribute is an object reference as in the case of tickerSymbol, a copy method (if available) or a copy constructor of the appropriate class (in this example, String) must be invoked.

Consider the following statements:

```
Stock stockOne, stockTwo;
stockOne = new Stock(200, "ABC", 1.60);
stockTwo = new Stock(300, "XYZ", 4.20);
```

The situation at this point can be visualized as follows:

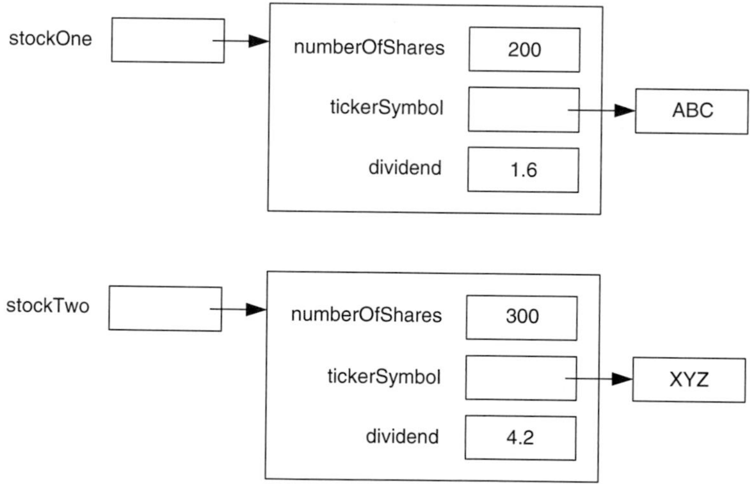

The invocation of the copy constructor in the following statement

```
stockTwo = new Stock(stockOne);
```

results in creating a new `Stock` object and leaving previously allocated memory of the object referenced by `stockTwo` as unused, as shown in the following figure:

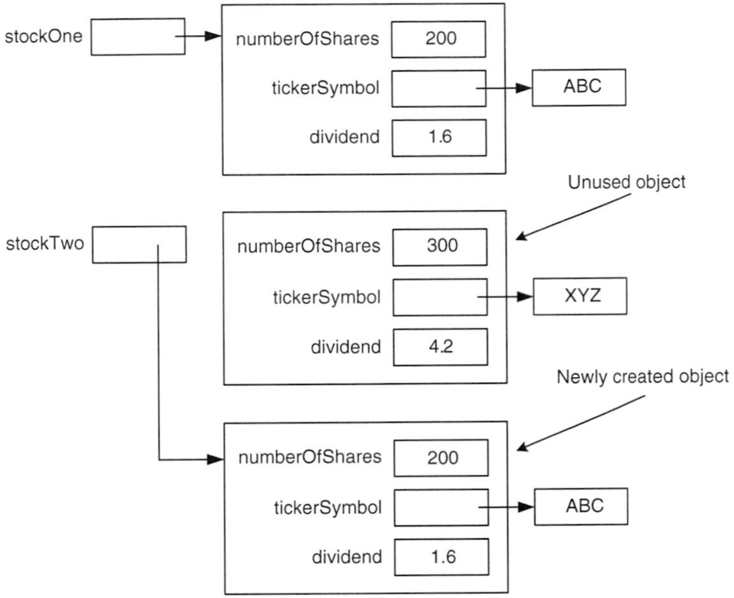

However, invoking the `copy` method as follows:

```
stockTwo.copy(stockOne);
```

results in the following:

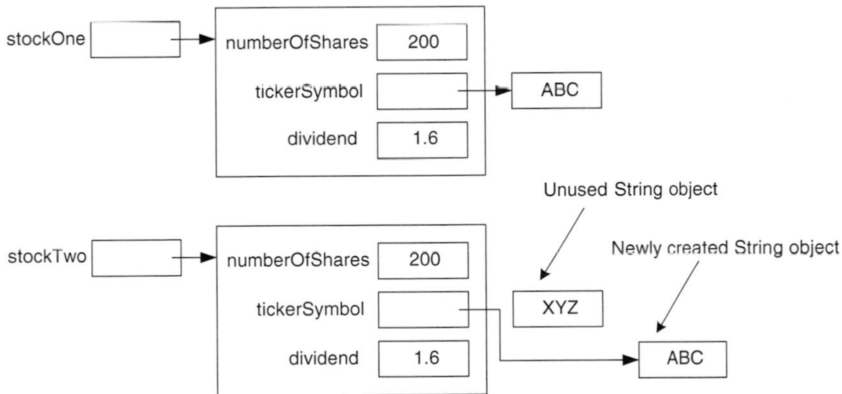

Note 6.8 A slightly different variation of the `copy` method is addressed in Exercise 10e.

`equals` Method

The `equals` method can be explained through the following *pseudo code*:

```
if (two objects have identical instance variable values)
    return true;
```

```
else
    return false;
```

The syntax template of the equals method is

```
public boolean equals(ClassName obj)
{
    [statements]
}
```

Observe that if stockOne and stockTwo are two objects of the class Stock and match is a boolean variable then the equals method is invoked as follows:

```
match = stockOne.equals(stockTwo);
```

If stockOne and stockTwo references are identical, then they both refer the same object and there is no need to compare individual instance variables. Thus, we have the following:

```
if (this != obj)
{
    //compare individual attributes
}
```

As in the case of copy constructor, you need to treat attributes differently based on whether or not they are of primitive data types. In the case of primitive data types, attributes can be compared using == operator. However, reference variables need to be compared using equals method of the appropriate class.

Example 6.16

The equals method of the Stock class can be written as follows:

```
public boolean equals(Stock obj)
{
    boolean theSame = true;
    if (this != obj)
{
        theSame
                = theSame && (numberOfShares == obj.
                                                numberOfShares);
        theSame
                = theSame && (tickerSymbol.equals(obj.
                                                tickerSymbol));
```

```
            theSame = theSame && (dividend == obj.dividend);
    }
        return theSame;
    }
```

Recall that numberOfShares and dividend are primitive data types. Thus == operator is used to compare data values for equality. Since tickerSymbol is a reference variable, the method equals of the corresponding class (in this example, String) is used to compare data values for equality.

Consider the following segment of code:

```
Stock stockOne;
Stock stockTwo;
stockOne = new Stock(200, "ABC", 1.60);
stockTwo = new Stock(stockOne);
```

In this case, we have the following:

```
stockOne.equals(stockTwo);
```

returns true and

```
(stockOne == stockTwo)
```

returns false.

You can test both copy constructor and the equals method using the following application. Note that you need to add both copy constructor and equals method to the class Stock.

```
public class StockTestingTwo
{
    public static void main (String[] args)
    {

        boolean tempBool;
        Stock stockOne;
        Stock stockTwo;

        //Invoke the constructor with int, String, double
        stockOne = new Stock(200, "ABC", 1.60);
        System.out.println("The stockOne is " + stockOne);
        System.out.println("\n");
```

```
        //Invoke the copy constructor with stockOne
        System.out.println
        ("Copy Constructor: Stock(stockOne)");
        stockTwo = new Stock(stockOne);
        System.out.println("The stockTwo is " + stockTwo);
        System.out.println("\n");


        //Invoke the equals method
        System.out.print("stockOne.equals(stockTwo) returns");
        tempBool = stockOne.equals(stockTwo);
        System.out.println(tempBool);
        System.out.println("\n");


        //Invoke the == operator
        System.out.print("(stockOne == stockTwo) is");
        tempBool = (stockOne == stockTwo);
        System.out.println(tempBool);
        System.out.println("\n");
    }
}
```

Output

```
The stockOne is 200 ABC 1.6


Copy Constructor : Stock(stockOne)
The stockTwo is 200 ABC 1.6


stockOne.equals(stockTwo) returns true


(stockOne == stockTwo) is false
```

Note 6.9 A slightly different variation of the `equals` method is addressed in Exercise 11e.

Advanced Topic 6.2: Finalizer and Garbage Collection

In Java, there is a method for performing the final cleaning up before the object goes out of scope. By cleaning up, what we really mean is releasing the resources held by the class. The most common resource is memory used by the object. Each class can have only one finalizer and its syntax template is as follows:

```
public void finalize()
{
    [statements]
}
```

Note that unlike the constructor, finalizer is a `void` method and the name of the finalizer is `finalize`. It is a common practice among Java programmers not to include the finalizer in the class definition. Instead, Java programmers rely on the garbage collection service provided by Java.

There is a simple way to mark an object as "garbage" or not useful, and make a request to the system to perform the necessary memory reclaiming. To release the memory used by an object, all that is required is to assign the keyword `null` to all reference variables that reference the object. As long as at least one variable references an object, the object is not released. In Java, `null` is a keyword.

Periodically, Java system reclaims all memory used by objects no longer referenced, through a system method `gc`. You can make a request for immediate garbage collection by invoking the `gc` method in your program. The syntax for invoking `gc` is

```
System.gc();
```

Example 6.17

Assume the following statements:

```
Stock stockOne;                //stockOne is a reference variable
stockOne = new Stock();        //a new Stock object is created
...
stockOne = null;               //The object referenced by stockOne
                               //is marked as garbage.
System.gc()                    //Garbage collection method is
                               //explicitly requested.
```

Example 6.18

Assume the following statements:

```
Stock stockOne;
Stock stockTwo;
stockOne = new Stock();        // a new Stock object is created
...
stockTwo = stockOne
```

```
stockOne = null;          // The object referenced by stockOne
                          // is marked as garbage.
```

The situation at this point can be visualized as follows:

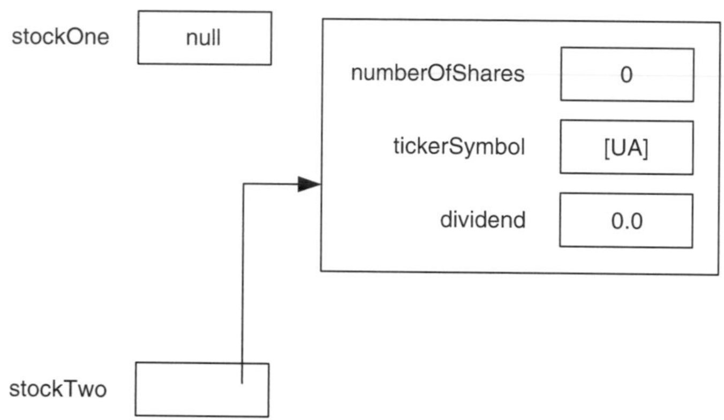

Since `stockTwo` references the object previously referenced by `stockOne`, the object will not be reclaimed by the system garbage collection method `gc`.

Advanced Topic 6.3: Class Variable

Recall that each object has its own copy of instance variables. However, there are many situations where it is worthwhile to have an attribute common to all instances of a class. For example, all humans have 23 chromosomes. This information is common to all humans unlike first name, last name that are instance specific. An attribute that is shared by all instances is called a *class variable*.

These are some of the facts on class variable:

- A class variable is shared by all instances of the class.
- A class variable exists even if there is no instance of the class ever created using a `new` operator.
- A class variable is initialized along with its declaration.
- A class variable can be modified inside the constructor.
- A class variable is marked `static`. Therefore, a class variable is also known as `static` variable or `static` field.

Example 6.19

For certain application, it is worth knowing the average salary of all employees. Such information is class specific and is not instance specific. So there is no need to assign an attribute for each employee. It makes perfect sense to keep an attribute for the entire class. Thus, we have the following attributes for the class `Employee`:

```
private static double averageSalary = 1265.43;

private String firstName;
private char middleInitial;
private String lastName;
private double salary;
```

Assume that `Employee` class has a constructor with the following header:

```
public Employee(String fName, char mIni, String lName, double
                                                          sal)
```

that creates an instance of `Employee` and assigns `fName`, `mIni`, `lName`, and `sal` as `firstName`, `middleInitial`, `lastName`, and `salary`, respectively. Note that even before creating any object of the class `Employee`, the class variable `averageSalary` exists and is initialized to `1265.43`.

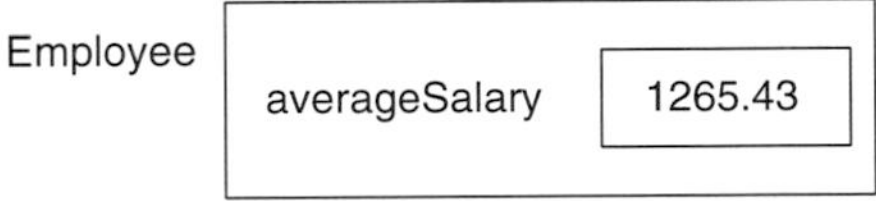

Consider the following declarations:

```
Employee empOne;
Employee empTwo;
```

These declarations create two reference variables of type `Employee`.

Employee

averageSalary 1265.43

empOne

empTwo

After executing the statement

```
empOne = new Employee("Chris", 'R', "Cox", 2468.57)
```

we have the following:

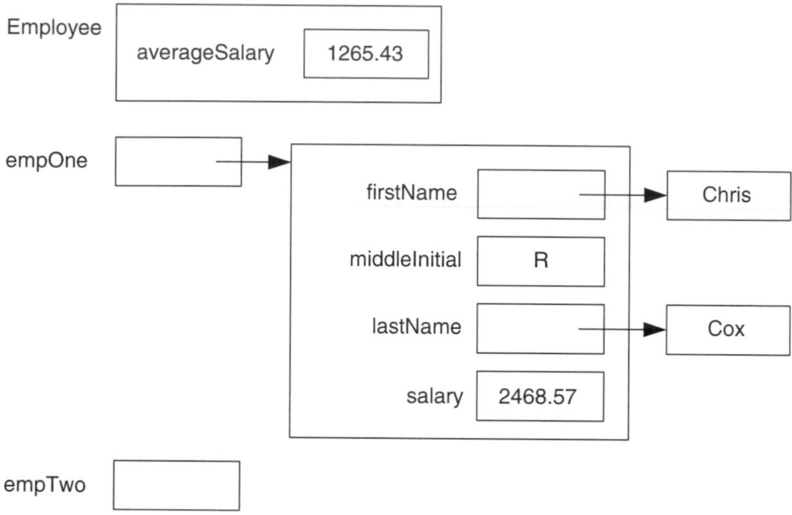

Note that the constructor has not allocated any memory location for the class variable `averageSalary`. Similarly, an assignment statement that assigns an object reference to another object reference has no impact on the class variable `averageSalary` either. Thus, after executing the assignment statement

```
empTwo = empOne;
```

the situation can be visualized as follows:

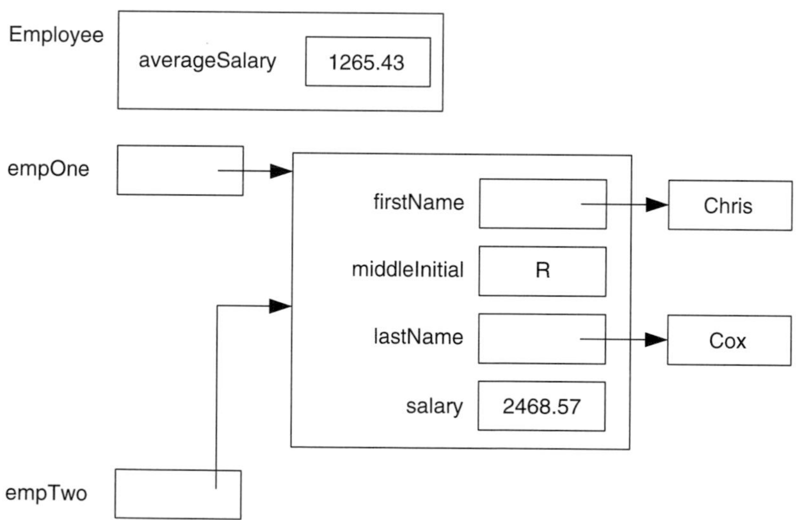

Observe that if you execute the following statements:

```
empOne = null;
empTwo = null;
```

the system will reclaim the memory used by the object. However, memory allocated to any class variable is not reclaimed.

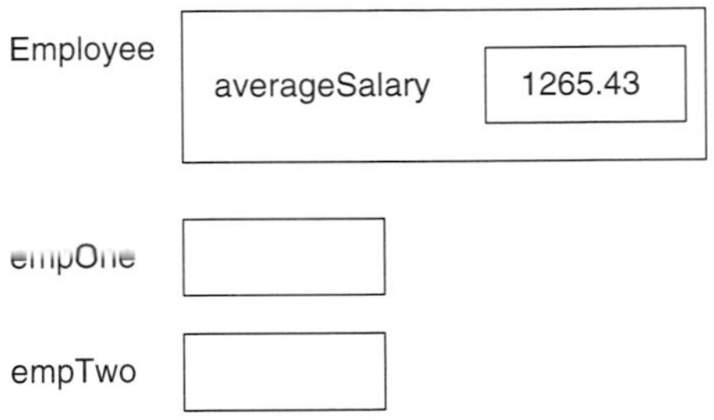

Employee averageSalary 1265.43

empOne

empTwo

static Methods

As mentioned before, a `static` method can be invoked using the class name or using a reference variable as implicit parameter. This is illustrated in Example 6.20. In this book, except in Example 6.20, we consistently invoke a `static` method using the class name.

There are two categories of `static` methods. The first category of `static` method deals with `static` variables. In Java, it is perfectly legal to write a method without marking it as `static` that modifies or accesses a class variable. Further, it is perfectly legal to modify a class variable within a constructor. Although a `static` method can only access other `static` methods or class variables, there is a definite advantage in marking a method `static`. In particular, it is possible to access and modify the `static` attribute even when there is no instance of the class. Therefore, throughout this book, any method that modifies or accesses a `static` attribute is maintained as a `static` method. We illustrate a `static` method that deals with class attributes in Example 6.20.

The second category of `static` methods is utility methods. They do not depend on any attribute of the class or an instance of the class. For instance, you have seen the class `Math` of the `java.language package` with a collection of utility methods. All those methods are defined as `static` methods and they do not depend on any attribute of the class or on any object of the class. Consider the method `pow` of the `Math` class. The method `pow` invoked as `Math.pow(x, y)` returns x^y. The method does not depend on any other information explicit or implicit. Therefore, such methods must be marked `static`. We illustrate `static` method that fall under this category in Example 6.21.

In contrast, consider the `computeSalary` method you have encountered in Chapter 4. Observe that `computeSalary` method needs the values of instance variables and thus an implicit parameter is essential. Therefore, such a method cannot be marked as `static`.

Example 6.20

In this example, we continue with Example 6.19. Note that `averageSalary` is a `static` attribute of the class. Therefore, we mark both accessor and mutator methods as `static`.

```
public class Employee
{
    private static double averageSalary = 1265.43;

    private String firstName;
    private char middleInitial;
    private String lastName;
    private double salary;

    //constructor
    public Employee(String inFirstName, char inMiddleInitial,
                    String inLastName, double inSalary)
    {
        firstName = inFirstName;
        middleInitial = inMiddleInitial;
        lastName = inLastName;
        salary = inSalary;
    }

    public static double getAverageSalary()
    {
        return averageSalary;
    }

    public static void setAverageSalary(double inAverageSalary)
    {
        averageSalary = inAverageSalary;
    }

    //toString method
    public String toString()
    {
        String str;
        str = firstName + " " + middleInitial + "." +
        lastName + " " + salary + "\nAverage Salary is " +
                                        averageSalary;
        return str;
    }
}
```

```
public class EmployeeTesting
{
    public static void main (String[] args)
    {
        Employee empOne;

        System.out.print("Average Salary before any objects
                                          are" + "instantiated is"),
        System.out.println(Employee.getAverageSalary());

        empOne = new Employee("Jack", 'E', "Smith", 3256.45);

        System.out.print("empOne.getAverageSalary() returns");
        System.out.println(empOne.getAverageSalary());
        System.out.println("empOne is " + empOne);

        Employee.setAverageSalary(2010.78);
        System.out.println("\n\nInvoked:"+
                    "Employee.setAverageSalary(2010.78)");

        System.out.print("Employee.getAverageSalary() returns");
        System.out.println(Employee.getAverageSalary());
        System.out.print("empOne.getAverageSalary() returns");
        System.out.println(empOne.getAverageSalary());
        System.out.println("empOne is " + empOne);

        empOne.setAverageSalary(2546.99);
        System.out.println("\n\nInvoked: empOne.
                                    setAverageSalary(2010.78)");

        System.out.print("Employee.getAverageSalary() returns");
        System.out.println(Employee.getAverageSalary());
        System.out.print("empOne.getAverageSalary() returns");
        System.out.println(empOne.getAverageSalary());
        System.out.println("empOne is " + empOne);

        empOne = null;
        System.out.println("\n\nNo objects of the type
                                              Employee exists");
        System.out.print("Employee.getAverageSalary() returns");
        System.out.println(Employee.getAverageSalary());
    }
}
```

Output

```
Average Salary before any objects are instantiated is 1265.43
empOne.getAverageSalary() returns 1265.43
empOne is Jack E. Smith 3256.45
Average Salary is 1265.43


Invoked: Employee.setAverageSalary(2010.78)
Employee.getAverageSalary() returns 2010.78
empOne.getAverageSalary() returns 2010.78
empOne is Jack E. Smith 3256.45
Average Salary is 2010.78


Invoked: empOne.setAverageSalary(2010.78)
Employee.getAverageSalary() returns 2546.99
empOne.getAverageSalary() returns 2546.99
empOne is Jack E. Smith 3256.45
Average Salary is 2546.99


No objects of the type Employee exists
Employee.getAverageSalary() returns 2546.99
```

Example 6.21

In this example, we create a class with a `static` method and illustrate its usage. The `static` method returns the greatest common divisor (gcd) of two integers. Since the gcd depends only on two integers, gcd must be defined as a `static` method.

Some of you may not remember the gcd and Euclid's algorithm to compute gcd. Here is a quick overview. Consider the integers 24 and 30. The divisors of 24 are 1, 2, 3, 4, 6, 8, and 12. Similarly, the divisors of 30 are 1, 2, 3, 5, 6, 10, and 15. Therefore, 6 is the gcd of 24 and 30.

An *algorithm* is a specific set of instructions to be performed to solve a problem with the requirement that it terminates for every set of data values that satisfies all preconditions. There is a simple algorithm to compute the gcd that appeared in 300 BC in Euclid's *Elements*. Probably this algorithm may be the oldest one still in use. The Euclid's algorithm proceeds as follows. Between 30 and 24, 30 is the larger one. So divide 30 by 24. Note that it yields a remainder 6. The main observation behind Euclid's algorithm is the fact that gcd of 30 and 24 is the same as the gcd of

24 and 6. Therefore, next you attempt to determine the gcd of 24 and 6. Note that 6 divides 24. Therefore, remainder is 0. Thus, gcd(30, 24) = gcd(24, 6) = gcd(6, 0). Now, gcd of 6 and 0 is 6. The algorithm terminates by determining 6 as the gcd of 30 and 24. These steps can be formally stated as follows:

```
Euclid's Algorithm (version 1)
Step 1: Take larger number as firstNumber;
Step 2: Take smaller number as secondNumber;
Step 3: If secondNumber is zero then
           return firstNumber;
        else
        {
            thirdNumber = firstNumber % secondNumber;
            firstNumber = secondNumber;
            secondNumber = thirdNumber;
            repeat Step 3
        }
```

From Step 3, it may be observed that you need to repeat certain steps as long as the secondNumber is not zero. Therefore, a while loop is required to implement the algorithm. Eventually, as the secondNumber becomes zero, the firstNumber is the gcd. Thus, we can rewrite the above algorithm as follows:

```
Euclid's Algorithm (version 2)
Input: a, b ; two integers.

Step 1: Preprocessing
     1.1 if (a < 0) then a = -a; // make number positive
     1.2 if (b < 0) then b = -b; // make number positive

Step 2: Initialization
     if (a > b) then
     {
         firstNumber = a;  // firstNumber > secondNumber
         secondNumber = b;
     }
     else
     {
         firstNumber = b;  // firstNumber > secondNumber
         secondNumber = a;
     }
```

```
Step 3: Find gcd
    while (secondNumber > 0)
    {
        thirdNumber = firstNumber % secondNumber;
        firstNumber = secondNumber;
        secondNumber = thirdNumber;
    }
Step 4: Return gcd
    return firstNumber;
```

The gcd is implemented as a static method of the class Utility.

```
/**
    A set of common utility static methods
    The list of method(s): gcd
*/
public class Utility
{
    /**
        Computes gcd of two integers
        @param a one of the integers
        @param b the second integer
        @return gcd of a and b
    */
    public static int gcd(int a, int b)
    {
        int firstNumber;
        int secondNumber;
        int thirdNumber;
        if (a < 0)
              a = -a;
        if (b < 0)
              b = -b;
        if (a > b)
            {
                firstNumber = a;
                secondNumber = b;
            }
        else
        {
            firstNumber = b;
            secondNumber = a;
        }
```

```
        while (secondNumber > 0)
        {
             thirdNumber = firstNumber % secondNumber;
             firstNumber = secondNumber;
             secondNumber = thirdNumber;
        }
        return firstNumber;
    }
}

/**
    An application to test the gcd method
*/
public class GcdTesting
{
    public static void main (String[] args)
    {

        int numOne;
        int numTwo;

        int numThree;

        numOne = 91;
        numTwo = 98;
        numThree = Utility.gcd(numOne, numTwo);

        System.out.println("numOne = " + numOne);
        System.out.println("numTwo = " + numTwo);
        System.out.println("numThree = Utility.gcd(numOne,
                                                 numTwo)");
        System.out.println("numThree = " + numThree);

        numTwo = 48;
        numThree = Utility.gcd(74, -numTwo);

        System.out.println("\nnumTwo = " + numTwo);
        System.out.println
        ("numThree = Utility.gcd(74, -numTwo)");
        System.out.println("numThree = " + numThree);

        numOne = -48;
        numThree = Utility.gcd(numOne, 21);
```

```
            System.out.println("\nnumOne = " + numOne);
            System.out.println("numThree = Utility.gcd(numOne, 21)");
            System.out.println("numThree = " + numThree);

            numThree = Utility.gcd(0, 17 + 8);
            System.out.println
            ("\nnumThree = Utility.gcd(0, 17 + 8)");
            System.out.println("numThree = " + numThree);

            numThree = Utility.gcd(-73, 0);
            System.out.println("\nnumThree = Utility.gcd(-73, 0)");
            System.out.println("numThree = " + numThree);
    }
}
```

Output

```
numOne = 91
numTwo = 98
numThree = Utility.gcd(numOne, numTwo)
numThree = 7

numTwo = 48
numThree = Utility.gcd(74, -numTwo)
numThree = 2

numOne = -48
numThree = Utility.gcd(numOne, 21)
numThree = 3

numThree = Utility.gcd(0, 17 + 8)
numThree = 25

numThree = Utility.gcd(-73, 0)
numThree = 73
```

Advanced Topic 6.4: Creating and Using Packages

This section explains the creation and usage of user-defined classes and packages. There are two ways to use a class.

Option 1

The simplest way is to keep the class you want to use in the same directory as the application program. In this case, you need not import the class. In fact you have been using classes created by you this way.

Option 2

This is the most general option. In this case, you create a package. Once the package is created, you can import the classes in various applications. However, creating and using a package involves five distinct steps. The formal description of each of these five steps along with illustrative examples are presented next.

Step 1

In this step you specify the package name you would like to use. For example, the package name chosen for the package created for this chapter is given the following name:

```
edu.creighton.cs1.ch06
```

In this package name, `edu.creighton` is in fact the author's domain name `creighton.edu` written backward. Even though you can choose any name, it is a well-accepted practice among Java programmers to start their package name with domain name written backward. Next `cs1` stands for Java book for CS 1 course, that is, this book itself and finally, `ch06` represents chapter 6 of the book. This convention allows the author to group the classes based on book and chapters of the book.

The general syntax for specifying the package is as follows:

```
package packageName;

//class definition
```

Example 6.22

This example illustrates the specification of the package name `edu.creighton.cs1.ch06` in the java file `Utility.java`.

```
package edu.creighton.cs1.ch06;

public class Utility
{
    public static int gcd(int a, int b)
    {
        int firstNumber;
        int secondNumber;
        int thirdNumber;

        if (a < 0)
            a = -a;
        if (b < 0)
            b = -b;

        if (a > b)
```

```
            {
                firstNumber = a;
                secondNumber = b;
            }
            else
            {
                firstNumber = b;
                secondNumber = a;
            }

            while (secondNumber > 0)
            {
                thirdNumber = firstNumber % secondNumber;
                firstNumber = secondNumber;
                secondNumber = thirdNumber;
            }
           return firstNumber;
       }
   }
```

Step 2

In this step, you will actually create a package and place it in a directory of your choice. In the author's computer, the following directory

```
C:\Program Files\Java\j2sdk1.6.0_02\jre\lib
```

exists. Assume that we want to put all packages in the following directory:

```
C:\Program Files\Java\j2sdk1.6.0_02\jre\lib\classes
```

Then, you must create the subdirectory classes first. Once the directory is created, you can create the package using the following command:

```
javac -d C:\Progra~1\Java\j2sdk1.6.0 _ 02\jre\lib\classes Utility.java
```

Step 3

The purpose of this step is to update the CLASSPATH environment variable. Your CLASSPATH must include the following:

```
.;C:\Progra~1\Java\j2sdk1.6.0_02\jre\lib\classes
```

Items in the CLASSPATH are separated by semicolons (for windows and colons for unix). The very first period stands for current directory. Thus, the above String indicates the fact that classes can be found either in the current directory or in C:\Progra~1\Java\j2sdk1.6.0 _ 02\jre\lib\classes.

Step 4

To use a class in one of the packages created, you must use appropriate `import` statement. For instance, to use class `Utility` of the package `edu.creighton.cs1.ch06`, you may use either of the following `import` statements:

```
import edu.creighton.cs1.ch06.Utility;

import edu.creighton.cs1.ch06.*;
```

Step 5

In this step you compile your application specifying location of the `package`. For instance, to compile a Java file `GcdTesting.java` you need the following command:

```
javac -classpath C:\Progra~1\Java\j2sdk1.6.0_02\jre\lib\classes
                                                    GcdTesting.java
```

Note that the above statement can be replaced by

```
javac GcdTesting.java
```

provided, in Step 4, you have the following `import` statement

```
import edu.creighton.cs1.ch06.Utility;
```

that explicitly identifies the class.

CASE STUDY 6.1: FRACTION CALCULATOR

In this section we present a fraction calculator to illustrate the various concepts presented in this text book so far. The fraction calculator depends on the Fraction class that maintains a fraction as a pair of integers with gcd value 1.

The fraction calculator program has two menus. The top-level menu takes you to either first or second operand set and display menu. The top-level menu also allows you to perform basic calculations such as add, subtract, multiply, and divide. The second-level menu allows you to set an operand and display it.

In the case study in Chapter 5, reader is challenged to create the test program. In this case study, the reader is challenged to read and understand the code written by someone else. Although you would like to write the code yourself, more often than not, you may be forced to read, understand, and maintain the code written by someone else. Therefore, the code in this section is intentionally left with minimum comments. Further, the UML 2 diagrams are also not presented for the same reason. In fact, we do not even provide a sample run to simulate the real life situation. Of course, an interested reader can easily perform a test run and learn about the program.

```
import java.util.Scanner;

/**
    Maintains a fraction as pair of integers
```

```
*/
public class Fraction
{
    private int numer;
    private int denom;

    /**
        Constructor with no arguments; creates a 0
    */
    public Fraction()
    {
        setFraction(0, 1);
    }

    /**
        Constructor that initializes numerator and denominator
        @param numerator
        @param denominator
    */
    public Fraction(int inNumer, int inDenom)
    {
        setFraction(inNumer, inDenom);
    }

    /**
        Constructor; integer to fraction
        @param numerator; denominator is set to 1
    */
    public Fraction(int inNumer)
    {
        setFraction(inNumer);
    }

    /**
        Constructor; real to fraction
        @param a real number
    */
    public Fraction(double inWholeNumber)
    {
        setFraction(inWholeNumber);
    }
```

```
/**
    Adds implicit and explicit parameters
    @param fraction to be added
    @return the sum of implicit and explicit parameters
*/
public Fraction add(Fraction inFraction)
{
    Fraction returnValue = new Fraction(0,1);

    returnValue.numer = numer * inFraction.denom
                               + denom * inFraction.numer;
    returnValue.denom = denom * inFraction.denom;
    returnValue.normalize();
    return returnValue;

}

/**
    Subtracts explicit parameter from the implicit
                                       parameter
    @param fraction to subtract
    @return the difference of implicit and explicit
                                       parameters
*/
public Fraction sub(Fraction inFraction)
{
    Fraction returnValue = new Fraction(0,1);

    returnValue.numer = numer * inFraction.denom - denom *
                                       inFraction.numer;
    returnValue.denom = denom * inFraction.denom;
    returnValue.normalize();
    return returnValue;
}

/**
    Multiplies implicit and explicit parameters
    @param multiplier
    @return the product of implicit and explicit
                                    parameters
*/
public Fraction mul(Fraction inFraction)
{
    Fraction returnValue = new Fraction(0,1);
```

```
        returnValue.numer = numer * inFraction.numer;
        returnValue.denom = denom * inFraction.denom;

        returnValue.normalize();
        return returnValue;
    }

    /**
        Divides implicit parameter by explicit parameter
        @param divisor
        @return implicit divided by explicit parameter
    */
    public Fraction div(Fraction inFraction)
    {
        Fraction returnValue = new Fraction(0, 1);

        returnValue.numer = numer * inFraction.denom;
        returnValue.denom = denom * inFraction.numer;

        returnValue.normalize();
        return returnValue;
    }

    /**
        Computes the additive inverse    of the implicit
                                              parameter
    */
    public Fraction minus()
    {
        Fraction returnValue = new Fraction(0, 1);

        returnValue.numer = - numer;
        returnValue.denom = denom;

        return returnValue;
    }

    /**
        Computes the multiplicative inverse of the implicit
                                                  parameter
    */
```

```
public Fraction inverse()
{
    Fraction returnValue = new Fraction(0, 1);
    if (numer > 0)
    {
         returnValue.denom = numer;
         returnValue.numer = denom;
    }
    else
    {
         returnValue.denom = - numer;
         returnValue.numer = - denom;
    }
    return returnValue;
}


/**
    Checks whether or not the explicit parameter is an
                                            integer value.
*/


private boolean isInteger(double inValue)
{
    if (Math.round(inValue) * 10 == Math.round(10 *
                                              inValue))
         return true;
    else
        return false;
}


/**
    Normalize the implicit parameter
*/
private void normalize()
{
    int gcdValue;


    if (denom < 0)
```

```
        {
            numer = -numer;
            denom = -denom;
        }
        gcdValue = Utility.gcd(Math.abs(numer), denom);
        numer = numer/gcdValue;
        denom = denom/gcdValue;
    }
    /**
        Copy method
    */
    public void setFraction(Fraction inFraction)
    {
        numer = inFraction.numer;
        denom = inFraction.denom;
    }


    /**
        Sets both numerator and denominator
        @param numerator
        @param denominator
    */
    public void setFraction(int inNumer, int inDenom)
    {
        int gcdValue;

        numer = inNumer;
        denom = inDenom;

        if (denom == 0)
              denom = 1;
        this.normalize();
    }


    /**
        Sets both numerator; denominator is 1.
        @param numerator an integer
    */
    public void setFraction(int inNumer)
```

```
{
    numer = inNumer;
    denom = 1;
}

/**
    Sets both numerator and denominator
    @param a real value, get the best approximate fraction
                                                   possible
*/
public void setFraction(double inWholeNumber)
{
    double tempNumer;
    int tempDenom;

    tempNumer = inWholeNumber;
    tempDenom = 1;
    while (!isInteger(tempNumer))
    {
        tempNumer = tempNumer * 10;
        tempDenom = tempDenom * 10;
    }
    numer = (int) tempNumer;
    denom = tempDenom;
    this.normalize();
}

/**
    Returns the real value
    @return real value obtained by division.
*/
public double valueOf()
{
    return numer/denom;
}

/**
    Accessor method for numerator
    @return numerator
*/
public double getNumer()
```

```
    {
        return numer;
    }

    /**
        Accessor method for denominator
        @return denominator
    */
    public double getDenom()
    {
        return denom;
    }

    /**
        toString method for neat printing
        @return string representation
    */
    public String toString()
    {
        String returnStr;
        int wholeNum;
        int tempNumer;
        boolean negative;

        if (numer < 0)
        {
            tempNumer = -numer;
            negative = true;
        }
        else
        {
            tempNumer = numer;
            negative = false;
        }
        wholeNum = tempNumer/denom;
        tempNumer = tempNumer % denom;

        if (negative)
        {
            if (wholeNum == 0)
                tempNumer = -tempNumer;
```

```
            else
                wholeNum = -wholeNum;
        }
        if (tempNumer == 0)

            returnStr = wholeNum + " ";

        else if (wholeNum == 0)

            returnStr = " [" + tempNumer + " / " +
                                    denom + "]";
        else

            returnStr = wholeNum + "[" + tempNumer + " / " +
                                             denom + "]";

        return returnStr;
    }
}


import java.io.*;
import java.util.*;

//import edu.creighton.cs1.ch06.*;

/**
    Fraction Calculator
*/
public class FractionCalculator
{
    static Scanner scannedInfo = new Scanner(System.in);

    public static void main(String[] args) throws IOException
    {
        Fraction firstOp, secondOp, result;
         firstOp = new Fraction(0,1);
         secondOp = new Fraction(0,1);
         result = new Fraction(0,1);

        int topSelection; //holds top level the selection
```

```
System.out.println("Welcome to Fraction Calculator\n");
displayTopMenu();
topSelection = scannedInfo.nextInt();

while(topSelection != 10)
{
    switch(topSelection)
    {
        case 0:
            System.out.println("\n\n\t\t\tDisplay or Set
                                    1st Operand\n\n");
            setOperand(firstOp,result);
            break;
        case 1:
            System.out.println("\n\n\t\t\tDisplay or
                                Set 2nd Operand\n\n");
            setOperand(secondOp, result);
            break;
        case 2:
            result = firstOp.minus();
            System.out.println("The result is " + result);
            break;
        case 3:
            result = secondOp.minus();
            System.out.println("The result is " + result);
            break;
        case 4:
            result = firstOp.inverse();
            System.out.println("The result is " + result);
            break;
        case 5:
            result = secondOp.inverse();
            System.out.println("The result is " + result);
            break;
        case 6:
            result = firstOp.add(secondOp);
            System.out.println("The result is " + result);
            break;
        case 7:
            result = firstOp.sub(secondOp);
            System.out.println("The result is " + result);
            break;
```

```
            case 8:
                result = firstOp.mul(secondOp);
                System.out.println("The result is " + result);
                break;
            case 9:
                result = firstOp.div(secondOp);
                System.out.println("The result is " + result);
                break;
            case 10:
                System.out.println("Good Bye");
                return;
            default:
                System.out.println("Select a number between 0
                                                    and 10");
        }//end switch

        displayTopMenu();
        topSelection = scannedInfo.nextInt();
        }//end while (choice != 10)
    }//end main

/**
    Displays the top menu
*/
private static void displayTopMenu()
{
    System.out.println("\t\t\tThe Main menu");
    System.out.println("Set the operands and specify the
                                        operation\n");
    System.out.println("Enter 0 to set or display the first
                                               operand");
    System.out.println("Enter 1 to set or display the
                                   second operand");
    System.out.println("Enter 2 to compute negative first
                                              operand");
    System.out.println("Enter 3 to compute negative second
                                               operand");
    System.out.println("Enter 4 to inverse first operand");
    System.out.println("Enter 5 to inverse the second
                                         operand");
    System.out.println("Enter 6 add : operand one + operand
                                                  two");
```

```
        System.out.println("Enter 7 subtract : operand one -
                                                operand two");
        System.out.println("Enter 8 multiply : operand one *
                                                operand two");
        System.out.println("Enter 9 divide : operand one /
                                              operand two");
        System.out.println("Enter 10 to exit");
    }//end displayTopMenu

    /**
        Displays operand display and set menu
    */
    private static void displayOpMenu()
    {
        System.out.println("Operand display and set menu");
        System.out.println("Enter 0 Whole number");
        System.out.println("Enter 1 Numerator, Denominator");
        System.out.println("Enter 2 Decimal value");
        System.out.println("Enter 3 Use the result of last
                                                    operation");
        System.out.println("Enter 4 Display the result of last
                                                    operation");
        System.out.println("Enter 5 Display the operand");
        System.out.println("Enter 6 Exit this menu");
    }//end displayFirstOpMenu


    /**
        Helper method to set an operand
        @param a new fraction
        @param result of the last computation
    */
    private static void setOperand(Fraction operand, Fraction
                                                    lastResult)
    {
        int wholeNumber, numerator, denominator;
        double value;

        int secondLevelSelection;//holds top level the selection

        displayOpMenu();
        secondLevelSelection = scannedInfo.nextInt();
```

```
    while(secondLevelSelection != 6)
    {
        switch(secondLevelSelection)
        {
            case 0:
                System.out.println("Enter the whole number");
                wholeNumber = scannedInfo.nextInt();
                operand.setFraction(wholeNumber);
                return;
            case 1:
                System.out.println("Enter Numerator, Denominator");
                numerator = scannedInfo.nextInt();
                denominator = scannedInfo.nextInt();
                operand.setFraction(numerator,denominator);
                return;
            case 2:
                System.out.println("Enter a decimal value");
                value = scannedInfo.nextDouble();
                operand.setFraction(value);
                return;
            case 3:
                operand.setFraction(lastResult);
                return;
            case 4:
                System.out.println("The last result is " +
                                              lastResult);
                break;
            case 5:
               System.out.println("The operand is " + operand);
                break;
            case 6:
                return;
            default:
                System.out.println("Select a number between 0
                                                    and 6");
        }//end switch
        displayOpMenu();
        secondLevelSelection = scannedInfo.nextInt();
    }//end while(secondLevelSelection != 5)
}// end setOperand

}
```

REVIEW

1. In order for an object X to invoke a method (or to request a service) of another object Y, two conditions need to be met: The method must be public; the object Y must be accessible to the object X.
2. Methods can be classified into two categories: methods with no parameter and methods with parameters.
3. Methods can be classified into two categories: void methods and value returning methods.
4. It is a syntax error to invoke a method using the syntax ClassName.methodName unless it is static.
5. Java API is grouped under different units called packages.
6. The signature of a method consists of the name of the method along with the list of all data types in the order they appear in the formal parameter list of the method.
7. The signature of a method distinguishes one method from the other.
8. The programming language feature that allows the programmer to create more than one method within a class having the same name is known as method overloading. Java allows method overloading.
9. It is a compilation error to have two methods within a class having identical signatures.
10. The return type is not part of the signature.
11. Parameter passing mechanism in Java is known as call by value. That is, at the beginning of a method invocation, actual parameters are copied to formal parameters.
12. A change in the formal parameter value during method execution has no impact on the value of the actual parameter.
13. In the case of parameter passing, actual parameters play the role similar to the expression on the right-hand side and formal parameters play the role of the variable on the left-hand side of an assignment statement.
14. The call by value of a reference variable is exactly the same. The value of an object reference in the actual parameter is copied to formal parameter during the parameter passing. Thus, the object reference alone is copied and any changes made in the object by the method are persistent.
15. The name of the constructor is the same as the class name.
16. It is legal to overload a constructor. The compiler selects a constructor based on the signature (as is the case for a method).
17. The Java compiler, by default, will provide a constructor if no constructor is included in the class definition. Such a constructor is known as default constructor. The default constructor has no parameters.
18. Once a constructor is included as part of the class definition, Java compiler does not provide the default constructor.

19. The self-reference `this` is maintained by the compiler. As a programmer, you can access it; however, you cannot change it.
20. A class variable is shared by all instances of the class and exists even if there is no instance of the class ever created.
21. A class variable is initialized along with its declaration and can be modified inside the constructor.
22. A class variable is marked `static`.
23. A `static` method can be invoked using the class name or using a reference variable as implicit parameter.
24. It is perfectly legal to write a method without marking it as `static` that modifies or accesses a class variable.

EXERCISES

1. Mark the following statements as true or false.
   a. A `static` method has no explicit parameter.
   b. A `static` method must be invoked using the class name.
   c. Only a `static` method can modify a class variable.
   d. Only a `static` method can access a class variable.
   e. Every method need not be `public`.
   f. In order for a class variable to exist there need not be any instances.
   g. Every method has a signature.
   h. The return type is part of the signature.
   i. Method overloading refers to the use of the same method name in two different classes.
   j. It is not an error to have two methods with identical signature in two different classes.
   k. During the execution of a method, any change in formal parameter will be reflected in the actual parameter.
   l. A constructor has `void` as its return type.
   m. A constructor can have `return` statement.
   n. A default constructor has no formal parameters.
   o. Throughout the execution of a method the self-reference remains the same.
   p. If == returns `true`, then `equals` will also return `true`.
   q. If == returns `false`, then `equals` will also return `false`.
   r. If `equals` returns `true`, then == will also return `true`.
   s. If `equals` returns `false`, then == will also return `false`.

t. A class variable can be initialized in a constructor.

u. A class variable can be updated in a constructor.

v. Since `random` method of the class `Math` is invoked as `Math.random()`, it is a `static` method.

w. A class can have two methods that are identical in there heading except in their return type specification.

2. Check whether or not the signatures match. If not, explain. Assume the following declarations: **short** `sh, sp;` **int** `i, j;` **long** `lone, ltwo;` **double** `x, y;` **char** `c1;` `String str1;`

a. **void** `trial(`**int** `x,` **int** `y)` and `trial(sh, 5)`

b. **void** `trial(`**int** `x,` **int** `y)` and `trial(sh, lone)`

c. **void** `try(`**int** `x,` **double** `y,` **char** `c)` and `try(10, 2.5, c1)`

d. **void** `try(`**int** `x,` **double** `y,` **char** `c)` and `try(10.0, 2, str1)`

e. **void** `try(`**int** `x,` **double** `y,` **char** `c)` and `try(10.0, 2)`

f. **void** `try(`**int** `x,` **double** `y,` **char** `c)` and `try(i,j, '<')`

g. **void** `try(`**int** `x,` **double** `y,` **char** `c)` and `try(i,j, q)`

h. **void** `try(`**int** `x,` **double** `y,` **char** `c)` and `try(ltwo,j, '<')`

i. **void** `try(`**int** `x,` **double** `y,` **char** `c)` and `try(2, 17.5, "J")`

j. **void** `try(`**int** `x,` **double** `y,` **char** `c1)` and `try(x, y, c1)`

k. **double** `track()` and `track(10.5)`

l. **double** `track()` and `track(7)`

m. **long** `trace(String s,` **int** `i)` and `trace("r", sh)`

n. **long** `trace(String s,` **int** `i)` and `trace( , 4)`

o. **long** `trace(String s,` **int** `i)` and `trace("Q", 8)`

p. **long** `trace(String s,` **int** `i)` and `trace(8)`

q. **long** `trace(String s,` **int** `i)` and `trace("Take a look")`

r. **long** `trace(String s,` **int** `i)` and `trace("Take "+ "a look", 7)`

s. **long** `trace(String s,` **int** `i)` and `trace("Hello ", 7 - 3)`

t. **long** `trace(String s,` **int** `i)` and `trace("Welcome ", sh * 5 + sp)`

3. Write a method or a constructor heading as specified:

a. A method named `cashflow` that returns a `double` and has three formal parameters of the following types: `int`, `double`, `char`.

b. A method named `isFull` that returns a `boolean` and has three formal parameters of the following types: `boolean`, `double`, `String`.

c. A method named `countVal` that returns an `int` and has three formal parameters of the following types: `boolean`, `long`, `String`.

d. A method named `displayInfo` that returns nothing and has two formal parameters of the following types: `int`, `String`.

e. A method named `getStatus` that returns a `boolean` and has no formal parameters.

f. A method named `getVal` that returns nothing and has no formal parameters.

g. A default constructor for the class `Student`.

h. A constructor for the class `Student` having three formal parameters of the following types: `boolean, double, String`.

i. A copy constructor for the `Student` class.

j. An `equals` method for the `Student` class.

4. Consider the following method:

```
public static int modify(int a, b)
{
        int one, two;

        one = a + b;
        two = a - b;

        if (one > two)
        {
            a = one;
            b = two;
        }
        else
        {
            a = two;
            b = one;
        }
        return a/b;
}
```

a. Correct any syntax errors. For the rest of this question, assume that syntax errors are corrected.

b. What will be value returned if the first actual parameter is 7 and second actual parameter is 12.

c. What will be value returned if the first actual parameter is 12.0 and second actual parameter is 7.

d. If the static method is a member of the class `DataValues`, how will you invoke it with the first actual parameter 7 and the second actual parameter 12.

e. Assuming the declarations `int x` = 20, `y` = 40; if `modify` is invoked with the first actual parameter `x` and second actual parameter `y`, what will be the values of `x` and `y` once modify is completed.

5. Consider the following method:

```
static public trial(int a, b; double c);
{
    if (a > b)
        c = a + b / c;
    else
        c = a - b / c;
    return c;
}
```

a. Correct any syntax errors. For the rest of this question, assume that syntax errors are corrected.

b. What will be the value returned if the first actual parameter is 1, the second actual parameter is 2, and the third actual parameter is 3, respectively.

c. What will be the value returned if the first actual parameter is 7, the second actual parameter is 3, and the third actual parameter is 4.0, respectively.

d. If the `static` method is a member of the class `Useful`, how will you invoke it with the first actual parameter 9, the second actual parameter 4, and the third actual parameter 12.5.

e. Assuming the declarations `int x` = 20, `y` = 40; `double z` = 18.2; if trial is invoked with the first actual parameter `x`, the second actual parameter `y` and the third actual parameter `z`, what will be the values of `x, y`, and `z` once the trial is completed.

6. Consider the following method:

```
public static double testing(int a; int b)
{
    int one, two, three;

    one = a + b - c;
    two = one + 10;

    if (a + b < c)
    {
        c = one + 7;
    return c;
    }
    else
```

```
    {
        c = two + 7;
    return c;
    }
}
```

a. Can the data member c and the method testing be members of the same class? If such a c is not allowed, assume that the heading

```
public static double testing(int a; int b)
```

is replaced by

```
public static double testing(int a; int b; int c)
```

If such a c is allowed, clearly explain if there are any restrictions. In either case, assume c is 0.

b. Correct any syntax errors. For the rest of this question, assume that syntax errors are corrected.

c. What will be the value returned if actual parameters are 1 and 3?

d. What will be the value returned if the first actual parameters are 3 and 2?

e. If the static method is a member of the class DataValues, how will you invoke it with the first actual parameter 4 and the third actual parameter 5?

f. Assuming the declarations int x = 2, y = 6, if testing is invoked with the first actual parameter x and the second actual parameter y, what will be the value of c once the method is completed.

7. Consider the following method with no formal parameters:

```
public void int nextValue()
{

    if (n % == 2)
    {
        n = n/2;
    }
    else
    {
        n = 3 * n + 1;
    }

}
```

a. Observe that n is an instance variable of the class that has nextValue as a method. Correct any syntax errors. For the rest of this question, assume that syntax errors are corrected.

b. What will be the value returned by the method `nextValue` (assuming that method returned `n`), if `n` is 10?

c. What will be the value returned by the method `nextValue` (assuming that method returned `n`), if `n` is 7?

8. Consider a class `Name` with two instance variables `fName` and `lName` of the type `String`.

    a. List the signatures of at least four different constructors.

    b. Is it possible to have two different constructors, each with exactly one formal parameter of the type `String` such that one of them sets `fName` as the actual parameter and the other constructor sets `lName` as the actual parameter.

    c. Write a copy constructor.

    d. Write an `equals` method.

    e. Write a `compareTo` method similar to the `compareTo` method of the `String` class.

9. Write Java statements that accomplish the following tasks for a class `Item`:

    a. Declare two instance variables `height` and `weight` of type `double`.

    b. Write at least three different constructors.

    c. Write a copy constructor.

    d. Write an `equals` method.

    e. Write a `compareTo` method (assume that `Item` objects are ordered based on the value of `height` times `weight`).

10. Consider the `copy` method presented in this book.

    a. Identify the header of the method.

    b. Identify the body.

    c. What are the explicit parameters?

    d. What is the signature of the method?

    e. If the `copy` method presented in this book is modified by changing the return type to `Stock`, what other changes you need to make so that the `copy` method will compile error-free.

11. Consider the `equals` method presented in this book.

    a. Identify the header of the method.

    b. Identify the body.

    c. What are explicit parameters?

    d. What is the signature of the method?

    e. Rewrite the `equals` method without any local variables.

12. Consider the class `Utility` presented in this book. What will be the gcd produced according to the code in the `Utility` class
    a. `numOne` is 225 and `numTwo` is 45?
    b. `numOne` is 220 and `numTwo` is 25?
    c. `numOne` is −100 and `numTwo` is 8?
    d. `numOne` is −13 and `numTwo` is −17?
    e. `numOne` is 0 and `numTwo` is 21?
    f. `numOne` is 38 and `numTwo` is 0?
    g. `numOne` is 0 and `numTwo` is 0?

PROGRAMMING EXERCISES

1. Create an `Employee` class and test it. The instance variables are used to store first name, last name, annual salary, and number of dependents. There are three constructors: the default constructor, a constructor with first name and last name as parameters, and a constructor with first name, last name, salary, and number of dependents as parameters. There are two application-specific methods. The first returns the monthly salary. The second returns bonus calculated as the maximum of percentage of the annual salary or the minimum bonus for the year. The percentage and minimum bonus for the year are explicit parameters of this method.
2. Write a program to estimate the profit from a particular product for a month. Information such as product name, unit cost, sale price, and average number of items sold per month are available. Note that product name may consist of many words such as "Hunter Miller 56in Ceiling Fan." Provide at least four constructors.
3. Design a class `Student` with six data members to store first name, last name, and four test scores. Provide three constructors: the default constructor, a constructor with first name and last name as formal parameters, and a constructor with first name, last name, and four test scores as formal parameters. Create a method `validateTestScore` with one formal parameter that returns the formal parameter if it is between 0 and 100 and 0 otherwise. Use `validateTestScore` method in all mutator methods that deal with a test score. Create another method `setAllData` that sets all attributes and has six formal parameters. Use `setAllData` in all constructors. Provide a method `getLetterGrade()` that returns the letter grade based on the policy outlined in Example 4.21.
4. Create and test a class `Vehicle` with attributes to store information on model, year, cost basis and sale price, and used. The attribute `used` indicates whether or not it is a used vehicle. Create a method `validateCost` with one formal parameter that returns the formal parameter if it is greater than or equal to 1000.00 and 0 otherwise. Use `validateCost` method in all mutator methods that deal with a cost.

Create another method `validatePrice` with one formal parameter that returns the formal parameter if it is greater than 10% the cost of the vehicle and 10% cost of the vehicle otherwise. Also include a method `setAllData` that sets all attributes. Use `setAllData` in all constructors. Provide at least four different constructors.

5. Create a class `Order` with three static methods: `max`, `middle`, and `min`. All three methods have three formal parameters and they return maximum, middle, and minimum values, respectively.
6. Create a class `Conversion` with two static methods: `toCentigrade` and `toFahrenheit`.
7. Create a class `Name` with two static methods: `toShorter` and `toInitials`. The `toShorter` returns a shorter version of the name and `toInitials` returns initials from the full name. For example, a name such as Meera S. Nair will have a shorter name M.S.Nair and MSN as initial.
8. Write a static method to convert a `String` into corresponding telephone number. If it is a uppercase letter or a lowercase letter, the program will substitute it with the corresponding digit. If it is already a digit, no substitution is done. Thus, "GOODCAR," "gooDCar," and "go6DC2r" will be translated to 4663227.
9. Consider the following sequence:

   1, 1, 2, 3, 5, 8, 13, 21, ...

   In this sequence, from the third number onward, next number in the sequence is the sum of the previous two numbers. For example, 2 = 1 + 1; 3 = 2 + 1; 5 = 3 + 2; and 8 = 5 + 3. This sequence is known as Fibonnaci sequence. In fact, you can start a Fibonacci sequence with any two values. As an example, Fibonacci sequence starting with 3 and 4 is as follows:

   3, 4, 7, 11, 18, 29, ...

   Create a static method `fibonacci` having three formal parameters. The first two integer parameters correspond to the first and the second values of the sequence and the third parameter specifies the requested term in the sequence. Thus, we have the following:

   `fibonacci`(3, 4, 1) returns 3

   `fibonacci`(3, 4, 2) returns 4

   `fibonacci`(3, 4, 3) returns 7

   `fibonacci`(3, 4, 4) returns 11

10. Modify the `Employee` class of the Programming Exercise 1 so that it has a static data member `noOfEmployees` to keep track of the total number of employees.
11. Modify the `Product` class of Programming Exercise 2 so that it has a static data member `noOfProducts` to keep track of the total number of products.
12. Redo Programming Exercise 3 by introducing an additional static variable to keep track of the number of students.

13. Redo Programming Exercise 4 by introducing an additional static variable to maintain the number of vehicles.
14. Modify the `Employee` class of Programming Exercise 1 or 10 by adding a `copy` constructor, a `copy` method, and an `equals` method.
15. Modify the `Product` class of Programming Exercise 2 or 11 by adding a `copy` constructor, a `copy` method, and an `equals` method.
16. Modify the `Student` class of Programming Exercise 3 or 12 by adding a `copy` constructor, a `copy` method, and an `equals` method.
17. Modify the `Car` class of Programming Exercise 4 or 13 by adding a `copy` constructor, a `copy` method, and an `equals` method.

ANSWERS TO SELF-CHECK

1. object reference, method name
2. `static`
3. value returning
4. `void`
5. comma
6. data type
7. False
8. True
9. call by value
10. value
11. True
12. name of the class
13. False
14. one
15. implicit
16. `myStock`

CHAPTER 7

Object-Oriented Software Design

In this chapter you learn

- Object-oriented concepts
  - Encapsulation, information hiding, interface, service, message passing, responsibility, delegation, late binding, inheritance hierarchy, composition, and abstract class
- Java concepts
  - Subclass, superclass, reference super, access modifiers, final class, abstract method, abstract class, and interface
- Programming skills
  - Design, create, execute, and test Java programs having many classes related through superclass/subclass relationship or composition

Reliable, error-free, maintainable, and flexible software is very difficult to produce. Today's software systems are quite complex and no level of abstraction can eliminate the complexity completely. However, certain abstractions are more natural to human thinking compared to other forms of abstractions. In the case of object-oriented paradigm, real-world entities are modeled as objects. This form of abstraction enables the software developer to divide the software into a collection of mutually collaborating objects working toward achieving a common goal of solving the problem. In this chapter, you will explore object-oriented paradigm in a more comprehensive fashion. Further, examples and analogies presented in this chapter will help you understand the object-oriented way of developing Java programs.

OBJECTS

The most fundamental concept of the object-oriented paradigm is that of an object. An object can be perceived in three different ways. In fact, these are not contradictory views; rather, they complement each other to enhance our ability to model the real world.

Three perspectives of an object are

- Data-centric view
- Client–server view
- Software design view

Data-Centric View

From a data-centric perspective, an object is a collection of attributes and operations that manipulate the data. In Chapter 1, you have seen that a computer is a general-purpose information-processing machine. Thus, every task performed by a computer is based on some information. So it is natural to view an application as a collection of data items that needs to be processed to produce the desired output.

Practically anything can be modeled as an object. You could model your e-mail system as an object. Your cable service provider can be an object. The local public library can be modeled as an object. You could also model a geometric shape as an object. In Chapter 6 you have seen how fractions can be modeled as objects. In data-centric view, a class is a template of objects with identical attributes and operations.

Self-Check

1. Practically anything can be modeled as an ________.
2. A class is a ________ of objects with identical attributes and operations.

Attribute

An attribute is an internal variable that captures some characteristics of the object. Therefore, an attribute has a name, a data type, and a value. For instance, in the case of an `Employee`, `employeeName` can be an attribute of type `String`. The attribute `employeeName` can have a value such as James Smith. Similarly, `monthlySalary` can be another attribute of `Employee`. It is of the type `double` and keeps the present monthly salary of the employee.

Self-Check

3. ________ are used to keep data of the object.
4. The data type of `monthlySalary` is ________.

Operation

In an object, an attribute is kept `private`. Therefore, to manipulate an attribute, there must be `public` operations. Without such operations, a class is not useful.

Thus, from a data-centric view, an object is a collection of closely related attributes along with operations on them. In other words, closely related attributes and operations on those attributes are bundled together as one semantic entity. For example, if you need to manipulate hours, minutes, and seconds, you can group all three attributes together into one semantic entity called clock. Next, you include necessary operations such as increment hour, increment minute, and increment second. This process of grouping data and operations on data is called *encapsulation*. In addition to encapsulation, all attributes of an object are hidden from the user. In fact, how the operations are implemented is also completely hidden from the user. The fact that implementation is hidden from the user is known as *information hiding*.

Encapsulation combined with information hiding helps reduce the complexity both at the user and at the designer level. To better understand the impact, consider the primitive data type `int`. Suppose you want to use the `int` data type. You know the valid data values that can be represented as `int`. You also know the set of operations that can be performed on an `int`. However, you need not know how an `int` value is stored in a memory location. Further, you need not know how each of the valid operations is implemented. Therefore, you, as a user of `int` data type, need to be concerned with the complexities associated with integer representations and integer operations. Now let us look from the designer's perspective. If you are responsible for implementing `int` data type, you have complete freedom to choose the representation scheme and methods to implement various operations on `int`. Whether you use binary system or decimal system to represent the integers is completely hidden from the user. Thus, as a designer, you can focus on efficient implementation under the current technology.

Self-Check

5. As a general rule, attributes are maintained as ________ and operations are maintained as ________.
6. True or false: Information hiding reduces the complexity of the system.

Client–Server View

From a client–server perspective, attributes are not the most important part of an object. Rather, the focus is on the services an object could provide for other objects. For example, in the case of a public library, as a user or client you are interested in the services such as borrowing a book, reserving a book, and searching the catalog for a book. In fact, the primary mission of the library is to provide these services to its patrons. In this case, you are a client and the service provider, your local public library, is the server.

Every object in the application provides some service. It is the service that makes them relevant to the program. Similar to human behavior, objects communicate with each other by sending messages. These messages are in fact requests for service. In this view, if the client object needs the service of a server object, it sends a message to the server. The server object, upon receiving a valid service request from a client, performs the requested service. In this model of computation, a Java program is a collection of cooperating objects carrying out various service requests made by other objects.

Thus, each object is a server or provider of some service. A service is a `public` method of a class. Similarly, message passing refers to method `invocation`. For instance, consider two objects, Ms. Jane Olsen and her puppy Nacho. Nacho provides certain services for Jane. One of such services is to fetch the tennis ball. So any time Jane wants Nacho to fetch a tennis ball, she sends a message to Nacho. Thus, Jane makes a request for the service fetch to Nacho. Thus, Jane is a client of Nacho since Nacho provides the fetch service. The object-oriented way of expressing this idea is as follows:

```
nacho.fetch(ball);
```

Note that in the Java programming environment,

- `nacho` is a reference variable of the type `Puppy`.
- `fetch` is a `public` method of the class `Puppy`.
- `ball` is an actual parameter that references an instance of `FetchableItem` class.

Note that server and client are role designations. Assume that Jane has a service `feed`. As Nacho becomes hungry, Nacho sends a message to the feeding service by producing a special grunt sound. Jane interprets the grunt as a request for the feeding service and places Nacho's favorite puppy food in the bowl. Note that in this situation, Jane is the server and Nacho is the client.

Mr. Jones and Mr. Clark are very good friends. One day, Mr. Jones decided to send a surprise gift to Mr. Clark through an overnight delivery service Air America Overnight Inc. (AAOI). Mr. Jones takes the package to the nearest AAOI kiosk. In this case, AAOI is the server and Mr. Jones is the client. Once Mr. Jones gives the package to an employee of AAOI, the safe and on-time delivery of the package to Mr. Clark is the *responsibility* of the AAOI. By offering to provide delivery service, AAOI also assumes the responsibility associated with it. Now, AAOI may or may not hire other local subcontractors to complete the task. However, as far as Mr. Jones is concerned, AAOI is responsible for the package. The way in which AAOI is going to accomplish the delivery service is not at all important to Mr. Jones.

Let us assume that AAOI has made some agreement with Freight Services (FS) to make the delivery. In that case, FS is a server for AAOI and AAOI is a client of FS. Once again, as far as AAOI is concerned, FS is responsible for the delivery of Mr. Jones' package and AAOI is not interested in the way in which FS is going to accomplish the service. Observe that Mr. Jones is completely unaware of the fact that AAOI depends on FS to carry out the delivery service.

Thus, in a client–server environment, clients can request any of the advertised services (i.e., `public` methods) of the server. The server itself may use other servers to accomplish the task. This *delegation* of task by a server is completely hidden to the client. As far as the client–server relationship is concerned, the responsibility rests with the server.

Observe that the concepts "responsibility" and "delegation" allow us to reduce the complexity. For instance, consider the case of Mr. Jones. Once the package is delivered to an AAOI personnel, Mr. Jones need not be concerned with the safe delivery of the package. AAOI took complete responsibility from that point on. Whether or not AAOI delegates the task in no way reduces AAOI's responsibility. However, using FS reduces the complexity for AAOI.

Self-Check

7. The object that requests the service is a _________ and the object that provides the service is a _________.
8. True or false: Delegation does not absolve responsibility.

Software Design View

From a design perspective, you consider an object as an instance of a class. For example, in the case of a payroll program, you may start with a class `Employee`. In this view all employees are instances of the class `Employee`. Each class is an abstraction of some real-life entity or concept. The class definition is similar to a blueprint. The definition as such does not create an instance or object of the class. Rather, individual objects are instantiated by applying the `new` operator on a constructor of the class. Each of the instances is an object created from the specification given in the class definition.

Example 7.1

Consider the problem of computing the area and perimeter of a rectangle. To compute the required information, you must know the length and width of the rectangle. Therefore, you decide to have objects with two attributes representing the length and width. Further, the application needs methods such as area and perimeter. If you proceed along these lines, you are in fact viewing in a data-centric way. However, if you start out identifying services required, say `computeArea` and `computePerimeter`, and then identify length and width as necessary data items, the approach taken is that of a client–server view. Finally, from the problem statement you could identify `Rectangle` as a class and then make the `Rectangle` class appropriate for this application by properly selecting the attributes and operations; this approach is more in tune with the software design view.

In all three views, you need to identify classes, attributes of each class, and services of each class. To a great extent, the above three views just assign different priorities in arriving at the final product. In the case of data-centric view, the priority is attribute, then service followed by class. However, in the client–server view, the priority is service, then attribute followed by class. Finally, in the software design view, the top priority is on identifying the classes. The order in which you may decide on attributes and services for each of those classes is not that important.

Self-Check

9. Irrespective of the view, you need to identify ________, identify ________ of each class, and identify ________ of each class.
10. In the software design view, the top priority is on identifying the ________.

SUBCLASS

Consider two classes `Person` and `Employee`. Observe that the following statements are true:

1. An instance of `Employee` is *always* an instance of `Person`.
2. An instance of `Person` is *sometimes* an instance of `Employee`.

We say `Employee` is a subclass of `Person`. In general, if you have two classes `ClassOne` and `ClassTwo` such that an instance of `ClassTwo` is *always* an instance of ClassOne and an instance of `ClassOne` is *sometimes* an instance of `ClassTwo`, then `ClassTwo` is a subclass of `ClassOne`. Alternatively, `ClassOne` is a superclass of `ClassOne`.

Using the unified modeling language (UML 2) you can show the subclass/superclass relationship between the two classes as shown in Figure 7.1.

Example 7.2

Consider three classes `Student`, `GradStudent`, and `UnderGradStudent`. Note that a `GradStudent` is *always* a `Student` and a `Student` is *sometimes* a `GradStudent`. Therefore, `GradStudent` is a subclass of `Student` class. Similarly, `UnderGradStudent` is a subclass of `Student` class.

Observe that a `GradStudent` is *never* an `UnderGradStudent` and an `UnderGradStudent` is *never* a `GradStudent`. Therefore, `GradStudent` is not a subclass of `UnderGradStudent` or `UnderGradStudent` is not a subclass of `GradStudent`. The UML 2 diagram in Figure 7.2 shows the relationship that exists among the three classes `Student`, `GradStudent`, and `UnderGradStudent`.

Example 7.3

Consider five classes `Triangle`, `Square`, `Rectangle`, `Circle`, and `Ellipse`. The relationship among these classes can be summarized in the form of a table as shown Table 7.1.

Thus, `Square` is a subclass of `Rectangle`. Similarly, `Circle` is a subclass of `Ellipse`. Thus, we have the UML 2 diagram shown in Figure 7.3.

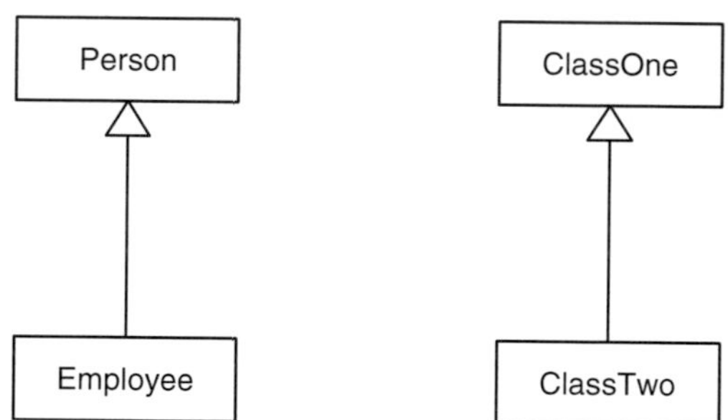

FIGURE 7.1 Subclass/superclass relationship.

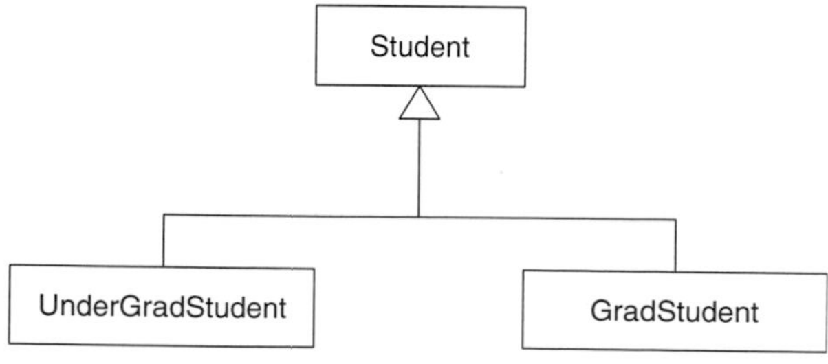

FIGURE 7.2 Relationship between `Student` and its subclasses.

TABLE 7.1 Superclass/Subclass Determination Scheme

	Triangle	Square	Rectangle	Circle	Ellipse
Triangle	×	Never	Never	Never	Never
Square	Never	×	Always	Never	Never
Rectangle	Never	Sometimes	×	Never	Never
Circle	Never	Never	Never	×	Always
Ellipse	Never	Never	Never	Sometimes	×

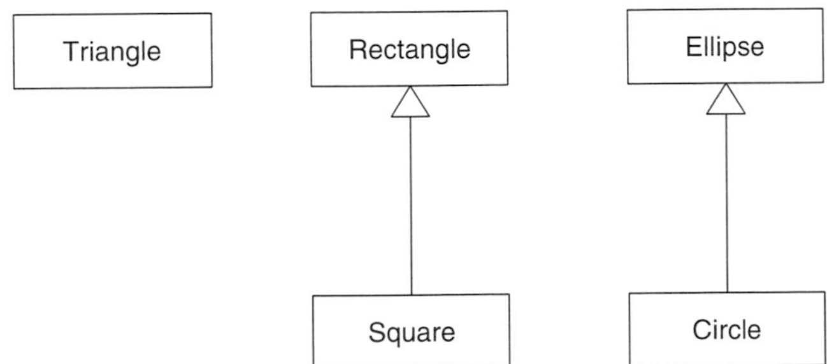

FIGURE 7.3 Relationship among geometric shapes.

Self-Check

11. True or false: The `Apple` class is a subclass of the `Fruit` class.
12. True or false: The `Keyboard` class is a subclass of the `Computer` class.

Inheritance

Consider Mr. Jones, an instance of the class `Employee`. Since every instance of `Employee` class is an instance of `Person` class, Mr. Jones, in particular, is an instance of the `Person` class. Thus, Mr. Jones has all the attributes of `Employee` class and `Person` class. Similarly, Mr. Jones provides all the services of `Employee` as well as `Person`. Due to this reason, there is no need to repeat the same attribute or service in a subclass.

Consider the attribute `name`. Since every instance of a `Person` has a name, the attribute `name` is kept as an attribute of the superclass `Person`. In this case, the subclass `Employee` inherits the attribute `name` from its superclass `Person`. However, consider the attribute `salary` of `Employee`. Note that every instance of a `Person` need not be an instance of `Employee` and as such may not have a `salary`. Therefore, `salary` is an attribute of the `Employee` class. The `Person` class does not inherit the attribute `salary` from `Employee`. Thus, in the case of a superclass/subclass relationship, subclass inherits all the attributes of the superclass. Superclass does not inherit any attribute of a subclass.

Services also follow the same pattern. For instance, `talk` is a service of `Person` class. Thus, every instance of `Employee` is also capable of providing that service. However, `prepareTimesheet` is a service of `Employee`. The `Person` class does not inherit `prepareTimesheet` service of `Employee` class. Thus, the superclass/subclass relationship establishes an *inheritance hierarchy*.

As mentioned above, subclass inherits all services of the superclass. In a subclass, if necessary, you could *override* a service. While all animals eat, their eating methods are different. For example, dogs and humans have different eating methods. This form of inheritance is known as *inheritance with polymorphism*. Thus, there are two forms of inheritance: inheritance with polymorphism and *inheritance without polymorphism*.

Examples 7.4 and 7.5 illustrate inheritance with polymorphism. An example for inheritance without polymorphism can be found in Example 7.6.

Example 7.4

Consider the inheritance hierarchy between an `Ellipse` and a `Circle`. One of the services of the `Ellipse` class is `rotate`. The service `rotate` tilts an ellipse by the angle specified in the counterclockwise direction with the center fixed. Note that in the case of a circle, rotating it by any angle makes no difference at all. Therefore, in the subclass `Circle`, you can override the `rotate` method such that it does nothing at all. Thus, you can have a more efficient implementation of the service `rotate` in the subclass `Circle`.

Example 7.5

Consider the inheritance hierarchy between `Student` and `GradStudent`. Let `studentData` be a method of the `Student` class that returns a `String` containing all the basic information about a `Student`. If `GradStudent` has additional attributes such as `thesisTitle`, you may override the `studentData` in the `GradStudent` class to include this information. In this case, the `studentData` method of the `GradStudent` class can be implemented in such a way that it invokes the `studentData` method of the superclass `Student` first; and then appends more information to produce the desired `String`.

Example 7.6

An `Employee` is still a `Person`. Being an `Employee` does not change the common services of `Person` such as `getFirstName`, `getLastName`, `setFirstName`, and `setLastName`. Therefore, the superclass/subclass relationship between `Person` and `Employee` establishes an inheritance hierarchy without polymorphism.

Note that establishing an inheritance hierarchy has many advantages. Some of them are as follows:

1. *Complexity reduction*. Typically you can start with a very simple class. Once that class is completely implemented and tested, you can add more attributes and services. This incremental approach greatly reduces the complexity.
2. *Ability to add new attributes and services*. Once a class is created, you cannot add more services to that class. The best way to add more services is by creating a new subclass. The newly created subclass inherits all the services of the superclass. Therefore, the newly created subclass can be used in place of the

superclass with the added advantage of new services, which were not available in the superclass.

3. *Code reusage*. The services of a subclass can make use of services of the superclass. This feature eliminates the need to repeat the same code both in superclass and in subclass. This ability to reuse the code indirectly promotes consistency and reduces complexity.
4. *Ability to modify services*. Applying the principle of polymorphism, you can override any method provided by the superclass. This feature helps to custom tailor the method for the subclass. Thus, it is possible to start with a most general superclass and specialize into various subclasses.

Self-Check

13. The subclass ________ all the services and attributes of its superclass.
14. A subclass can ________ any method of a superclass.

Creating Subclass

The syntax template for creating a new class `SubClassName` from an existing class `SuperClassName` through inheritance is as follows:

```
[classModifiers] class SubClassName extends SuperClassName
                                                [modifiers]
{
    [attributes and methods of SubClassName alone]
}
```

Note that `extends` is a reserved word in Java.

Note 7.1 From an object-oriented design perspective, a class can be created through single or multiple inheritance. In the case of single inheritance a subclass has a unique superclass. In multiple inheritance a class has more than one superclass. The programming language C++ allows multiple inheritance. However, Java does not permit multiple inheritance. Instead, Java introduces the concept of an interface and allows multiple interfaces. You will be introduced to interfaces later in this chapter.

Example 7.7

Suppose the class `Person` exists. Then a new class `Employee` can be created through inheritance as follows:

```
public class Employee extends Person
{
    //attributes and operations
}
```

Similarly, suppose the class `Student` exists. Then a new class `GradStudent` can be created through inheritance as follows:

```
public class GradStudent extends Student
{
    //attributes and operations
}
```

As stated during the general discussion, all attributes and services of the superclass are inherited by the subclass. However, those are not members of the subclass. All `private` members (attributes and operations) of a superclass are not directly accessible to any other class, including the subclass. Similarly, all `public` members (data as well as services) of a superclass are directly accessible to any other class, including the subclass. However, there is a third option possible, namely, `protected` access. All `protected` members of a superclass are directly accessible in the subclass but not to any other class. These facts can be summarized in the form of a table as shown in Table 7.2.

Recall that there is a restriction on `static` methods. A `static` method can access only other `static` methods and class variables. For example, a `public` instance variable of a superclass is not accessible inside a `static` method of the subclass.

Next we address three major issues that arise in the context of a superclass/subclass relationship. Once these issues are addressed, we will be ready to present our first example of subclass creation.

1. Due to polymorphism, the same service is available in both the superclass and the subclass. Therefore, how to distinguish between these two services inside a method of the subclass. In other words, how to invoke the service of a superclass inside a method of the subclass.
2. Both `public` and `protected` members of the superclass are directly accessible in the subclass. However, `private` members (attributes and operations) of the superclass are not accessible in the subclass. Therefore, how to access `private` attributes of the superclass inside a method of the subclass.
3. Both `public` and `protected` attributes of the superclass are directly accessible in the subclass. However, `private` members attributes of the superclass are not accessible in the subclass. Therefore, how to initialize attributes of superclass in the subclass constructor. In other words, how to invoke the constructor of the superclass inside a subclass constructor.

TABLE 7.2 Inheritance and the Role of Access Modifier

Access Modifier of a Member in Superclass	Accessibility Inside Methods of a Subclass	Accessibility Inside Methods of any Class Other Than a Subclass
`public`	Accessible	Accessible
`protected`	Accessible	Not accessible
`private`	Not accessible	Not accessible

Self-Check

15. True or false: Java allows multiple inheritance.
16. A __________ member of the superclass is not directly accessible in the subclass.

Invoking Method of Superclass

Consider two classes `Student` and `GradStudent` such that the class `GradStudent` is a subclass of the class `Student`. Now the class `GradStudent` inherits all `public` and `protected` methods of the class `Student`. Due to polymorphism, it is possible to override a method of the class `Student` in the subclass `GradStudent`. For instance, `computegpa` can be a method of both classes. However, the way grade point average (gpa) is computed may be totally different for a graduate student as opposed to other students. If that is the case, the method `computegpa` needs to be overridden in the class `GradStudent`. Thus, in general, a method of the class `SuperClass` may be overridden in the class `SubClass`. In that case, both methods have the same signature. Thus, there are two methods with identical signature, one in the `SuperClass` and the other in the `SubClass`. Recall that you can have two methods with identical names but different signatures within a class. In that case, it is called *method overloading*. You can have two methods with identical names and identical signatures, one in the superclass and the other in the subclass. This is known as method overriding (or polymorphism).

Note 7.2 Two overloaded methods have different signatures. Thus, a compiler can uniquely identify the method that needs to be invoked. However, overridden methods have identical signatures. Therefore, within the subclass, there needs to be a mechanism to distinguish between the one that is the member of the subclass and the one that is inherited from the superclass.

Note 7.3 The issues mentioned in Note 7.2 are not relevant to `static` methods. Recall that a `static` method can be invoked using the class name. Therefore, the class name will uniquely identify the method involved.

Java provides an elegant solution through a reference variable `super`. In Java, `super` is a keyword. The reference variable `super` can be used inside all methods of a subclass that are not marked `static` and it references the implicit parameter as an instance of its superclass. Recall that the keyword `this` can be used in all methods of a class that are not marked `static` and it references the implicit parameter as an instance of the class.

Note 7.4 The keyword `super` is a reference similar to the self-reference `this`, with the difference that while `this` references the implicit parameter as an instance of the class, `super` references the implicit parameter as an instance of the superclass.

The template

```
super.methodName([actualParameterList])
```

can be used inside all methods of a subclass that are not marked `static` to invoke the method `methodName` of a superclass that is not marked `static`.

Good Programming Practice 7.1

It is a good programming practice to invoke all methods of the superclass using the reference `super`.

Self-Check

17. Having two methods with identical signature, one in a superclass and other in the subclass is known as method ____________.
18. Having two methods with different signatures in the same class is known as method ____________.

Accessing Private Attribute of Superclass

Recall that `private` attributes of the superclass remain `private` even under inheritance. Therefore, the subclass cannot directly access `private` attributes of the superclass. Just as any other class, subclass also needs to use `public` services provided by the superclass. Therefore, if `getDataMember` and `setDataMember` are two methods provided by the superclass to access and mutate an attribute `dataMember`, you can access and mutate `dataMember` inside all methods of the subclass that are not marked `static` using the reference `super`. Thus,

```
super.getDataMember()
```

and

```
super.setDataMember(actualParameterList)
```

can be used inside all methods of a subclass that are not marked `static` to access and modify the `private` attribute `dataMember` of the superclass.

Self-Check

19. To access a private attribute of a superclass, a public method of the __________ is required.
20. Both __________ and __________ attributes of a superclass are directly accessible in the subclass.

Invoking Constructor of Superclass

Inside the constructor of the subclass, you cannot use the `new` operator to initialize the superclass attributes. For example, consider the superclass/subclass relationship between `Student` and `GradStudent`. The following code is illegal:

```
public GradStudent(argListOne)
```

```
{
    super = new Student(argListTwo);        // illegal
    // more statements
}
```

Instead, the correct way of invoking the constructor of the superclass inside the constructor of a subclass is as follows:

```
public GradStudent(argListOne)
{
    super(argListTwo);   //Assume that superclass has
                         //a constructor whose signature
                         //matches argListTwo

    // more statements
}
```

Note that if the constructor of the superclass is invoked inside the subclass constructor using the keyword `super`, it must be the first executable statement of the subclass constructor.

We conclude this section with Example 7.8 that illustrates all the concepts introduced so far.

Example 7.8

In this example, we first create a class `Circle`. This class has one `private` attribute, `radius` of type `double`. There are two application-specific methods `area` and `circumference` that compute and return the values of area and circumference, respectively. Two constructors are also included in the class `Circle`.

We use inheritance to create a new class `CylinderInherited`. The `CylinderInherited` class has one new `private` attribute `height` of type `double`. There are two methods `area` and `volume` that compute and return the values of surface area and volume, respectively. The class `CylinderInherited` has two constructors and one `boolean` method `isTall`.

Now consider the `CylinderInherited` class created from `Circle` class through inheritance. The following points merit special mention:

1. Methods `setRadius` or `getRadius` need not be defined again. `CylinderInherited` inherits those methods.
2. Every time a method of the superclass is invoked, the reference `super` is explicitly used. The methods `circumference`, `getRadius`, and `setRadius` can be invoked without the reference `super`.
3. The very first executable statement in each of the constructors is `super` with appropriate parameters to invoke a constructor of the superclass.
4. For the sake of illustrating the mechanism of accessing a `private` attribute inside a method that is not marked `static`, we introduce a method `isTall`.

Let us define a cylinder to be tall if its height is at least four times its radius. Note that you must use `getRadius` method to access the `private` attribute `radius` of the class `Circle`.

5. The class `CylinderInherited` overrides methods `area` and `toString` of the class `Circle`. In fact, `toString` is defined in a Java system class `Object` and every class with no explicit superclass such as the `Circle` class is implicitly a subclass of the `Object` class. Thus, the `toString` method of the `Circle` class is in fact overriding the `toString` method of the `Object` class. The `Object` class is presented later in this chapter.
6. The method `area` is defined in the class `Circle` first and overridden in the class `CylinderInherited`.
7. The method `circumference` is not overridden in the class `CylinderInherited`, and therefore the method `circumference` has identical behavior in both classes.
8. The method `volume` is defined in the class `CylinderInherited` only. This method is not available in the class `Circle`.

```java
/**
    Circle class computes area and circumference
*/
public class Circle
{
    private double radius;
    /**
        Constructor with no parameters
    */
    public Circle()
    {
        radius = 0;
    }
    /**
        Constructor to creates a circle of given radius
        @param inRadius the radius of the circle
    */
    public Circle(double inRadius)
    {
        setRadius(inRadius);
    }
    /**
        Computes and returns the area
        @return area of the circle
    */
    public double area()
```

```
    {
        return (Math.PI * radius * radius);
    }

    /**
        Computes and returns the circumference
        @return circumference of the circle
    */
    public double circumference()
    {
        return (2 * Math.PI * radius);
    }

    /**
        Accessor method for the radius
        @return radius

    */
    public double getRadius()
    {
        return radius;
    }

    /**
        Mutator method for the radius
        @param inRadius the new value of the radius
    */
    public void setRadius(double inRadius)
    {
        if(inRadius >= 0)
            radius = inRadius;
        else
            radius = 0;
    }

    /**
        toString method
        @return radius as a String
    */
    public String toString()
    {
        String str;
        str = "Radius is "  + radius ;
        return str;
    }
}
```

```
/**
    Cylinder inherited from Circle
*/
public class CylinderInherited extends Circle
{
    private double height;

    /**
        Constructor initializes radius and height
        @param inRadius
        @param inHeight
    */
    public CylinderInherited(double inRadius, double
                                                 inHeight)
    {
        super(inRadius);
        setHeight(inHeight);
    }

    /**
        Constructor initializes radius and height with
                                          default values
    */
    public CylinderInherited()
    {
        super();
        setHeight(0);
    }

    /**
        Computes the surface area
        @return area
    */
    public double area()
    {
        return (super.circumference() * height + 2 *
                                         super.area());
    }

    /**
        Computes the volume
        return volume
    */
    public double volume()
```

```
    {
        return (super.area() * height);
    }

    /**
        Checks whether or not height is >= 4 times the
                                                  radius
        @return boolean value
    */
    public boolean isTall()
    {
        return (height >= 4 * super.getRadius());
    }

    /**
        Accessor method for the height
        @return height
    */
    public double getHeight()
    {
        return height;
    }

    /**
        Mutator method for the height
        @param inHeight new value for height
    */
    public void setHeight(double inHeight)
    {
        height = inHeight;
    }

    /**
        toString method
        @return a String with radius and height
    */
    public String toString()
    {
        String str;
        str = super.toString() + "; Height is "  +
                                           height;
        return str;
    }
}
```

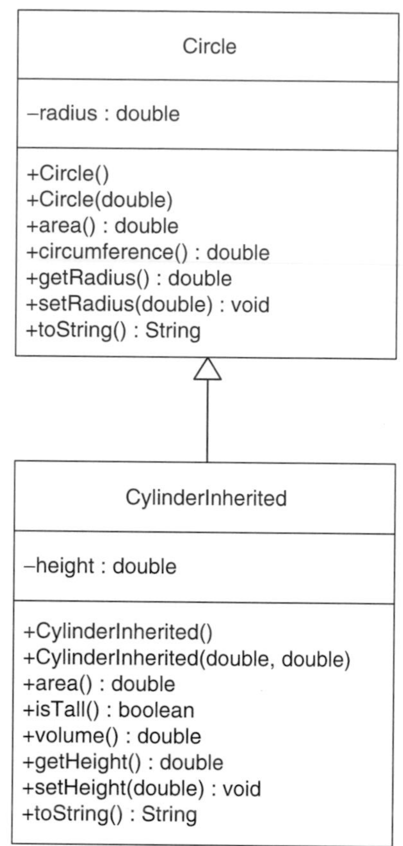

FIGURE 7.4 Class diagram of `Circle` and `CylinderInherited`.

The class diagram in UML 2 notation is given in Figure 7.4.

Consider the following statements that create one instance of `Circle` and one instance of `CylinderInherited`:

```
Circle round = new Circle(8.74);

CylinderInherited roller = new CylinderInherited(4.23, 20.45);
```

The reference variables `round` and `roller` along with the objects created can be visualized as shown in Figure 7.5.

In Figure 7.5, members that are accessible through the reference variable are shown as small rectangles projecting outward. Thus, using the `round` reference variable, you can invoke five methods `area`, `circumference`, `getRadius`, `setRadius`, and `toString` of the class `Circle`. The attribute `radius` is not accessible through the reference variable `round`. Similarly, using the `roller` reference variable, you can invoke nine methods. They are three methods `circumference`, `getRadius`, and `setRadius` of the class `Circle` and six methods `area`, `volume`, `isTall`, `getHeight`, `setHeight`,

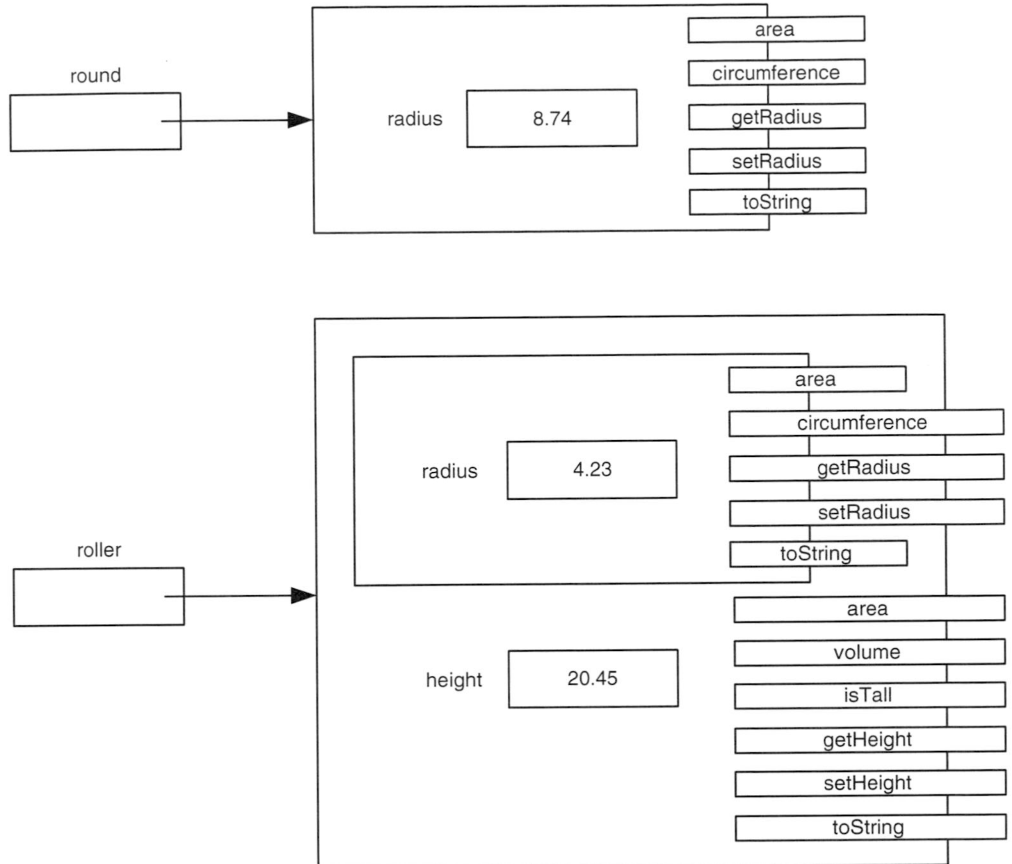

FIGURE 7.5 Visualization of objects `round` and `roller`.

and `toString` of the class `CylinderInherited`. Note that none of the attributes are accessible. Further, since methods `area` and `toString` of the class `Circle` are overridden in the class `CylinderInherited`, those two methods cannot be accessed using the reference variable `roller`.

Note that the Java statements

```
System.out.println(round);

System.out.println(roller);
```

will produce the following output lines, respectively, by invoking the corresponding `toString` methods:

```
Radius is 8.74

Radius is 4.23; Height is 20.45
```

We conclude this example by the following Java program that illustrates the use of all methods:

```
import java.text.DecimalFormat;

public class CircleCylinderInherited
{
    public static void main(String[] args)
    {
        DecimalFormat twoDecimalPlaces = new
                     DecimalFormat("0.00");                   // 1
        Circle roundOne = new Circle();                       // 2
        Circle round = new Circle(8.74);                      // 3

        CylinderInherited rollerOne
                = new CylinderInherited();                    // 4
        CylinderInherited roller
                    = new CylinderInherited(4.23, 20.45);     // 5

        System.out.println("(6)  roundOne data: " +
                                        roundOne);            // 6
        System.out.println("(7)  round data: " + round);      // 7
        System.out.println("\nMethods of Circle");            // 8
        System.out.println("(8) Area of round: "
            + twoDecimalPlaces.format(round.area()));         // 9
        System.out.println("(10) Circumference of
            round: " + twoDecimalPlaces.format(round.
                                  circumference()));          // 10
        roundOne.setRadius(6.98);                             // 11
        System.out.println("(12) Radius of roundOne : "
                              + roundOne.getRadius());        // 12
        System.out.println();                                 // 13
        System.out.println("(14) rollerOne data: "
                                   + rollerOne);              // 14
        System.out.println("(15) roller data: " +
                                        roller);              // 15
        System.out.println("\nMethods of CylinderInherited");
                                                              // 16
        System.out.println("(17) Area of roller: "
            + twoDecimalPlaces.format(roller.area()));        // 17
```

```
        System.out.println("(18) Volume of roller: "
            + twoDecimalPlaces.format(roller.volume()));    // 18
        if (roller.isTall())
            System.out.println("(19) The roller is tall."); // 19
        else
            System.out.println("(20) The roller is not tall.");
                                                            // 20
        rollerOne.setHeight(12.7);                          // 21
        System.out.println("(22) Height of rollerOne : " +
                               rollerOne.getHeight());      // 22

        System.out.println("\nMethods Inherited from Circle");
                                                            // 23
        System.out.println("(24) Circumference of roller: "
                      + twoDecimalPlaces.format(roller.
                                  circumference())); // 24
        rollerOne.setRadius(6.98);                          // 25
        System.out.println("(26) Radius of rollerOne : " +
                               rollerOne.getRadius());      // 26
    }
}
```

Output

```
(6)  roundOne data: Radius is 0.0
(7)  round data: Radius is 8.74

Methods of Circle
(8)  Area of round: 239.98
(10) Circumference of round: 54.92
(12) Radius of roundOne: 6.98
(14) rollerOne data: Radius is 0.0; Height is 0.0
(15) roller data: Radius is 4.23; Height is 20.45

Methods of CylinderInherited
(17) Area of roller: 655.94
(18) Volume of roller: 1149.54
(19) The roller is tall.
(22) Height of rollerOne: 12.7

Methods Inherited from Circle
(24) Circumference of roller: 26.58
(26) Radius of rollerOne: 6.98
```

Self-Check

21. Inside the constructor of the subclass, you cannot use the ________ operator to initialize the superclass attributes.
22. If the constructor of the superclass is invoked inside the subclass constructor using the keyword ________, it must be the first executable statement of the subclass constructor.

Subclass Objects as Superclass Instance

We begin our discussion through a real-life analogy.

Example 7.9

Consider a class `Pet` and its two subclasses `Cat` and `Dog`. Assume the following variable declarations:

```
Pet petOne;
Cat catOne;
Cat catTwo;
Dog dogOne;
```

Now

```
catOne = new Cat("Snowball");
```

creates an instance of `Cat`. Since `Cat` is a subclass of `Pet` class, the following statement is legal:

```
petOne = catOne;
```

This allows us to treat `catOne` as a member of the `Pet` class. In other words, you can assign a subclass reference to a superclass reference.

However, you cannot assign a superclass reference to a subclass reference. Thus,

```
catTwo = petOne;                // illegal
```

You can correct the above statement as follows:

```
catTwo = (Cat) petOne;          // is legal
```

Note that in this case, `petOne` in fact references an instance of `Cat` and hence the cast operation was legal. Further note that the following statement is illegal:

```
dogOne = (Dog) petOne;          // illegal : petOne references an
                                // instance of Cat.
```

From an object-oriented perspective, an instance of a subclass is always an instance of the superclass. Therefore, you can assign a reference variable of the subclass to a reference variable of the superclass.

Consider the following declarations:

```
Circle roundOne;                                              // 1
Circle round = new Circle(4.23);                              // 2

CylinderInherited rollerOne;                                  // 3
CylinderInherited roller = new CylinderInherited(4.23, 20.45);
                                                              // 4
```

where `Circle` and `CylinderInherited` are classes as in Example 7.8. Thus, `CylinderInherited` is a subclass of the class `Circle` and the following assignment statement is legal:

```
roundOne = roller;
```

However, the following two statements are illegal:

```
rollerOne = round;  //illegal
rollerOne = (CylinderInherited) round;  //illegal
```

The reason being `round` refers to a `Circle` object only.

However, the following segment of code is legal:

```
roundOne = roller;          //a reference variable of type Circle
                            // references a CylinderInherited Object.

rollerOne = (CylinderInherited) roundOne;  // is legal.
```

Observe that even though `roundOne` is a reference variable of the type `Circle`, `roundOne` references an instance of `CyliderInherited` class.

Self-Check

23. True or false: A subclass object reference can be assigned to a superclass object reference.
24. True or false: A superclass object reference can be assigned to a subclass object reference.

Polymorphic Behavior

Consider the following declarations:

```
Circle round;

CylinderInherited roller = new CylinderInherited(4.23, 20.45);
```

where `Circle` and `CylinderInherited` are classes as in Example 7.8. As you have seen,

```
round = roller;
```

is legal. Note that round is a reference variable of type Circle. Therefore, you can use round to invoke any of the methods of the Circle class. Further, if a method is overridden in the subclass CylinderInherited, the overridden method is invoked rather than the method in the superclass Circle. Thus,

```
round.area();
```

will invoke the method area of CylinderInherited class. Similarly,

```
roller.toString();
```

and

```
round.toString();
```

invoke the method toString of CylinderInherited class. This is the polymorphism in action. Even though round is a reference variable of type Circle, since it references an object of the subclass CylinderInherited, the method in the subclass is invoked instead of the one in the Circle class. Observe that compiler cannot determine the actual method to be executed. The decision has to be made at the run time. This is known as *dynamic binding* or *late binding* in object-oriented terminology.

However,

```
roller.volume();
```

invokes the method volume of CylinderInherited class and

```
round.volume();      //Compilation error
                     //Even when round has the reference of an
                     //object of the class CircleInherited,
                     //it is a compile time error, since volume is
                     //not a method of Circle.
```

results in a compile time error. Observe that volume is not a method of the Circle class.

The above discussion can be summarized as follows. If a superclass reference references an object of a subclass, then

- Superclass reference can invoke any of the methods defined in the superclass and if the method is overridden in the subclass, the method in the subclass is invoked.
- Superclass reference cannot invoke any of the methods defined in the subclass but not in the superclass.

Example 7.10 illustrates the concepts presented in this section.

Example 7.10

```
import java.text.DecimalFormat;

/**
    Application program to illustrate various concepts related
                                                  to inheritance
*/
public class ObjectReference
{
    public static void main(String[] args)
    {
        DecimalFormat twoDecimalPlaces = new
                     DecimalFormat("0.00");                     // 1
        Circle roundOne;                                        // 2
        Circle round = new Circle(4.23);                        // 3

        CylinderInherited rollerOne;                            // 4
        CylinderInherited roller
            = new CylinderInherited(4.23, 20.45);               // 5
        System.out.println("(6)  round data: " + round);        // 6
        System.out.println("\nMethod invocation:
                             Circle reference");                // 7
        System.out.println("(8)  Area of round: "
            + twoDecimalPlaces.format(round.area()));           // 8
        System.out.println("(9)  Circumference of round: "
                  + twoDecimalPlaces.format(round.
                                 circumference()));             // 9
        System.out.println("\n(10) roller data: " +
                                              roller);          // 10
        roundOne = roller;
        System.out.println("(11) roundOne data: " +
                                            roundOne);          // 11
        System.out.println("\nMethod invocation: Superclass
                                            reference");  // 12
        System.out.println("(13) Area of roundOne: "
            + twoDecimalPlaces.format(roundOne.area()));
                                                                // 13
        System.out.println("(14) Circumference of roundOne: "
                  + twoDecimalPlaces.format(roundOne.
                                 circumference()));  // 14
        //The volume cannot be invoked; it is not defined in
                                                   Circle class.
```

```
        //The next commented line will generate compile time
                                                          error
        //System.out.println("(15) Volume of roundOne: "
        //   + twoDecimalPlaces.format(roundOne.volume()));
                                                          // 15
        System.out.println("\nMethod invocation: Subclass
                                        reference");  // 16
        System.out.println("(17) Area of roller: "
            + twoDecimalPlaces.format(roller.area()));  // 17
        System.out.println("(18) Circumference of roller: "
                   + twoDecimalPlaces.format(roller.
                                  circumference()));    // 18
        System.out.println("(19) Volume of roller: "
            + twoDecimalPlaces.format(roller.volume()));
                                                          // 19
    }
}
```

Output

```
(6)  round data: Radius is 4.23

Method invocation: Circle reference
(8)  Area of round: 56.21
(9)  Circumference of round: 26.58

(10) roller data: Radius is 4.23; Height is 20.45
(11) roundOne data: Radius is 4.23; Height is 20.45

Method invocation: Superclass reference
(13) Area of roundOne: 655.94
(14) Circumference of roundOne: 26.58

Method invocation: Subclass reference
(17) Area of roller: 655.94
(18) Circumference of roller: 26.58
(19) Volume of roller: 1149.54
```

Observe that `toString` method of `CylinderInherited` is invoked in both Lines 10 and 11. Similarly, `area` computed in Lines 13 and 17 are also identical. Once again, the method `area` of `CylinderInherited` is invoked. Observe that method `area` of `Circle` class will return a value identical to one on Line 8. As a final observation, note that volume can only be invoked using a reference variable of type `CylinderInherited`.

Self-Check

25. True or false: If a superclass reference references an object of a subclass then superclass reference can invoke any of the methods defined in the superclass and if the method is overridden in the subclass, the method in the subclass is invoked.
26. True or false: Irrespective of the object it references, a superclass reference cannot invoke any of the methods defined in the subclass but not in the superclass.

Advanced Topic 7.1: `instanceof` Operator

You can use the `instanceof` operator to determine whether or not a reference variable currently references an object of a particular class. Consider the following statements:

```
Student student = new Student();
Person person = student;
Employee employee;
```

Now the expression `person instanceof Student` evaluates to `true` and the expression `person instanceof Employee` evaluates to `false`.

Good Programming Practice 7.2

Avoid the use of `instanceof` operator unless it is absolutely necessary.

Advanced Topic 7.2: Use of `protected` Attributes

The purpose of the next example is to illustrate the use of `protected` attributes. This example will illustrate the pros and cons of declaring an attribute as `protected`.

Example 7.11

In this example, once again we create a circle class and then extend it to a cylinder class. The names `Circle` and `CylinderInherited` were used in Example 7.8. Therefore, `CirclePro` and `CylinderProInherited` are used to name circle and cylinder classes of this example. The attribute `radius` of the `CirclePro` has `protected` access. There are two design choices:

1. Methods `getRadius` and `setRadius` can be omitted from the `CirclePro`. In other words, `protected` attribute need not have associated accessor and mutator methods. However, you may decide to keep the associated mutator as a `private` method. We illustrate such an approach in this example.
2. The subclass has direct access to the `protected` attribute. Thus, attribute `radius` can be accessed in the `CylinderProInherited` class. However, the protected attribute is not accessible to any class that uses `CylinderProInherited`. Therefore, you may decide to keep the associated accessor and mutator methods as `public` method.

The classes CirclePro and CylinderProInherited are as follows. The changes compared to Circle and CylinderInherited are shown in comments.

```
/**
    Circle class with radius as a protected member
*/
public class CirclePro              //Circle :> CirclePro
{
    protected double radius;        //private :> protected
    /**
        Constructor creates a circle with radius 0
    */
    public CirclePro()              //Circle :> CirclePro
    {
        radius = 0;
    }
/**
    Constructor creates a circle of given radius
    @param inRadius radius of the circle
*/
public CirclePro(double inRadius) //Circle :> CirclePro
{
    setRadius(inRadius);
    @return area of the circle
}
/**
    Computes area
*/
public double area()
{
    return (Math.PI * radius * radius);
}
/**
    Computes circumference
    @return circumference of the circle
*/
public double circumference()
{
    return (2 * Math.PI * radius);
}
                                            //getRadius deleted
```

```
/**
    Mutator method for the radius
    @param inRadius new value for the radius
*/
private void setRadius(double inRadius)
                                            //public :> private
{
    if(inRadius >= 0)
        radius = inRadius;
    else
        radius = 0;
}
/**
    toString method
    @return radius as a String
*/
public String toString()
{
    String str;
        str = "Radius is "  + radius ;
        return str;
    }
}
/**
    Cylinder created by inheriting a circle with protected
                                                    attribute
*/
public class CylinderProInherited extends CirclePro
{                                           //class names changed
    private double height;
    /**
        Constructor creates a cylinder with default values
    */
    public CylinderProInherited()    //constructor name changed
    {
        super();
        setHeight(0);
    }
    /**
        Constructor creates a cylinder with given radius and
                                                      height
```

```
        @param inRadius radius of the cylinder
        @param inHeight height of the cylinder
    */
    public CylinderProInherited(double inRadius, double
            inHeight)                      //constructor name changed
    {
        super(inRadius);
        setHeight(inHeight);
    }
    /**
        Computes and returns surface area
        @return area of the cylinder
    */
    public double area()
    {
        return (super.circumference() * height + 2 *
                                            super.area());
    }
    /**
        Computes and returns volume
        @return volume of the cylinder
    */
    public double volume()
    {
        return (super.area() * height);
    }
    /**
        Determines whether or not height > 4 * radius
        @return true if cylinder is tall
    */
    public boolean isTall()
    {
        return (height >= 4 * radius); //getRadius() :> radius
    }
    /**
        Mutator method for radius
        @param inRadius new value for radius
    */
    public void setRadius(double inRadius)
    {
        radius = inRadius;
    }
```

```
    /**
        Mutator method for height
        @param inHeight new value for height
    */
    public void setHeight(double inHeight)
    {
        height = inHeight;
    }
    /**
        Accessor method for radius
        @return radius
    */
    public double getRadius()
    {
        return radius;
    }
    /**
        Accessor method for height
        @return height
    */
    public double getHeight()
    {
        return height;
    }
    /**
        toString mcthod
        @return radius and height as a String
    */
    public String toString()
    {
        String str;
        str = super.toString() + "; Height is "  + height;
        return str;
    }
}
```

Observe that inside isTall method, the protected attribute radius of the class CirclePro is directly accessed. As mentioned before, both radius and super.radius mean the attribute radius of the superclass. However, super.radius highlights the fact that radius is an attribute of the superclass.

The task of creating a class CircleProCylinderInherited modifying the class CircleCylinderInherited is left as Programming Exercise 1.

Advanced Topic 7.3: Design Options

protected Operations

As in the case of attributes, operations can also be declared protected. For instance, you could have defined circumference as protected method of the class Circle. The major advantage of such a design decision is to limit the method circumference to classes that inherit Circle. In Example 7.11, CylinderInherited can treat circumference similar to any other public service of the Circle. However, the class CircleCylinderInherited cannot invoke the method circumference.

package Access

You have already seen the access modifiers public, protected, and private. They can be specified in connection with attributes, services, and classes. You can opt for no access modifier at all. This results in package access. Note that every class in the same package has access to an item with package access and all classes not in the package have no access.

Modifier final

You have seen the keyword final in Chapter 2. The keyword final is used to declare constants. You can also use final modifier in the context of a method or a class. A method marked as final cannot be overridden in a subclass.

The syntax template of a value-returning static method is as follows:

```
[accessModifier] final [static] returnType methodName
                                ([formalParameterList])

{
    [statements]
}
```

Similarly, a class marked final cannot have any subclass. The syntax template of a class definition is as follows:

```
[accessModifier] final ClassName modifiers
{
    [statements]
}
```

Three concepts protected methods, package access, and final modifier, though important, are quite easy to assimilate without a full-blown example and hence such an example is omitted. However, the next concept is a major design concept. Therefore, we introduce it in the next section.

ABSTRACT CLASSES AND METHODS

A class is marked `abstract` if any instance of it is an instance of one of its subclasses. Recall that an instance of a subclass is always an instance of the superclass. For example, consider the `Circle` and `CylinderInherited` class of Example 7.8. Every instance of the `CylinderInherited` class is an instance of the `Circle` class. However, it is possible to have an instance of `Circle` class that is not an instance of `CylinderInherited`. In other words, it is possible to have an instance of `Circle` class that is not an instance of any of its subclasses. Therefore, `Circle` is not an `abstract` class.

So far in this chapter, we have discussed the need to create new subclasses from existing classes. In some situations, from a design perspective, we would like to create a new superclass from existing classes. For example, consider the following situation. The public library carries various items that can be borrowed by its patrons. These items can be classified into four categories: books, journals, CDs, and DVDs. Therefore, you may start with four classes: `Book`, `Journal`, `CompactDisc`, and `DigitalVideoDisc`. However, these classes have common attributes and common services. Therefore, from an object-oriented design perspective, a new class `Item` can be defined that abstracts the common attributes and common services. Note that every instance of the `Item` class is an instance of one of the subclasses. Therefore, `Item` is an `abstract` class.

As another example, consider the patrons of the library. All patrons fall under two categories: children and adults. Note that there are many attributes and methods common to both groups. Therefore, it is reasonable to create three classes: `Patron`, `Child`, and `Adult` such that both `Child` and `Adult` classes are subclasses of the class `Patron`. Observe that `Patron` is an `abstract` class since every instance of the `Patron` is an instance of one of its subclasses.

The concept of an `abstract` method is quite similar. Let us look at a familiar example. Consider three classes `Cat`, `Dog`, and `Person`. We can treat all three classes as subclasses of an `abstract` class `Animal`. Note that `eat` is an operation for all animals. However, the way an animal eats depends upon the class it belongs. In other words, even though `eat` is a common operation for all animals, it is impossible to implement the `eat` method in the `Animal` class. Instead, the implementation is done at each of the subclasses.

For the rest of this section we explain the Java way of implementing `abstract` methods and `abstract` classes.

A method in Java is specified `abstract` by explicitly including the keyword `abstract` in its heading. Further, an `abstract` method has no body; rather, the method heading ends with a semicolon.

The following are some examples of `abstract` methods:

```
public abstract double computeFine();
public abstract void borrow(Item item);
```

Here are some important points worth noticing about `abstract` classes:

1. An `abstract` class is declared by placing the keyword `abstract` immediately after the access modifier.
2. If a class has an `abstract` method, the class becomes `abstract`.
3. An `abstract` class may or may not have an `abstract` method.
4. It is legal to declare a reference variable of an `abstract` class type.
5. It is illegal to instantiate a reference variable of an `abstract` class type using its constructor.
6. Subclass or subclasses are created by overriding all `abstract` methods of the `abstract` class. Such a subclass is no longer `abstract`. Thus, a reference variable of an `abstract` class type is instantiated through a constructor of one of its subclasses.

To illustrate these concepts, consider the following problem. CTN University has two types of students: undergraduates and graduates. Registrar services decided to printout a slim down status report of each student for academic monitoring purposes. In the case of undergraduate students, the most crucial parameter is the gpa. However, for a graduate student, the number of years spent is the most important parameter.

From an object-oriented software development approach, the above problem calls for three classes: `Student` (an `abstract` class) and two subclasses `GradStudent` and `UnderGradStudent` of `Student` class (see Figure 7.6).

Design decisions can be explained as follows:

1. The first name and last name are common to all students. Therefore, they must be kept as attributes of the `abstract` class.
2. The service `StudentInfo` is common to all students and is kept as member of the `abstract` class. However, this operation cannot be implemented at `Student` class. Therefore, it is marked as an `abstract` operation.
3. Our decision to keep first name as `private` and last name as `protected` is purely for the purpose of illustrating both `private` and `protected` attributes of a class.

The three classes `Student`, `UnderGradStudent`, and `GradStudent` along with the application program and the output are as follows:

```
/**
    Superclass for all students
*/
```

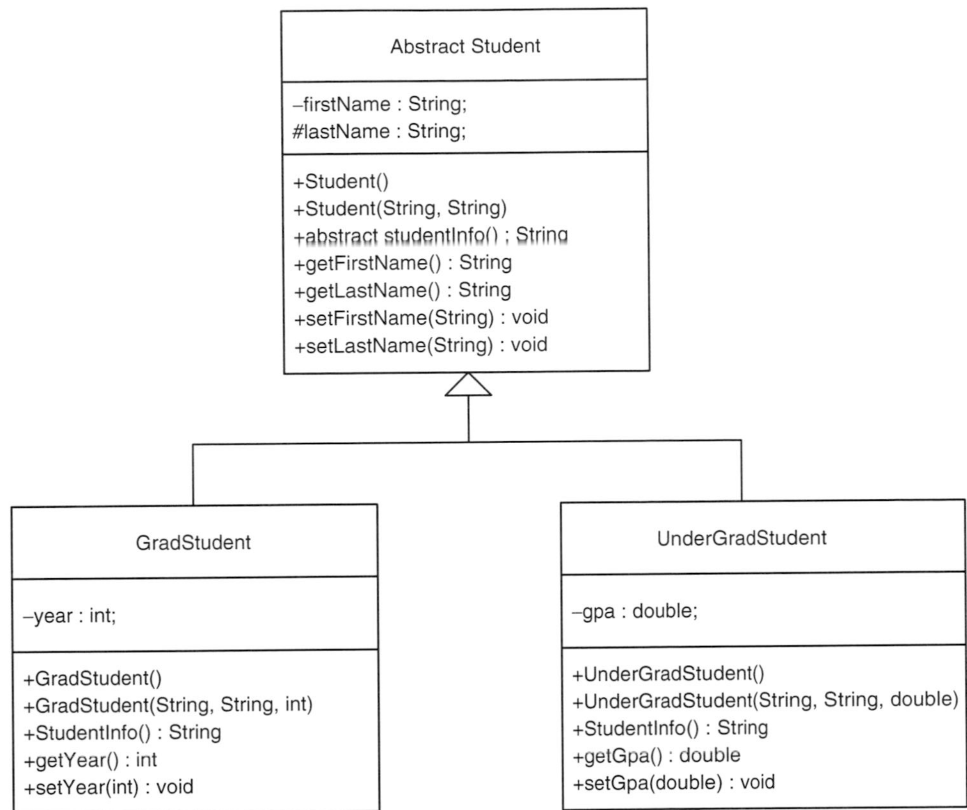

FIGURE 7.6 Class diagram of `Student` and its two subclasses.

```
public abstract class Student
{
    private String firstName;
    protected String lastName;
    /**
        Constructor creates a Student with name null
    */
    public Student()
    {
        firstName = null;
        lastName = null;
    }
    /**
        Constructor creates a Student with given values for
                                                       name
        @param inFirstName first name
```

```
        @param inLastName last name
    */
    public Student(String inFirstName, String inLastName)
    {
        firstName = inFirstName;
        lastName = inLastName;
    }
    /**
        Abstract method creates a String with relevant
                                            student info
        @return relevant student info as a String
    */
    public abstract String StudentInfo();
    /**
        Accessor method for first name
        @return first name
    */
    public String getFirstName()
    {
        return firstName;
    }
    /**
        Accessor method for last name
        @return last name
    */
    public String getLastName()
    {
        return lastName;
    }
    /**
        Mutator method for last name
        @param inLastName new value for last name
    */
    public void setLastName(String inLastName)
    {
        lastName = inLastName;
    }
    /**
        Mutator method for first name
        @param inFirstName new value for first name
    */
    public void setFirstName(String inFirstName)
```

```
    {
        firstName = inFirstName;
    }
    /**
        toString method
        @return relevant student information
    */
    public String toString()
    {
        return StudentInfo();
    }
}
/**
    Undergrad student class keeps gpa
*/
public class UnderGradStudent extends Student
{
    private double gpa;
    /**
        Constructor creates an undergrad student with
                                            default values
        @param inFirstName first name
        @param inLastName last name
    */
    public UnderGradStudent()
    {
        super();
        gpa = 0;
    }
    /**
        Constructor creates an undergrad student with given
                                                   values
        @param inFirstName first name
        @param inLastName last name
        @param inGpa gpa
    */
    public UnderGradStudent(String inFirstName, String
                            inLastName, double inGpa)
    {
        super(inFirstName, inLastName);
        gpa = inGpa;
    }
```

```
    /**
        String with relevant student info
        @return relevant student info as a String
    */
    public String StudentInfo()
    {
        return (getFirstName() + " " + lastName + "; gpa = " +
                                                            gpa);
    }
    /**
        Accessor method for gpa
        @return gpa
    */
    public double getGpa()
    {
        return gpa;
    }
    /**
        Mutator method for gpa
        @param inGpa new value of gpa
    */
    public void setGpa(double inGpa)
    {
        gpa = inGpa;
    }
}
/**
    Gradstudent class keeps year
*/
public class GradStudent extends Student
{
    private int year;
    /**
        Constructor creates a Gradstudent with default values
    */
        public GradStudent()
    {
        super();
        year = 1;
    }
    /**
        Constructor creates a Gradstudent with given values
```

```
        @param inFirstName first name
        @param inLastName last name
        @param inYear year
    */
    public GradStudent(String inFirstName, String inLastName,
                                                    int inYear)
    {
        super(inFirstName, inLastName);
        year = inYear;
    }
    /**
        String with relevant student info
        @return relevant student info as a String
    */
    public String StudentInfo()
    {
        return (getFirstName() + " " + lastName + "; year = "
                                                     + year);
    }
    /**
        Accessor method for year
    */
    public int getYear()
    {
        return year;
    }
    /**
        Mutator method for year
        @param inYear new value for year
    */
    public void setYear(int inYear)
    {
        year = inYear;
    }
}
import java.io.*;
import java.util.*;
/**
    The testing class for Student and its subclasses
*/
public class AbstractClassTesting
{
```

```
static Scanner scannedInfo = new Scanner(System.in);
public static void main(String[] args) throws IOException
{
    Student studentRef;
    String fname, lname;
    double gradePtAvg;
    int  yearAtSchool;
    int topSelection;  //holds top level the selection
    System.out.println
           ("\t\t\tWelcome to Student Info. Service");
    displayTopMenu();
    topSelection = scannedInfo.nextInt();
    while(topSelection != 0)
    {
        switch(topSelection)
            {
                case 1:
                    System.out.println
                          ("tEnter first name, last name and gpa");
                    fname = scannedInfo.next();
                    lname = scannedInfo.next();
                    gradePtAvg = scannedInfo.nextDouble();
                    studentRef = new UnderGradStudent
                                   (fname, lname, gradePtAvg);
                    System.out.println
                                  (studentRef StudentInfo());
                    break;
                case 2:
                    System.out.println("tEnter first
                    name, last name and year at school");
                    fname = scannedInfo.next();
                    lname = scannedInfo.next();
                    yearAtSchool = scannedInfo.nextInt();
                    studentRef = new GradStudent(fname,
                                        lname, yearAtSchool);
                    System.out.println
                                  (studentRef.StudentInfo());
                    break;
                case 0:
                    System.out.println("Good Bye");
                    return;
                default:
```

```
                        System.out.println
                            ("Select a number between 0 and 2");
                    }//end switch
            displayTopMenu();
            topSelection = scannedInfo.nextInt();
        }//end while (choice != 10)
    }//end main
    /**
        This top menu displayed as the program runs
    */
    private static void displayTopMenu()
    {
        System.out.println("Select from choices given below:");
        System.out.println("Enter 1 for undergraduate student");
        System.out.println("Enter 2 for graduate student");
        System.out.println("Enter 0 to exit");
    }//end displayTopMenu
}
```

Output

```
                Welcome to Student Info. Service
Select from choices given below:
Enter 1 for undergraduate student
Enter 2 for graduate student
Enter 0 to exit
1
    Enter first name, last name and gpa
Mark Lloyd 3.27
Mark Lloyd; gpa = 3.27
Select from choices given below:
Enter 1 for undergraduate student
Enter 2 for graduate student
Enter 0 to exit
2
    Enter first name, last name and year at school
Eliza Downy 2
Eliza Downy; year = 2
Select from choices given below:
Enter 1 for undergraduate student
Enter 2 for graduate student
```

```
Enter 0 to exit
0
```

Example 7.12

In this example, we illustrate the power of polymorphism. For this purpose, we introduce two new classes: `CorrespondentStudent` and `PartTimeStudent`. To make this example more illustrative, we create `CorrespondentStudent` as a subclass of `Student` and `PartTimeStudent` as a subclass of `UnderGradStudent`, respectively. Thus, we have the class diagram shown in Figure 7.7.

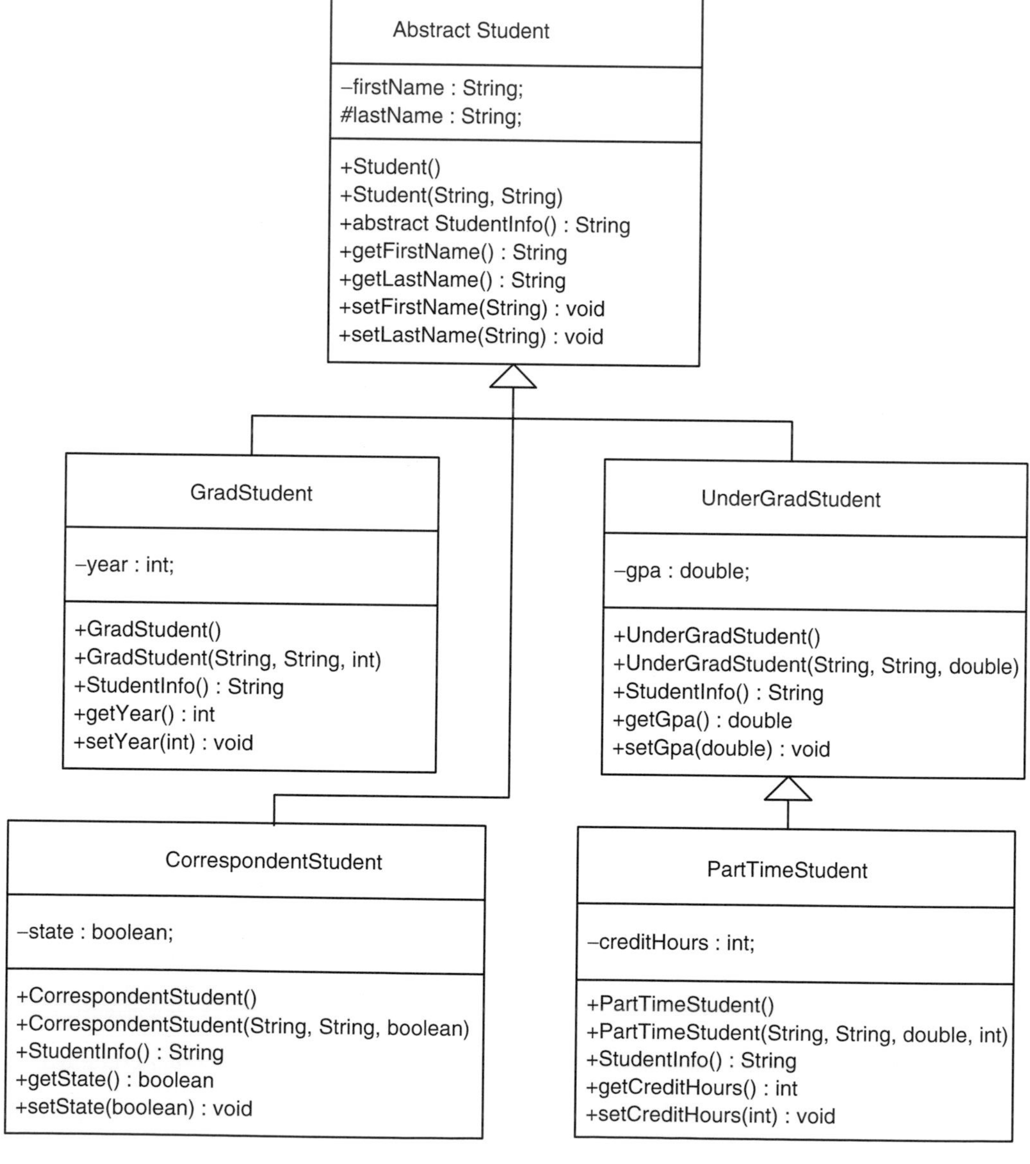

FIGURE 7.7 Class diagram of the `Student` class and its subclasses.

```
/**
    Correspondent student class keeps state
*/
public class CorrespondentStudent extends Student
{
    private boolean state;

    /**
        Constructor creates a corres. student with default values
    */
    public CorrespondentStudent()
    {
        super();
        state = false;
    }

    /**
        Constructor creates a corres. student with given values
        @param inFirstName first name
        @param inLastName last name
        @param inState in or out of state
    */
    public CorrespondentStudent(String inFirstName, String
                                inLastName, boolean inState)
    {
        super(inFirstName, inLastName);
        state = inState;
    }

    /**
        String with relevant student info
        @return relevant student info as a String
    */
    public String StudentInfo()
    {

        return (getFirstName() + " " + lastName + ";
                                in state = " + state);
    }

    /**
        Accessor method for state
        @return state
    */
```

```
    public boolean getState()
    {
        return state;
    }

    /**
        Mutator method for state
        @param inState new value of state
    */
    public void setState(boolean inState)
    {
        state = inState;
    }
}

/**
    Part-time student class keeps credit hours
*/
public class PartTimeStudent extends UnderGradStudent
{
    private int creditHours;

    /**
        Constructor creates a part-time student with default values
        @param inFirstName first name
        @param inLastName last name
        @param inGpa gpa
        @param inCreditHours credit hours
    */
    public PartTimeStudent()
    {
        super();
        creditHours = 0;
    }

    /**
        Constructor creates a part-time student with given values
    */

    public PartTimeStudent(String inFirstName, String
          inLastName, double inGpa, int inCreditHours)
    {
        super(inFirstName, inLastName, inGpa);
        creditHours = inCreditHours;
    }
```

```
    /**
        String with relevant student info
        @return relevant student info as a String
    */
    public String StudentInfo()
    {
        return (super.StudentInfo() + ";
                                credits hours = " + creditHours);
    }

    /**
        Accessor method for credit hours
        @return credit hours
    */
    public boolean getCreditHours()
    {
        return creditHours;
    }

    /**
        Mutator method for credit hours
        @param inCreditHours new value of credit hours
    */
    public void setCreditHours(int inCreditHours)
    {
        creditHours = inCreditHours;
    }
}

import java.io.*;
import java.util.*;

/**
    Application program for student and its subclasses
*/
public class AbstractClassTestingModified
{
    static Scanner scannedInfo = new Scanner(System.in);

    public static void main(String[] args) throws IOException
    {
        Student studentRef;

        String fname, lname;
        double gradePtAvg;
```

```
int yearAtSchool;
int stateInfo;
int creditsEnrolled;

int topSelection; //holds top level the selection

System.out.println
            ("\t\t\tWelcome to Student Info. Service");
displayTopMenu();
topSelection = scannedInfo.nextInt();

while(topSelection != 0)
{
    switch(topSelection)
    {
    case 1:
        System.out.println
              ("tEnter first name, last name and gpa");
        fname = scannedInfo.next();
        lname = scannedInfo.next();
        gradePtAvg = scannedInfo.nextDouble();
        studentRef = new UnderGradStudent(fname,
                            lname, gradePtAvg);
        System.out.println(studentRef.StudentInfo());
    break;

    case 2:
        System.out.println
                        ("tEnter first name, last name
                                and year at school");
        fname = scannedInfo.next();
        lname = scannedInfo.next();
        yearAtSchool = scannedInfo.nextInt();
        studentRef = new GradStudent(fname, lname,
                                yearAtSchool);
        System.out.println(studentRef.StudentInfo());
        break;

    case 3:
        System.out.println
                   ("tEnter first name, last name and
                                   in state info.");
        System.out.println("\t\t\t(0: out of state)");
```

```
            System.out.println("\t\t\t(1: in state)");
            fname = scannedInfo.next();
            lname = scannedInfo.next();
            stateInfo = scannedInfo.nextInt();
            if (stateInfo == 0)
                studentRef = new CorrespondentStudent
                               (fname, lname, false);
            else
                studentRef = new CorrespondentStudent
                                (fname, lname, true);
            System.out.println(studentRef.StudentInfo());
            break;

        case 4:
            System.out.println
                     ("tEnter first name, last name, gpa and
                                         credits enrolled");
            fname = scannedInfo.next();
            lname = scannedInfo.next();
            gradePtAvg = scannedInfo.nextDouble();
            creditsEnrolled = scannedInfo.nextInt();
            studentRef = new PartTimeStudent(fname,
                lname,gradePtAvg, creditsEnrolled);
            System.out.println(studentRef.StudentInfo());
            break;

        case 0:
            System.out.println("Good Bye");
            return;
            default:
            System.out.println
                        ("Select a number between 0 and 4");
            }//end switch

        displayTopMenu();
        topSelection = scannedInfo.nextInt();
    }//end while (topSelection != 0)
}//end main

/**
    Displays the top level menu
*/
```

```
    private static void displayTopMenu()
    {

        System.out.println
                         ("nSelect from choices given below:");
        System.out.println
                         ("Enter 1 for undergraduate student");
        System.out.println("Enter 2 for graduate student");
        System.out.println
                         ("Enter 3 for correspondent student");
        System.out.println("Enter 4 for part-time student");
        System.out.println("Enter 0  to exit");

    }//end displayTopMenu
}
```

Output

```
                Welcome to Student Info. Service

Select from choices given below:

Enter 1 for undergraduate student
Enter 2 for graduate student
Enter 3 for correspondent student
Enter 4 for part-time student
Enter 0 to exit
3

    Enter first name, last name and in state info.
        (0: out of state)
        (1: in state)
Malissa Price 0
Malissa Price; in state = false

Select from choices given below:

Enter 1 for undergraduate student
Enter 2 for graduate student
Enter 3 for correspondent student
Enter 4 for part-time student
Enter 0 to exit
4

    Enter first name, last name, gpa and credits enrolled
Darren McGill 2.71 6
Darren McGill; gpa = 2.71; credits hours = 6
```

```
Select from choices given below:

Enter 1  for undergraduate student
Enter 2  for graduate student
Enter 3  for correspondent student
Enter 4  for part-time student
Enter 0  to exit
0
```

Self-Check

27. If a class has at least one abstract method, it must be marked ________.
28. True or false: An abstract class has at least one abstract method.

Advanced Topic 7.4: `Object` Class

In Java, all classes are subclasses of a class `Object`. If a class is not derived from an existing class, then the class you define is implicitly derived from the `Object` class. Thus,

```
public class Student
{
    //members of the class
}
```

is equivalent to the following:

```
public class Student extends Object
{
    // members of the class
}
```

The `Object` class is the topmost class in the inheritance hierarchy. Every other class in Java inherits all services of the `Object` class. Some of the constructors and services of the `Object` class are presented in Table 7.3.

Notice that `toString` is a method of the `Object` class. Recall that `toString` method is invoked implicitly by the system during the invocation of `print` and `println` methods. Therefore, you are encouraged to provide a `toString` method in every class. Thus, you have been overriding the method `toString` without actually realizing it. You have been using inheritance with polymorphism all along!

Advanced Topic 7.5: Composition

Composition is another way to use an existing class to create a new class. In fact all application programs we developed so far in this book have used composition. In the case of superclass/subclass relationship, an instance of the subclass is an ("is-a") instance of the superclass. In the case of composition, an instance of the newly created class has an ("has-a") instance of the existing class as an attribute.

TABLE 7.3 Some of the Constructors and Services of the `Object` Class

Operation	Explanation
`public Object()`	Constructor

`public boolean equals(Object ob)`	Returns `true` if implicit parameter and the explicit parameter `ob` have identical attributes; returns `false` otherwise
	Example:
	`Object obOne, obTwo;`
	`...`
	`obOne.equals(obTwo)`
	is `true` if the `obOne` and `obTwo` objects have identical attributes; and is `false` otherwise

`protected String toString()`	Returns a `String` describing the object
`protected void finalize()`	The garbage collector invokes this method once it determines there is no existing reference

Earlier in this chapter you have seen creation of a new class `CylinderInherited` from an existing class `Circle`. In this section, we create a new class `CylinderComposed` from the same existing class `Circle`. Thus, in this section, we view a cylinder as having two attributes: `base` of the type `Circle` and `height` of the type `int`. Thus, we have the following:

```
public class CylinderComposed
{
    private Circle base;
    private double height;

    //constructors and services
}
```

Observe that `base` is an attribute. The reference `super` cannot be used to invoke services of `Circle`. Similarly, the keyword `super` cannot be used inside a constructor of the `CylinderComposed` class to instantiate the `base`. Each of these issues is addressed next.

Accessor and Mutator Methods

Accessor and mutator methods of the `height` attribute are as in the case of `CylinderInherited` class. However, `radius` is not an attribute of this class. Further, `base` being a `private` attribute, none of the methods of the `Circle` are available to the application program. Therefore, if the application program of the `CylinderComposed` class needs to set the radius, such a method needs to be included in the class `CylinderComposed`.

Note 7.5 While composition allows the designer of the new class to block access to services (such as `getRadius` and `setRadius` in this case), it necessitates additional coding if the designer would like to provide those services. In the case of inheritance, the designer of the new class has no such option. All the `public` methods of the superclass are directly available to the application program.

Thus, we have the following accessor and mutator methods:

```
public double getHeight()
{
    return height;
}

public double getBaseRadius()
{
    return base.getRadius();
}

public void setHeight(double inHeight)
{
    height = inHeight;
}

public void setBaseRadius(double inBaseRadius)
{
    base.setRadius(inBaseRadius);
}
```

Observe that accessor and mutator methods for the `radius` attribute of the `base` need to invoke the `getRadius` and `setRadius` methods of the `Circle` class.

Constructor

In the case of composition, the constructor must instantiate every attribute that happens to be an object reference through `new` operator. Thus, we have the following:

```
public CylinderComposed(double inRadius, double inHeight)
{
    base = new Circle(inRadius);
    setHeight(inHeight);
}

public CylinderComposed()
{
    base = new Circle();
    setHeight(0);
}
```

Application-Specific Services

We had included three services: area, volume, and isTall in the CylinderInherited class. To illustrate the differences and similarities, we provide these three services next.

```
public double area()
{
    return (base.circumference() * height + 2 * base.area());
}

public double volume()
{
    return (base.area() * height);
}

public boolean isTall()
{
    return (height >= 4 * base.getRadius());
}
```

Note that the only difference is that the attribute base is used instead of the keyword super. Similar comments apply to toString method as well. Thus, we have the following:

```
public String toString()
{

    String str;
    str = base.toString() + "; Height is "  + height;
    return str;
}
```

The complete listing of CylinderComposed class, UML 2 diagram, and visual representations are presented next.

```
/**
    Cylinder composed from Circle
*/
public class CylinderComposed
{
    private Circle base;
    private double height;
    /**
        Constructor initializes radius and height with default
                                                         values
    */
    public CylinderComposed()
```

```
{
    base = new Circle();
    setHeight(0);
}
/**
    Constructor initializes radius and height
    @param inRadius
    @param inHeight
*/
public CylinderComposed(double inRadius, double inHeight)
{
    base = new Circle(inRadius);
    setHeight(inHeight);
}
/**
    Computes the surface area
    @return area
*/
public double area()
{
    return (base.circumference() * height
                                    + 2 * base.area());
}
/**
    Computes the volume
    @return volume
*/
public double volume()
{
    return (base.area() * height);
}
/**
    Checks whether or not height is >= 4 times the radius
    @return boolean value
*/
public boolean isTall()
{
    return (height >= 4 * base.getRadius());
}
/**
    Accessor method for the height
    @return height
```

```
    */
    public double getHeight()
    {
        return height;
    }

    /**
        Accessor method for the radius
        @return radius
    */
    public double getBaseRadius()
    {
        return base.getRadius();
    }
    /**
        Mutator method for the height
        @param inHeight new value for height
    */
    public void setHeight(double inHeight)
    {
        height = inHeight;
    }

    /**
        Mutator method for the radius
        @param inBaseRadius new value for radius
    */
    public void setBaseRadius(double inBaseRadius)
    {
        base.setRadius(inBaseRadius);
    }

    /**
        toString method
        @return a String with radius and height
    */
    public String toString()
    {
        String str;

        str = base.toString() + "; Height is "  + height;
        return str;
    }
}
```

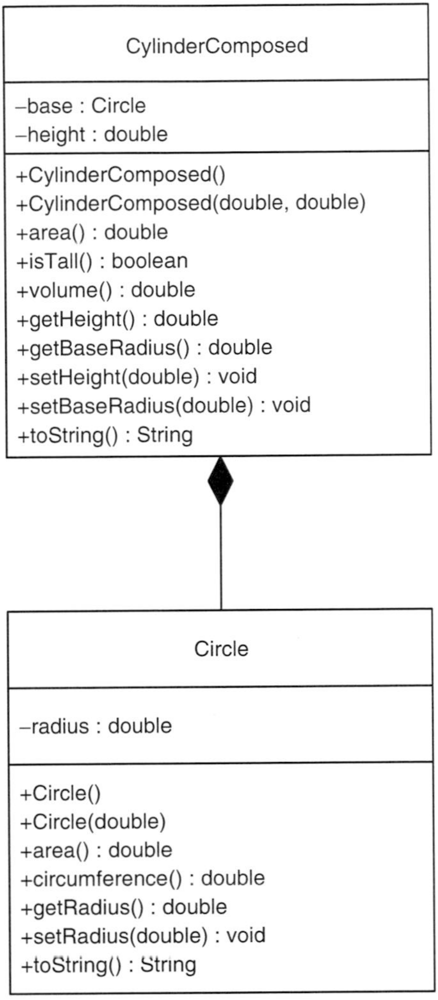

FIGURE 7.8 Class diagram of `CylinderComposed` and `Circle`.

The class diagram in UML 2 notation is shown in Figure 7.8.

Consider the following statements that create one instance of `Circle`, one instance of `CylinderInherited`, and one instance of `CylinderComposed`. An instance of `CylinderInherited` is included here for easy comparison.

```
Circle round = new Circle(8.74);
CylinderInherited roller = new CylinderInherited(4.23, 20.45);
CylinderComposed solidTube = new CylinderComposed(4.23, 20.45);
```

The reference variables `round`, `roller`, and `solidTube` along with the objects created can be visualized as shown in Figure 7.9.

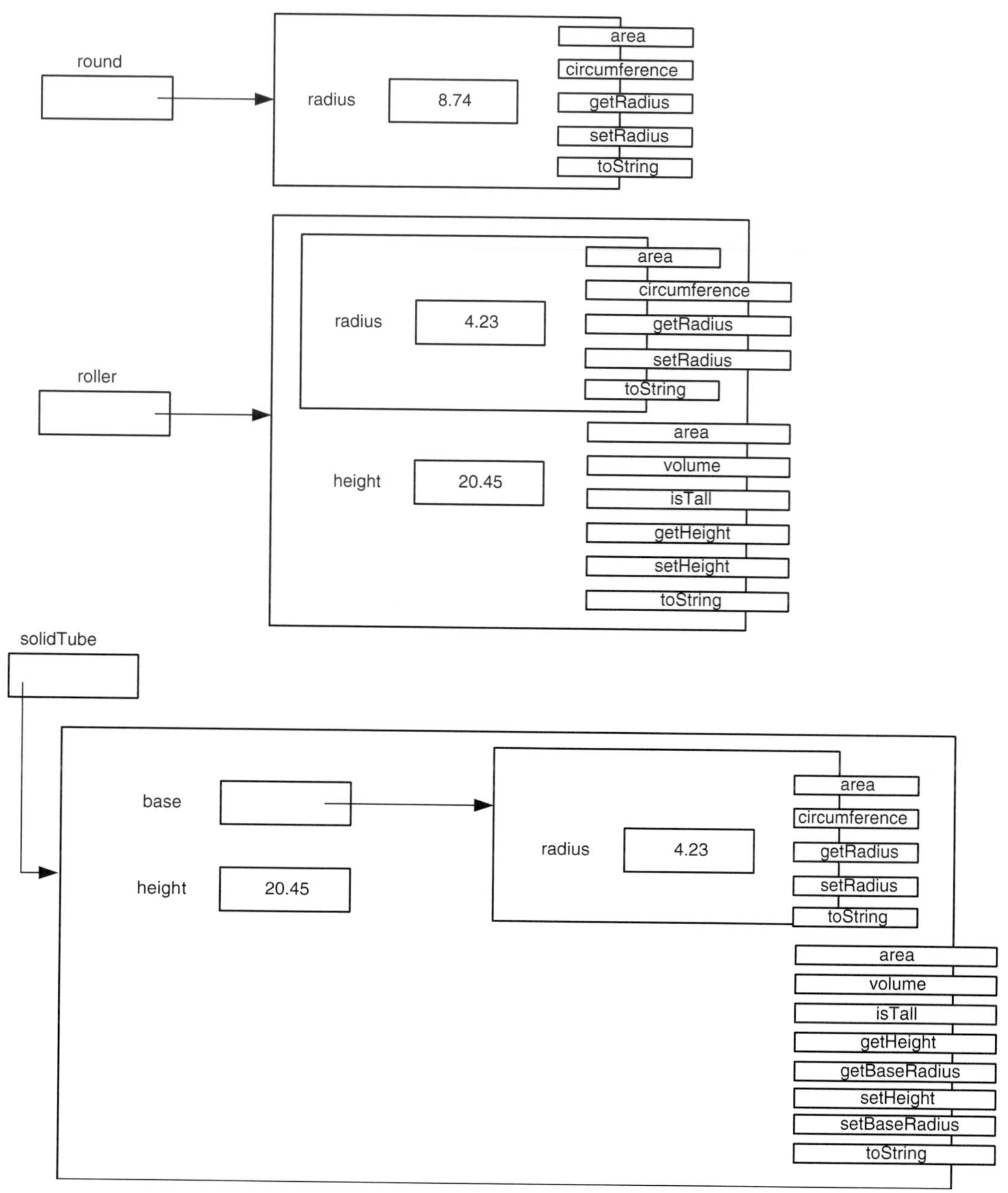

FIGURE 7.9 Visualization of inheritance and composition.

Methods that are accessible through the reference variable `solidTube` are the only services of the class `CylinderComposed`. Note that since `base` is a `private` attribute, all services associated with `base` are hidden from the user.

Note that Java statements

```
System.out.println(roller);
System.out.println(solidTube);
```

will produce the same output as

```
Radius is 4.23; Height is 20.45
```

We conclude this example by the following Java program that illustrates the use of all methods of the class `CylinderComposed`:

```
import java.text.DecimalFormat;

/**
    Application class to test the Cylinder created by composition
*/
public class CircleCylinderComposed
{
    public static void main(String[] args)
    {
        DecimalFormat twoDecimalPlaces
                                  = new  DecimalFormat("0.00"); // 1
        CylinderComposed solidTubeOne
                                      = new CylinderComposed(); // 2
        CylinderComposed solidTube
                             = new CylinderComposed(4.23,20.45); // 3
        System.out.println("(4) solidTubeOne data: "
                                             + solidTubeOne); // 4
        System.out.println("(5) solidTube data: "
                                                + solidTube); // 5
        System.out.println("\nMethods of CylinderComposed"); // 6
        System.out.println("(7)  Area of solidTube: "
              + twoDecimalPlaces.format(solidTube.area())); // 7
        System.out.println("(8)  Volume of solidTube: "
            + twoDecimalPlaces.format(solidTube.volume())); // 8
        if (solidTube.isTall())
            System.out.println("(9)  The solidTube is tall.");
                                                              // 9

        else
            System.out.println("(10) The solidTube is not
                                                tall.");    // 10
            solidTubeOne.setHeight(12.7);                   // 11
        System.out.println("(12) Height of solidTubeOne : "
                                 + solidTubeOne.getHeight()); // 12
                            solidTubeOne.setBaseRadius(6.98); // 13
        System.out.println("(14) Radius of solidTubeOne : "
                              + solidTubeOne.getBaseRadius()); // 14
    }
}
```

Output

```
(4)  solidTubeOne data: Radius is 0.0; Height is 0.0
(5)  solidTube data: Radius is 4.23; Height is 20.45

Methods of CylinderComposed
(7)  Area of solidTube: 655.94
(8)  Volume of solidTube: 1149.54
(9)  The solidTube is tall.
(12) Height of solidTubeOne: 12.7
(14) Radius of solidTubeOne: 6.98
```

INTERFACE

In Java, an `interface` can be thought of as an `abstract class` with no attribute. As mentioned before, Java does not allow multiple inheritance. Thus, Java allows single inheritance along with multiple interfaces. We will be using interfaces in Chapter 8. The following points are worth noting:

1. An `interface` can have named constants.
2. An `interface` cannot be instantiated (as in the case of an `abstract` class).
3. It is legal to declare reference variable of the `interface` type (as in the case of an `abstract` class).

As you have seen in this chapter, in the case of inheritance, we use the keyword `extends`. Similarly, in the case of interfaces, we use the keyword `implements`. We will cover all these ideas in detail in Chapter 8.

Self-Check

29. True or false: An `interface` cannot be instantiated.
30. True or false: It is legal to declare reference variable of the `interface` type.

CASE STUDY 7.1: PAYROLL FOR SMALL BUSINESS: REDESIGNED

Having learned inheritance, Ms. Smart has decided to redesign her payroll program for small business. The program specification remains the same.

Last time we saw Ms. Smart, who had three classes: `FullTimeEmp`, `PartTimeEmp`, and `SalesEmp`. Ms. Smart has decided to revisit the create classes step.

The UML 2 diagram of `FullTimeEmp`, `PartTimeEmp`, and `SalesEmp` classes is shown in Figure 7.10.

There are two attributes that are common to all three employee types: first name and last name. Every employee presently in the system or who may join in future must have a name. If there happens to be a new class of employees, they also will have first name

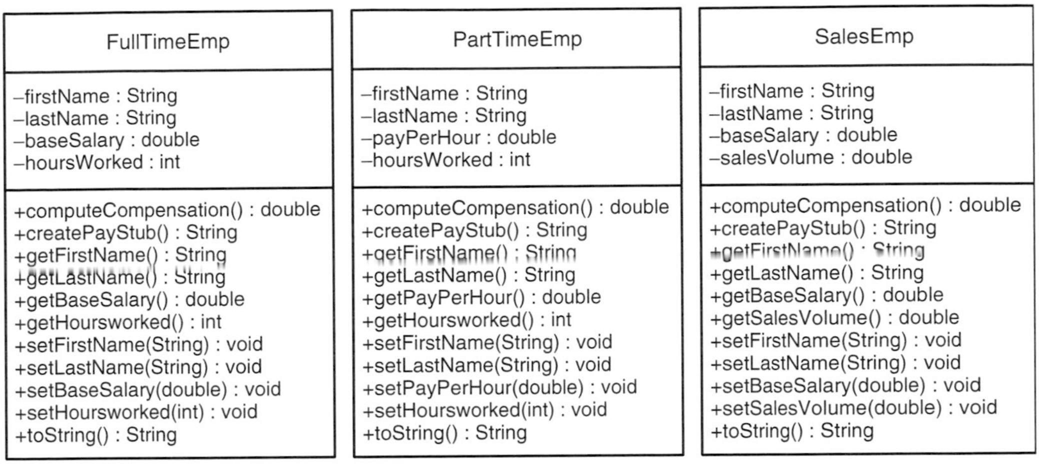

FullTimeEmp
–firstName : String –lastName : String –baseSalary : double –hoursWorked : int
+computeCompensation() : double +createPayStub() : String +getFirstName() : String +getLastName() : String +getBaseSalary() : double +getHoursworked() : int +setFirstName(String) : void +setLastName(String) : void +setBaseSalary(double) : void +setHoursworked(int) : void +toString() : String

PartTimeEmp
–firstName : String –lastName : String –payPerHour : double –hoursWorked : int
+computeCompensation() : double +createPayStub() : String +getFirstName() : String +getLastName() : String +getPayPerHour() : double +getHoursworked() : int +setFirstName(String) : void +setLastName(String) : void +setPayPerHour(double) : void +setHoursworked(int) : void +toString() : String

SalesEmp
–firstName : String –lastName : String –baseSalary : double –salesVolume : double
+computeCompensation() : double +createPayStub() : String +getFirstName() : String +getLastName() : String +getBaseSalary() : double +getSalesVolume() : double +setFirstName(String) : void +setLastName(String) : void +setBaseSalary(double) : void +setSalesVolume(double) : void +toString() : String

FIGURE 7.10 Three types of employees.

and last names as attributes. Thus, first name and last name are attributes for any type of employee, present or future.

Now consider the services. Since it is a payroll program, every employee must have an operation to compute the compensation. Similar is the case for creating pay stub. Therefore, both `computeCompensation` and `createPayStub` are services of any employee. However, these methods vary by the employee type. Therefore, for a general employee these methods are to be declared `abstract`. Recall that once a method is `abstract`, the class itself becomes `abstract`. Therefore, a new `abstract` class `Employee` is introduced. Existing three classes `FullTimeEmp`, `PartTimeEmp`, and `SalesEmp` change accordingly and become the subclasses of the `abstract` class `Employee`. In all four classes we include constructors instead of depending on the system-provided default constructor. The UML 2 diagram for all four classes is shown in Figure 7.11 and the corresponding Java code is as follows:

```
/**
    Abstract class Employee
*/
public abstract class Employee
{
    private String firstName;
    private String lastName;

/**
    Constructor initializes name with default values
*/
public Employee()
```

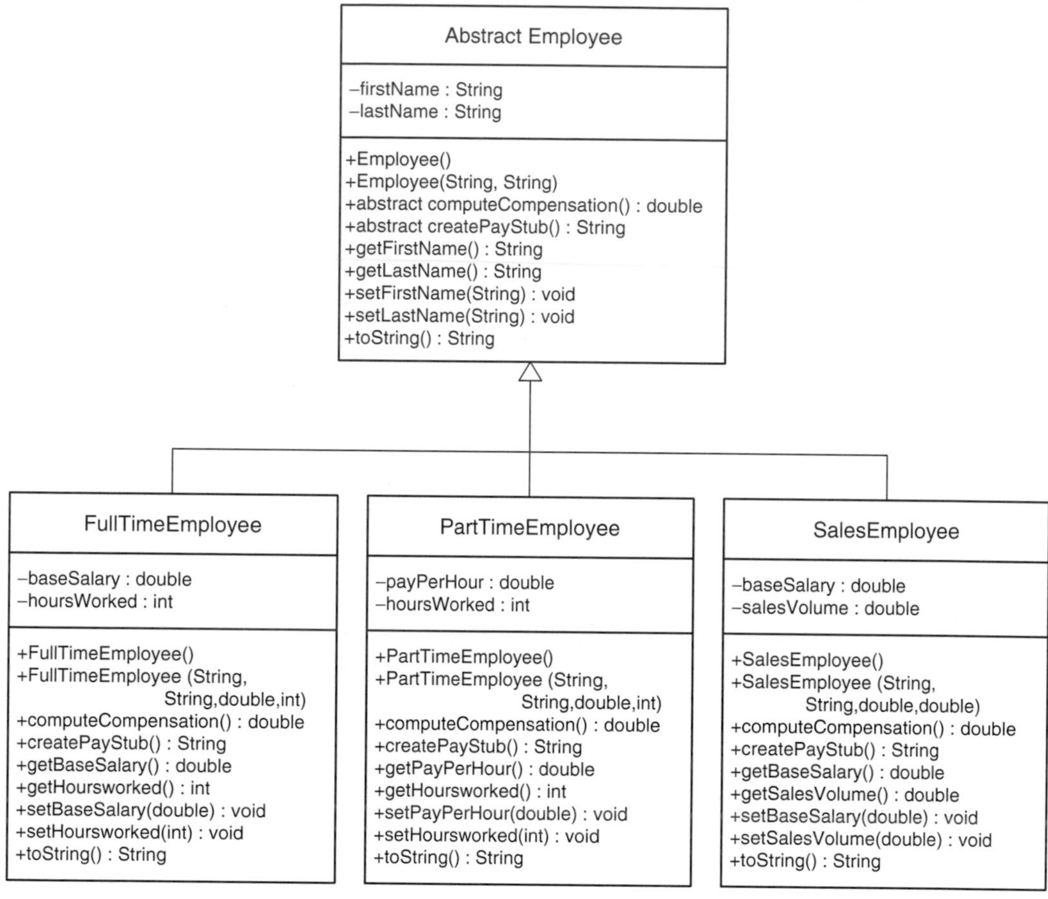

FIGURE 7.11 Employee and its subclasses.

```
{
    firstName = null;
    lastName = null;
}
/**
    Constructor initializes first and last name
    @param inFirstName first name
    @param inLastName last name
*/
public Employee(String inFirstName, String inLastName)
{
    firstName = inFirstName;
    lastName = inLastName;
}
/**
    Computes compensation
```

```
    @return compensation
*/
public abstract double computeCompensation();
/**
    Creates pay stub
    @return pay stub
*/
public abstract String createPayStub();
/**
    Accessor method for the first name
    @return first name
*/
public String getFirstName()
{
    return firstName;
}
/**
    Accessor method for the last name
    @return last name
*/
public String getLastName()
{
    return lastName;
}
/**
    Mutator method for the first name
    @param inFirstName new value for first name
*/
public void setFirstName(String inFirstName)
{
    firstName = inFirstName;
}
/**
    Mutator method for the last name
    @param inLastName new value for last name
*/
public void setLastName(String inLastName)
{
    lastName = inLastName;
}
/**
    toString method
```

```
        @return a String with name
    */
    public String toString()
    {
        String  str;
        str = firstName + " " + lastName;
        return str;
    }
}
/**
    FullTimeEmployee inherited from Employee
*/
public class FullTimeEmployee extends Employee
{
    private double baseSalary;
    private int hoursWorked;
    /**
        Constructor initializes with default values
    */
    public FullTimeEmployee()
    {
        super();
        baseSalary = 0.0;
        hoursWorked = 0;
    }
    /**
        Constructor initializes all values
        @param inFirstName the first name
        @param inLastName the last name
        @param inBaseSalary base salary
        @param inHoursWorked hours worked
    */
    public FullTimeEmployee(String inFirstName, String
              inLastName, double inBaseSalary, int inHoursWorked)
    {
        super(inFirstName, inLastName);
        baseSalary = inBaseSalary;
        hoursWorked = inHoursWorked;
    }
    /**
        Computes compensation
        @return compensation
```

```
*/
public double computeCompensation()
{
    double compensation, payPerHour;

    payPerHour = baseSalary /80;
    if (hoursWorked > 80)
    {
        compensation = baseSalary +
                        (hoursWorked - 80) * 1.5 * payPerHour;
    }
    else
    {
        compensation = baseSalary;
    }

    return compensation;
}

/**
    Creates pay stub
    @return pay stub
*/
public String createPayStub()
{
    DecimalFormat twoDecimalPlaces = new
                 DecimalFormat("0.00");
    double salary;

    salary = computeCompensation();
    String outStr;

    outStr = "\t\t\t" +
             "HEARTLAND CARS OF AMERICA" +
             "\n\n\t" +
             getFirstName() + " " + getLastName() +
             "\n\n" +
             "\n\tBasic Salary \t$" +
             twoDecimalPlaces.format(baseSalary) +
             "\n\tHours Worked \t " + hoursWorked +
             "\n\tPay          \t$" +
             twoDecimalPlaces.format(salary) +
             "\n\n";
    return outStr;
}
```

```
    /**
        Accessor method for the base salary
        @return base salary
    */
    public double getBaseSalary()
    {
        return baseSalary;
    }

    /**
        Accessor method for the hours worked
        @return hours worked
    */
    public int getHoursWorked()
    {
        return hoursWorked;
    }

    /**
        Mutator method for base salary
        @param inBaseSalary new value of base salary
    */
    public void setBaseSalary(int inBaseSalary)
    {
        baseSalary = inBaseSalary;
    }

    /**
        Mutator method for hours worked
        @param inHoursWorked new value of hours worked
    */
    public void setHoursWorked (int inHoursWorked)
    {
        hoursWorked = inHoursWorked;
    }

    /**
        toString method
        @return a String with name
    */
    public String toString()
    {
        String  str;
        str = "Full time employee : " + super.toString();
```

```
        return str;
    }
}

/**
    PartTimeEmployee inherited from Employee
*/
public class PartTimeEmployee extends Employee
{
    private double payPerHour;
    private int hoursWorked;

    /**
        Constructor initializes with default values
    */
    public PartTimeEmployee()
    {
        super();
        payPerHour = 0.0;
        hoursWorked = 0;
    }

    /**
        Constructor initializes all values
        @param inFirstName the first name
        @param inLastName the last name
        @param inPayPerHour pay per hour
        @param inHoursWorked hours worked
    */
    public PartTimeEmployee(String inFirstName, String inLastName,
                     double inPayPerHour, int inHoursWorked)
    {
        super(inFirstName, inLastName);
        payPerHour = inPayPerHour;
        hoursWorked = inHoursWorked;
    }

    /**
        Computes compensation
        @return compensation
    */
    public double computeCompensation()
```

```
{
    double compensation;
    compensation = payPerHour * hoursWorked;
    return compensation;
}

/**
    Creates pay stub
    @return pay stub
*/
public String createPayStub()
{
    DecimalFormat twoDecimalPlaces = new
                  DecimalFormat("0.00");
    double salary;

    salary = computeCompensation();
    String outStr;
    outStr = "\t\t\t" +
                "HEARTLAND CARS OF AMERICA" +
                "\n\n\t" +
                getFirstName() + " " + getLastName() +
                "\n\n" +
                "\n\tSalary/Hour \t$" +
                twoDecimalPlaces.format(payPerHour) +
                "\n\tHours worked \t " + hoursWorked +
                "\n\tPay          \t$" +
                twoDecimalPlaces.format(salary) +
                "\n\n";
    return outStr;
}

/**
    Accessor method for the pay per hour
    @return pay per hour
*/
public double getPayPerHour()
{
    return payPerHour;
}

/**
    Accessor method for the hours worked
    @return hours worked
*/
```

```
    public int getHoursWorked()
    {
        return hoursWorked;
    }

    /**
        Mutator method for pay per hour
        @param inPayPerHour new value of pay per hour
    */
    public void setPayPerHour (double inPayPerHour)
    {
        payPerHour = inPayPerHour;
    }

    /**
        Mutator method for hours worked
        @param inHoursWorked new value of hours worked
    */
    public void setHoursWorked (int inHoursWorked)
    {
        hoursWorked = inHoursWorked;
    }

    /**
        toString method
        @return a String with name
    */
    public String toString()
    {
        String  str;
        str = "Part-time employee : " + super.toString();
        return str;
    }
}

/**
    SalesEmployee inherited from Employee
*/
public class SalesEmployee extends Employee
{
    private double baseSalary;
    private double salesVolume;
```

```
    /**
        Constructor initializes with default values
    */
    public SalesEmployee()
    {
        super();
        baseSalary = 0.0;
        salesVolume = 0.0;
    }

    /**
        Constructor initializes all values
        @param inFirstName the first name
        @param inLastName the last name
        @param inBaseSalary base salary
        @param inSalesVolume sales volume
    */
    public SalesEmployee(String inFirstName, String inLastName,
                    double inBaseSalary, double inSalesVolume)
    {
        super(inFirstName, inLastName);
        baseSalary = inBaseSalary;
        salesVolume = inSalesVolume;
    }

    /**
        Computes compensation
        @return compensation
    */
    public double computeCompensation()
    {
        double compensation;

        compensation = baseSalary + 0.02 * salesVolume;

        return compensation;
    }

    /**
        Creates pay stub
        @return pay stub
    */
    public String createPayStub()
```

```
{
    DecimalFormat twoDecimalPlaces = new
                    DecimalFormat("0.00");
    double  salary;

    salary = computeCompensation();
    String outStr;

    outStr = "\t\t\t" +
                 "HEARTLAND CARS OF AMERICA" +
                 "\n\n\t" +
                 getFirstName( )+ " " + getLastName() +
                 "\n\n" +
                 "\n\tBasic Salary \t$" +
                 twoDecimalPlaces.format(baseSalary) +
                 "\n\tSales Volume \t$" +
                 twoDecimalPlaces.format(salesVolume) +
                 "\n\tPay          \t$" +
                 twoDecimalPlaces.format(salary) +
                 "\n\n";
    return outStr;
}

/**
    Accessor method for the base salary
    @return base salary
*/
public double getBaseSalary()
{
    return baseSalary;
}

/**
    Accessor method for the sales volume
    @return sales volume
*/
public double getSalesVolume()
{
    return salesVolume;
}

/**
    Mutator method for the base salary
    @param inBaseSalary new value of base salary
*/
```

```
    public void setBaseSalary(int inBaseSalary)
    {
        baseSalary = inBaseSalary;
    }

    /**
        Mutator method for the sales volume
        @param inSalesVolume new value of sales volume
    */
    public void setSalesVolume (double inSalesVolume)
    {
        salesVolume = inSalesVolume;
    }

    /**
        toString method
        @return a String with name
    */
    public String toString()
    {
        String  str;
        str = "Sales employee : " + super.toString();
        return str;
    }
}
import java.util.*;
import java.io.*;

/**
    Application program for Heartland Cars of America
*/
public class HeartlandCarsOfAmericaEmployeePayRoll
{
    //public static void main (String[] args)
    public static void main (String[] args) throws
                FileNotFoundException, IOException
    {
        //Create reference variable of all three employee types
        Employee employee = null;

        //Declare variables to input

        char inputEmployeeType;
        String inputFirstName;
```

```
String inputLastName;
double inputBaseSalary;
double inputPayPerHour;
int inputSalesVolume;
int inputHoursWorked;

//Get two input values
//    Scanner scannedInfo = new Scanner(System.in);
Scanner scannedInfo = new Scanner(
                     new File("C:\\Employee.dat"));
PrintWriter outFile = new PrintWriter(
                new FileWriter("C:\\payroll.dat"));

while (scannedInfo.hasNext())
{

    inputEmployeeType
                    = scannedInfo.next().charAt(0);
    switch (inputEmployeeType)
    {
    case 'F' :
    case 'f' :

        inputFirstName = scannedInfo.next();
        inputLastName = scannedInfo.next();
        inputBaseSalary
                     = scannedInfo.nextDouble();
        inputHoursWorked = scannedInfo.nextInt();

        //create an object
        employee = new FullTimeEmployee
               (inputFirstName, inputLastName,
           inputBaseSalary, inputHoursWorked);
        break;

    case 'P' :
    case 'p' :

        inputFirstName = scannedInfo.next();
        inputLastName = scannedInfo.next();
        inputPayPerHour
                     = scannedInfo.nextDouble();
        inputHoursWorked = scannedInfo.nextInt();

        //create an object and initialize attributes
```

```
                employee = new PartTimeEmployee
                  (inputFirstName, inputLastName,
                  inputPayPerHour, inputHoursWorked);
                break;

            case 'S' :
            case 's' :

                inputFirstName = scannedInfo.next();
                inputLastName = scannedInfo.next();
                inputBaseSalary
                             = scannedInfo.nextDouble();
                inputSalesVolume = scannedInfo.nextInt();

                //create an object and initialize
                                       attributes
                employee = new SalesEmployee
                         (inputFirstName,inputLastName,
                    inputBaseSalary, inputSalesVolume);

                break;

            default:
                System.out.println("Check data file.");
                return;

            }        // End of switch

            //invoke the createPayStub method
            outFile.println(employee.createPayStub());

        }            // End of while

        outFile.close();
    }                // End of main
}                    // End of class
```

Input File Content

```
F Adam Smith 2450.00 87
F Joyce Witt 3425.67 80
F Mike Morse 1423.56 75
P Chris Olsen 34.56 34
S Patrick McCoy 1040.57 856985
```

Output

```
                    HEARTLAND CARS OF AMERICA

Adam Smith

Basic Salary    $2450.00
Hours Worked     87
Pay             $2771.56

                    HEARTLAND CARS OF AMERICA

Joyce Witt

Basic Salary    $3425.67
Hours Worked     80
Pay             $3425.67

                    HEARTLAND CARS OF AMERICA

Mike Morse

Basic Salary    $1423.56
Hours Worked     75
Pay             $1423.56

                    HEARTLAND CARS OF AMERICA

Chris Olsen

Salary/Hour     $34.56
Hours worked     34
Pay             $1175.04

                    HEARTLAND CARS OF AMERICA

Patrick McCoy

Basic Salary    $1040.57
Sales Volume    $856985.00
Pay             $18180.27
```

REVIEW

1. From a data-centric perspective, an object is a collection of attributes and operations that manipulate the data.
2. Grouping data and operations on data is called encapsulation.

3. The fact that implementation is hidden from the user is known as information hiding.
4. From a client–server perspective, every object in the application provides some service.
5. The delegation of task by a server is completely hidden to the client and as far as the client is concerned, the responsibility rests with the server.
6. From a design perspective, you consider an object as an instance of a class.
7. Each class is an abstraction of some real-life entity or concept.
8. The superclass/subclass relationship establishes an inheritance hierarchy.
9. In a subclass a service can be overridden. This form of inheritance is known as inheritance with polymorphism.
10. There are two forms of inheritance: inheritance with polymorphism and inheritance without polymorphism.
11. Java does not permit multiple inheritance.
12. All `private` members of a superclass are not directly accessible to any other class, including the subclass.
13. All `public` members of a superclass are directly accessible to any other class, including the subclass.
14. All `protected` members of a superclass are directly accessible to subclass but not to any other class.
15. Method overloading refers to two or more methods with identical names but different signatures within a class.
16. Method overriding refers to two methods with identical names and identical signatures, one in the superclass and the other in the subclass.
17. Use `super` as an implicit parameter inside all methods of a subclass that are not marked `static` to invoke a method of a superclass that is not marked `static`.
18. Invoking the constructor of the superclass is the first executable statement inside the constructor of the subclass.
19. An instance of a subclass is always an instance of the superclass. Therefore, a subclass reference can be assigned to a superclass reference.
20. A method marked as `final` cannot be overridden in a subclass.
21. A class marked as `final` cannot have any subclass.
22. A class is marked `abstract` if any instance of it is an instance of one of its subclasses.
23. If a class has an `abstract` method, the class becomes `abstract`.
24. An `abstract` class need not contain any `abstract` method.
25. An `abstract` class can have reference variables.
26. The `Object` class is the topmost class in the inheritance hierarchy.

27. Every other class in Java inherits all services of the `Object` class.
28. In the case of composition, the constructor must instantiate every attribute that happens to be an object reference through a `new` operator.
29. In Java, interfaces can be thought of as `abstract` classes with no attribute.

EXERCISES

1. Mark the following statements as true or false:
   a. You, as a user, need not understand the technical details inside your iPhone is an example of encapsulation.
   b. In a program, an object will be either a client or a server. It cannot be both.
   c. The subclass inherits all attributes and services of the superclass.
   d. Rectangle is a subclass of square.
   e. Polymorphism refers to the presence of a method having the same name but possibly different signatures in a superclass and subclass.
   f. The keyword `extends` is used to create a subclass from a superclass.
   g. A `protected` attribute is accessible in the subclass.
   h. The keyword `super` references the implicit attribute as an instance of the superclass.
   i. It is legal to modify an inherited `private` attribute.
   j. The `new` operator is not allowed inside a constructor.
   k. For a class to be `abstract`, it must have at least one `abstract` method.
   l. It is legal to create instances of an `abstract` class.
   m. It is legal to create reference variables of an `abstract` class.

2. Fill in the blanks
   a. As a designer of the class you may mark a class ______________ to prevent creation of any subclass.
   b. A ______________ member is available only in the class and its subclass.
   c. ______________ data member of a subclass is inherited by a superclass.
   d. Let `B` be a subclass of `A`. Let `try()` be a method of `A`. Inheritance without polymorphism means method `try()` is ______________ implemented in `B`.
   e. Let `B` be a subclass of `A`. Let `try()` be a method of both classes. In class `B`, you invoke `try` of class `A` using the keyword ______________.
   f. If you see the code "`X extends Y`" in a Java program, `X` is a ______________ of `Y`.
   g. In a subclass constructor, you invoke the default constructor of the superclass as ______________.
   h. If no access modifier is specified, the class has ______________ access.

3. Consider three classes `A`, `B`, and `C`. Class `A` is a superclass of `B` and `C`. The class `A` has two methods: `void alpha()` and `void beta()`. The class `B` has two methods: `void beta()` and `void gamma()`. The class `C` has two methods: `void alpha()` and `void gamma()`. Further assume that `a` is an instance of `A`, `b` is an instance of `B`, and `c` is an instance of `C`. Indicate whether or not the segment of code has any error. If there is an error that can be corrected through proper casting, then correct it. Once the segment of code is error free, identify the method that is being invoked.

   i. `a = b; a.beta();`

   ii. `a = b; ...; b = a; b.beta();`

   iii. `a = b; ...; c = a; c.alpha();`

   iv. `a = b; ...; b = a; b.gamma();`

   v. `b.alpha();`

   vi. `a = b; ...; b = (C) a; b.beta();`

   vii. `a = b; ...; a.alpha();`

   viii. `a = c; ...; a.alpha();`

4. Determine whether or not each of the following is a superclass/subclass pair. If so, identify the superclass. Can the superclass be an `abstract` class? If not, suggest a possible superclass that is `abstract`.

   a. Person, car
   b. Vehicle, car
   c. Course, test
   d. Country, USA
   e. Car, truck
   f. Employee, person
   g. House, building
   h. Part-time employee, full-time employee
   i. Animal, dog
   j. Doctor, health care professional

5. Consider three classes `BankAccount`, `SavingsAccount`, and `CheckingAccount` maintained by your local bank. A person having a savings account receives interest at the end of each month based on his average balance. However, the savings account allows only three withdrawals for each month. The checking account receives no interest and it allows unlimited number of withdrawals per month. Based on these facts, answer the following:

   a. Is there any superclass/subclass relationship(s) among `BankAccount`, `SavingsAccount`, and `CheckingAccount`?

   b. Suggest the best possible candidate to be an `abstract` class. For the remaining questions, assume that the class you have suggested is chosen `abstract`.

c. Suggest an attribute for the `abstract` class.

d. Suggest a method for the `abstract` class.

e. Suggest an `abstract` method for the `abstract` class.

f. Suggest a method for one of the classes that is not a member of the other two classes, including the `abstract` class.

6. Let `ClassOne` be superclass of `ClassTwo`. Assume that `ClassOne` has a method with the following heading:

```
public void testing()
```

and it is overridden in class `ClassTwo`. Answer the following:

a. How to invoke `testing` of `ClassTwo` inside another method of `ClassTwo`?

b. How to invoke `testing` of `ClassOne` inside another method of `ClassTwo`?

c. How to invoke `testing` of `ClassOne` inside the method `testing` of `ClassTwo`?

d. How to invoke `testing` of `ClassTwo` inside a method of `ClassThree`.

e. How to invoke `testing` of `ClassOne` inside a method of `ClassThree`.

f. Assume that `ClassThree`, one of the methods, have a local variable `z` of type `ClassOne`. Is the statement `z = new ClassTwo();` legal?

g. Assume that `ClassThree` has an instance variable `z` of type `ClassOne`. Is the statement `z = new ClassTwo();` legal?

h. Assume that `ClassThree`, one of the methods, has a local variable `z` of type `ClassTwo`. Is the statement `z = new ClassOne();` legal?

i. Assume that `ClassThree` has an instance variable `z` of type `ClassTwo`. Is the statement `z = new ClassOne();` legal?

7. Write Java method headings. In all cases assume that the method name is `trial`, returns an `int`, and has two formal parameters of `String` and `double`, respectively. If a certain case is not possible, explain the reason.

a. Marked as `abstract`; but not `static`; not `final`

b. Marked as `abstract` and `static`; but not `final`

c. Marked as `final` and `static`; but not `abstract`

d. Marked as `abstract` and `final`; but not `static`

8. Given two classes, `ClassOne` and `ClassTwo`. Consider the three different ways a new class, `ClassThree`, can be created from `ClassOne` and `ClassTwo`: (a) `ClassThree` is a subclass of `ClassOne` and there is an attribute of the type `ClassTwo`, (b) `ClassThree` is a subclass of `ClassTwo` and there is an attribute of the type `ClassOne`, and (c) `ClassThree` has two attributes: one of the type `ClassOne` and the other of the type `ClassTwo`. Explain, as a designer, which option is the most appropriate?

PROGRAMMING EXERCISES

1. Create the class CircleProCylinderInherited by modifying the class CircleCylinderInherited.
2. Create a class Rectangle having two attributes: length and width. Keep length as private and width as protected instance variables of type int. The Rectangle class has two services: area and perimeter. Extend the class to ThreeDRectangle by adding a protected int attribute height. The ThreeDRectangle overrides the method area of the Rectangle class. Further, it has a method volume.
3. Redesign the Employee class of Programming Exercise 1 of Chapter 6. Create a class Name with two attributes and then create Employee class by extending the Name class.
4. Modify the Name and Employee classes of the Programming Exercise 3 by adding a copy constructor, copy method, equals method, and compareTo method to both classes. Two employees are ordered based on their names.
5. Extend the Employee class of Programming Exercise 3 or 4 to Boss class by introducing one more attribute noOfEmployees to keep track of the number of employees to supervise. If you extend Employee class of Programming Exercise 4, add a copy constructor, copy method, equals method, and compareTo method to Boss class.
6. Redo Programming Exercise 3, applying composition instead of inheritance.
7. Redo Programming Exercise 4, applying composition instead of inheritance.
8. Redo Programming Exercise 5, applying composition instead of inheritance.
9. Consider the class CircularCounter of Chapter 3. Create a class SecondClock by extending the CircularCounter. The second clock counts up to 60. There are two methods, tick() and tick(int s). tick() increments by 1 second and tick(int s) increments by s seconds. Create a MinuteClock by extending SecondClock and create Clock by extending MinuteClock. (*Hint*: You may need to override tick and tick(int s) in one or both of the classes, MinuteClock and Clock.)
10. Consider the class CircularCounter of Chapter 3. Create a class Clock having three attributes minute, second, and hour. Each one is of the type CircularCounter. Provide methods tick and tick(int s) that will increment the clock by 1 and s seconds, respectively.
11. Create a class Person from two other classes: Name and Address. The Name class has two attributes: first name and last name. The Address class has four attributes: street, city, state, and zip code. Create Person by extending Name.
12. Redo Programming Exercise 11, using composition instead of inheritance.
13. Create two subclasses Student and Faculty of the Person class of either Programming Exercise 11 or 12 or start with a Person class with six attributes: first

name, last name, street, city, state, and zip code. Student has two new attributes: major and year of graduation. The faculty has two more attributes: specialty and salary.

14. Create an abstract class `GeometricFigure` with one attribute `dimension` and two abstract methods `area` and `magnify`. Create each of the following classes. The area of the ellipse can be calculated as `Math.PI * a * b`, where `a` and `b` are major and minor axes of the ellipse, respectively.
    a. Create `Point` as a subclass of `GeometricFigure`. `Point` has two attributes `x` and y, both of type `int`.
    b. Create `Ellipse` as a subclass of `GeometricFigure`. Ellipse has two attributes `a` and `b`, both of type `int` for major and minor axes.
    c. Create `Circle` as a subclass of `Ellipse`. No additional attribute required.
    d. Create `Rectangle` as a subclass of `GeometricFigure`. Rectangle has two attributes `a` and `b`, both of type `int` for length and width.
    e. Create `Square` as a subclass of `Rectangle`. No additional attribute required.
15. Redo Programming Exercise 14, by creating an interface `GeometricFigure` with two methods, area and magnify.

ANSWERS TO SELF-CHECK

1. object
2. template
3. Attributes
4. `double`
5. `private`, `public`
6. True
7. client, server
8. True
9. classes, attributes, services
10. classes
11. True
12. False
13. inherits
14. override
15. False
16. `private`
17. overriding

18. overloading
19. superclass
20. `public`, `protected`
21. `new`
22. `super`
23. True
24. False
25. True
26. True
27. abstract
28. False
29. True
30. True

CHAPTER 8

GUI Applications, Applets, and Graphics

In this chapter you learn

- Java concepts
  - Principles of event-driven programming, event-interface model, interfaces, inner classes, anonymous inner classes, listener interfaces, GUI components, graphics, color, and font classes
- Programming skills
  - Design and create three different types of Java GUI programs: applications, applets, and applet applications

Graphical user interfaces (GUIs) have revolutionized the world of computers and made computers useful to ordinary people. Internet became a very useful tool mainly due to its sophisticated GUI. One no longer needs to type strange character combinations of yesteryears. One can perform almost any task by just clicking the mouse.

In this chapter, you learn the fundamentals of GUI programming. A GUI program is more attractive, intuitive, and user-friendly. You will be introduced to six Java classes: `Component`, `Container`, `JFrame`, `JLabel`, `JTextField`, and `JButton`. A GUI program is an event-driven program. User-generated events, such as clicking the mouse button or pressing the Enter key, determine the next task performed by a GUI program. This chapter explains the principles behind event-driven programming. With the help of the six Java classes and the principles introduced in this chapter, you can create a wide variety of application programs and applets. Further, Programming Exercises at the end of this chapter introduce additional GUI components. Graphics `package` presented in this chapter will enable you to use colors, fonts, and drawing services.

COMMON THEME BEHIND ALL GUI APPLICATION PROGRAMS

Creating a GUI program involves two stages of separate but interlinked activities. The first stage is creating the desired appearance on the screen. This is done by creating instances of appropriate GUI classes. For the sake of simplicity, let us refer this stage as *creating the application window.* The second stage involves writing a program so that the application window created in stage 1 becomes a functional *user interface.* Let us refer to this stage as *event-driven programming.*

Creating a GUI program can be summarized as follows:

1. *Creating the application window.* Create the desired appearance on the screen by creating instances of appropriate GUI classes.
2. *Event-driven programming.* Add necessary Java code that turns the application window created in stage 1 into a functional user interface.

We cover these two stages of GUI program development in sequence.

Example 8.1

Consider the problem of computing the sum of first N integers. Creating the application window begins with deciding on inputs and outputs. In this case, all you need is one input and one output. Next, you need to decide on control buttons. You need one to calculate the sum and another to exit the application. Therefore, in the first stage we create an application window as shown in Figure 8.1.

Recall that the control buttons `Calculate` and `Exit` will not work at the end of the first stage. The second stage of the application development makes controls `Calculate` and `Exit` behave as planned.

Example 8.2

This example identifies various GUI components required to create the application window of Example 8.1 (Figure 8.2).

Here, a `JFrame` object contains all other GUI components. There are two `JTextField` objects. We use one of them for the input and another for the output. The

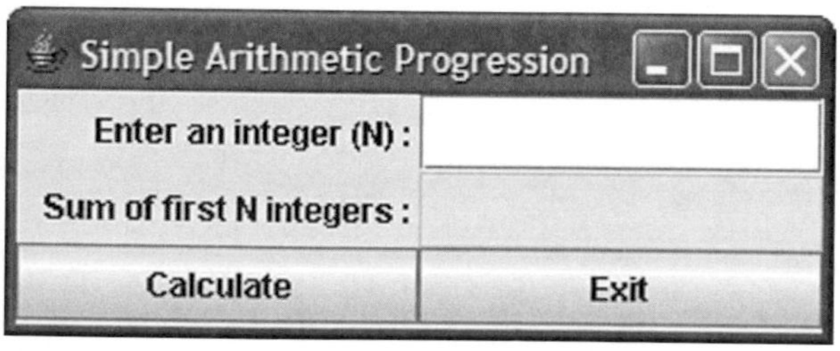

FIGURE 8.1 Application window for simple arithmetic progression.

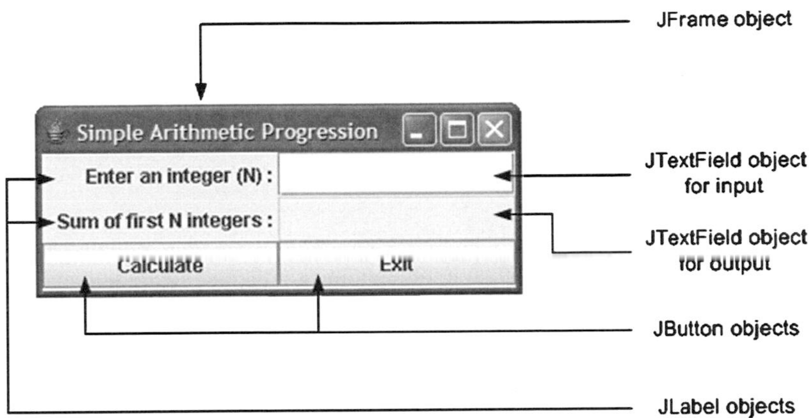

FIGURE 8.2 Identification of GUI components.

JTextField object intended for the output is kept as noneditable to prevent the user from entering any value. We use JLabel objects to properly label inputs and outputs. Finally, JButton objects are used as controls. In this example, eventually, after entering an integer, clicking the JButton object labeled Calculate computes the sum and displays it in the output JTextField. Clicking the JButton Exit terminates the application.

Self-Check

1. Creating the application window involves creating the desired appearance on the screen by creating instances of appropriate ________ classes.
2. Event-driven programming involves adding necessary ________ that turns the application window created in stage 1 into a functional user interface.

CREATING APPLICATION WINDOW

The application window of all applications is created by extending the JFrame class of the Java package javax.swing. All other GUI components are contained inside the borders of the JFrame object. To be more specific, the interior area of a JFrame object is called a *content pane*. The content pane itself is an instance of the Container class. The Container class has a service add that can be used to add any instance of Component class. In particular, we use add service of the Container class to place various GUI objects in the content pane of the JFrame object. Thus, the process of creating the application window has three steps:

1. Create a new application class by extending the JFrame class
2. Get the reference of the content pane
3. Create and place necessary GUI components in the content pane

Each one of the above steps is explained below.

Self-Check

3. The application window of all applications is created by extending the _________ class of the Java package _________.
4. The content pane itself is an instance of the _________ class.

Creating New Application Class

We have been creating applications by creating a new class or classes and using them in an application program. The application program had a single method, the `main`. A GUI application also follows the same pattern, except that the new class is created by extending `JFrame` class. Thus, we have the following:

```
import javax.swing.*;

public class StartingPoint extends JFrame
{
    //private data members

    //constructors and methods

}

public class StartingPointApplication
{
    //main method
    public static void main(String[] args)
    {
        StartingPoint sp = new StartingPoint();
    }
}
```

Unlike the applications you have seen so far, the `main` of a GUI application is going to be quite short. Therefore, instead of creating two classes, we keep `main` as part of the new class created. Thus, we have the following:

```
import javax.swing.*;
//other import statements

public class StartingPoint extends JFrame
{
    //private data members
```

```
    //constructors and methods

    //main method
    public static void main(String[] args)
    {
        StartingPoint sp = new StartingPoint();
    }
}
```

Observe that we need to create the necessary graphics only once. Therefore, we make use of the constructor to create the application window. Thus, the general structure of all GUI applications in this chapter is as follows:

```
import javax.swing.*;
//other import statements

public class StartingPoint extends JFrame
{
    //private data members

    public StartingPoint()
    {
        //Java statements to create
        //application window
    }

    //additional constructors and methods

    //main method
    public static void main(String[] args)
    {
        StartingPoint sp = new StartingPoint();
    }
}
```

Recall that the application starts by executing the `main`. The `main` invokes the constructor of the class. The constructor in turn creates the application window. Therefore, the necessary code to create and display the application window is part of the constructor of the class. Therefore, at the very minimum, the constructor needs to perform the following tasks:

1. Invoke the constructor of the superclass
2. Define the size of the `JFrame`

TABLE 8.1 Constructors of the `JFrame` Class

Constructor	Explanation
`public JFrame();`	`Constructor with no arguments. Creates a JFrame object with no title. Both width and height are zero. The object is not visible`
`public JFrame(String str);`	`Constructor with a String argument. Creates a JFrame object with str as title. Both width and height are zero. The object is not visible`

3. Make the `JFrame` visible
4. Provide a graceful way to exit the application

We address each of these issues in detail.

Self-Check

5. Every Java application has a ________ method.
6. Every Java application starts by executing the ________ method.

Invoking Constructor of Superclass

Two constructors of the `JFrame` class are shown in Table 8.1. Throughout this chapter, we use the second constructor. This allows us to provide a meaningful name to our application in the title bar of the `JFrame`. For instance, to give a title such as `"Simple Arithmetic Progression"` you invoke the constructor of `JFrame` as shown below:

```
super("Simple Arithmetic Progression");
```

Self-Check

7. Write the necessary Java code to create an instance of `JFrame` with title `"Welcome to Java"`.
8. True or false: The `JFrame` constructor with zero arguments creates an application window with no title.

Define Size of `JFrame`

One of the ways to custom design a GUI application is to define the size explicitly. All GUI components occupy a certain rectangular area in your monitor's screen. Thus, every GUI component has a width (the horizontal measure) and height (the vertical measure). The unit of measure is a *pixel.* The term pixel stands for picture element and is the smallest unit on your screen that you can control. The pixel size depends on the current monitor setting. For instance if our current monitor resolution is 1280 by 1024, then there are 1280 pixels in each horizontal line and 1024 pixels in each vertical line.

In this case, if you want your frame to occupy 1/4 horizontally, you select width as 1280/4 = 320. Similarly, if you want your frame to occupy 1/2 vertically, you select height as 1024/2 = 512.

To set the size, invoke the `setSize` method of the superclass `Component`. Since the program is not changing the values of width and height, we can keep them as named constants. Thus, we have the following:

```
private static final int WIDTH = 300;
private static final int HEIGHT = 120;

super.setSize(WIDTH, HEIGHT);
```

Self-Check

9. The unit of measure of a GUI component is a _________.
10. The term pixel stands for _________.

Make `JFrame` Visible

You can control the visibility of a GUI component through `setVisible` method of the superclass `Component`. This method has one formal parameter of the type `boolean`. If the actual parameter is `true`, the GUI is visible; otherwise, GUI remains invisible. Thus, you need the following statement to make the application window visible:

```
super.setVisible(true);
```

Self-Check

11. The method `setVisible` belongs to _________ class.
12. The `setVisible` method has one formal parameter of the type _________.

Provide Graceful Way to Exit Application

As you may be well aware of, a window comes with a close button. The user can click on the close button to terminate the application program. To enable this feature, we include the following statement:

```
setDefaultCloseOperation(EXIT_ON_CLOSE);
```

Thus, we have the following application program that can be compiled and executed just like any other application. You can close the window by clicking the "close" button, ×, on the upper right-hand corner of the window (in Microsoft Windows). (Other platforms may have different locations and different looks.)

```
import javax.swing.*;

public class StartingPoint extends JFrame
```

```
{
    private static final int WIDTH = 300;
    private static final int HEIGHT = 120;

    public StartingPoint()
    {
        super("Simple Arithmetic Progression");

        super.setSize(WIDTH, HEIGHT);

        //add other GUI components here.

        super.setVisible(true);
        super.setDefaultCloseOperation(EXIT_ON_CLOSE);

    }

    public static void main(String[] args)
    {
        StartingPoint sp = new StartingPoint();
    }
}
```

Output

Many of the useful services of `JFrame` are inherited from two superclasses, the `Component` and the `Container` (Table 8.2). In Java, every GUI component is derived from the `Component` class. Some of the components are such that they can contain other components. These components are members of the `Container` subclass. The simple arithmetic progression is shown in Figure 8.3 and the inheritance hierarchy is shown in Figure 8.4.

Self-Check

13. In Java, every GUI component is derived from the __________ class.
14. The method __________ determines the action taken as the user clicks the close button, ×, appearing on the top right-hand corner of the window.

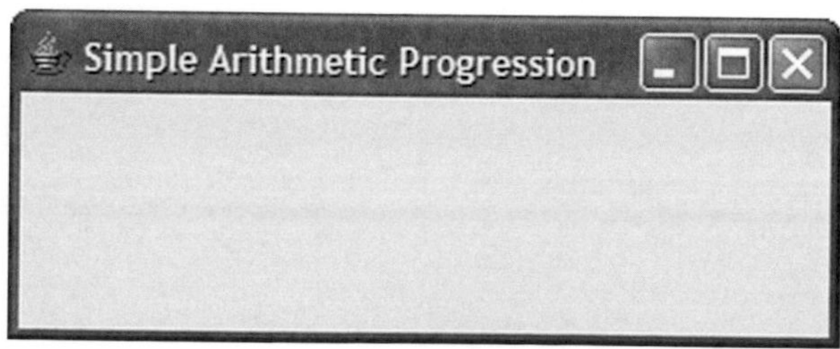

FIGURE 8.3 Application window with a visible `JFrame`.

TABLE 8.2 Services of `Component`, `Container`, and `JFrame` Classes

Service	Explanation
java.awt. Component	
public void setSize(**int** width, **int height**)	Sets both width and height
public void setVisible(**boolean** bool);	The Component object is visible if bool is true. The object remains invisible otherwise
java.awt.Container	
public void add(Component comp)	Adds the Component comp to the Container object
java.awt.JFrame	
public Container getContentPane()	Returns the reference of the content pane
public void setDefaultCloseOperation (**int** action)	Determines the action taken as the user clicks the close button, ×, appearing on the top right-hand corner of the window. There are four choices for the parameter action. They are the constants: EXIT_ON_CLOSE, DISPOSE_ON_CLOSE, HIDE_ON_CLOSE, and DO_NOTHING_ON_CLOSE. The constant EXIT_ON_CLOSE is defined in the JFrame class and other constants are defined in javax.swing.WindowConstants

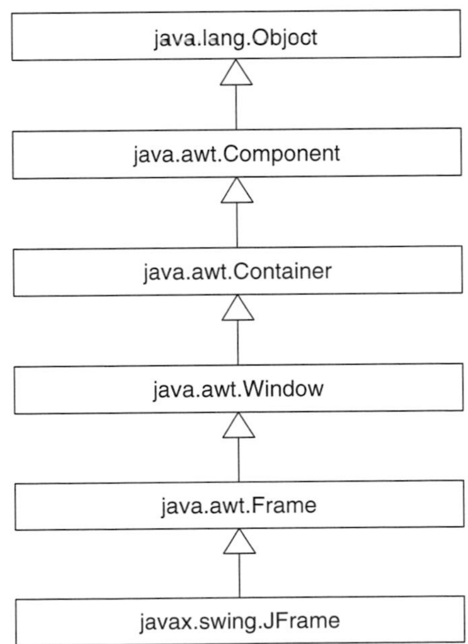

FIGURE 8.4 The inheritance hierarchy of `JFrame`.

Get Reference of Content Pane

Before jumping into coding, keep in mind that all the code we develop in the first stage is part of the constructor of the class we are developing.

Recall that `JFrame` has a *content pane* and GUI components are placed on the content pane and not on the `JFrame`. The content pane can be thought of as the interior part of the `JFrame`. The content pane is an instance of the `Container` class. The following statement creates a reference variable `conInterior` of type `Container` and initializes with the reference of the content pane of the `JFrame`:

```
Container conInterior = super.getContentPane();
```

Once the content pane is accessible, `add` service of the `Container` class can be used to place other components. However, before we can place other GUI components, a layout manager has to be set for the content pane. The layout manager determines the size and location of components placed inside a container. We use the `GridLayout` manager in this chapter (Table 8.3).

For example, the following Java statement creates a `GridLayout` manager with three rows and two columns:

```
GridLayout gridlayout;                          // (1)

gridlayout = new GridLayout(3,2);               // (2)
```

Now, the layout manager of the content pane `conInterior` can be set as `gridlayout` through the `setLayout` service of the `Container` class as shown below:

```
conInterior.setLayout(gridlayout);              // (3)
```

Observe that the only reason we declared a reference variable `gridlayout` in Line 1 is to use it as an actual parameter in Line 3. There is no need to store the reference returned by

TABLE 8.3 Constructors of the `GridLayout` Class

Constructor	Explanation
`public GridLayout()`	Creates a one row, one column grid layout
`public GridLayout(int r, int c)`	Creates a grid layout with r rows and c columns. All grids have the same dimension
`public GridLayout(int r, int c, int rsep, int csep)`	Creates a grid layout with r rows and c columns. Rows are separated by rsep pixels, and columns are separated by csep pixels. All grids have the same dimension

new GridLayout(3,2) in a reference variable such as gridlayout as in Line 2. Instead, you could use the expression **new** GridLayout(3,2) as the actual parameter. Thus, Lines 1–3 can be replaced by Line 4.

```
conInterior.setLayout(new GridLayout(3,2));        // (4)
```

Self Check

15. JFrame has a _________ and GUI components are placed on it.
16. The content pane is an instance of the _________ class.

Create and Place GUI Components in Content Pane

In this section, you learn how to create and place various GUI components in the content pane. This step can be further split into two smaller steps:

1. Component creation. Create a reference variable of the component type and use one of the constructors to instantiate the reference variable.
2. Component placement. Place the object created in Step 1 in the content pane using the add service of the Container class.

We illustrate these concepts by three different types of GUI components: JLabel, JtextField, and JButton.

Component creation

JLabel As the name suggests, an instance of JLabel is used to label various GUI components of an application. The primary use of JLabel objects is to label other GUI components. The four constructors of the JLabel class are listed in Table 8.4.

TABLE 8.4 Constructors of JLabel Class

Constructor	Explanation
public JLabel(String str)	Constructor with a String argument. Creates a JLabel object with str as a left-aligned label
public JLabel(Icon ic)	Constructor with an Icon argument. Creates a JLabel object with Icon ic
public JLabel(String str, **int** halign)	Constructor with a String argument and a horizontal alignment specification. The halign can be SwingConstants.LEFT, SwingConstants.RIGHT, SwingConstants.CENTER for left, right, or center alignment of the label str
public JLabel(String str, Icon ic, **int** halign)	Constructor with all three arguments. Icon ic will appear left of the label str

The following segment of code creates two `JLabel` objects:

```
JLabel jLNumber;
JLabel jLSumToNumber;

jLNumber = new JLabel("Enter an integer (N) : ",
                              SwingConstants.RIGHT);
jLSumToNumber = new JLabel("Sum of first N integers : ",
                              SwingConstants.RIGHT);
```

Note that the string appearing on both of these `JLabel` objects are right-justified.

Self-Check

17. Write a Java statement to declare a `JLabel` reference named "title."
18. Instantiate the `JLabel` reference created with a `JLabel` object with label "Java for Game Development." Make sure that the label is centered.

`JTextField` `JTextField` can be used for input and output. If a `JTextField` is used for output alone, then it is advisable to make it noneditable. The data entered in a `JTextField` object by the user is treated as one `String`, and similarly the output displayed in a `JTextField` is also a `String`. The `String` itself appears as a single line. If you want multiple lines in your graphical display, you can use `JTextArea` class. Both `JTextField` and `JTextArea` inherit many operations from their common superclass `JTextComponent`. Some of the constructors of `JTextField` and some of the services of its superclass `JTextComponent` are presented in Tables 8.5 and 8.6, respectively.

Thus the following segment of code creates two instances of `JTextField`, each with 10 columns:

```
JTextField jTFNumber;
JTextField jTFSumToNumber;

jTFNumber = new JTextField(10);
jTFSumToNumber = new JTextField(10);
```

TABLE 8.5 Constructors of `JTextField` Class

Constructor	Explanation
`public JTextField(int c)`	Constructor with an int argument. Creates a JTextField object with c columns
`public JTextField(String str)`	Constructor with a String argument. Creates a JTextField object initialized with str
`public JTextField(String str, int c)`	Creates a JTextField object initialized with str and c columns

TABLE 8.6 Services of `JTextField` Inherited from `JTextComponent`

Services	Explanation
`public String getText()`	Get method. Returns String contained in the JTextComponent used in the context of input
`public void setEditable(boolean b)`	The default value is true. If the value is set false, user can no longer enter data. Thus used mainly to designate a JTextComponent as output only
`public void setText(String str)`	Set method. The String str becomes the new String. Used in the context of output

Further, the Java statement

```
jTFSumToNumber.setEditable(false);
```

makes the `JTextField` instance `jTFSumToNumber` as noneditable. Note that by default, a `JTextField` is editable. Thus, there is no need to include the statement

```
jTFNumber.setEditable(true);     // this statement is not required.
```

in your program.

Self-Check

19. Write a Java statement to declare and instantiate a `JTextField` reference named "message."
20. Make the `JTextField` object created noneditable.

JButton

We use `JButton` in a GUI to start an action. For instance, you can use a `JButton` to start a computation once all the data values are entered. A `JButton` can also be used to terminate the application. Some of the constructors of `JButton` are presented in Table 8.7.

The following segment of code creates two `JButton` objects with labels `Calculate` and `Exit`:

```
JButton jBCompute;
JButton jBExit;
jBCompute = new JButton("Calculate");
jBExit = new JButton("Exit");
```

TABLE 8.7 Constructors of `JButton` Class

Constructor	Explanation
`public JButton(Icon ic)`	Constructor with an Icon argument. Creates a JButton object with Icon ic
`public JButton(String str)`	Constructor with a String argument. Creates a JButton object with String str as its label
`public JButton(String str, Icon ic)`	Creates a JButton object with an icon and a label

Self-Check

21. Write a Java statement to declare a `JButton` reference named "startButton."
22. Instantiate the `JButton` reference created with a `JButton` object with the label "Start."

Component `placement`

Recall that we have chosen `GridLayout` manager as the layout manager. In the case of `GridLayout` manager, all we need to do to place various components in the content pane is to use the `add` service of the `Container` class. Components are placed in the grid from left to right in each row, and rows are filled from top to bottom. For instance, in the case of a 3 by 2 grid, there are three rows and two columns. The components are placed in the order (1, 1), (1, 2), (2, 1), (2, 2), (3, 1), and (3, 2).

The ordered pair (i, j) stands for row i, column j position in the grid. Thus components are placed in the order row 1, column 1, followed by row 1, column 2. This completes row one from left to right. Therefore, the next component will be placed in row 2, column 1, followed by row 2, column 2. This completes row 2 from left to right. The next component will be placed in row 3, column 1, followed by row 3, column 2. This completes row 3 from left to right and rows from top to bottom.

The following Java statements will place the six components we created in the content pane `conInterior`:

```
conInterior.add(jLNumber);          // place at (1, 1)
conInterior.add(jTFNumber);         // place at (1, 2)
conInterior.add(jLSumToNumber);     // place at (2, 1)
conInterior.add(jTFSumToNumber);    // place at (2, 2)
conInterior.add(jBCompute);         // place at (3, 1)
conInterior.add(jBExit);            // place at (3, 2)
```

This completes the first stage of the GUI application development.

```
/**
    Computes the sum of first N numbers
*/
import javax.swing.*;
import java.awt.*;

public class StageOne extends JFrame
{
    private JLabel jLNumber;
    private JLabel jLSumToNumber;
    private JTextField jTFNumber;
    private JTextField jTFSumToNumber;
    private JButton jBCompute;
    private JButton jBExit;

    private static final int WIDTH = 300;
    private static final int HEIGHT = 120;

    /**
        Constructor with no arguments
    */
    public StageOne()
    {
        //Invoke JFrame constructor and set the size
        super("Simple Arithmetic Progression");
        super.setSize(WIDTH,HEIGHT);

        // Create two labels
        jLNumber = new JLabel("Enter an integer (N) : ",
                              SwingConstants.RIGHT);
        jLSumToNumber = new JLabel("Sum of first N integers : ",
                              SwingConstants.RIGHT);

        //Create two JTextfields
        jTFNumber = new JTextField(10);
        jTFSumToNumber = new JTextField(10);
        jTFSumToNumber.setEditable(false);

        //Create compute  JButton
        jBCompute = new JButton("Calculate");

        //Create Exit JButton
        jBExit = new JButton("Exit");
```

```
        Container conInterior = super.getContentPane();

        //Set the layout
        conInterior.setLayout(new GridLayout(3,2));

        //Place components in the container
        //left to right; top to bottom order
        conInterior.add(jLNumber);
        conInterior.add(jTFNumber);
        conInterior.add(jLSumToNumber);
        conInterior.add(jTFSumToNumber);
        conInterior.add(jBCompute);
        conInterior.add(jBExit);

        //Make the frame visible and allow graceful exit
        super.setVisible(true);
        super.setDefaultCloseOperation(EXIT_ON_CLOSE);
    }

    public static void main( String args[] )
    {
        StageOne stageOne = new StageOne();
    }
}
```

This will produce the application window as shown in Figure 8.1. You can type in the `JTextField`; however, buttons `Compute` and `Exit` will not work.

Self-Check

23. True or false: In the case of GridLayout manager components are placed in the grid from left to right in each row, and rows are filled from top to bottom.
24. The add method is inherited from the ________ class.

EVENT-DRIVEN PROGRAMMING

In the first stage of our GUI application development we created `JButton` objects: `jBCompute` and `jBExit`. This section implements the behavior of these buttons as the user clicks them using a mouse. Note that the behavior as such is application-specific. For instance, in this application, we would like the application to compute the desired sum and display it in the appropriate `JTextField` as the user clicks the `Compute` `JButton` and terminate the application as the user clicks the `Exit` `JButton`. However, the underlying principles remain the same for all GUI applications. In this section, we explain and illustrate those principles.

Every time a JButton object is clicked, Java virtual machine (JVM) creates an instance of the class ActionEvent. JVM provides you with an actionListener interface. An interface can be thought of as a class with one or more abstract methods, but no data. From a programmer's perspective, for a JButton object to carry out an action you must implement the actionListener interface in your program. Implementing an interface means, defining all the abstract methods of the interface. In the case of the actionListener interface, the only method that needs to be implemented is actionPerformed, which has an ActionEvent as its formal parameter. During the program execution, as the user clicks a JButton object, an ActionEvent object is created. The associated actionListener invokes the actionPerformed method with this newly created ActionEvent object as the actual parameter.

As you click a JButton object, creating an ActionEvent object as well as invoking the corresponding actionPerformed method with this newly created ActionEvent object as actual parameter is performed by the JVM. As a GUI application programmer, you only need to do the following:

1. Implement the actionListener interface. For each JButton object, an actionListener interface needs to be implemented. This amounts to providing the necessary code for the actionPerformed method.
2. Register the actionListener interface. For each JButton object, its actionListener needs to be specified. In Java terminology, this is called registering the actionListener.

Event-Driven Model of Computation

You have seen that clicking a mouse in JButton is an ActionEvent. There is an interface associated with ActionEvent class, called ActionListener. The ActionListener interface has one method actionPerformed and is defined as follows:

```
public interface ActionListener
{
    public void actionPerformed(ActionEvent e);
}
```

Therefore, implementing the ActionListener interface amounts to creating the method ActionPerformed as a member of a class. Finally, you need to register the interface implemented.

Our discussion so far on the event-driven model of computation can be summarized as follows:

1. Every user-generated event falls under some event class.
2. Corresponding to each event class, there is an associated listener interface.

3. An `interface` may contain one or more abstract methods.
4. Implementing a listener `interface` involves defining all the methods of the `interface` as part of some class in your program.
5. Once the listener `interface` is implemented, it needs to be registered.

Implementing Listener interface

Implementing a listener `interface` involves defining all the methods of the `interface` as members of some class in your program. Therefore let us begin by creating a class, `ComputeJButtonInterface`, to implement the `ActionListener` associated with `jBCompute` object as follows:

```
private class ComputeJButtonInterface implements ActionListener
{
    public void actionPerformed(ActionEvent e)
    {
            //application specific code
    }
}
```

Observe that the class `ComputeJButtonInterface` is declared as `private`. Our application alone needs access to this class and that can be achieved by keeping this as an *inner class* (or a class within a class) of our program. Recall that in the case of inheritance we use the keyword `extends` and in the case of `interface` we use the keyword `implements`. We now proceed to provide the application-specific code for the method `actionPerformed`.

As the user clicks the `Compute JButton`, he/she must have already entered an integer value in the input `JTextField`. Therefore, `actionPerformed` method must first get that value as an integer. The `JTextField` class has a method `getText` that returns the input entered by the user as a `String`. Thus

```
jTFNumber.getText();
```

returns a `String`. In Java, there is a class `Integer`. One of the `static` methods of the `Integer` class is `parseInt`, which has `String` as its formal parameter and returns the `int` value if in fact the actual parameter is a `String` representation of an integer. Therefore,

```
int n;
n = Integer.parseInt(jTFNumber.getText());
```

converts the integer value entered by the user into an `int` and stores it in the variable `n`. Now the sum can be computed using the mathematical formula

$$\text{sum of first n integers} = \frac{n(n+1)}{2}$$

Since either n or n + 1 is always even, we could code the above formula as shown below:

```
if (( n % 2) == 0)
     sum = (n / 2) * (n + 1);
else
     sum = ((n + 1) / 2) * n;
```

Now all that remains is to display the computed sum in the designated `JTextField`. We use the service `setText` of the `JTextField` class as shown below:

```
jTFSumToNumber.setText("" + sum);
```

Observe that `setText` requires a `String` as its argument and one of the simple ways to convert an `int` into a `String` is to use a `null` `String` along with the concatenation operation.

Thus we have the following:

```
private class ComputeJButtonInterface implements ActionListener
{
     public void actionPerformed(ActionEvent e)
     {
          int n;
          int sum;

          n = Integer.parseInt(jTFNumber.getText());

          if (( n % 2) == 0)
               sum = (n / 2) * (n + 1);
          else
               sum = ((n + 1) / 2) * n;

          jTFSumToNumber.setText("" + sum);

     }
}
```

We could similarly create an ActionListener interface for the Exit JButton as shown below. Note that all that needs to be done is to make use of the exit service of the System class. The method exit can be used to indicate an error code. Since our program is exiting without any error, we shall use 0. Thus, we have the following:

```
private class ExitJButtonInterface implements ActionListener
{
    public void actionPerformed(ActionEvent e)
    {
        System.exit(0);
    }
}
```

Self-Check

25. Implementing a listener interface involves defining ________ the methods of the interface as members of some class in your program.
26. We use the keyword ________ to create a subclass from an existing class and we use the keyword ________ to implement an interface.

Registering Listener interface

In this section, we illustrate how to register a listener interface. The central idea involves using the appropriate registering method of the class and passing an instance of the listener interface as an actual parameter. For instance, to register an ActionListener interface, you need to use the service addActionListener. Therefore,

```
ComputeJButtonInterface computeJBInterface;
computeJBInterface = new ComputeJButtonInterface();
```

creates an instance of ComputeJButtonInterface. The following statement completes the registration step:

```
jBCompute.addActionListener(computeJBInterface);
```

Note that the above three lines of code can be replaced by the following single Java statement:

```
jBCompute.addActionListener(new ComputeJButtonInterface());
```

Similarly,

```
jBExit.addActionListener(new ExitJButtonInterface());
```

registers the listener interface ExitJButtonInterface as the ActionListener of the Exit JButton object.

The `ActionListener interface` is part of `java.awt.event package`. Therefore, we must include one of the following `import` statements in our program:

```
import java.awt.event.*;
```

or

```
import java.awt.event.ActionListener;
```

This completes the second and the final stage of GUI application development. The complete program listing and sample output are given as follow:

```
/**
    Computes the sum of first N numbers
*/
import javax.swing.*;
import java.awt.*;
import java.awt.event.*;

public class SumToN extends JFrame
{
    private JLabel jLNumber;
    private JLabel jLSumToNumber;
    private JTextField jTFNumber;
    private JTextField jTFSumToNumber;
    private JButton jBCompute;
    private JButton jBExit;

    private static final int WIDTH = 300;
    private static final int HEIGHT = 120;

    /**
        Constructor with no arguments
    */
    public SumToN()
    {
        //Invoke JFrame constructor and set the size
        super("Simple Arithmetic Progression");
        super.setSize(WIDTH,HEIGHT);

        // Create two labels
        jLNumber = new JLabel("Enter an integer (N) : ",
                                      SwingConstants.RIGHT);
```

```
        jLSumToNumber = new JLabel("Sum of first N integers : ",
                                            SwingConstants.RIGHT);

        //Create two JTextfields
        jTFNumber = new JTextField(10);
        jTFSumToNumber = new JTextField(10);
        jTFSumToNumber.setEditable(false);

        //Create compute  JButton
        jBCompute = new JButton("Calculate");
        jBCompute.addActionListener(new ComputeJButton Interface() );

        //Create Exit JButton
        jBExit = new JButton("Exit");
        jBExit.addActionListener(new ExitJButtonInterface());

        Container conInterior = super.getContentPane();

        //Set the layout
        conInterior.setLayout(new GridLayout(3,2));

        //Place components in the container
        //left to right; top to bottom order
        conInterior.add(jLNumber);
        conInterior.add(jTFNumber);
        conInterior.add(jLSumToNumber);
        conInterior.add(jTFSumToNumber);
        conInterior.add(jBCompute);
        conInterior.add(jBExit);

        //Make the frame visible and allow graceful exit
        super.setVisible(true);
        super.setDefaultCloseOperation(EXIT_ON_CLOSE);
    }

    /**
        Implements action listener interface for compute button
    */
    private class ComputeJButtonInterface implements
                                    ActionListener
    {
        public void actionPerformed(ActionEvent e)
```

```
        {
            int n, sum;

            n = Integer.parseInt(jTFNumber.getText());

            if (( n % 2) == 0)
                sum = (n / 2) * (n + 1);
            else
                sum = ((n + 1) / 2) * n;
            jTFSumToNumber.setText("" + sum);

        }
    }

    /**
        Implements action listener interface for exit button
    */
    private class ExitJButtonInterface implements ActionListener
    {
        public void actionPerformed(ActionEvent e)
        {
            System.exit(0);
        }
    }

    public static void main( String args[] )
    {
        SumToN sumTN = new SumToN();
    }
}
```

Output

Figure 8.5 shows the sample run of simple arithmetic progression.

Simple Arithmetic Progression	
Enter an integer (N) :	100
Sum of first N integers :	5050
Calculate	Exit

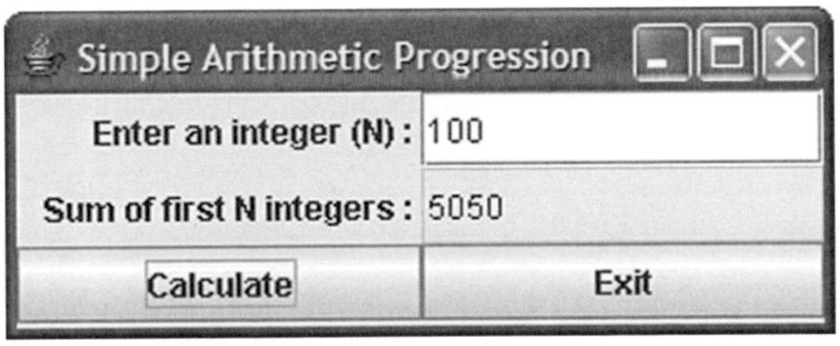

FIGURE 8.5 Sample run of simple arithmetic progression.

Self-Check

27. To register an `ActionListener interface`, you need to use the service ________.
28. The `ActionListener interface` is part of ________ package.

For ease of reference, we summarize our discussion so far as follows.

The common theme behind all GUI application programs involve the following:

1. Create the application window
   a. Create a new application class by extending `JFrame`
   b. Invoke the constructor of the superclass
   c. Define the size of the `JFrame`
   d. Make the `JFrame` visible
   e. Provide a graceful way to exit the application
   f. Get the reference of the content pane
   g. Create a layout manager
   h. Create and place GUI components in the content pane
2. Create code for event-driven programming
   a. Implement the `actionListener interface`
   b. Register the `actionListener interface`

METRIC CONVERSION HELPER

In this section, we create a GUI application to convert between the following pairs of units of measure:

1. Miles and kilometers
2. Pounds and kilograms
3. Gallons and liters
4. Fahrenheit and Centigrade

There are eight units of measure. Therefore, you need eight instances of `JTextField` and eight instances of `JLabel`. The question is do we really need any instance of `JButton`. As you might have observed, the `Exit JButton` is not really necessary. The user can always close the application by clicking the close button. What about the `Compute JButton`? Not really! In fact, `JTextField` object can also generate an `ActionEvent` object every time the Enter key is pressed. Therefore, we can design the program so that as the user enters any one of the eight possible values and presses the Enter key, the value is converted into the corresponding unit in the other system of measurement. Thus, all you need is eight instances of `JTextField` and eight instances of `JLabel` corresponding to each of the units of measure, and they can be organized as shown in Figure 8.6.

Metric Conversion Assistant			
Mile :		Kilometer :	
Pound :		Kilogram :	
Gallon :		Liter :	
Fahrenheit :		Centigrade :	

FIGURE 8.6 Application window for metric conversion assistant.

If the user enters a certain value in one of the JTextField objects and presses the Enter key, the equivalent value in the other system of measurement will be displayed in the adjacent JTextField object. For instance, if the user enters distance in JTextField object adjacent to the label Mile and presses the Enter key, the equivalent distance in kilometers will appear in the JTextField object adjacent to the label Kilometer. Similarly, when the user enters the volume in JTextField object adjacent to the label Liter and presses the Enter key, the equivalent volume in gallons will appear in the JTextField adjacent to the label Gallon.

Observe that there are eight JLabel objects and eight JTextField objects. Thus, there are 16 components and they are placed in a 4 by 4 grid.

Recall that pressing the Enter key in a JTextField object generates an ActionEvent. Therefore, you had to create and register action listeners to each of the JTextField objects. The action performed methods get the data entered through the getText service of the JTextComponent (superclass of JTextField), convert it into the other unit of measurement, and set it in the designated JTextField through the setText service of the JTextComponent (superclass of JTextField). Thus, we have the following:

```
import java.awt.*;
import java.awt.event.*;
import javax.swing.*;
import java.text.DecimalFormat;

/**
    Metric conversion class
*/
public class MetricConversion extends JFrame
{
    private JLabel jLMile;
    private JLabel jLKilometer;
    private JLabel jLPound;
    private JLabel jLKilogram;
    private JLabel jLGallon;
    private JLabel jLLiter;
```

```
    private JLabel jLFahrenheit;
    private JLabel jLCentigrade;

    private JTextField jTFMile;
    private JTextField jTFKilometer;
    private JTextField jTFPound;
    private JTextField jTFKilogram;
    private JTextField jTFGallon;
    private JTextField jTFLiter;
    private JTextField jTFFahrenheit;
    private JTextField jTFCentigrade;

    private static final int WIDTH = 400;
    private static final int HEIGHT = 150;

    private static final double MILE_KM = 1.6; //Mile to
                                                 Kilometer

    private static final double LB_KG = 0.454; // Pound to
                                                  Kilogram
    private static final double GL_LT = 3.7;    // Gallon to
                                                      Liter

    private static final double CENT_FAHR = 1.8;
                                                // Centigrade to
                                                      Fahrenheit

    private static final double FREEZING_POINT = 32.0;

    public MetricConversion()
    {
        //Invoke JFrame constructor and set the size
        super("Metric Conversion Assistant");
        super.setSize(WIDTH,HEIGHT);

        // Create labels
        jLMile = new JLabel("Mile : ", SwingConstants.RIGHT);
        jLKilometer = new JLabel("Kilometer : ",
                             SwingConstants.RIGHT);
        jLPound = new JLabel("Pound : ", SwingConstants.
                                                 RIGHT);
        jLKilogram = new JLabel("Kilogram : ", SwingConstants.
                                                       RIGHT);
```

```
jLGallon = new JLabel("Gallon : ", SwingConstants.
                                          RIGHT);
jLLiter = new JLabel("Liter : ", SwingConstants.
                                          RIGHT);
jLFahrenheit = new JLabel("Fahrenheit : ",
                    SwingConstants.RIGHT);
jLCentigrade = new JLabel("Centigrade : ",
                    SwingConstants.RIGHT);

//Create JTextfields
jTFMile = new JTextField(10);
jTFKilometer = new JTextField(10);
jTFPound = new JTextField(10);
jTFKilogram = new JTextField(10);
jTFGallon = new JTextField(10);
jTFLiter = new JTextField(10);
jTFFahrenheit = new JTextField(10);
jTFCentigrade = new JTextField(10);

jTFMile.addActionListener(new
               MileJTextFieldInterface());
jTFKilometer.addActionListener(new
               KilometerJTextFieldInterface());
jTFPound.addActionListener(new
               PoundJTextFieldInterface());
jTFKilogram.addActionListener(new
               KilogramJTextFieldInterface());
jTFGallon.addActionListener(new
               GallonJTextFieldInterface());
jTFLiter.addActionListener(new
               LiterJTextFieldInterface());
jTFFahrenheit.addActionListener(new
               FahrenheitJTextFieldInterface());
jTFCentigrade.addActionListener(new
               CentigradeJTextFieldInterface());

Container conInterior = super.getContentPane();

//Set the layout
conInterior.setLayout(new GridLayout(4,2));

//Place components in the container
//left to right; top to bottom order
```

```
        conInterior.add(jLMile);
        conInterior.add(jTFMile);
        conInterior.add(jLKilometer);
        conInterior.add(jTFKilometer);
        conInterior.add(jLPound);
        conInterior.add(jTFPound);
        conInterior.add(jLKilogram);
        conInterior.add(jTFKilogram);
        conInterior.add(jLGallon);
        conInterior.add(jTFGallon);
        conInterior.add(jLLiter);
        conInterior.add(jTFLiter);
        conInterior.add(jLFahrenheit);
        conInterior.add(jTFFahrenheit);
        conInterior.add(jLCentigrade);
        conInterior.add(jTFCentigrade);

        //Make the frame visible and allow graceful exit
        super.setVisible(true);
        super.setDefaultCloseOperation(EXIT_ON_CLOSE);
    }

    /**
        Implements action listener interface for MileJTextField
    */
    private class MileJTextFieldInterface implements
                                            ActionListener
    {
        public void actionPerformed(ActionEvent e)
        {
            double mile;
            double kilometer;
            DecimalFormat fourPlaces =    new
                     DecimalFormat("0.0000");

            mile = Double.parseDouble(jTFMile.getText());
            kilometer = mile * MILE_KM;

            jTFKilometer.setText(""+ fourPlaces.
                                format(kilometer));
        }
    }
```

```
/**
    Implements action listener interface for
                        KilometerJTextField
*/
private class KilometerJTextFieldInterface implements
                                      ActionListener
{
    public void actionPerformed(ActionEvent e)
    {
        double mile;
        double kilometer;

        DecimalFormat fourPlaces = new
              DecimalFormat("0.0000");

        kilometer = Double.parseDouble(jTFKilometer.
                                      getText());
        mile = kilometer / MILE_KM;

        jTFMile.setText(""+ fourPlaces.format(mile));
    }
}

/**
    Implements action listener interface for PoundJTextField
*/
private class PoundJTextFieldInterface implements
                                    ActionListener
{
    public void actionPerformed(ActionEvent e)
    {
        double pound;
        double kilogram;

        DecimalFormat fourPlaces = new
              DecimalFormat("0.0000");

        pound = Double.parseDouble(jTFPound.getText());
        kilogram = pound * LB_KG;

        jTFKilogram.setText(""+ fourPlaces.
                          format(kilogram));
    }
}
```

```
/**
    Implements action listener interface for
                          KilogramJTextField
*/
private class KilogramJTextFieldInterface implements
                                      ActionListener
{
    public void actionPerformed(ActionEvent e)
    {
        double pound;
        double kilogram;

        DecimalFormat fourPlaces = new
              DecimalFormat("0.0000");

        kilogram = Double.parseDouble(jTFKilogram.getText());
        pound = kilogram / LB_KG;

        jTFPound.setText(""+ fourPlaces.format(pound));
    }
}

/**
    Implements action listener interface for GallonJTextField
*/
private class GallonJTextFieldInterface implements
                                      ActionListener
{
    public void actionPerformed(ActionEvent e)
    {
        double gallon;
        double liter;

        DecimalFormat fourPlaces = new
              DecimalFormat("0.0000");

        gallon = Double.parseDouble(jTFGallon.
                                  getText());
        liter = gallon * GL_LT;

        jTFLiter.setText(""+ fourPlaces.format(liter));
    }
}
```

```
/**
    Implements action listener interface for LiterJTextField
*/
private class LiterJTextFieldInterface implements
                                    ActionListener
{
    public void actionPerformed(ActionEvent e)
    {
        double gallon;
        double liter;

        DecimalFormat fourPlaces = new
              DecimalFormat("0.0000");

        liter = Double.parseDouble(jTFLiter.getText());
        gallon = liter / GL_LT;

        jTFGallon.setText(""+ fourPlaces.
                        format(gallon));
    }
}

/**
    Implements action listener interface for
                        FahrenheitJTextField
*/
private class FahrenheitJTextFieldInterface implements
                                        ActionListener
{
    public void actionPerformed(ActionEvent e)
    {
        double centigrade;
        double fahrenheit;

        DecimalFormat fourPlaces = new
                DecimalFormat("0.0000");

        fahrenheit = Double.parseDouble(jTFFahrenheit.
                                    getText());
        centigrade = (fahrenheit - FREEZING_POINT) /
                                    CENT_FAHR;

        jTFCentigrade.setText(""+ fourPlaces.
                        format(centigrade));
    }
}
```

```
    /**
        Implements action listener interface for
                            CentigradeJTextField
    */
    private class CentigradeJTextFieldInterface implements
                                           ActionListener
    {
        public void actionPerformed(ActionEvent e)
        {
            double centigrade;
            double fahrenheit;

            DecimalFormat fourPlaces = new
                    DecimalFormat("0.0000");

            centigrade = Double.parseDouble(jTFCentigrade.
                                             getText());
            fahrenheit = centigrade * CENT_FAHR +
                                   FREEZING_POINT;

            jTFFahrenheit.setText(""+ fourPlaces.
                                 format(fahrenheit));
        }
    }

    public static void main( String[] args)
    {
        MetricConversion metricConversion = new
                                MetricConversion();
    }
}
```

Advanced Topic 8.1: Programming Options for Implementing Event Listeners

In this chapter, so far, we followed a simple programming style for creating and registering action listeners. To begin our discussion, let us refer to the style of programming we have followed so far as Option A. In Option A, for each GUI component that needs an event listener, an inner class that `implements` the necessary `interface` is created. However, there are many other programming options available to a Java programmer in terms of coding. This section illustrates some of the most common programming styles. This will allow you to choose a style you may prefer for your GUI programs. Further, you may encounter these styles as you read Java programs developed by other programmers. You may omit this section completely if you choose to do so.

Option B

In this option, you create one class that `implements` the necessary `interface` for all GUI components that needs an event listener. For instance, we can replace the following eight inner classes:

```
MileJTextFieldInterface, KilometerJTextFieldInterface,
PoundJTextFieldInterface, KilogramJTextFieldInterface,
GallonJTextFieldInterface, LiterJTextFieldInterface,
FahrenheitJTextFieldInterface, CentigradeJTextFieldInterface
```

of Option A by a single inner class.

```
private class ActionListenerInterface implements ActionListener
{
    public void actionPerformed(ActionEvent e)
    {
        double mile, kilometer;
        double pound, kilogram;
        double gallon, liter;
        double centigrade, fahrenheit;

        DecimalFormat fourPlaces = new
              DecimalFormat("0.0000");

        if (e.getSource() == jTFMile)
        {
            mile = Double.parseDouble(jTFMile.getText());
            kilometer = mile * MILE_KM;
            jTFKilometer.setText(""+ fourPlaces.
                            format(kilometer));
        }
        else if (e.getSource() == jTFKilometer)
        {
            kilometer = Double.parseDouble(jTFKilometer.
                                        getText());
            mile = kilometer / MILE_KM;
            jTFMile.setText(""+ fourPlaces.format(mile));
        }
        else if (e.getSource() == jTFPound)
```

```
{
    pound = Double.parseDouble(jTFPound.getText());
    kilogram = pound * LB_KG;
    jTFKilogram.setText(""+ fourPlaces.
                    format(kilogram));
}
else if (e.getSource() == jTFKilogram)
{
    kilogram = Double.parseDouble(jTFKilogram.
                              getText());
    pound = kilogram / LB_KG;
    jTFPound.setText(""+ fourPlaces.format(pound));
}
else if (e.getSource() == jTFGallon)
{
    gallon = Double.parseDouble(jTFGallon.
                         getText());
    liter = gallon * GL_LT;
    jTFLiter.setText(""+ fourPlaces.format(liter));
}
else if (e.getSource() == jTFLiter)
{
    liter = Double.parseDouble(jTFLiter.getText());
    gallon = liter / GL_LT;
    jTFGallon.setText(""+ fourPlaces.
                    format(gallon));
}
else if (e.getSource() == jTFFahrenheit)
{
    fahrenheit = Double.parseDouble(jTFFahrenheit.
                                getText());
    centigrade = (fahrenheit - FREEZING_POINT) /
                              CENT_FAHR;
    jTFCentigrade.setText(""+ fourPlaces.
                    format(centigrade));
}
else if (e.getSource() == jTFCentigrade)
{
    centigrade = Double.parseDouble(jTFCentigrade.
                                getText());
    fahrenheit = centigrade * CENT_FAHR +
                       FREEZING_POINT;
```

```
            jTFFahrenheit.setText(""+ fourPlaces.
                                    format(fahrenheit));
        }
    }
}
```

Observe that there is only one actionPerformed method. Thus, it is imperative that you determine the GUI component associated with the action event inside the actionPerformed method. The source that generated the event can be identified through the getSource service of the ActionEvent class. Thus instead of creating eight separate classes, we just create one inner class. Further, instead of registering eight different objects, you could register the same object. In other words, the following eight statements:

```
jTFMile.addActionListener(new
        MileJTextFieldInterface());
jTFKilometer.addActionListener(new
        KilometerJTextFieldInterface());
jTFPound.addActionListener(new
        PoundJTextFieldInterface());
jTFKilogram.addActionListener(new
        KilogramJTextFieldInterface());
jTFGallon.addActionListener(new
        GallonJTextFieldInterface());
jTFLiter.addActionListener(new
        LiterJTextFieldInterface());
jTFFahrenheit.addActionListener(new
        FahrenheitJTextFieldInterface());
jTFCentigrade.addActionListener(new
        CentigradeJTextFieldInterface());
```

of Option A needs to be replaced by the following nine statements:

```
ActionListenerInterface instanceALI =
        new ActionListenerInterface();

jTFMile.addActionListener(instanceALI);
jTFKilometer.addActionListener(instanceALI);
jTFPound.addActionListener(instanceALI);
jTFKilogram.addActionListener(instanceALI);
jTFGallon.addActionListener(instanceALI);
jTFLiter.addActionListener(instanceALI);
jTFFahrenheit.addActionListener(instanceALI);
jTFCentigrade.addActionListener(instanceALI);
```

Option C

In this option, you need not create any inner class at all. Instead, the application class itself implements the interface. Thus, the actionPerformed method is a member of the application class. For instance, we can replace the following eight inner classes:

```
MileJTextFieldInterface, KilometerJTextFieldInterface,
PoundJTextFieldInterface, KilogramJTextFieldInterface,
GallonJTextFieldInterface, LiterJTextFieldInterface,
FahrenheitJTextFieldInterface, CentigradeJTextFieldInterface
```

of Option A by a single actionPerformed method. Observe that the actionPerformed method in this case is a member of the application program itself and not a member of an inner class as in the case of Option B. In other words, in both Option B and Option C, there is only one actionPerformed method. In the case of Option B, the actionPerformed method is a member of an inner class whereas in the case of Option C, it is a member of the application class itself.

The actionPerformed method is identical to the one presented in Option B. Since it is the application class that implements the action listener, you need to modify the class declaration. For instance, you need to modify the following class declaration in Option A (or Option B)

```
public class MetricConversionThree extends JFrame
```

as shown below:

```
public class MetricConversionThree extends JFrame
                              implements ActionListener
```

During registering, you need to use an instance of the application class itself. Therefore, we use the keyword this. In other words, the following eight statements of Option A

```
jTFMile.addActionListener(new
        MileJTextFieldInterface());
jTFKilometer.addActionListener(new
        KilometerJTextFieldInterface());
jTFPound.addActionListener(new
        PoundJTextFieldInterface());
jTFKilogram.addActionListener(new
        KilogramJTextFieldInterface());
jTFGallon.addActionListener(new
        GallonJTextFieldInterface());
```

```
jTFLiter.addActionListener(new
        LiterJTextFieldInterface());
jTFFahrenheit.addActionListener(new
        FahrenheitJTextFieldInterface());
jTFCentigrade.addActionListener(new
        CentigradeJTextFieldInterface());
```

are to be replaced by the following eight statements:

```
jTFMile.addActionListener(this);
jTFKilometer.addActionListener(this);
jTFPound.addActionListener(this);
jTFKilogram.addActionListener(this);
jTFGallon.addActionListener(this);
jTFLiter.addActionListener(this);
jTFFahrenheit.addActionListener(this);
jTFCentigrade.addActionListener(this);
```

Option D

In this option, you specify the action to be taken through an *anonymous inner class*. Thus the code appears as part of the registering mechanism. For instance, we can replace the following inner class

```
private class MileJTextFieldInterface implements
                                        ActionListener
{
    public void actionPerformed(ActionEvent e)
    {
        double mile, kilometer;
        DecimalFormat fourPlaces =  new
                DecimalFormat("0.0000");

        mile = Double.parseDouble(jTFMile.getText());
        kilometer = mile * MILE_KM;

        jTFKilometer.setText(""+ fourPlaces.
                        format(kilometer));
    }
}
```

and the Java statement in Option A

```
jTFMile.addActionListener(new MileJTextFieldInterface());
```

by the following segment of code:

```
jTFMile.addActionListener(
    new ActionListener()
    {
        public void actionPerformed(ActionEvent e)
        {
            double mile, kilometer;
            DecimalFormat fourPlaces = new
                  DecimalFormat("0.0000");

            mile = Double.parseDouble(jTFMile.getText());
            kilometer = mile * MILE_KM;

            jTFKilometer.setText(""+ fourPlaces.
                                 format(kilometer));
        }
    });
```

Similar is the case of the remaining seven inner classes in Option A.

Advanced Topic 8.2: Applets

Java allows you to create small applications, commonly known as applets, which can be executed inside a web browser. Since an applet runs inside a web browser, it is subject to certain restrictions. One of the major restrictions is that an applet cannot access the local disk. However, on the bright side, you need not worry about your applet causing damage to someone else's computer! You create an applet by inheriting the `JApplet` class of the `package javax.swing` (Tables 8.8 and 8.9).

In the case of a Java application, program execution begins by executing the very first executable statement in the method `main`. Therefore, all application programs must have a method `main`. Java applets do not start by executing the method `main`. However, as the

TABLE 8.8 Commonly Used Methods of the `JApplet`

Method	Explanation
public Container getContentPane()	Returns the reference of the contentPane
public void setContentPane(Container c)	Sets c as the contentPane of the JApplet
public JMenuBar getJMenuBar()	Returns the reference of the JMenuBar
public void setJMenuBar(JMenuBar menuBar)	Sets menuBar as the JMenuBar of the JApplet
public void update(Graphics g)	Updates the graphics by invoking the paint method

TABLE 8.9 Inherited Methods of the `JApplet` from `Applet`

Method	Explanation
public void init()	Automatically invoked by the browser or applet viewer to initialize the applet
public void start()	Automatically invoked by the browser or applet viewer to start its normal execution. It is invoked after the init method and every time the applet moves into sight in a web browser
public void stop()	Automatically invoked by the browser or applet viewer to stop its normal execution. It is invoked every time the applet moves out of sight in a web browser. It is also invoked just before destroy
public void destroy()	Automatically invoked by the browser or applet viewer as applet is no longer required
public void resize(**int** w, **int** h)	Width and height are changed to w and h, respectively
public void showStatus(String str)	Displays the String str in the status bar of the web browser window
public URL getDocumentBase()	Returns the Uniform Resource Locator (URL) of the document that has the applet
public URL getCodeBase()	Returns the URL of the applet

web browser loads the web page, it loads the applet and invokes methods `init` followed by `start`. Hence from a programming perspective, creating an applet involves inheriting a `JApplet` class and overriding methods `init` and `start`.

Creating Applet from GUI Application

In this section, an applet version of the metric conversion application is created. In fact, we start with the GUI application program and convert it into a Java applet. This process will illustrate major similarities as well as differences between an applet and an application. An applet is a GUI program and as such has many features in common with a stand-alone GUI application. However, there are many differences and those differences are listed as follows:

1. An applet is created by extending `JApplet` class. A GUI application can be created from `JFrame` either by inheritance or through composition.

2. Applets have no main method. Web browser invokes init, start, stop, and dispose methods in sequence. A GUI application must have a main method. The application program starts by executing the first executable statement of the main.
3. Applet initialization code is kept in init method. In a GUI application, the initialization code is kept in a constructor.
4. An applet is visible so long as the web page is visible. Thus, you need not invoke set Visible method. In the case of a GUI application, you must invoke setVisible method with true as its actual parameter.
5. Applets have no title. Therefore, you cannot invoke setTitle method. The HTML file is used to set the title. In a GUI application, you invoke setTitle method to display a string in the title bar.
6. Applets have a method resize. However, the size of an applet is specified in the HTML file. A GUI application invokes setSize method to set the dimensions.
7. Applets need no code for closing. Typically, an application provides mechanisms to gracefully end the execution.

Therefore, the following *seven* steps broadly outline the process of converting a GUI application into an applet:

1. Create your applet by inheriting JApplet class
2. Replace the constructor with init method
3. Delete Java statements that invoke the superclass constructor super and setTitle method
4. Delete the setSize and setVisible methods
5. Delete main method
6. Delete all the code corresponding to an Exit JButton
7. Delete statements, such as setDefaultCloseOperation, that refer to a window

We illustrate these steps by modifying the metric conversion GUI application presented earlier in this chapter. We have included the comments to indicate the change. For the sake of clarity, we have intentionally removed all other comments.

```
import java.awt.*;
import java.awt.event.*;
import javax.swing.*;
import java.text.DecimalFormat;
/*
    Step 1

    public class MetricConversion extends JFrame
```

```
    is replaced by the following: */

public class MetricConversionApplet extends JApplet
{
    private JLabel jLMile;
    private JLabel jLKilometer;
    private JLabel jLPound;
    private JLabel jLKilogram;
    private JLabel jLGallon;
    private JLabel jLLiter;
    private JLabel jLFahrenheit;
    private JLabel jLCentigrade;

    private JTextField jTFMile;
    private JTextField jTFKilometer;
    private JTextField jTFPound;
    private JTextField jTFKilogram;
    private JTextField jTFGallon;
    private JTextField jTFLiter;
    private JTextField jTFFahrenheit;
    private JTextField jTFCentigrade;

    /*

        Step 4
        Delete the following. These are used in connection
                                             with setSize only
    private static final int WIDTH = 400;
    private static final int HEIGHT = 150;
                                                           */

    private static final double MILE_KM = 1.6;
    private static final double LB_KG = 0.454;
    private static final double GL_LT = 3.7;
    private static final double CENT_FAHR = 1.8;
    private static final double FREEZING_POINT = 32.0;

    /*

        Step 2
        The following constructor heading
```

```
public MetricConversion()

    is replaced by the following: */

public void init()
{
    /*

        Step 3
        The following invocation of super is deleted

        super("Metric Conversion Assistant");
    */

    /*

        Step 4
        Delete invocation of setSize

        super.setSize(WIDTH,HEIGHT);
    */

    jLMile = new JLabel("Mile : ", SwingConstants.RIGHT);
    jLKilometer = new JLabel("Kilometer : ",
                    SwingConstants.RIGHT);
    jLPound = new JLabel("Pound : ", SwingConstants.
                                        RIGHT);
    jLKilogram = new JLabel("Kilogram : ", SwingConstants.
                                              RIGHT);
    jLGallon = new JLabel("Gallon : ", SwingConstants.
                                          RIGHT);
    jLLiter = new JLabel("Liter : ", SwingConstants.
                                        RIGHT);
    jLFahrenheit = new JLabel("Fahrenheit : ",
                      SwingConstants.RIGHT);
    jLCentigrade = new JLabel("Centigrade : ",
                      SwingConstants.RIGHT);

    jTFMile = new JTextField(10);
    jTFKilometer = new JTextField(10);
    jTFPound = new JTextField(10);
    jTFKilogram = new JTextField(10);
```

```
jTFGallon = new JTextField(10);
jTFLiter = new JTextField(10);
jTFFahrenheit = new JTextField(10);
jTFCentigrade = new JTextField(10);

jTFMile.addActionListener(new
      MileJTextFieldInterface());
jTFKilometer.addActionListener(new
      KilometerJTextFieldInterface());
jTFPound.addActionListener(new
      PoundJTextFieldInterface());
jTFKilogram.addActionListener(new
      KilogramJTextFieldInterface());
jTFGallon.addActionListener(new
      GallonJTextFieldInterface());
jTFLiter.addActionListener(new
      LiterJTextFieldInterface());
jTFFahrenheit.addActionListener(new
      FahrenheitJTextFieldInterface());
jTFCentigrade.addActionListener(new
      CentigradeJTextFieldInterface());

Container conInterior = super.getContentPane();

conInterior.setLayout(new GridLayout(4,2));

conInterior.add(jLMile);
conInterior.add(jTFMile);
conInterior.add(jLKilometer);
conInterior.add(jTFKilometer);
conInterior.add(jLPound);
conInterior.add(jTFPound);
conInterior.add(jLKilogram);
conInterior.add(jTFKilogram);
conInterior.add(jLGallon);
conInterior.add(jTFGallon);
conInterior.add(jLLiter);
conInterior.add(jTFLiter);
conInterior.add(jLFahrenheit);
```

```
        conInterior.add(jTFFahrenheit);
        conInterior.add(jLCentigrade);
        conInterior.add(jTFCentigrade);

        /*

            Step 4
            Invocation of setVisible is deleted

            super.setVisible(true);

                                                        */
        /*

            Step 7
            Invocation of setDefaultCloseOperation is
                                                deleted

            super.setDefaultCloseOperation(EXIT_ON_CLOSE);

        */
    }

    private class MileJTextFieldInterface implements
                                    ActionListener
    {
        public void actionPerformed(ActionEvent e)
        {
            double mile;
            double kilometer;

            DecimalFormat fourPlaces =    new
                        DecimalFormat("0.0000"

            mile = Double.parseDouble(jTFMile.getText());
            kilometer = mile * MILE_KM;

            jTFKilometer.setText(""+ fourPlaces.
                            format(kilometer));
        }
    }

    private class KilometerJTextFieldInterface implements
                                        ActionListener
    {
        public void actionPerformed(ActionEvent e)
```

```
        {
            double mile;
            double kilometer;

            DecimalFormat fourPlaces = new
                   DecimalFormat("0.0000");

            kilometer = Double.parseDouble(jTFKilometer.
                                           getText());
            mile = kilometer / MILE_KM;

            jTFMile.setText(""+ fourPlaces.format(mile));
        }
    }

    private class PoundJTextFieldInterface implements
                                       ActionListener
    {
        public void actionPerformed(ActionEvent e)
        {
            double pound;
            double kilogram;

            DecimalFormat fourPlaces = new
                   DecimalFormat("0.0000");

            pound = Double.parseDouble(jTFPound.getText());
            kilogram = pound * LB_KG;

            jTFKilogram.setText(""+ fourPlaces.
                                format(kilogram));
        }
    }

    private class KilogramJTextFieldInterface implements
                                          ActionListener
    {
        public void actionPerformed(ActionEvent e)
        {
            double pound;
            double kilogram;

            DecimalFormat fourPlaces = new
                   DecimalFormat("0.0000");
```

```
            kilogram = Double.parseDouble(jTFKilogram.
                                                getText());
            pound = kilogram / LB_KG;

            jTFPound.setText(""+ fourPlaces.format(pound));
        }
    }

    private class GallonJTextFieldInterface implements
                                             ActionListener
    {
        public void actionPerformed(ActionEvent e)
        {
            double gallon;
            double liter;

            DecimalFormat fourPlaces = new
                  DecimalFormat("0.0000");

            gallon = Double.parseDouble(jTFGallon.
                                             getText());
            liter = gallon * GL_LT;

            jTFLiter.setText(""+ fourPlaces.format(liter));
        }
    }

    private class LiterJTextFieldInterface implements
                                             ActionListener
    {
        public void actionPerformed(ActionEvent e)
        {
            double gallon;
            double liter;

            DecimalFormat fourPlaces = new
                  DecimalFormat("0.0000");

            liter = Double.parseDouble(jTFLiter.getText());
            gallon = liter / GL_LT;

            jTFGallon.setText(""+ fourPlaces.
                                 format(gallon));
        }
    }
```

```
private class FahrenheitJTextFieldInterface implements
                                        ActionListener
{
    public void actionPerformed(ActionEvent e)
    {
        double centigrade;
        double fahrenheit;

        DecimalFormat fourPlaces = new
              DecimalFormat("0.0000");

        fahrenheit = Double.parseDouble(jTFFahrenheit.
                                         getText());
        centigrade = (fahrenheit - FREEZING_POINT) /
                                        CENT_FAHR;

        jTFCentigrade.setText(""+ fourPlaces.
                           format(centigrade));
    }
}

private class CentigradeJTextFieldInterface implements
                                        ActionListener
{
    public void actionPerformed(ActionEvent e)
    {
        double centigrade;
        double fahrenheit;

        DecimalFormat fourPlaces = new
              DecimalFormat("0.0000");

        centigrade = Double.parseDouble(jTFCentigrade.
                                         getText());
        fahrenheit = centigrade * CENT_FAHR +
                             FREEZING_POINT;

        jTFFahrenheit.setText(""+ fourPlaces.
                           format(fahrenheit));
    }
}

/*

    Step 5
    method main is deleted
```

```
    public static void main( String[] args)
    {
        MetricConversion metricConversion = new
                                  MetricConversion();
    }
    */
}
```

Just like an application, to run an applet, you must first compile it and create a .class file. Next you need to embed the applet (the .class file) inside a web page. If you have never created an HTML file, you can use the following simple HTML file. Once the contents are typed into file using a text editor, remember to save it as a .html or .htm file. For instance, you can save as `MetricConversionJApplet.html`.

```
<HTML>
<HEAD>
<TITLE>Metric Conversion Helper</TITLE>
</HEAD>
<BODY>
<APPLET code = "MetricConversionJApplet.class" width = "400" height = "120">
</APPLET>
</BODY>
</HTML>
```

Finally, you run your applet either by opening HTML file `MetricConversionJApplet.html` with a web browser, or running the appletviewer (an HTML viewer with limited capabilities) at command line prompt as shown below:

```
appletviewer MetricConversionJApplet.html
```

Output

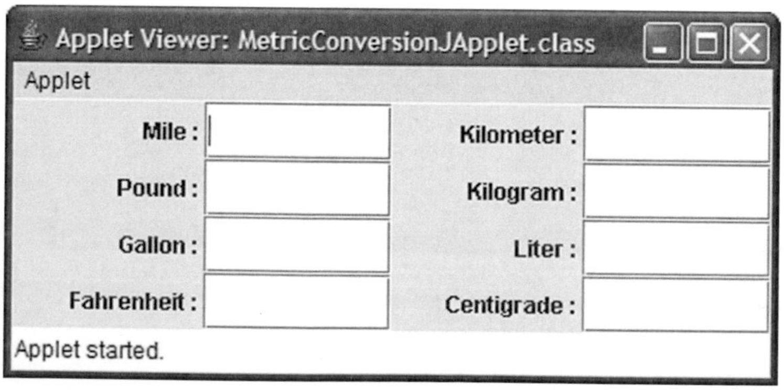

FIGURE 8.7 Metric conversion applet.

Advanced Topic 8.3: Applet and GUI Application

You have learned how to create GUI applications and applets. You may be wondering whether or not it is necessary to keep two versions of the same program, one to execute as a stand-alone application and the other to execute embedded inside a web page as an applet. Fortunately, you can write a Java program that is both an application as well as an applet. The major steps are as follows:

1. Create an applet class by extending `JApplet` class
2. Add a main method to the class created in step 1
3. Create an instance of the `applet` class and `JFrame` class
4. Add the applet instance created in step 3 to the content pane of the `JFrame` instance
5. Invoke the `setSize` method of the `JFrame` class
6. Invoke the `init` and the `start` methods of the `applet` class
7. Invoke `setVisible` method to make the `JFrame` instance visible and invoke `setDefaultCloseOperation` method to gracefully close the application window

We illustrate these steps by converting the `MetricConversionJApplet` presented in the previous section. We have included the comments to indicate the change. For the sake of clarity, we have intentionally removed all other comments.

```
import java.awt.*;
import java.awt.event.*;
import javax.swing.*;
import java.text.DecimalFormat;

public class MetricConversionJAppletApp extends JApplet
{
    private JLabel jLMile;
    private JLabel jLKilometer;
    private JLabel jLPound;
    private JLabel jLKilogram;
    private JLabel jLGallon;
    private JLabel jLLiter;
    private JLabel jLFahrenheit;
    private JLabel jLCentigrade;

    private JTextField jTFMile;
    private JTextField jTFKilometer;
    private JTextField jTFPound;
    private JTextField jTFKilogram;
    private JTextField jTFGallon;
    private JTextField jTFLiter;
    private JTextField jTFFahrenheit;
    private JTextField jTFCentigrade;
```

```
    /*
        Step 5
        Next two lines are added as part of Step 5.
    */

    private static final int WIDTH = 400;
    private static final int HEIGHT = 150;

    private static final double MILE_KM = 1.6;
    private static final double LB_KG = 0.454;
    private static final double GL_LT = 3.7;
    private static final double CENT_FAHR = 1.8;
    private static final double FREEZING_POINT = 32.0;

    public void init()
    {
        jLMile = new JLabel("Mile : ", SwingConstants.RIGHT);
        jLKilometer = new JLabel("Kilometer : ",
                                 SwingConstants.RIGHT);
        jLPound = new JLabel("Pound : ", SwingConstants.
                                                     RIGHT);
        jLKilogram = new JLabel("Kilogram : ", SwingConstants.
                                                     RIGHT);
        jLGallon = new JLabel("Gallon : ", SwingConstants.
                                                     RIGHT);
        jLLiter = new JLabel("Liter : ", SwingConstants.
                                                     RIGHT);
        jLFahrenheit = new JLabel("Fahrenheit : ",
                                  SwingConstants.RIGHT);
        jLCentigrade = new JLabel("Centigrade : ",
                                  SwingConstants.RIGHT);

        jTFMile = new JTextField(10);
        jTFKilometer = new JTextField(10);
        jTFPound = new JTextField(10);
        jTFKilogram  = new JTextField(10);
        jTFGallon = new JTextField(10);
        jTFLiter = new JTextField(10);
        jTFFahrenheit = new JTextField(10);
        jTFCentigrade = new JTextField(10);

        jTFMile.addActionListener(new
                          MileJTextFieldInterface());
        jTFKilometer.addActionListener(new
                          KilometerJTextFieldInterface());
```

```
        jTFPound.addActionListener(new
                        PoundJTextFieldInterface());
        jTFKilogram.addActionListener(new
                        KilogramJTextFieldInterface());
        jTFGallon.addActionListener(new
                        GallonJTextFieldInterface());
        jTFLiter.addActionListener(new
                        LiterJTextFieldInterface());
        jTFFahrenheit.addActionListener(new
                        FahrenheitJTextFieldInterface());
        jTFCentigrade.addActionListener(new
                        CentigradeJTextFieldInterface());

        Container conInterior = super.getContentPane();

        conInterior.setLayout(new GridLayout(4,2));

        conInterior.add(jLMile);
        conInterior.add(jTFMile);
        conInterior.add(jLKilometer);
        conInterior.add(jTFKilometer);
        conInterior.add(jLPound);
        conInterior.add(jTFPound);
        conInterior.add(jLKilogram);
        conInterior.add(jTFKilogram);
        conInterior.add(jLGallon);
        conInterior.add(jTFGallon);
        conInterior.add(jLLiter);
        conInterior.add(jTFLiter);
        conInterior.add(jLFahrenheit);
        conInterior.add(jTFFahrenheit);
        conInterior.add(jLCentigrade);
        conInterior.add(jTFCentigrade);

    }

    private class MileJTextFieldInterface implements
                                        ActionListener
    {
        public void actionPerformed(ActionEvent e)
        {
            double mile;
            double kilometer;

            DecimalFormat fourPlaces =  new
                    DecimalFormat("0.0000");
```

```
            mile = Double.parseDouble(jTFMile.getText());
            kilometer = mile * MILE_KM;

            jTFKilometer.setText(""+ fourPlaces.
                                format(kilometer));
        }
    }

    private class KilometerJTextFieldInterface implements
                                            ActionListener
    {
        public void actionPerformed(ActionEvent e)
        {
            double mile;
            double kilometer;

            DecimalFormat fourPlaces = new
                  DecimalFormat("0.0000");

            kilometer = Double.parseDouble(jTFKilometer.
                                            getText());
            mile = kilometer / MILE_KM;

            jTFMile.setText(""+ fourPlaces.format(mile));
        }
    }

    private class PoundJTextFieldInterface implements
                                        ActionListener
    {
        public void actionPerformed(ActionEvent e)
        {
            double pound;
            double kilogram;

            DecimalFormat fourPlaces = new
                  DecimalFormat("0.0000");
            pound = Double.parseDouble(jTFPound.getText());
            kilogram = pound * LB_KG;

            jTFKilogram.setText(""+ fourPlaces.
                                format(kilogram));
        }
    }
```

```
private class KilogramJTextFieldInterface implements
                                        ActionListener
{
    public void actionPerformed(ActionEvent e)
    {
        double pound;
        double kilogram;

        DecimalFormat fourPlaces = new
              DecimalFormat("0.0000");

        kilogram = Double.parseDouble
             (jTFKilogram.getText());
        pound = kilogram / LB_KG;

        jTFPound.setText(""+ fourPlaces.format(pound));
    }
}

private class GallonJTextFieldInterface implements
                                        ActionListener
{
    public void actionPerformed(ActionEvent e)
    {
        double gallon;
        double liter;

        DecimalFormat fourPlaces = new
              DecimalFormat("0.0000");

        gallon = Double.parseDouble(jTFGallon.getText());
        liter = gallon * GL_LT;

        jTFLiter.setText(""+ fourPlaces.format(liter));
    }
}

private class LiterJTextFieldInterface implements
                                        ActionListener
{
    public void actionPerformed(ActionEvent e)
    {
        double gallon;
        double liter;
```

```
            DecimalFormat fourPlaces = new
                  DecimalFormat("0.0000");

            liter = Double.parseDouble(jTFLiter.getText());
            gallon = liter / GL_LT;

            jTFGallon.setText(""+ fourPlaces.
                                 format(gallon));
        }
    }

    private class FahrenheitJTextFieldInterface implements
                                             ActionListener
    {
        public void actionPerformed(ActionEvent e)
        {
            double centigrade;
            double fahrenheit;

            DecimalFormat fourPlaces = new
                  DecimalFormat("0.0000");

            fahrenheit = Double.parseDouble(jTFFahrenheit.
                                              getText());
            centigrade = (fahrenheit - FREEZING_POINT) /
                                             CENT_FAHR;

            jTFCentigrade.setText(""+ fourPlaces.
                                 format(centigrade));

        }
    }

    private class CentigradeJTextFieldInterface implements
                                             ActionListener
    {
        public void actionPerformed(ActionEvent e)
        {
            double centigrade;
            double fahrenheit;

            DecimalFormat fourPlaces = new
                  DecimalFormat("0.0000");
```

```
            centigrade = Double.parseDouble(jTFCentigrade.
                                            getText());
            fahrenheit = centigrade * CENT_FAHR +
                                    FREEZING_POINT;

            jTFFahrenheit.setText(""+ fourPlaces.
                                format(fahrenheit));
        }
    }

    /*

            Step 2
            method main is added
                                                        */

    public static void main( String[] args)
    {
        /*

            Step 3
            instance of applet and JFrame
                                                        */

        MetricConversionJAppletApp
        metricConversion = new
                        MetricConversionJAppletApp();
        JFrame appWindow = new JFrame("Metric Conversion
                                        Helper");

        /*

            Step 4
            add instance of applet to the
           content pane of the appWindow
                                                        */

        Container conInterior = appWindow.getContentPane();
        conInterior.add(metricConversion);

        /*

            Step 5
            invoke setSize for appWindow
                                                        */
```

```
        appWindow.setSize(WIDTH,HEIGHT);

        /*

              Step 6
              invoke init and start of the applet
                                                                  */

        metricConversion.init();
        metricConversion.start();

        /*

              Step 7
              invoke setVisible and
              setDefaultCloseOperation
                                                                  */

        appWindow.setVisible(true);
        appWindow.setDefaultCloseOperation(JFrame.
                                       EXIT_ON_CLOSE);

    }

}
```

The program can be executed either as an application or as an applet. The output produced will be similar to the one shown in Figure 8.6 or 8.7, respectively, and hence omitted.

Advanced Topic 8.4: Graphics

Event handlers discussed in previous sections are examples of a *callback* mechanism. In simple terms, Java system invokes certain methods based on the events occurring during the execution of a program. However, the developer of the application has the ability to override those methods. In this section, you will encounter a new callback method `paint`. The `paint` service is defined in the `Component` class and it renders contents of a `Component` object.

Throughout this section, you will be using the `paint` service of the applet to draw various objects in an instance of `Canvas` class that is part of the applet. The `Canvas` class is a subclass of the `Component` class.

To draw, a graphics context is required. A graphics context is an instance of `Graphics` class. Since `Graphics` class is an `abstract` class, it is impossible to create a graphics context using the `new` operator. As a programmer, there are two ways to get a graphics context for drawing on a component. First, the formal parameter to the paint method is a graphics context, and thus it is available inside a component's paint method. The second option is to use getGraphics method of a component. In this section, we will be using the first option. The reader is encouraged to try Programming Exercise 10 that requires getGraphics method.

The paint is invoked by the system in three different occasions:

1. The component is made visible on the screen for the first time.
2. The component is resized.
3. The component needs to be made visible again as some other component that previously obscured this component has moved.

The paint method can also be called by the user indirectly. For instance, the component determines that it needs to update its contents, due to a user action such as clicking a JButton object, by displaying a pushed-in look. As a programmer, you override the paint method of a component. However, you never invoke the paint method directly. Instead, you call repaint method with no parameters and repaint method in turn will call update method followed by the paint method of the component.

You draw graphics by overriding the paint method. We start with the following Java program that draws a string "Graphics and Java" starting at pixel location (10, 25) (Figure 8.8). The paint method has a single argument that is an instance of the Graphics class of the package java.awt. There are many methods in the Graphics class to produce various geometric shapes and writing strings. You can use these methods inside the paint method. Some of the most commonly used Graphics methods are presented in Table 8.10. For the present, we use drawString method of the Graphics class with three arguments. The first argument is the String to be drawn and the next two int values specify the starting pixel location. Therefore, the following paint method draws the String at pixel location (10, 25):

```
public void paint(Graphics g)
{
    super.paint(g);
    g.drawString("Graphics and Java", 10, 25);

}
```

The complete Java program listing is given below:

```
import java.awt.Graphics;
import javax.swing.*;
```

```
/**
    Illustration of drawString method
*/
public class GraphicsJAppletApp extends JApplet
{
    private static final int WIDTH = 200;
    private static final int HEIGHT = 100;
    public void paint(Graphics g)
    {
         super.paint(g);
         g.drawString("Graphics and Java", 10, 25);

    }

    public static void main( String[] args)
    {
         GraphicsJAppletApp appletApp = new
                      GraphicsJAppletApp();
         Jframe appWindow = new Jframe("Graphics and Java");

         appWindow.getContentPane().add(appletApp);
         appWindow.setSize(WIDTH,HEIGHT);

         appletApp.init();
         appletApp.start();

         appWindow.setVisible(true);
         appWindow.setDefaultCloseOperation(Jframe.EXIT_ON_CLOSE);
    }

}
```

Output (produced by application)

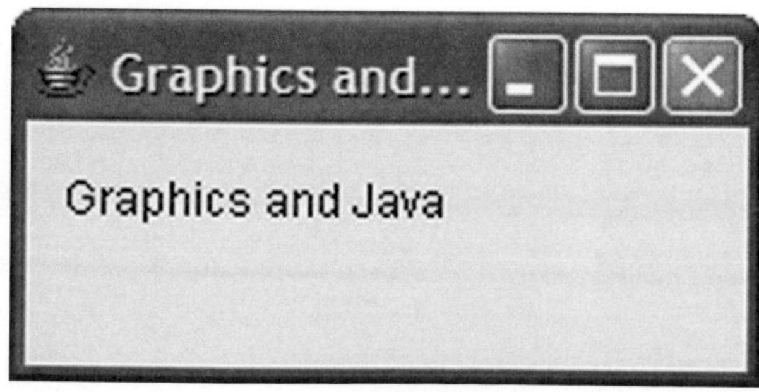

FIGURE 8.8 Demonstration of `drawString`.

TABLE 8.10 Methods For Creating Geometric Shapes

Method	Explanation
`public abstract void drawLine (int sx, int sy, int dx, int dy)`	Draws a line starting from (sx, sy) to (dx, dy)
`public abstract void drawArc (int x, int y, int w, int h, int sa, int aa)`	Draws an arc of the ellipse with top-left corner (x,y) and width w, height h. The arc starts at an angle sa (degree) and ends at an angle sa + aa (degree)
`public abstract void fillArc (int x, int y, int w, int h, int sa, int aa)`	Draws a filled arc of the ellipse with top-left corner (x,y) and width w, height h. The arc starts at an angle sa (degree) and ends at an angle sa + aa (degree)
`public abstract void drawRect (int x, int y, int w, int h)`	Draws a rectangle with top-left corner at (x,y) and width w, height h
`public abstract void fillRect (int x, int y, int w, int h)`	Draws a filled rectangle with top-left corner at (x,y) and width w, height h
`public abstract void draw3DRect (int x, int y, int w, int h, boolean r)`	Draws a 3-dimensional rectangle with top-left corner at (x,y) and width w, height h. Rectangle will appear raised if r is true
`public abstract void fill3DRect (int x, int y, int w, int h, boolean r)`	Draws a filled 3-dimensional rectangle with top-left corner at (x,y) and width w, height h. Rectangle will appear raised if r is true
`public abstract void drawRoundRect(int x, int y, int w, int h, int aw, int ah)`	Draws a corner-rounded rectangle with top-left corner at (x,y) and width w, height h. The arc width aw and arc height ah determine the shape of the corner
`public abstract void fillRoundRect(int x, int y, int w, int h, int aw, int ah)`	Draws a filled corner-rounded rectangle with top-left corner at (x,y) and width w, height h. The arc width aw and arc height ah determine the shape of the corner
`public abstract void drawOval (int x, int y, int w, int h)`	Draws an Oval with top-left corner at (x,y) and width w, height h
`public abstract void fillOval (int x, int y, int w, int h)`	Draws a filled Oval with top-left corner at (x,y) and width w, height h
`public abstract void drawPolygon(int[] x, int[] y, int n)`	Draws a polygon with n points: (x[0], y[0]), ...,(x[n-1], y[n-1])
`public abstract void fillPolygon(int[] x, int[] y, int n)`	Draws a filled polygon with n points: (x[0], y[0]), ...,(x[n-1], y[n-1])
`public abstract void drawPolygon(Polygon p)`	Draws the polygon p
`public abstract void fillPolygon(Polygon p)`	Draws the filled polygon p
`public abstract void drawString(Polygon p)`	Draws the String str starting at pixel location (x, y)

Before we introduce other methods of the Graphics class, let us explore the Color class and the Font class of the java.awt package.

Advanced Topic 8.5: Color

Every GUI component in Java has a background color and a foreground color. As you draw on a component, you see the foreground color. You can set both background color and foreground color of a Component using methods setBackground and setForeground, respectively.

Java employs RGB (Red Green Blue) color scheme. Every color in this scheme is obtained by mixing red, green, and blue hues in various proportions. For instance, yellow is obtained by mixing equal amounts of red and green with no blue at all. Similarly, the color magenta is created by mixing red and blue in equal amounts. The Color class has three constructors. The first constructor in Table 8.11 allows us to specify the RGB value as three separate int values. Each of these int values can be between 0 and 255. Therefore, using the first constructor, you can create red, green, blue, yellow, and magenta as shown below:

```
Color redColor = new Color(255, 0, 0);         //creates a red color
Color greenColor = new Color(0, 255, 0);       //creates a green
                                                          color
Color blueColor = new Color(0, 0, 255);        //creates a blue color
Color yellowColor = new Color(255, 255, 0);    //creates a yellow
                                                          color
Color magentaColor = new Color(255, 0, 255);   //creates a magenta
                                                          color
```

TABLE 8.11 Selected Constructors and Methods of Color Class

Constructor or Method	Explanation
Color(**int** r, **int** g, **int** b)	Creates a color object with red value r, green value g, and blue value b. Values r, g, and b are in the range 0-255
Color(**int** c)	Creates a color object with red value r, green value g, and blue value b such that r, g, and b are in the range 0-55; and c = r * 65536 + g * 256 + b
Color(**float** r, **float** g, **float** b)	Creates a color object with red value r, green value g, and blue value b. Values r, g, and b are in the range 0-1
public int getBlue()	Accessor method that returns blue component value in the range 0-255
public int getGreen()	Accessor method that returns green component value in the range 0-255
public int getRed()	Accessor method that returns red component value in the range 0-255
public int getRGB()	Accessor method that returns the RGB value c. c = r * 65536 + g * 256 + b, then r, g, and b are red, green, and blue components in the range 0-255

TABLE 8.12 Most Commonly Used Colors

	Primary colors of RGB		
Color.red	(255,0,0)	Color.green	(0,255,0)
Color.blue	(0,0,255)		
	Secondary colors		
Color.yellow	(255,255,0)	Color.magenta	(255,0,255)
Color.cyan	(0,255,255)	Color.orange	(255,200,0)
Color.pink	(255,175,175)		
	Shades of gray		
Color.black	(0,0,0)	Color.darkGray	(64,64,64)
Color.gray	(128,128,128)	Color.lightGray	(192,192,192)
Color.white	(255,255,255)		

Mixing all three colors in the same proportion produces different shades of gray. White color has RGB values 255, 255, 255, and black has RGB values 0, 0, 0. Note that an RGB value of 64, 64, 64 creates a 75% gray color and an RGB value of 192, 192, 192 creates a 25% gray color. Similarly, an RGB value of 255, 255, 0 produces the most light yellow color and an RGB value of 1, 1, 0 produces the most dark yellow color.

The `Color` class defines 15 most commonly used colors as constants. Table 8.12 lists these constants and their RGB values. Note that all color constants are spelled in lowercase.

We now create a simple applet application to demonstrate the `Color` class. Three instances of `JTextField`, `jTFRed`, `jTFGreen`, and `jTFBlue` are used to get the RGB values from the user. If the user fails to enter a value in any one of the textfields, we would like to treat it as zero. This can be accomplished by checking the length of the string entered by the user. If the length of the string happens to be zero, we will set the value as zero. If the length is not zero, then we use `parseInt` method of the `Integer` class to parse the string and get the integer value. Recall that each of the RGB component value has to be between `0` and `255`. We compute the remainder of the user-entered value upon division by 256. This will make sure that the RGB values are always between 0 and 255. Thus, we have the following code:

```
int r, g, b;
String str;

str = jTFRed.getText();
if (str.length() == 0)
    r = 0;
else
    r = Integer.parseInt(str) % 256;

str = jTFGreen.getText();
if (str.length() == 0)
    g = 0;
```

```
else
    g = Integer.parseInt(str) % 256;

str = jTFBlue.getText();
if (str.length() == 0)
    b = 0;
else
    b = Integer.parseInt(str) % 256;
```

The following Java statement will create a new `Color` and set it as the background color of the content pane `conInterior`:

```
conInterior.setBackground(new Color(r, g, b));
```

To make this change to take effect we need to call the `repaint` method of the `Component` class as shown below:

```
conInterior.repaint();
```

Thus, we have the following `actionPerformed` method:

```
public void actionPerformed(ActionEvent e)
{
    int r, g, b;
    String str;

    str = jTFRed.getText();
    if (str.length() == 0)
        r = 0;
    else
        r = Integer.parseInt(str) % 256;

    str = jTFGreen.getText();
    if (str.length() == 0)
        g = 0;
    else
        g = Integer.parseInt(str) % 256;

    str = jTFBlue.getText();
    if (str.length() == 0)
        b = 0;
    else
        b = Integer.parseInt(str) % 256;
```

```
        conInterior.setBackground(new Color(r, g, b));
        conInterior.repaint();
    }
```

Observe that reference variable conInterior needs to be available inside the action Performed. Therefore, it has to be declared as an attribute of the applet class. Since grid layout places components from left to right and top to bottom order, we create an extra JLabel object jLnoLabel and place it so that our JButton object jBsetColor appears at the center. The complete listing is given below:

```
import java.awt.*;
import java.awt.event.*;
import javax.swing.*;
import java.text.DecimalFormat;

/**
    Applet application illustrating the Color class
*/
public class ColorDisplayJAppletApp extends JApplet
{
    private JLabel jLRed, jLGreen, jLBlue, jLnoLabel;

    private JTextField jTFRed, jTFGreen, jTFBlue;

    private JButton jBsetColor;

    private Container conInterior;

    private static final int WIDTH = 300;
    private static final int HEIGHT = 150;

    /**
        init method creates the GUI
    */
    public void init()
    {
        jLRed = new JLabel("Red", SwingConstants.CENTER);
        jLGreen = new JLabel("Green", SwingConstants.CENTER);
        jLBlue = new JLabel("Blue", SwingConstants.CENTER);
        jLnoLabel = new JLabel();

        //Create JTextfields
        jTFRed = new JTextField(3);
```

```
        jTFGreen = new JTextField(3);
        jTFBlue = new JTextField(3);

        //Create JButton
        jBsetColor = new JButton("Set Color" );

        jBsetColor.addActionListener(new
            SetColorJButtonInterface());

        conInterior = super.getContentPane();

        //Set the layout
        conInterior.setLayout(new GridLayout(3,3));

        //Place components in the container
        //left to right; top to bottom order
        conInterior.add(jLRed);
        conInterior.add(jLGreen);
        conInterior.add(jLBlue);
        conInterior.add(jTFRed);
        conInterior.add(jTFGreen);
        conInterior.add(jTFBlue);
        conInterior.add(jLnoLabel);
        conInterior.add(jBsetColor);

    }

    /**
        Implements the action listener class
    */
    private class SetColorJButtonInterface implements
                                          ActionListener
    {
        public void actionPerformed(ActionEvent e)
        {
            int r, g, b;
            String str;

            str = jTFRed.getText();
            if (str.length() == 0)
                r = 0;
            else
                r = Integer.parseInt(str) % 256;

            str = jTFGreen.getText();
            if (str.length() == 0)
                g = 0;
```

```
            else
                 g = Integer.parseInt(str) % 256;

            str = jTFBlue.getText();
            if (str.length() == 0)
                 b = 0;
            else
                 b = Integer.parseInt(str) % 256;

            conInterior.setBackground(new Color(r, g, b));
            conInterior.repaint();
        }
    }

    public static void main( String[] args)
    {

        ColorDisplayJAppletApp appletApp = new
                      ColorDisplayJAppletApp();
        JFrame appWindow = new JFrame("Color Creator");

        appWindow.getContentPane().add(appletApp);
        appWindow.setSize(WIDTH,HEIGHT);

        appletApp.init();
        appletApp.start();

        appWindow.setVisible(true);
        appWindow.setDefaultCloseOperation(JFrame.EXIT_ON_CLOSE);
    }
}
```

Output (produced by application)

The color creator is as shown in Figure 8.9.

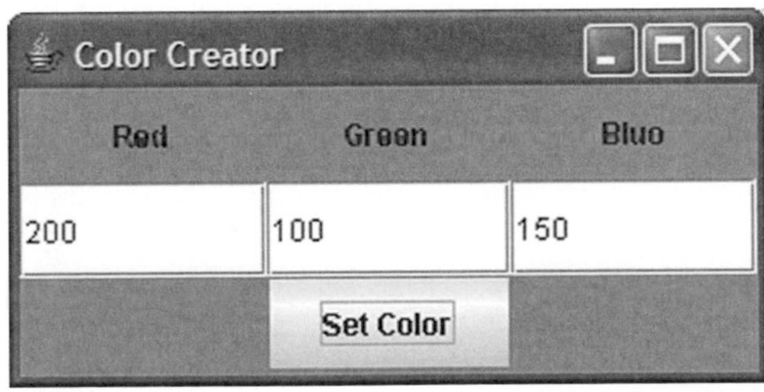

FIGURE 8.9 Color creator.

Advanced Topic 8.6: Font

In the previous section, you have learned about the Color class in Java. In this section, we explore the Font class in Java. Using the Font class you could create a Font object and use it in your GUI programs and applets to add more visual effects. The Font class is part of java.awt package.

The Font class has a constructor that takes three arguments as shown below:

```
public Font(String fontName, int fontStyle, int pointSize)
```

fontName: A string to indicate the Font face name or font name. Examples are

- Serif
- Monospaced Sanserif
- Dialog
- DialogInput

These fonts will be available in all Java systems.

fontStyle: Font class has three int constants to indicate the font style:

- Font.PLAIN for plain style text
- Font.BOLD for bold style text
- Font.ITALIC for italic style text

You can also use Font.BOLD + Font.ITALIC for bold and italic style text.

pointSize: An int value to specify the size of the character in points. An inch is 72 points. Thus, 12 points is one-sixth of an inch.

Before we start creating fonts, it is a good idea to find out the fonts available in your system. The following Java application can be used for that purpose. The program uses graphics environment and arrays that we have not introduced. The graphics environment will be introduced later in this chapter. The arrays are presented in Chapter 9. For the present, execute this program and observe the output. Since output depends on the machine, it is not presented.

```
import java.awt.*;

/**
    Lists all fonts available in the system.
    Applet application version
*/
public class FontAvailable
{
    public static void main(String[] args)
```

```
    {
          int j;

         String[] FontNameList =
         GraphicsEnvironment.getLocalGraphicsEnvironment()
                          .getAvailableFontFamilyNames();

         for (j = 0; j < listOfFontNames.length; j++)
                System.out.println(FontNameList[j]);
    }
}
```

The following example illustrates the use of Font class inside a paint method.

Example 8.3

```
import java.awt.*;
import java.awt.event.*;
import javax.swing.*;
import java.text.DecimalFormat;

/**
    Illustration of Font class
    Applet application version
*/
public class FontDisplayJAppletApp extends JApplet
{

    private static final int WIDTH = 600;
    private static final int HEIGHT = 160;

    /**
         Overridden paint method
    */
    public void paint( Graphics g)
    {
         super.paint(g);

         g.setFont(new Font("Monospaced Sanserif ", Font.PLAIN,
                                                          10));
         g.drawString(
               "Sea to Shinning Sea : Monospaced Sanserif plain
                                                  10pt",110,20);
```

```
        g.setFont(new Font("Courier New", Font.ITALIC, 15));
        g.drawString(
                    "Sea to Shinning Sea : Courier New italic
                                                      15pt",80,40);

        g.setFont(new Font("Arial", Font.BOLD, 20));
        g.drawString("Sea to Shinning Sea : Arial bold
                                              20pt",50,65);

        g.setFont(new Font("Times Roman",
                              Font.ITALIC + Font.BOLD, 25));
        g.drawString(
              "Sea to Shinning Sea : Times Roman italic bold
                                                   25pt", 20,95);
    }

    public static void main( String[] args)
    {

        FontDisplayJAppletApp appletApp = new
                    FontDisplayJAppletApp();
        JFrame appWindow = new JFrame("Font Creator");
        appWindow.getContentPane().add(appletApp);
        appWindow.setSize(WIDTH,HEIGHT);
        appletApp.init();
        appletApp.start();
        appWindow.setVisible(true);
        appWindow.setDefaultCloseOperation(JFrame.
                                      EXIT_ON_CLOSE);
    }
}
```

Output (produced by application)

Figure 8.10 shows the demonstration of the Font class.

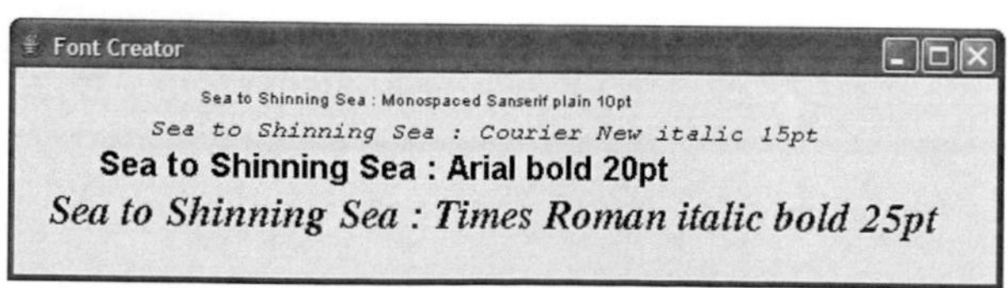

FIGURE 8.10 Demonstration of the Font class.

Advanced Topic 8.7: Drawing Services

The `Graphics` class of `java.awt package` has many services to create a wide variety of geometric shapes and drawing strings. The basic shapes supported are

- Line
- Arc
- Rectangle
- Three-dimensional rectangle
- Rectangles with round corners
- Oval
- Polygon

All these shapes except line, come with a filled version as well. Observe that in the case of line, there is no need for a filled version.

Example 8.4

In this example, we illustrate some of the drawing methods. Program is quite self-explanatory.

```
import java.awt.*;
import javax.swing.*;

/**
    Illustration of various drawing methods
*/
public class DrawingJAppletApp extends JApplet
{
    private static final int WIDTH = 400;
    private static final int HEIGHT = 400;
    public void paint(Graphics g)
    {
         super.paint(g);
         g.setColor(Color.red);
         g.drawOval(100, 100, 60, 60);

         g.setColor(Color.green);
         g.fillOval(100, 100, 30, 30);

         g.setColor(Color.blue);
         g.drawOval(200, 100, 60, 40);

         g.setColor(Color.pink);
         g.fillOval(215, 115, 30, 10);

         g.setColor(Color.green);
         g.drawRect(100, 200, 60, 60);
```

```
            g.setColor(Color.red);
            g.fillRect(160, 230, 25, 25);

            g.setColor(Color.pink);
            g.drawRoundRect(200, 200, 60, 40,10,20);

            g.setColor(Color.green);
            g.drawRoundRect(260, 260, 30, 10, 10,20);

            g.setColor(Color.cyan);
            g.fill3DRect(150, 300, 80, 40,true);
        }
        public static void main( String[] args)
        {
            DrawingJAppletApp appletApp = new DrawingJAppletApp();
            JFrame appWindow = new JFrame("Drawing");

            appWindow.getContentPane().add(appletApp);
            appWindow.setSize(WIDTH,HEIGHT);

            appletApp.init();
            appletApp.start();

            appWindow.setVisible(true);
            appWindow.setDefaultCloseOperation(JFrame. EXIT_ON_CLOSE);
        }
    }
```

Output (produced by applet)

Figure 8.11 shows various drawing services to create geometric shapes.

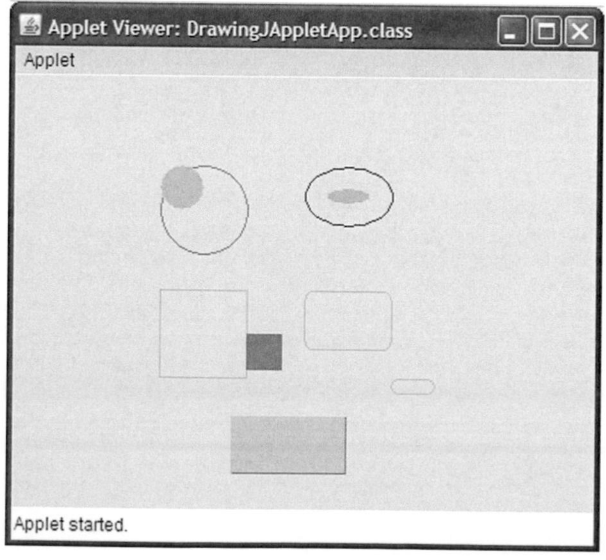

FIGURE 8.11 Drawing services.

REVIEW

1. The application window of all applications is created by extending the `JFrame` class of the Java `package javax.swing`.
2. In particular, we use the `add` service of the `Container` class to place various GUI objects in the content pane of the `JFrame` object.
3. The term pixel stands for picture element and is the smallest unit on your screen that you can control.
4. You can control the visibility of a GUI component through `setVisible` method of the superclass `Component`.
5. `JFrame` has a content pane and GUI components are placed on the content pane and not on the `JFrame`.
6. The layout manager determines size and location of components placed inside a container.
7. Components are placed in the grid from left to right in each row and rows are filled from top to bottom.
8. Every user-generated event falls under some event class.
9. Corresponding to each event class, there is an associated listener `interface`.
10. An `interface` may contain one or more abstract methods.
11. Implementing a listener `interface` involves defining all the methods of the `interface` as part of some class in your program.
12. Pressing the Enter key in a `JTextField` object generates an `ActionEvent`.
13. Java allows you to create small applications, commonly known as applets that can be executed inside a web browser.
14. An applet cannot access the local disk.
15. Java application program begins by executing the very first executable statement in the method `main`.
16. An applet is created by extending `JApplet` class.
17. Applets have no `main` method. Web browser invokes `init`, `start`, `stop`, and `dispose` methods in sequence.
18. Applets have no title.
19. An applet is executed by opening an HTML file with a web browser or running the appletviewer.
20. An appletviewer is a scaled down web browser.
21. The service `paint` method is defined in the `Component` class and the method renders contents of a `Component` object.
22. To draw, a graphics context is required.
23. A graphics context is an instance of `Graphics` class. Since `Graphics` class is an `abstract` class, it is impossible to create a graphics context using the `new` operator.

24. The formal parameter to the paint method is a graphics context and it is available inside a component's paint method.
25. The paint method is not invoked directly. Instead, you call repaint method with no parameters and repaint method in turn will call update method followed by the paint method.
26. Every GUI component in Java has a background color and a foreground color.
27. Java employs RGB (Red Green Blue) color scheme. Every color in this scheme is obtained by mixing red, green, and blue hues in various proportions.
28. The Font class is part of java.awt package.
29. An inch is 72 points. Thus, 12 points is one-sixth of an inch.
30. The Graphics class of java.awt package has many services to create a wide variety of geometric shapes.

EXERCISES

1. Mark the following statements as true or false:
   a. A GUI program responds to events generated by the user.
   b. GUI components are measured in points.
   c. In the case of GridLayout manager, components are placed from left to right in each row and rows are filled from top to bottom.
   d. There are interfaces with no abstract methods.
   e. parseInt is a method of the String class.
   f. Applet is a small application that can be executed inside a web browser.
   g. An applet is created as a subclass of JFrame.
   h. The setTitle method can be used to set the title of an applet.
   i. The setVisible method is used to make the applet visible.
   j. To draw, a graphics context is required.
   k. The paint is invoked by the user to draw various geometric shapes.
   l. The primary colors of the Java color system are red, blue, and yellow.
   m. There are user-generated events that do not fall under some event class.
   n. If all three components of a color are the same, then it is some shade of gray, including black and white.
2. Fill in the blanks.
   a. The two stages of creating a GUI program are ____________ and ____________.
   b. An instance of the class ____________ is used to label a GUI component.
   c. The getText method of a JTextField returns a ____________.

d. To get the reference of the content pane of a `JFrame`, use the method ________________________.

e. Both `JTextField` and `JTextArea` inherit common methods from its immediate superclass ________________________.

f. A component is placed in a container using the method ________________________ of ________________________ class.

g. Every time a `JButton` is clicked, JVM creates an instance of the class ________________________.

h. An applet starts with ________________________ method followed by ________________________ method.

i. The size of an applet is determined by ________________________.

j. A `Font` has ________________, ________________, and ________________.

k. The `actionListener interface` has an `abstract` method ________________.

l. The RGB values of the color Red are ________________, ________________, ________________.

m. The RGB values of the color Black are ____________, ____________, ____________.

n. The formal parameter of the `paint` method is of the type ____________.

3. Write the Java statement or statements to accomplish the following tasks:
   a. To make the `JFrame` of your application 300 by 500 pixel
   b. To make the `JFrame` visible
   c. To get the content pane of the JFrame
   d. To create a button labeled "Ok"
   e. To place a button in a content pane
   f. To create a new color with red, green, and blue in the ratio 2:3:4 approximately
   g. To create a new font with name "Courier new," style "italic," and 48 points in size
   h. To draw a circle of radius 100 and center (250, 400)
   i. To draw a filled rectangle of height 200, width 300, and center (500, 600) and color blue
4. In finding sum of first n integers program, the book has used the fact that either n or n + 1 is going to be an integer. Explain the advantages and disadvantages of this approach.
5. Suggest the appropriate GUI component
   a. To input data
   b. To output data
   c. To input and output data
   d. To identify other GUI components
   e. To generate an event

6. Using a `GridLayout` manager, six identical GUI components can be placed in four different ways: 1 by 6, 2 by 3, 3 by 2, and 6 by 1. List all possible ways one can place 24 identical GUI components using a `GridLayout` manager.
7. List all the methods automatically invoked by the system during the execution of an applet in the order they are invoked.
8. What are the primary colors in Java? Name at least four secondary colors.
9. Name the fonts that are available in Java independent of the machine.

PROGRAMMING EXERCISES

1. Create a user interface similar to the one shown in Figure 8.12. The user can enter any number and press the Enter key to see the running average. The user can reset the running average to zero using the reset button.
2. Create a GUI program to display maximum, minimum, average, and standard deviation of all the numbers input. (*Hint*: You can compute standard deviation using the formula square root of (average of (x * x) – average of x * average of x)).
3. Write a GUI program to assign a letter grade based on four test scores (use the grading policy of Chapter 4).
4. Design and implement GUI program that can convert a `String` into corresponding telephone number. If it is an uppercase letter or a lowercase letter, the program will substitute it with the corresponding digit. If it is already a digit, no substitution is done. Thus, "GOODCAR," "gooDCar," and "go6DC2r" will be translated to 4663227. Your program should first create a class `TelephoneNumber` that has a method to perform the conversion and use it in your GUI program.
5. Create an applet to determine the monthly payment of a mortgage. If the principal is p, and the interest rate is r, then amount to be paid back is $a = p * (1 + r/100)^n$ where n is the number of years. Now, the monthly payment is obtained by dividing the amount by the number of months.
6. This book has used the `GridLayout` manager for layout. Use `BorderLayout` manager to place five buttons North, South, Center, East, and West on an applet. If the user clicks any one of them, all others will disappear. If the user clicks again, then all the buttons will reappear.

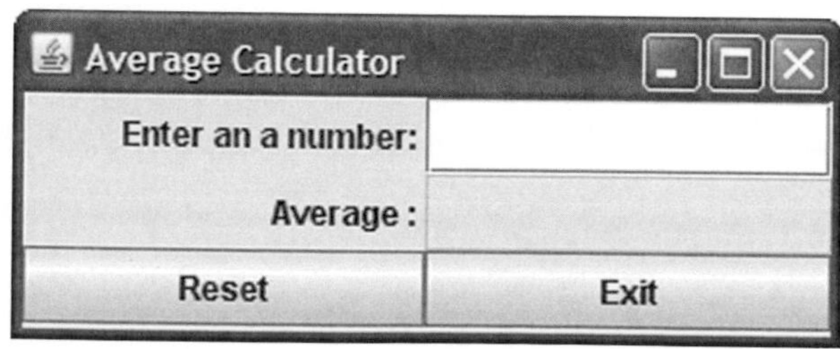

FIGURE 8.12 Average calculator.

7. This book has used the `GridLayout` manager for layout. Use `FlowLayout` manager to place five labels A, B, C, D, and E on an applet. If the user clicks any one of them, all others will disappear. If the user clicks again, then all the buttons will reappear.
8. In the book examples, components are always added to the content pane. Instead, you can create add components to different panels. Then add the panels to the content pane. Create an application that will produce the GUI shown in Figure 8.13. As the user clicks on a button, either it will disappear or all buttons will be restored. The aim of this game is to clear all buttons. (*Hint*: Use a random number generator to produce a number between 1 and 9. If the number is <7, let the button disappear. Otherwise let all buttons appear.)
9. In this program you are asked to create a GUI with two `JTextArea` objects. Unlike a `JTextField`, `JTextArea` can have multiple lines. So the string you get as well as you set in a `JTextArea` object can contain new line characters. In addition to two `JTextArea` objects, there is a `JTextField` to input an integer value. The objective of this program is to encrypt and decrypt the text in one `JTextArea` and display it in the other `JTextArea`. The encryption scheme used is called a shift method. For example, if the input is 5, the character "A" will be encrypted to "F". (*Hint*: Use the panel introduced in Programming Exercise 8.)
10. Create a GUI program to magnify and reduce the text by a given percentage. Use `JTextField` for the text. Use the panel introduced in Programming Exercise 8. Use two panels: one to keep all GUI components and the other for drawing. Get the graphic context of the panel to draw.

FIGURE 8.13 Panel exercise.

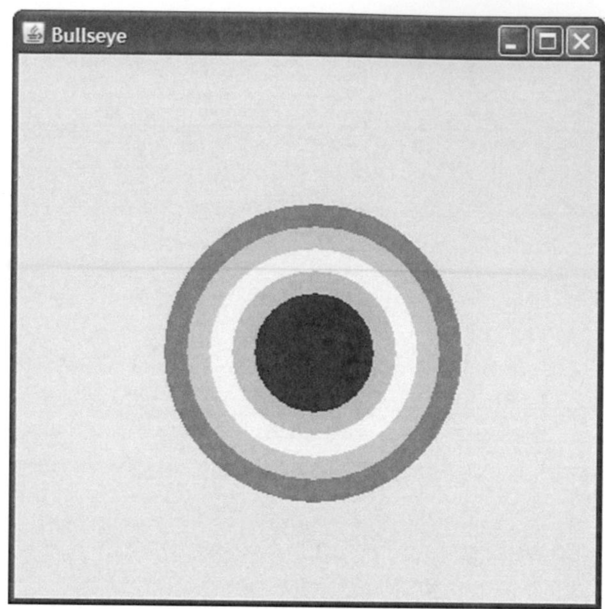

FIGURE 8.14 Bullseye.

11. Create an application applet to produce a bullseye similar to the one shown in Figure 8.14. The order of colors from the outside is red, orange, yellow, green, and blue.
12. Create a traffic light system. Every time the user pushes a button, the signal cycles through the familiar pattern of green on, yellow on, red on, and green on.
13. This book has illustrated four different programming options to implement an interface. For this exercise, create a "hybrid" version of the `Metric Conversion Helper` showing all four different programming options.

ANSWERS TO SELF-CHECK

1. GUI
2. Java code
3. `JFrame, javax.swing`
4. `Container`
5. `main`
6. `main`
7. `super("Welcome to Java");`
8. True
9. pixel
10. picture element
11. `Component`

12. `boolean`
13. `Component`
14. `setDefaultCloseOperation`
15. content pane
16. `Container`
17. `JLabel title;`
18.
```
title = new JLabel("Java for Game Development",
                                SwingConstants.CENTER);
```
19. `JTextField message = new JTextField(10);`
20. `message.setEditable(false);`
21. `JButton startButton;`
22. `startButton = new JButton("Start");`
23. True
24. `Container`
25. all
26. `extends`, `implements`
27. `addActionListener`
28. `java.awt.event`

CHAPTER 9

Simple Data Structures

In this chapter you learn

- Object-oriented concepts
  - Use classes `Vector` and `ArrayList`, members of Java Collection Framework
- Java concepts
  - One-dimensional arrays, two-dimensional arrays, multidimensional arrays, ragged arrays, enhanced `for` loop, wrapper classes, auto-boxing, and auto-unboxing
- Programming skills
  - Create classes with arrays of various data types, develop simple array processing algorithms

You have seen primitive data types and classes. Primitive data types allow us to store only one value at a time. An object, however, is capable of encapsulating more than one data values. As you have seen in Chapter 5, the ability to perform the same task again and again, or repetition, is an important control structure. This chapter introduces its data structure counterparts: arrays, `ArrayList` class, and `Vector` class.

Let us visit Heartland Cars of America and see the project Ms. Smart is currently working on. Now Heartland Cars of America is well established with many employees. To promote competition among sales personnel, the manager has decided to introduce a "Best sales person award" to the most successful sales person at the end of each pay period. Thus, the most successful sales person is awarded an extra $500.00. Ms. Smart is currently working on her program to incorporate this new change in the payroll program.

As she analyzed the problem, she noticed the following facts. She needs to process all sales person information to find out the most successful sales person for the period. Therefore, she can no longer compute the compensation and print the pay stub as she used to do. Instead, she needs to determine the most successful sales employee first. Ms. Smart decided to use arrays to simplify the computation.

TABLE 9.1 Mr. Grace's New Grading Policy

Points Range	Final Grade
More than 30% below class average	F
Between 10 and 30% below the class average	D
Between 10% below and 10% above class average	C
Between 10 and 30% above the class average	B
More than 30% above the class average	A

Observe that similar problem can arise in many situations. Suppose that Mr. Grace has decided on the following grading scheme. A student's final grade no longer depends solely on the points he or she receives. Rather it is directly linked to the class average as shown in Table 9.1.

Note that in this case, Mr. Grace first needs to process the entire data to determine the class average before assigning final grades. Mr. Grace could use an array to simplify the computation involved.

ONE-DIMENSIONAL ARRAY

Array is a contiguous fixed-length structure for storing multiple values of the same type. Array is a directly supported language feature. Thus, their performance is on par with primitive data types. Further, they have a unique syntax that is different from other objects. Each element of an array is referred to by the array name along with its position or *index*. For instance, if an array is named `pointsEarned`, then the individual elements are named `pointsEarned[0]`, `pointsEarned[1]`, `pointsEarned[2]`, and so on. Note that index of the first location is `0` and the first element is `pointsEarned[0]`. You can use any integer expression as an index.

Self-Check

1. Array is a __________ structure for storing multiple values of the __________.
2. Each element of an array is referred to by the __________ along with its __________.
3. The first index value is __________.
4. You can use any __________ as an index.

Declaring Array

The syntax template for declaring an array is as follows:

```
dataType[] arrayName;
```

In the above declaration, `dataType` specifies the data type of the elements that can be kept at each of the array locations, `arrayName` is the name of the array and the pair `[]` distinguishes this as a one-dimensional array declaration. Note that a declaration of the form

```
dataType arrayName;
```

is a valid statement in Java and declares a single variable of the type `dataType`.

Example 9.1

Consider the following array declarations:

```
char[] line;                // 1 dimensional array of
                                              characters
int[] gameScores;           // 1 dimensional array of integers
double[] taxRate;           // 1 dimensional array of double
                                                      values
String[] page;              // 1 dimensional array of String
                                                  references
Employee[] EmployeeList; // 1 dimensional array of Employee
                                                  references
```

Self-Check

5. Declare a one-dimensional array named `points` of `double` values.
6. Let `Student` be a class. Declare a one-dimensional array named `classList` of `Student`.

Instantiating Array

A declaration such as

```
char[] line;
```

creates a reference variable `line`. You need to explicitly create an array object using the `new` operator and instantiate the reference variable. The syntax template for instantiating a reference variable `arrayName` is:

```
arrayName = new dataType[IntegerExpression];
```

where `IntegerExpression` is any constant expression that evaluates to a nonnegative integer value, and it specifies the number contiguous storage locations or the *length* of the array. Since the very first location has an index value `0`, the index of the last location is length − `1` or `IntegerExpression` − `1`.

Example 9.2

The following statement

```
line = new char[20]; //instantiation of line
```

creates an array of characters of length `20`. The individual array locations are as follows:

```
line[0], line[1], line[2], ..., line[19].
```

Observe that each of these 20 locations can store one character. The statement

```
line = new char[0];
```

is legal in Java; however, no memory location is allocated. Note that you can also declare and instantiate an array using the following syntax template:

```
dataType[] arrayName = new dataType[IntegerExpression];
```

In the case of primitive data types, the array locations are initialized with default values. Thus, integral arrays are initialized by `0`, floating point arrays by `0.0`, and `boolean` arrays by `false`. In the case of object references, the array locations are initialized by `null`.

Self-Check

7. Instantiate array `points` of Self-Check 5 so that the array has 25 locations.
8. In the case of primitive data types, the array locations are initialized with __________ values.

Example 9.3

Consider the following Java statements:

```
double[] itemPrice;           //(1)
itemPrice = new double[4];  //(2)
```

Statement 1 creates a reference variable `itemPrice` as shown in Figure 9.1.

Statement 2 first allocates four contiguous memory locations, each capable of storing a `double` value, and then initializes the variable `itemPrice` with the reference of the array object created, as shown in Figure 9.2.

Observe that each of the array locations is initialized with the default value `0.0`. Statements 1 and 2 above can be combined as follows:

```
double[] itemPrice = new double[4];
```

The four locations created by the new operator are `itemPrice[0]`, `itemPrice[1]`, `itemPrice[2]`, and `itemPrice[3]`. Thus,

```
itemPrice[0] = 10.56;
itemPrice[1] = 34.12;
```

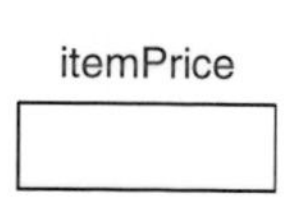

FIGURE 9.1 Primitive data type array reference variable.

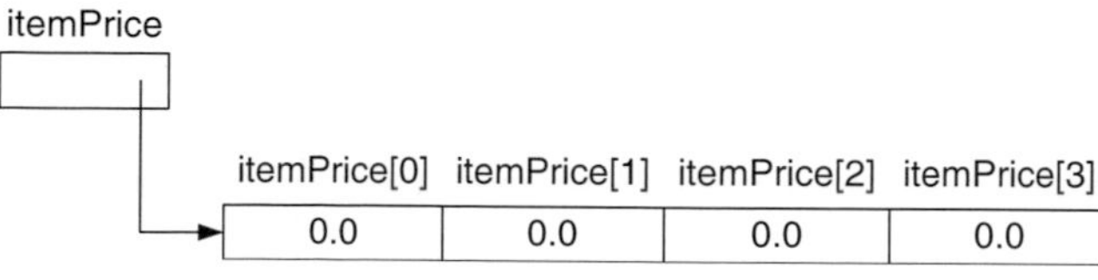

FIGURE 9.2 Primitive data type array instantiation.

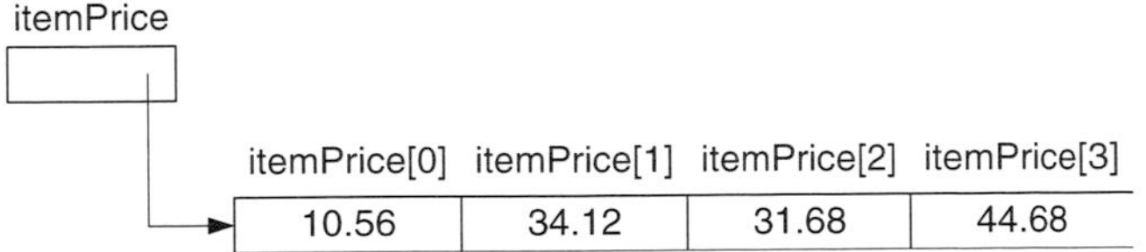

FIGURE 9.3 Primitive data type array location initialization.

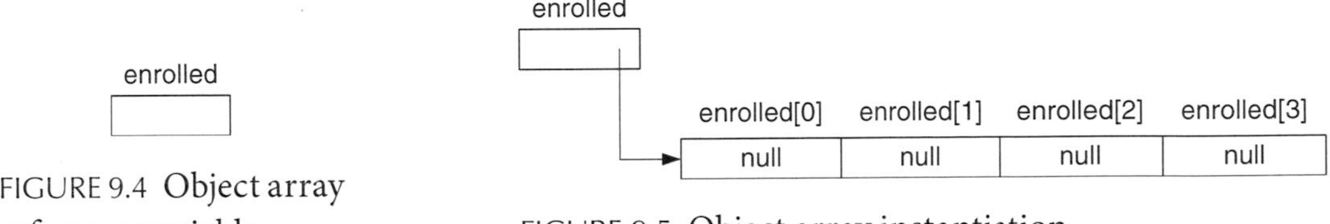

FIGURE 9.4 Object array reference variable.

FIGURE 9.5 Object array instantiation.

assigns `10.56` and `34.56` to first and second locations of the array `itemPrice` (see Figure 9.3). You can assign three times the first location value to the third location as follows:

```
itemPrice[2] = 3 * itemPrice[0];
```

The fourth location can be assigned with the sum of first two locations as follows:

```
itemPrice[3] = itemPrice[0] + itemPrice[1];
```

Example 9.4

This example illustrates the creation of array of object references. Let `Student` be a class. Now consider the following two Java statements:

```
Student[] enrolled;          //(3)
...
enrolled = new Student[4];  //(4)
```

Statement 3 creates a reference variable `enrolled` as shown in Figure 9.4.

Statement 4 first allocates four contiguous memory locations, each one of them capable of storing a `Student` reference, and then initializes the variable `enrolled` with the reference of the newly created array object, as shown in Figure 9.5.

Observe that each of the array locations is initialized with the default value `null`. Statements 3 and 4 can be combined as follows:

```
Student[] enrolled = new Student[4];
```

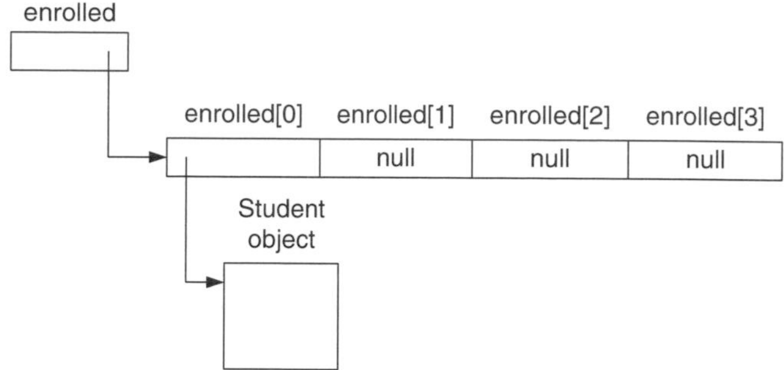

FIGURE 9.6 Object array location instantiation.

Recall that each of the array locations is a reference variable of the type `Student`. Therefore, you need to explicitly instantiate each of those locations. For instance, you can instantiate `enrolled[0]` using the default constructor as follows (see also Figure 9.6):

```
enrolled[0] = new Student();
```

Common Programming Error 9.1

Failure to instantiate array locations in the case of object references is a common programming error.

Self-Check

9. Assign 23.5 to the third location of the array `points` of Self-Check 7.
10. In the array `points`, assign five times the value at location 6 to location 8.

Advanced Topic 9.1: Programming Option

One programming option to avoid the common Programming Error 9.1 is to instantiate all array locations using a `for` statement similar to

```
for (int j = 0; j < enrolled.length; j++)
{
    enrolled[i] = new Student();
}
```

However, this approach has its drawbacks. See Exercise 12. A better approach is shown in `loadData` method of `Course` class that is part of Case Study 9.2.

Note 9.1 Observe the use of a constructor with no arguments inside the `for` loop. In Chapter 6, we made the observation that it is a good programming practice to make sure that every class has either a default constructor (i.e., one that is supplied by the compiler)

or a user-defined constructor with no formal parameters. Observe that without such a constructor the approach suggested in Advanced Topic 9.1 is not possible.

Advanced Topic 9.2: Alternate Syntax

Some of the most commonly used alternate syntax and their equivalent syntax are presented in Table 9.2.

Attribute length

The length of an array is defined as the number of locations and this value is available in an attribute `length`. You can access the `length` of an array in any expression using the array name and the dot operator. You cannot change the value of `length` through an assignment operator. In other words, if `itemPrice` is a `double` array,

```
itemPrice.length = 10;              // illegal
```

is an illegal statement. However,

```
int size = itemPrice.length;        // legal
```

is a legal statement. If the array reference variable is not instantiated using the `new` operator, `length` cannot be accessed. Once the array is instantiated with `new` operator, the `length` can be accessed in a Java program (see Table 9.3).

TABLE 9.2 Commonly Used Alternate Syntax

Alternate Syntax	Equivalent Syntax
`double itemPrice[];`	`double[] itemPrice;`
`double itemPrice[], salePrice[];`	`double[] itemPrice;` `double[] salePrice;`
`double[] itemPrice, salePrice;`	`double[] itemPrice;` `double[] salePrice;`
`double itemPrice[], salePrice;`	`double[] itemPrice;` `double salePrice;`

`double[] itemPrice`
`= {12.6, 78.45, 23.13, 0.0};`
	`double[] itemPrice = new double[4];`
	`itemPrice[0] = 12.6;`
	`itemPrice[1] = 78.45;`
	`itemPrice[2] = 23.13;`
	`itemPrice[3] = 0.0;`

TABLE 9.3 Attribute Length of an Array

Java Statements	Value of Length
`double[] itemPrice;`	It is illegal to access `itemPrice.length`
`double[] itemPrice = null;`	It is illegal to access `itemPrice.length`
`double[] itemPrice = new double[0];`	0
`double[] itemPrice = new double[10];`	10

Example 9.5

The following Java application illustrates the attribute `length`. Lines 7 and 11 will result in a syntax error and an execution error, respectively. They need to be kept commented for an error-free execution of the program.

```
/**
    Illustrates the length attribute
*/
public class LengthOfArray
{
    public static void main(String[] args)
    {

        double[] itemPrice;

        //System.out.println
               ("(7)  itemPrice.length = " + itemPrice.length);
        itemPrice = new double[10];
        System.out.println
               ("(9)  itemPrice.length = " + itemPrice.length);
        itemPrice = null;
        //System.out.println
               ("(11) itemPrice.length = " + itemPrice.length);
        itemPrice = new double[0];
        System.out.println
               ("(13) itemPrice.length = " + itemPrice.length);
        itemPrice = new double[20];
        System.out.println
               ("(15) itemPrice.length = " + itemPrice.length);
    }
}
```

Output

```
(9)  itemPrice.length = 10
(13) itemPrice.length = 0
(15) itemPrice.length = 20
```

Self-Check

11. The length of the array `points` of Self-Check 7 is available in __________.
12. The last index location of the array `points` is given by the expression __________.

PROCESSING ONE-DIMENSIONAL ARRAYS

Some of the most commonly performed operations on a one-dimensional array are

- To initialize the array with certain specific values
- To initialize the array using user input
- To output the array
- To perform various numeric computations (if the array data is numeric in nature) such as
  - Finding the sum
  - Finding the average
- To search for an item satisfying certain condition such as
  - The smallest item
  - The largest item

All of the above operations require the ability to process each one of the array elements. The repetition structures introduced in Chapter 5 are quite useful in processing each one of the array elements in a systematic manner. For example, consider the following declarations:

```
double[] itemPrice = new double[20];       // double array of
                                                    length 20
int index;                              // used for indexing
```

The repetition structure for

```
for(index = 0; index < itemPrice.length; index++)
{
    // Place the code for processing
    //
    // itemPrice[index]
    //
}
```

processes each element of the array itemPrice in the following order:

```
itemPrice[0], itemPrice[1], itemPrice[2], itemPrice[3], ...,
                                                itemPrice[19].
```

Observe that at first, index has a value 0. Therefore, during the first iteration of the above loop, itemPrice[index] is itemPrice[0]. In the next iteration, index has the value 1. Therefore, itemPrice[index] is itemPrice[1], and so on. Thus, as index takes each of the values from 0 to 19, itemPrice[index] becomes itemPrice[0], itemPrice[1], itemPrice[2], itemPrice[3], …, and itemPrice[19], respectively.

Initialize Array with Certain Specific Values

The following example illustrates how to assign the same value to all elements of an array.

Example 9.6

For the sake of discussion, let us assume that we need to initialize `itemPrice[0]`, `itemPrice[1]`, `itemPrice[2]`, `itemPrice[3]`, …, `itemPrice[19]` with `5.87`. That is,

```
itemPrice[0] = 5.87;
itemPrice[1] = 5.87;
itemPrice[2] = 5.87;
...
itemPrice[19] = 5.87;
```

Therefore, we have the following:

```
for (index = 0; index < itemPrice.length; index++)
{
    itemPrice[index] = 5.87;
}
```

Self-Check

13. Write the Java statement to initialize every location of the array `points` of Self-Check 7 by 50.67.

Enhanced for Statement

Java 5.0 introduced a new version of the `for` statement that can be used to process the elements of a collection such as an array, a `Vector`, an `ArrayList`, and so on. The `Vector` class and `ArrayList` class are covered later in this chapter.

The syntax of the enhanced `for` statement is

```
for (dataType elementVar : collectionName)
        statement
```

and is equivalent to

```
for (dataType elementVar, int j;j < collectionName.length;
                                                   j++)
{
    elementVar = collectionName[j];
    statement
}
```

The statement

```
for (index = 0; index < itemPrice.length; index++)
{
    System.out.println(itemPrice[index]);
}
```

can be written using the enhanced `for` statement as follows:

```
for (double p : itemPrice)
{
    System.out.println(p);
}
```

Note 9.2 In the enhanced `for` loop, the element variable `p` is assigned the value `itemPrice[0]`, `itemPrice[1]`, and so on. In the (ordinary) `for` loop, the index variable `index` is assigned the values `0`, `1`, `2`, and so on.

Note 9.3 The enhanced `for` loop cannot be used to assign value to individual members of an array. In other words, even though

```
for (double p : itemPrice)
{
    p = 10.5;
}
```

will not produce any syntax error, it will not assign `10.5` to `itemPrice[0]`, `itemPrice[1]`, and so on. To fully appreciate the underlying reason, consider the equivalent `for` statement.

```
for (double p, int j; j < itemPrice.length; j++)
{
    p = itemPrice[j];
    p = 10.5;
}
```

Observe that `p` and `itemPrice[j]` are two different variables and they have different memory locations. Therefore, any changes made to `p` will not change value at `itemPrice[j]`. In particular, assigning `10.5` to `p` will not change the value stored at `itemPrice[j]`.

Self-Check

14. True or false: Enhanced `for` loop can be used to assign values to array locations.

Initialize Array Locations with Different Values

The following example illustrates how to assign different values to all elements of an array.

Example 9.7

For the sake of discussion, let us assume that we need to initialize itemPrice[0], itemPrice[1], itemPrice[2], itemPrice[3], ..., itemPrice[19] with 5.87, 15.87, 25.87, and so on. That is,

```
itemPrice[0] = 5.87;
itemPrice[1] = 15.87; or itemPrice[1] = 5.87 + 10;
itemPrice[2] = 25.87; or itemPrice[1] = 5.87 + 20;
...
```

Considering Table 9.4 we have the following:

```
for (index = 0; index < itemPrice.length; index++)
{
    itemPrice[index] = 5.87 + 10 * index;
}
```

You could make use of the fact that value of itemPrice[index] is value of the previous location itemPrice[index-1] plus 10. In other words, the above code can be replaced by the following:

```
for (index = 0; index < itemPrice.length; index++)
{
    itemPrice[index] = itemPrice[index-1] + 10;
}
```

Self-Check

15. Write the Java statement to initialize every location of the array points of Self-Check 7 by the following sequence: 8.0, 13.1, 18.2, ...

TABLE 9.4 Array Values as a Function of Index

Index	5.87 + 10 * Index	Assignment Statement
0	5.87	itemPrice[0] = 5.87;
1	15.87	itemPrice[1] = 15.87;
2	25.87	itemPrice[2] = 25.87;
19	195.87	itemPrice[19] = 195.87;

Initialize Array Using User Input

The following example illustrates how to read input from the keyboard and store the array.

Example 9.8

Let `ScannedInfo` be a `Scanner` object created to receive standard input stream `System.in`. Then the expression

```
ScannedInfo.parseDouble();
```

returns the next `double` value. Thus, we have the following repetition structure:

```
for (index = 0; index < itemPrice.length; index++)
{
    itemPrice[index] = ScannedInfo.parseDouble();
}
```

Note that there is no equivalent enhanced `for` loop. See Note 9.3.

Self-Check

16. Write the Java statement to initialize every location of the array `points` of Self-Check 7 using user input. Assume that `ScannedInfo` is a `Scanner` object created to receive standard input stream `System.in`.

Output Array

The following example illustrates how to output the values in an array.

Example 9.9

Consider the array `itemPrice`.

```
for (index = 0; index < itemPrice.length; index++)
{
    System.out.print(itemPrice[index]+ " ");
}
```

The above segment will output all `20` values in one line. If you want to print `5` items per line, then you could include a statement such as the following as the first statement of the `for` structure:

```
if (index % 5 == 0)
    System.out.println();
```

However, above statement will print a new line for index value 0. To exclude 0, you could add the condition, index > 0. Thus, we have the following segment of code:

```
int final ITEMS_PER_LINE = 5;

...

for (index = 0; index < itemPrice.length; index++)
{
    if (index % ITEMS_PER_LINE == 0 && index > 0)
                              System.out.println();
    System.out.print(itemPrice[index]+ "  ");
}
```

Self-Check

17. Write the Java statement to output the array points of Self-Check 7, eight items per line.

Perform Various Numeric Computations

Next example illustrates numeric computations on an array.

Example 9.10

In this example, we compute the sum and average of all elements of an array.

```
sum = 0;
for (index = 0; index < itemPrice.length; index++)
{
    sum = sum + itemPrice[index];
}
```

To compute the average, you need to make sure that number of items is more than 0. Thus,

```
if (itemPrice.length > 0)
    average = sum / itemPrice.length;
else
    average = 0.0;
```

Note that in the above discussion, the array itemPrice is of type double. Therefore, the expression

```
sum / itemPrice.length;
```

is of type double. However, since the array elements are of type int and you have declared sum as int type, you need to cast either sum or the length to double before performing the division to avoid truncation due to integer division.

Self-Check

18. Write the Java statement to compute the average of all elements in `points` of Self-Check 7.

Search for Item

The following example illustrates searching for the smallest element of an array.

Example 9.11

In this example, we find the index of the smallest number in an array. In the case of multiple values, we find the index of the first smallest number in an array. Note that once the index value is known you can determine the smallest value. The algorithm to determine the index of the smallest element in an array can be described as follows. Keep a variable `minIndex` of type `int` to keep track of the index value of the smallest value seen so far in the array. Therefore, as we start inspecting the array, the first item is the smallest item seen so far. Hence, at the beginning, `minIndex` is initialized to `0`. We then compare the element at `minIndex` with all other elements in the array. As we compare, there are two possible outcomes: (1) the element being compared is smaller than the element at `minIndex` or (2) the element being compared is not smaller than the element at `minIndex`. In the first case, the element at `minIndex` is no longer the smallest value seen so far. We just found a new smaller value and hence we update `minIndex`. In the second case, the element at `minIndex` is still the smallest value seen so far. Therefore, `minIndex` remains the same. Thus, we have the following segment of code for determining the smallest value in the array `itemPrice`:

```
int minIndex;
...
minIndex = 0;
for (index = 1; index < itemPrice.length; index++)
    if (itemPrice[index]< itemPrice[minIndex])
        minIndex = index;
smallestValue = itemPrice[minIndex];
```

We illustrate the way above algorithm works using a sample data for the array `itemPrice`. Only first eight data values of the array and first four iterations are shown in Figure 9.7. As a challenge, a slightly different variation of this algorithm is implemented in `findMin` of Case Study 9.1. Even if you skip Case Study 9.1, it will be beneficial to check out `findMin` method and observe the differences.

Self-Check

19. Write the Java statement to find the largest element in the array `points` of Self-Check 7.

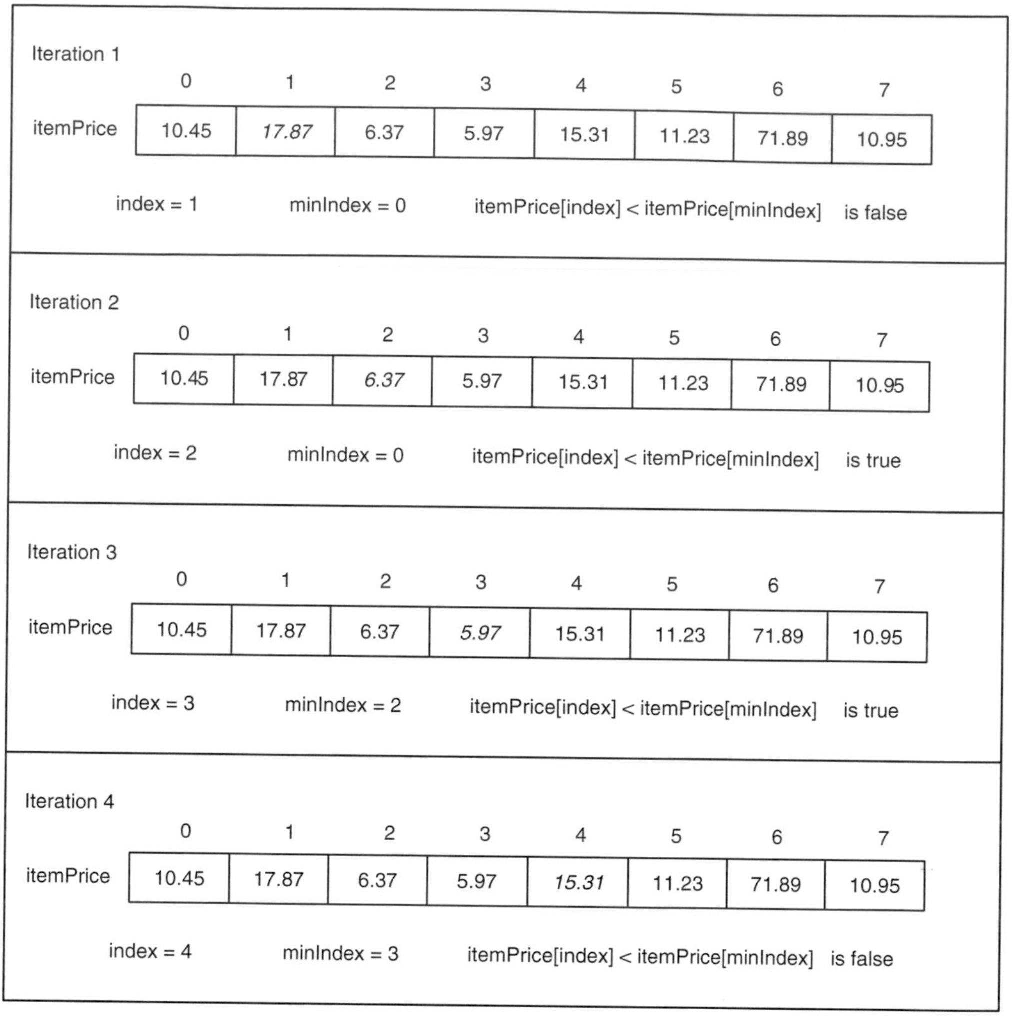

Iteration 1

	0	1	2	3	4	5	6	7
itemPrice	10.45	*17.87*	6.37	5.97	15.31	11.23	71.89	10.95

index = 1 minIndex = 0 itemPrice[index] < itemPrice[minIndex] is false

Iteration 2

	0	1	2	3	4	5	6	7
itemPrice	10.45	17.87	*6.37*	5.97	15.31	11.23	71.89	10.95

index = 2 minIndex = 0 itemPrice[index] < itemPrice[minIndex] is true

Iteration 3

	0	1	2	3	4	5	6	7
itemPrice	10.45	17.87	6.37	*5.97*	15.31	11.23	71.89	10.95

index = 3 minIndex = 2 itemPrice[index] < itemPrice[minIndex] is true

Iteration 4

	0	1	2	3	4	5	6	7
itemPrice	10.45	17.87	6.37	5.97	*15.31*	11.23	71.89	10.95

index = 4 minIndex = 3 itemPrice[index] < itemPrice[minIndex] is false

FIGURE 9.7 Finding the smallest value in an array.

CASE STUDY 9.1: MR. GRACE'S LATEST GRADING POLICY

Mr. Grace's latest grading policy is to conduct six tests and then take the average of top five scores to decide the final grade. He created a file in which each line has eight entries: first name, last name, and six test scores of the student. Mr. Grace decided to create a `Student` class with four attributes:

```
private static final int ARRAY_SIZE = 6;
private String firstName;
private String lastName;
private double testScores[] = new double[ARRAY_SIZE];
private double gradeScore;          // average of top ARRAY_SIZE-1
                                    // testScores
private String grade;
```

Instead of providing accessor and mutator methods for every attribute, Mr. Grace decided to provide three methods, and accessor and mutator methods only for attributes `gradeScore` and `grade`. (From a pedagogical point of view, our focus is not on get and set methods. Further, readers can easily add them in case they prefer.)

- Method `setStudentInfo` to set all attributes
- Method `computeGradeScore` to compute `gradeScore`
- Method `createGradeReport` to create a `String` that contains the necessary information about a student, including `gradeScore`

Assume that the `main` creates a `Scanner` object `ScannedInfo` and is used as an actual parameter in the `setStudentInfo` method. The `setStudentInfo` method reads one line from the file and sets all the attributes of one student. Thus, we have the following:

```
public void setStudentInfo(Scanner sc)
{
    int i;
    firstName = sc.next();
    lastName = sc.next() ;
    for (i = 0 ; i < testScores.length ; i++)
        testScores[i] = sc.nextDouble() ;
}
```

Now, the average of top five out of six scores can be computed in three steps:

1. Compute the sum of all six values
2. Compute the minimum of all six values
3. Compute the sum of top five scores by subtracting the minimum value from the sum of all six values and divide by `(testScores.length - 1)`

We implement each one of the above steps as a separate method. Since we need steps 1 and 2 for performing step 3 and not for any other purpose, we keep the associated methods `private`. These three methods are listed below:

```
private double computeSum()
{
    int i;
    double sum = 0;
    for (i = 0 ; i < testScores.length ; i++)
        sum = sum + testScores[i] ;
    return sum;
}
```

```
private double findMin()
{
    int i;
    double minimum = testScores[0];
    for (i = 1 ; i < testScores.length ; i++)
        if (testScores[i] < minimum)
            minimum = testScores[i];
    return minimum;
}

public void computeGradeScore()
{
    double adjustedTotal;
    adjustedTotal = computeSum() - findMin();
    gradeScore = adjustedTotal /( testScores.length - 1);
}
```

Our final method returns a `String` of necessary information about a student. The main application can use it to produce required output.

```
public String createGradeReport()
{

    int i;
    String str;
    DecimalFormat twoDecimalPlaces = new DecimalFormat("0.00");
    str = firstName + "\t"+ lastName + "\t";
    for (i = 0; i < testScores.length; i++)
        str = str + testScores[i] + "\t";
    str = str + twoDecimalPlaces.format(gradeScore);
    str = str + "\t" + grade;
    return str;
}
```

Now the application program creates an instance of the `Student`. For each student, the program sets the data values, computes the grade score, and prints the information into an output file. The program listing along with sample input and output follows. Observe that in the program listing we have intentionally used enhanced `for` statements to give the reader more examples of enhanced `for` statements. In particular, compare the `createGradeReport` presented in the programme listing.

```
import java.util.*;
import java.text.DecimalFormat;

/**
    Keeps name, six test scores, grade score and letter grade
*/
public class Student
{
    private static final int ARRAY_SIZE = 6;

    private String firstName;
    private String lastName;
    private double testScores[] = new double[ARRAY_SIZE];
    private double gradeScore;
    private String grade;

    /**
        Loads student data to an instance of Student
        @param a scanner instance
    */
    public void setStudentInfo(Scanner sc)
    {
        firstName = sc.next();
        lastName = sc.next() ;

        for (int i = 0; i < testScores.length; i++)
            testScores[i] = sc.nextDouble() ;
    }

    /**
        Computes the sum of all test scores
        @return sum of all test scores
    */
    private double computeSum()
    {
        double sum = 0;

        for (double ts : testScores)
            sum = sum + ts ;

        return sum;
    }
```

```
/**
    Computes the minimum of all test scores
    @return the minimum of all test scores
*/
private double findMin()
{
    double minimum = testScores[0];

    for (double ts : testScores)
        if (ts < minimum)
        minimum = ts;

    return minimum;
}

/**
    Computes the average test score ignoring the least
                                                 score
*/
public void computeGradeScore()
{
    double adjustedTotal;

    adjustedTotal = computeSum() - findMin();
    gradeScore = adjustedTotal /( testScores.length - 1);
}

/**
    Create a String with all information on a student
    @return String with all information on a student
*/
public String createGradeReport()
{

    String str;
    DecimalFormat twoDecimalPlaces = new
                     DecimalFormat("0.00");

    str = firstName + "\t"+ lastName + "\t";
    if (str.length() < 10) str = str+ "\t";
    for (double ts : testScores)
        str = str + ts + "\t";
    str = str + twoDecimalPlaces.format(gradeScore);
```

```
        if (grade != null) str = str + "\t" + grade;
        return str;
    }

    /**
        Returns the average test score after ignoring the
                                                      least
        @return average test score after ignoring the least
    */
    public double getGradeScore()
    {
        return gradeScore;
    }

    /**
        Returns the letter grade
        @return letter grade
    */
    public String getGrade()
    {
        return grade;
    }

    /**
        Mutator method for average test score
        @param inGradeScore new value of average test score
    */
    public void setGradeScore(double inGradeScore)
    {
        gradeScore = inGradeScore;
    }

    /**
        Mutator method for letter grade
        @param inGrade new value of letter grade
    */
    public void setGrade(String inGrade)
    {
        grade = inGrade;
    }
```

```
    /**
        toString method
        @return all information about student including
                                                letter grade
    */
    public String toString()
    {
        return createGradeReport();
    }

}



import java.io.*;
import java.util.Scanner;


/**
    Application program of Mr. Grace to assign grade
*/
public class StudentScores
{


    public static void main (String[] args) throws
                FileNotFoundException, IOException
    {
        Student student = new Student();
        Scanner ScannedInfo = new Scanner(new
                            File("C:\\ studentData.dat"));
        PrintWriter output = new PrintWriter(new
                    FileWriter("C:\\ studentData.out"));

        while (ScannedInfo.hasNext())
        {
            student.setStudentInfo(ScannedInfo);
            student.computeGradeScore();
            output.println(student.createGradeReport());
        }
        output.close();
    }
}
```

Input File Content

Kim Clarke	**70.50**	**69.85**	**90.25**	**100.00**	**81.75**	**100.00**
Chris Jones	**78.57**	**51.25**	**97.45**	**85.67**	**99.75**	**88.76**
Brian Wills	**85.08**	**92.45**	**67.45**	**71.57**	**50.92**	**72.00**
Bruce Mathew	**60.59**	**87.23**	**45.67**	**99.75**	**72.12**	**100.00**
Mike Daub	**56.60**	**45.89**	**78.34**	**64.91**	**66.12**	**70.45**

Output

```
Kim Clarke      70.5     69.85    90.25    100.0      81.75    100.0      88.50
Chris Jones     78.57    51.25    97.45      85.67    99.75      88.76    90.04
Brian Wills     85.08    92.45    67.45      71.57    50.92      72.0     77.71
Bruce Mathew    60.59    87.23    45.67      99.75    72.12    100.0      83.94
Mike Daub       56.6     45.89    78.34      64.91    66.12      70.45    67.28
```

Advanced Topic 9.3: Array Index Out of Bounds Exception

The following statement

```
double[] itemPrice = new double[5];
```

creates five contiguous memory locations

```
itemPrice[0], itemPrice[1], itemPrice[2], ..., itemPrice[4].
```

If `intExp` is an integer expression, then `itemPrice[intExp]` exists only if `intExp` is a value between the *lower bound* `0` and the *upper bound* `4`. If `intExp` is a value not between `0` and `4`, `itemPrice[intExp]` does not exist and index is said to be *out of bounds.* During program execution if an array index becomes out of bounds, Java throws an `ArrayIndexOutOfBoundException` exception. It is the programmer's responsibility to provide the necessary exception handling code in their programs. If the exception is not handled within the program, it terminates as shown in Example 9.12.

Example 9.12

```
/**
    Illustration of out of bounds exception
*/
public class ArrayIndexOutOfBounds
{

    public static void main (String[] args)
    {
        double[] sample = new double[5];
        int i;
```

```
        for (i = 0; i <= 5; i++)
        {
            sample[i] = (3*i + 1.0) / 2;
            System.out.println(sample[i] + "\t");
        }
    }
}
```

Output

```
0.5
2.0
3.5
5.0
6.5
Exception in thread "main" java.lang.ArrayIndexOutOfBounds
                                              Exception: 5
   at ArrayIndexOutOfBounds.main(ArrayIndexOutOfBounds.java:11)
```

In the above error message, (ArrayIndexOutOfBounds.java:11) indicates the exact line in the file where the exception had occurred in the format fileName:lineNumber. Further, the number 5 appearing in the very first line indicates the index value that caused the exception. Thus, from the above error message, you can infer that during the program execution the variable i became 5. The statement

```
sample[i] = (3*i + 1.0) / 2;
```

caused an error due to the fact that sample[5] is not a valid array location. Exception handling is covered in Chapter 11.

Advanced Topic 9.4: Assignment and Relational Operators

You can use the assignment operator = and relational operators == and != in the context of an array. Recall that

```
double[] itemPrice;
```

creates a reference variable itemPrice of type double array. Further,

```
itemPrice = new double[5];
```

creates 5 contiguous memory locations and initializes the reference variable itemPrice with the reference or *base address* (a term used in the context of arrays, especially in other programming languages) of the array object created.

Consider the following statements:

```
double[] arrayOne = new double[5];          //(1)
double[] arrayTwo = new double[5];          //(2)
double[] arrayThree;
int i;

for (i = 0; i < arrayOne.length)
{
    arrayOne[i] = (10 * i + 12.45);
};
```

The situation can be visualized as shown in Figure 9.8.

Now the statement

```
arrayTwo = arrayOne;
```

assigns the reference variable `arrayTwo` with the value contained in the reference variable `arrayOne`. Thus, both of them contain the reference of the same array (see Figure 9.9). Therefore, `arrayOne[i]` is same as `arrayTwo[i]` for `i = 0, 1, 2, 3, 4`. Note that no separate array is created. Recall from Chapter 6 that this type of copying is known as shallow copying.

Observe that the array locations previously referenced by `arrayTwo` are no longer accessible to the program. The garbage collector of Java will eventually reclaim the memory that became inaccessible to the program.

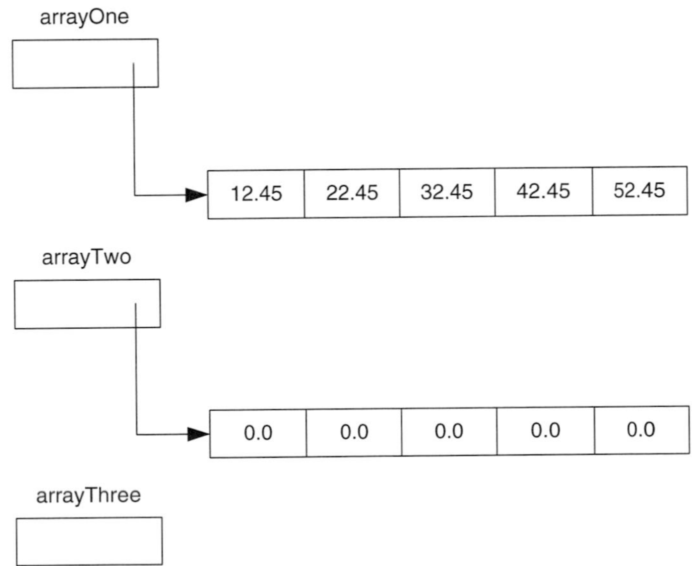

FIGURE 9.8 Copying an array—before.

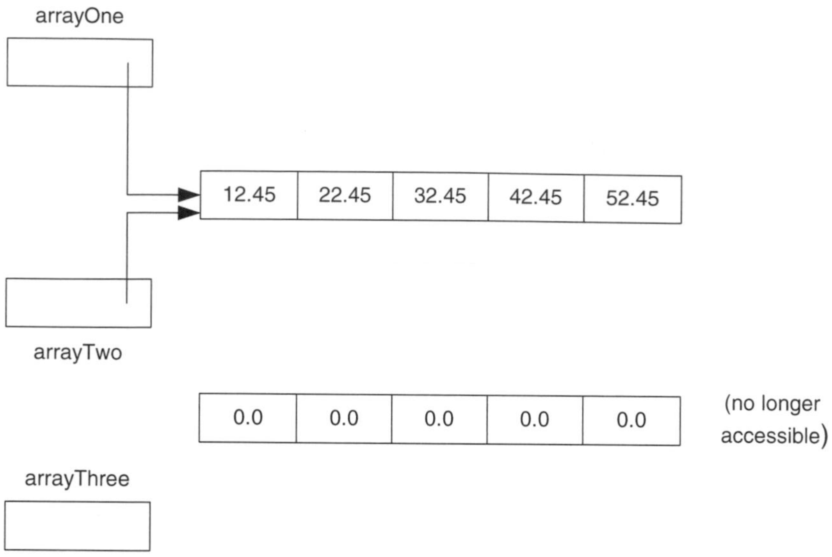

FIGURE 9.9 Shallow copying an array.

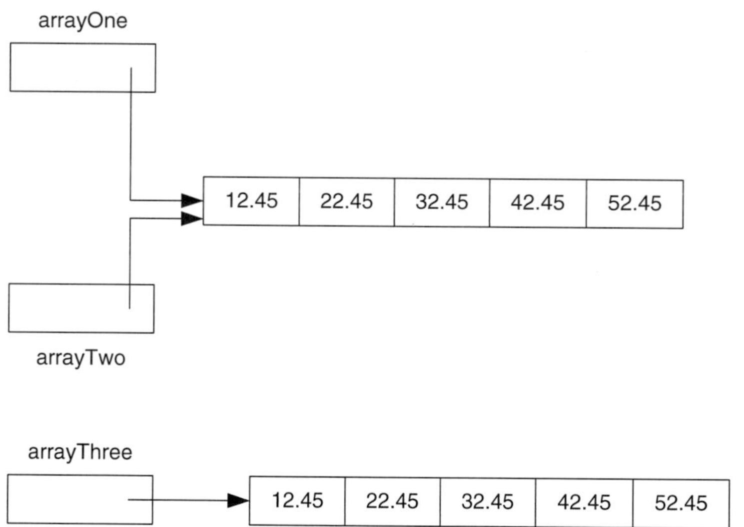

FIGURE 9.10 Deep copying an array.

If you want to make a copy of the array object referenced by `arrayOne`, you must allocate memory and perform an element-by-element copy using a repetition structure. As mentioned in Chapter 6, this form of copying is known as deep copying. Thus, to deep copy `arrayOne` to `arrayThree`, we need the following statements:

```
arrayThree = new double[arrayOne.length];

for (i = 0; i < arrayOne.length; i++)
    arrayThree[i] = arrayOne[i];
```

Once the deep copying is performed, we have the situation shown in Figure 9.10.

Similarly, comparing two arrays for equality in fact compares the reference variable. Thus, the expression (arrayOne == arrayTwo) is true, since both arrayOne and arrayTwo contain the reference of the same array. Note that the expression (arrayOne == arrayThree) evaluates to false. If you want to compare element by element for equality, you need to perform *deep comparison*. In this case, you start by checking whether or not both arrays have identical length. If that is the case, as in the case of deep copying, you need to use a repetition structure such as for to compare corresponding individual elements for equality.

Thus, the code to compare two int arrays arrayOne and arrayThree can be written as follows:

```
boolean equal;

equal = (arrayOne.length == arrayThree.length);
for (i = 0; (i < array.length && equal); i++)
    equal = (arrayOne[i] == arrayThree[i]);

if (equal)
    System.out.println("Arrays are identical");
else
    System.out.println("Arrays are not identical");
```

Advanced Topic 9.5: Role of Inheritance

So far in our presentation, we have insisted that all elements of an array must be of the same data type. Even though the above-mentioned statement is true, it is not a major restriction. For instance, let us assume you want to use an array to store three different types of employees: full-time, part-time, and sales. In this case, you could design a class Employee and three subclasses FullTimeEmp, PartTimeEmp, and SalesEmp and create an array currentEmployees of type Employee reference of length 100 as shown below:

```
Employee[] currentEmployee = new Employee[100];
```

If fullTimeEmp is an instance of FullTimeEmp the following assignment

```
currentEmployee[10] = fullTimeEmp;
```

is legal and stores the reference fullTimeEmp at currentEmployee[10]. Recall from Chapter 7 that you can always assign a subclass reference to a superclass reference. However,

```
fullTimeEmp = currentEmployee[20];
```

is illegal. Once again, as explained in Chapter 7, you cannot assign superclass reference to a subclass reference. If `currentEmployee[20]` is in fact referencing an object of the type `FullTimeEmp`, then the following statement is legal:

```
fullTimeEmp = (FullTimeEmp) currentEmployee[20];
```

The next example illustrates the use of inheritance in arrays.

Example 9.13

```
import java.util.*;
import java.io.*;

/**
    Illustrates the use of inheritance in an array
*/
public class HeartlandCarsOfAmericaEmployeePayRoll
{
    public static void main (String[] args) throws
                FileNotFoundException, IOException
    {

        //Create reference variable of all three employee
                                                  types
        Employee[] currentEmployees = new Employee[100];

        //Declare variables to input

        char inputEmployeeType;
        String inputFirstName;
        String inputLastName;
        double inputBaseSalary;
        double inputPayPerHour;
        int inputSalesVolume;
        int inputHoursWorked;
        int noOfEmployees;
        int idx;

        //Get two input values
        //      Scanner ScannedInfo = new Scanner(System.in);
        Scanner ScannedInfo = new Scanner(new
                        File("C:\\Employee.dat"));
        PrintWriter outFile = new PrintWriter(new
                    FileWriter("C:\\payroll.dat"));
```

```
index = 0;
while (ScannedInfo.hasNext())
{

    inputEmployeeType = ScannedInfo.next().
                                 charAt(0);
    switch (inputEmployeeType)
    {
        case 'F' :
        case 'f' :

            inputFirstName = ScannedInfo.next();
            inputLastName = ScannedInfo.next();
            inputBaseSalary = ScannedInfo.
                                   nextDouble();
            inputHoursWorked = ScannedInfo.nextInt();

            //create a FullTimeEmployee object
            currentEmployees[idx]
                = new FullTimeEmployee(inputFirstName,
                        inputLastName,inputBaseSalary,
                                    inputHoursWorked);

        break;

        case 'P' :
        case 'p' :

            inputFirstName = ScannedInfo.next();
            inputLastName = ScannedInfo.next();
            inputPayPerHour = ScannedInfo.
                                  nextDouble();
            inputHoursWorked = ScannedInfo.nextInt();

            //create a PartTimeEmployee object
            currentEmployees[idx]
                = new PartTimeEmployee(inputFirstName,
                        inputLastName,inputPayPerHour,
                                    inputHoursWorked);

        break;
```

```
                case 'S' :
                case 's' :


                    inputFirstName = ScannedInfo.next();
                    inputLastName = ScannedInfo.next();
                    inputBaseSalary = ScannedInfo.
                                      nextDouble();
                    inputSalesVolume = ScannedInfo.nextInt();


                    //create a SalesEmployee object
                    currentEmployees[idx]
                            = new SalesEmployee(inputFirstName,
                                 inputLastName,inputBaseSalary,
                                             inputSalesVolume);

                break;


                default:
                    System.out.println("Check data file.");
                return;

            }         // End of switch
            idx++;
        }             // End of while


        noOfEmployees = idx;


        for (idx = 0; idx < noOfEmployees; idx++)
        {
            //invoke the createPayStub method
            outFile.println(currentEmployees[idx].
                                     createPayStub());
        }

        outFile.close();

    }                          // End of main
}                              // End of class
```

Advanced Topic 9.6: Passing Arrays as Parameters in Methods

As in the case of objects, arrays can be passed as parameters in methods. The following method takes two `double` arrays and deep copies the second array into the first array:

```
public static void copy(double[] destination, double[] source)
{
    //works only if destination.length >= source.length
    //better option is let the copy method return destination
    int i;
    for (i = 0; i < source.length; i++)
        destination[i] = source[i];

}
```

Note 9.4 If a formal parameter is an array, then you need to include a pair of square brackets without any specification of size to distinguish it as a one-dimensional array.

Note 9.5 If an actual parameter is an array, then you do not use the pair of square brackets. You use the array name to pass the reference.

Sometimes you may be interested in only part of the array. Let us write a method to find the sum of all values between two index values `start` and end.

```
public static double sum(double[] arr, int start, int end)
{
    int i;
    double total = 0;

    for(i = start; i <= end; i++)
        total = total + arr[i];
    return total;
}
```

Similarly, you can write methods for inputting and outputting values between two indices.

```
public static void getData(Scanner sc, double[] arr, int start,
                                                        int end)

{
    int i;
    for(i = start;  i <= end; i++)
    arr[i] = sc.nextDouble();
}
```

```
public static void print(double[] arr, int start, int end)
{
    int i;
    for (i = start;  i <= end; i++)
        System.out.print(arr[i] + "\t");
}
```

The next example illustrates the method invocation if one of the parameters is of the type array. We do not provide many examples due to the following two reasons: (1) It is very rare in an object-oriented paradigm that you would need to use an array as such. Quite often, the array itself may be one of the attributes, and as such you may be passing objects as parameters. (2) Java provides two classes `ArrayList` and `Vector`. They are more flexible data structures and thus many Java programmers use them instead of arrays. Instances of `ArrayList` and `Vector` are objects and, once again, you are passing objects as parameters.

Example 9.14

The following program creates an `int` array of size `25` and initializes the array through user input. Then the program prints the entire array in five lines; each line containing `5` consecutive array values along with the sum in each of those rows.

```
import java.io.*;
import java.util.Scanner;
import java.text.DecimalFormat;

/**
    Illustration of array as a parameter
*/
public class ArrayAsParameter
{
    /**
        Gets data from a Scanner object and stores it in an
                                                        array
        @param sc input source; a scanner object
        @param arr the destination array name
        @param start the starting location of array
        @param end the starting location of array
    */
    public static void getData(
        Scanner sc, double[] arr, int start, int end)
    {
        int i;
        for (i = start;  i <= end; i++)
        arr[i] = sc.nextDouble();
    }
```

```
/**
    Computes the sum of values in an array
    @param arr the array name
    @param start the starting location of array
    @param end the starting location of array
*/
public static double sum(double[] arr, int start,
                                          int end)
{
    int i;
    double total = 0;

    for(i = start; i <= end; i++)
        total = total + arr[i];
    return total;
}

/**
    prints an array
    @param arr the array name
    @param start the starting location of array
    @param end the starting location of array
*/
public static void print(double[] arr, int start,
                                          int end)
{
    int i;
    for (i = start;  i <= end; i++)
        System.out.print(arr[i] + "\t");
}


public static void main (String[] args) throws
                                    IOException
{
    double[] sample = new double[25];
    int i;
    Scanner ScannedInfo = new Scanner(System.in);
    DecimalFormat twoDecimalPlaces = new
                    DecimalFormat("0.00");
    System.out.println("Enter 25 decimal values");
    getData(ScannedInfo, sample, 0, 24);
```

```
        System.out.println
                ("\nFive inputs per row and their sum
                                        follows:\n");
        for (i = 0; i < 5; i++)
        {
            print(sample, 5*i, 5*i + 4);
            System.out.println(
                    twoDecimalPlaces.format(sum(sample, 5*i,
                                                5*i + 4)));
        }
        System.out.println();
    }
}
```

Output

```
Enter 25 decimal values

12.34 45.12 65.67 23.00 78.56 34.51 87.01 20.70 19.45 52.92 17.37 65.17
45.60 73.09 18.50 31.17 27.11 88.99 91.54 29.25 74.39 56.14 72.05 18.36
50.07

Five inputs per row and their sum are as follows:

12.34    45.12    65.67    23.0     78.56    224.69
34.51    87.01    20.7     19.45    52.92    214.59
17.37    65.17    45.6     73.09    18.5     219.73
31.17    27.11    88.99    91.54    29.25    268.06
74.39    56.14    72.05    18.36    50.07    271.01
```

Next example solves the same problem in a more object-oriented fashion.

Example 9.15

This example solves the same problem as in Example 9.14. The approach shows a better design. This example is presented to illustrate the fact that quite often there is no need for passing an array as an actual parameter. We have kept the names of all variables and methods the same so that the interested reader can make an easy comparison of program presented in Example 9.14 with the one presented here.

```
import java.util.Scanner;

/**
    Object parameter passing instead of an array
*/
public class ArrayAsParameterClass
{
    private double[] arr;
```

```
/**
    Constructor with size specification
    @param size length of the array
*/
public ArrayAsParameterClass(int size)
{
    arr = new double[size];
}

/**
    Gets data from a Scanner object and stores it in an
                                                  array
    @param sc input source; a scanner object
    @param start the starting location of array
    @param end the starting location of array
*/
public void getData(
            Scanner sc, int start, int end)
{
    int i;
    for(i = start;  i <= end; i++)
    arr[i] = sc.nextDouble();
}

/**
    Computes the sum of values in an array
    @param start the starting location of array
    @param end the starting location of array
*/
public double sum(int start, int end)
{
    int i;
    double total = 0;

    for (i = start; i <= end; i++)
           total = total + arr[i];
    return total;
}

/**
    prints an array
    @param start the starting location of array
```

```
            @param end the starting location of array
        */
        public void print(int start, int end)
        {
            int i;
            for (i = start;  i <= end; i++)
                System.out.print(arr[i] + "\t");
        }
    }


    import java.io.*;
    import java.util.Scanner;
    import java.text.DecimalFormat;

    /**
        Application program to illustrate object instead of array
    */
    public class ArrayAsParameterClassApplication
    {
        public static void main (String[] args) throws IOException
        {
            ArrayAsParameterClass sample = new
                                         ArrayAsParameterClass(25);
            int i;
            Scanner ScannedInfo = new Scanner(System.in);
            DecimalFormat twoDecimalPlaces = new
                                         DecimalFormat("0.00");
            System.out.println("Enter 25 decimal values");
            sample.getData(ScannedInfo, 0, 24);

            System.out.println(
                         "\nFive inputs per row and their sum
                                                 follows:\n");
            for (i = 0; i < 5; i++)
            {
                sample.print(5*i, 5*i + 4);
                System.out.println(
                    twoDecimalPlaces.format(sample.sum(5*i,
                                                 *i + 4)));
            }
            System.out.println();
        }
    }
```

Advanced Topic 9.7: Returning Arrays in Method Invocation

As in the case of objects, arrays can be the return type of a method. The following method creates a new array, then deep copies the elements of source array between indices start and end:

```
public static double[] subArrayCopy
                    (double[] source, int start, int end)
{
    int i;
    double[] destination = null;
    destination = new double[end-start+1];
    for(i = 0; i < destination.length; i++)
        destination[i] = source[start + i];
    return destination;
}
```

Note 9.6 You need double[] to indicate the return type as a double array. However, in the return statement all you need is the array name.

The next example illustrates the use of array as a return type. However, through better design, returning an array can be completely eliminated and this is left as an exercise (see Programming Exercise 7).

Example 9.16

The following program creates an int array of size 25 and initializes the array through user input. Then the program makes a deep copy of the array, copying 5 consecutive values at a time into five new arrays.

```
import java.io.*;
import java.util.Scanner;
import java.text.DecimalFormat;
/**
    Program array as return type
*/
public class ArrayAsReturnType
{
    /**
        Gets data from a scanner object and stores it in an
                                                       array
        @param sc input source; a scanner object
        @param arr the destination array name
```

```
        @param start the starting location of array
        @param end the starting location of array
    */
    public static void getData(
            Scanner sc, double[] arr, int start, int end)
    {
        int i;
        for(i = start;  i <= end; i++)
            arr[i] = sc.nextDouble();

    }

    /**
        Copy a subarray of the source array
        @param arr the array name
        @param start the starting location of array
        @param end the starting location of array
    */
    public static double[] subArrayCopy
            (double[] source, int start, int end)
    {
        int i;
        double[] destination = null;
        destination = new double[end-start+1];
        for(i = 0; i < destination.length; i++)
            destination[i] = source[start + i];
        return destination;
    }

    /**
        prints an array
        @param arr the array name
        @param start the starting location of array
        @param end the starting location of array
    */
    public static void println(double[] arr, int start, int end)
    {
        int i;
        for(i = start;  i <= end; i++)
            System.out.print(arr[i] + "\t");

        System.out.println();
    }
```

```
    public static void main (String[] args) throws IOException
    {
        double[] sample = new double[25];
        double[] sampleOne;
        int i;
        Scanner ScannedInfo = new Scanner(System.in);
        DecimalFormat twoDecimalPlaces = new
                                    DecimalFormat("0.00");
        System.out.println("Enter 25 decimal values");
        getData(ScannedInfo, sample, 0, 24);
        System.out.println(
                    "\nInputs values are:\n");
        for (i = 0; i < 5; i++)
        {
            sampleOne = subArrayCopy(sample, 5*i, 5*i + 4);
            println(sampleOne, 0, 4);
        }
        System.out.println();
    }
}
```

Output

```
Enter 25 decimal values

12.34 45.12 65.67 23.00 78.56 34.51 87.01 20.70 19.45 52.92 17.37 65.17
45.60 73.09 18.50 31.17 27.11 88.99 91.54 29.25 74.39 56.14 72.05 18.36
50.07

Input values are

12.34    45.12    65.67    23.0     78.56
34.51    87.01    20.7     19.45    52.92
17.37    65.17    45.6     73.09    18.5
31.17    27.11    88.99    91.54    29.25
74.39    56.14    72.05    18.36    50.07
```

TWO-DIMENSIONAL ARRAY

In the previous section, you have seen one dimensional arrays. The data type of each of the elements of an array can be an array itself. Let us consider the following scenario. To study the effect of eating habits of rabbits, a zoologist has decided to store the daily food consumption of 40 rabbits for 180 days. Thus, there needs to be a one-dimensional array of size 180 to keep track of each of those day's data. Now each day's data in fact consists of 40 readings, one for each rabbit involved in the study. Therefore, each one of the 180 array locations contains an array of 40 elements. Thus, if a two-dimensional array named scores is used to store the data, then scores[0], scores[1], ..., scores[179] are themselves arrays, each storing an array of length 40. Therefore, it is meaningful to talk about the elements of scores[0], scores[1], ... scores[179]. Note that the elements of the array scores[0] are known as scores[0][0], scores[0][1], ..., scores[0][39]. Similarly, the elements of scores[1] are known as scores[1][0], scores[1][1], ..., scores[1][39].

Declaring and Instantiating Array

The syntax template for declaring a two-dimensional array is

```
dataType[][] arrayName;
```

In the above declaration, dataType specifies the type of data that can be kept in each of the array locations, arrayName is the name of the array and the pair [][] distinguishes this as a two-dimensional array declaration.

The following array declarations illustrate the use of various types in connection with a two-dimensional array declaration:

```
char[][]  page;
int[][] seasonScores;
double[][] scores;
String[][] section;
Employee[][] historyOfEmployees;
```

Recall that

```
double[][]   scores;
```

creates the reference variable scores. You need to explicitly create an array object using new operator and instantiate the reference variable scores.

The syntax template for creating a two-dimensional array object instantiating a reference variable arrayName is

```
arrayName = new dataType[IntExpOne][IntExpTwo];
```

where IntExpOne and IntExpTwo are the two expressions that evaluate to two non-negative integer values.

For example,

```
score = new double[180][40];
```

will create an array of 180 by 40 locations.

Example 9.17

Consider the following Java statements:

```
double[][] temperature;                    //(1)
temperature = new double[7][3];            //(2)
```

Statement 1 creates a reference variable `temperature` as shown in Figure 9.11. Statement 2 creates the structure shown in Figure 9.12.

Figure 9.12 shows the finer details of a two-dimensional array. Thus, this two-dimensional array is a one-dimensional array of one-dimensional arrays. Note that technically temperature is an array of size 7. Therefore, `temperature.length` has a value 7. Each of the one-dimensional arrays `temperature[0]`, `temperature[1]`, …, `temperature[6]` is a one-dimensional array of length 3. Consequently, `temperature[0].length`, `temperature[1].length`, …, `temperature[6].length` exists and each has value 3.

Figure 9.13 is a simplified view of Figure 9.12.

Recall that each of the array locations is initialized with value 0.0 (not shown in Figure 9.13 for clarity). Statements 1 and 2 can be combined as shown below:

```
double[][] temperature = new double[7][3];
```

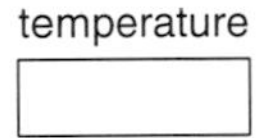

FIGURE 9.11 Two-dimensional array reference variable.

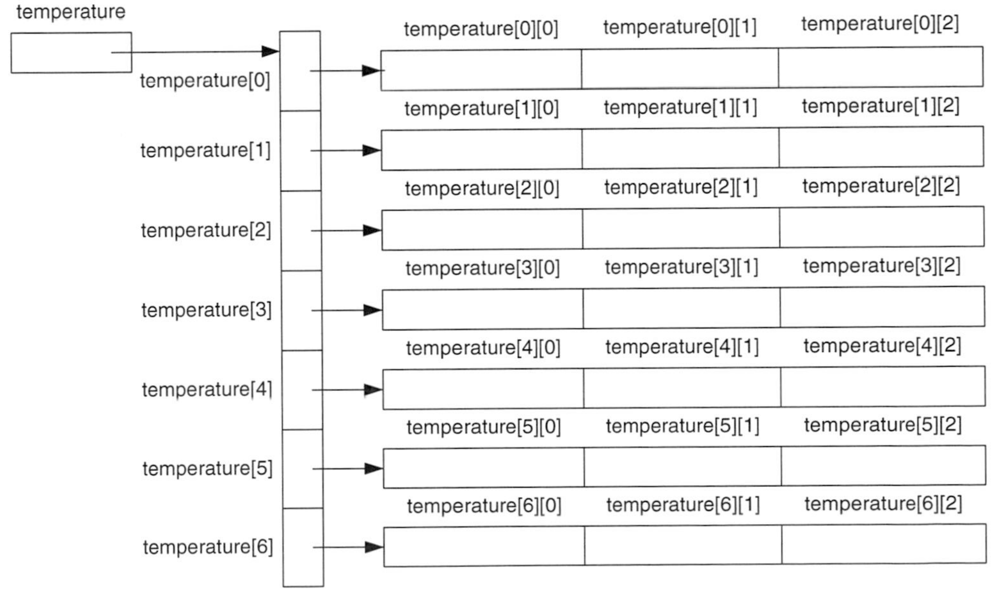

FIGURE 9.12 Two-dimensional array.

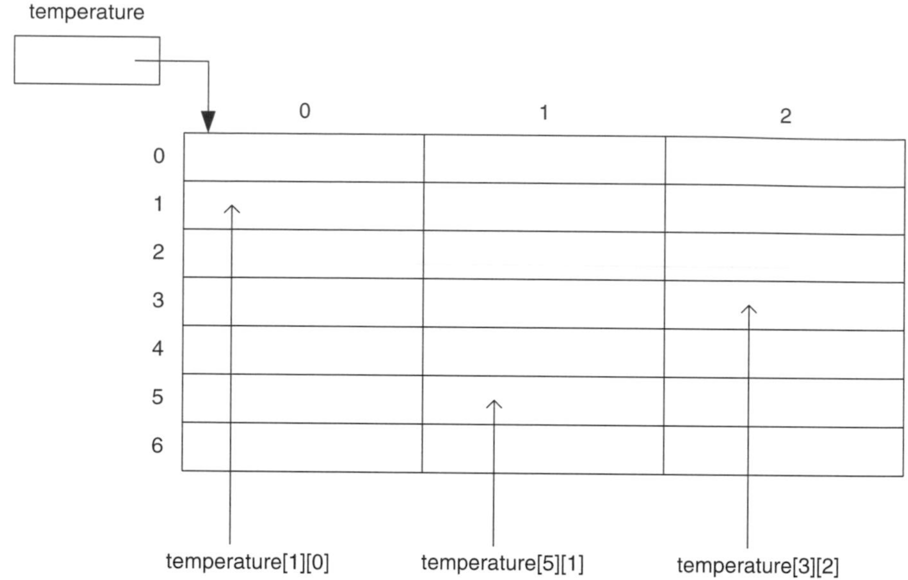

FIGURE 9.13 Simplified view of a two-dimensional array.

As shown in Figure 9.13, the 21 (7 times 3) locations created by the `new` operator are referenced using two indices. The first index can be thought of as the row and the second index as the column of the two-dimensional array. For instance, `temperature[1][0]` references the row 1, column 0 location of the two-dimensional array temperature. Similarly, `temperature[5][1]` references the row 5, column 1 location of the two-dimensional array temperature and `temperature[3][2]` references the row 3, column 2 location of the two-dimensional array `temperature`, respectively.

Now

```
temperature[1][0] = 70.56;
temperature[5][1] = 55.12;
temperature[3][2] = 68.12;
```

assigns `70.56, 55.12`, and `68.12` to three locations `temperature[1][0]`, `temperature[5][1]`, and `temperature[3][2]`, respectively.

Self-Check

20. Declare a two-dimensional array `variance` to store double values.
21. Instantiate `variance` so that the two-dimensional array has 7 rows and 12 columns.
22. Initialize the fifth row's seventh column element of `variance` with 134.53.

Advanced Topic 9.8: Alternate Syntax

Some of the most commonly used alternate syntax and their equivalent syntax are presented in Table 9.5.

TABLE 9.5 Alternate Syntax for a Two-Dimensional Array

Alternate Syntax	Equivalent Syntax
`double temperature[][];`	`double[][] temperature;`
`double temperature[][],` `taxTbl[][];`	`double[][] temperature;` `double[][] taxTbl;`
`double[][] temperature,` `taxTbl;`	`double[][] temperature;` `double[][] taxTbl;`

`double tbl[][], item[],`
`salePrice;`
	`double[][] tbl;`
	`double[] item;`
	`double salePrice;`

`double[][] twoDaytemp =`
`{{87.5, 65.2, 70.7},`
`{85.8, 63.9, 68.9}};`
	`double[]twoDaytemp`
	`= new double[2][3];`
	`twoDaytemp[0][0] = 87.5;`
	`twoDaytemp[0][1] = 65.2;`
	`twoDaytemp[0][2] = 70.7;`
	`twoDaytemp[1][0] = 85.8;`
	`twoDaytemp[1][1] = 63.9;`
	`twoDaytemp[1][2] = 68.9;`

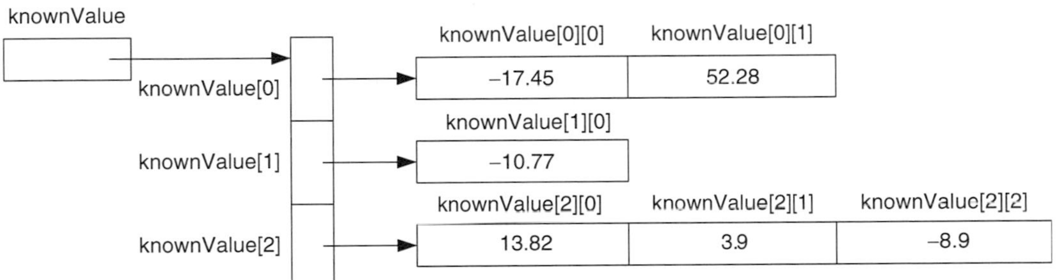

FIGURE 9.14 Ragged two-dimensional array.

Advanced Topic 9.9: Ragged Array

In this subsection, we present the most general definition of a two-dimensional array. Consider the following Java statement:

```
double[][] knownValue
        = {{-17.45, 52.28},{-10.77},{13.82, 3.9, -8.9}}
```

Note that in the above declaration, row `0` has two columns, row `1` has one column, and row `2` has three columns. In Java such a definition is legal. We can visualize the array `knownValue` as shown in Figure 9.14.

Observe that `knownValue.length` is `3`. The row `0` is of size `2` and thus `knownValue[0].length` is `2`. Similarly, the row `1` is of length `1` and thus `knownValue[1].length` is `1`. The row `2` has three elements and thus `knownValue[2].length` is `3`. Observe that all rows need not have same size and such an array is known as a ragged array.

You can create the ragged array shown in Figure 9.14 using the following Java statements as well:

```
double[][] knownValue;                  // (1)
knownValue = new double[3][];           // (2)
knownValue[0] = new double[2];          // (3)
knownValue[0][0] = -17.45;              // (3.0)
knownValue[0][1] = 52.28;               // (3.1)
knownValue[1] = new double[1];          // (4)
knownValue[1][0] = -10.77;              // (4.0)
knownValue[2] = new double[3];          // (5)
knownValue[2][0] = 13.82;               // (5.0)
knownValue[2][1] = 3.9;                 // (5.1)
knownValue[2][2] = -8.9;                // (5.2)
```

Figure 9.15 shows intermediate structures created after each of the Lines from 1–4. Observe that Lines (3.0), (3.1), (4.0), (5.0), (5.1), and (5.2) initialize the array as shown in Figure 9.14.

Advanced Topic 9.10: Processing Two-Dimensional Arrays

As you have noticed in our discussion on processing one-dimensional arrays, the central idea is to visit each element of the array in a systematic manner. In the case of a two-dimensional array, processing the entire array can be done either by processing row by row or column by column. Apart from processing the entire array, certain applications may need processing only certain specific rows or columns.

We assume the following declarations throughout this section:

```
int row;
int col;
double[][] rectMatrix = new double[4][5];
```

Processing Specific Row

Suppose you want to process all elements of row 2. Now elements of row 2 are

```
rectMatrix[2][0], rectMatrix[2][1], rectMatrix[2][2],
                  rectMarix[2][3], rectMatrix[2][4]
```

Observe that in this case, the first index remains the same and the second index assumes the values 0, 1, 2, ..., `rectMatrix[row].length - 1`. Therefore, the necessary code can be written as follows:

```
row = 2;
for (col = 0; col < rectMatrix[row].length; col++)
{
    //process rectMatrix[row][col];
}
```

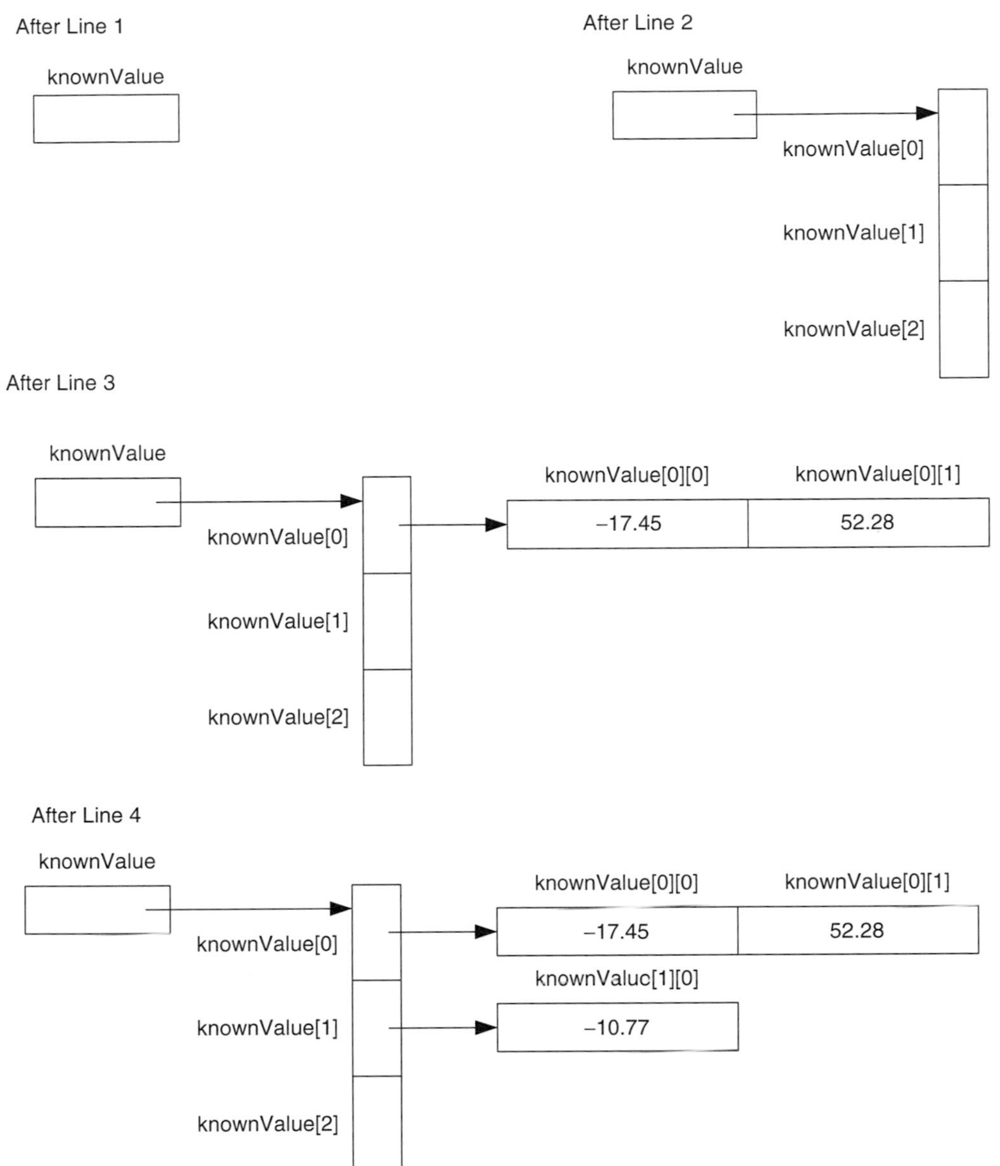

FIGURE 9.15 Intermediate stages of a ragged array creation.

You could adopt the above code to perform various row-processing tasks. For instance, the following code will output the row `1`:

```
row = 1;
for (col = 0; col < rectMatrix[row].length; col++)
{
    System.out.print(rectMatrix[row][col]);
}
System.out.println();
```

Similarly, the following code can compute the minColumn, the index value where the minimum occurs in row 3:

```
int minColumn;
...
row = 3;
minColumn = 0;
for (col = 1; col < rectMatrix[row].length; col++)
{
    if (rectMatrix[row][col] < rectMatrix[row][minColumn])
        minColumn = col;
}
```

Processing Entire Array Row by Row

Suppose you want to process all elements of the array. In this case, you may process row by row. Therefore, the necessary code can be written as follows:

```
for (row = 0; row < rectMatrix.length; row++)
{
    //place any statements that needs to be executed
    //before processing the row with index value row

    for (col = 0; col < rectMatrix[row].length; col++)
    {
        //process rectMatrix[row][col];
    }
    // place any statements that needs to be executed
    //after processing row with index value row
}
```

You could adopt the above code to perform various row-processing tasks. For instance, the following code will output the array row by row:

```
for (row = 0; row < rectMatrix.length; row++)
{
    for (col = 0; col < rectMatrix[row].length; col++)
    {
        System.out.print(rectMatrix[row][col]);
    }
    System.out.println();
}
```

Similarly, the following code can compute the minimum value in each row and place it in one-dimensional array rowMin. Observe that rowMin[0] is used to keep minimum value of row 0, rowMin[1] is used to keep minimum value of row 1, and so on. Note that rowMin is a double array of length rectMatrix.length.

```
for (row = 0; row < rectMatrix.length; row++)
{
    minColumn = 0;
    for (col = 1; col < rectMatrix[row].length; col++)
    {
        if (rectMatrix[row][col] < rectMatrix[row][minColumn])
            minColumn = col;
    }
    rowMin[row] = rectMatrix[row][minColumn];
}
```

Processing Specific Column

For the sake of example, assume that you want to process all elements of column 2. Now elements of the column 2 are

```
rectMatrix[0][2], rectMatrix[1][2], rectMatrix[2][2],
                                     rectMatrix[3][2]
```

Observe that in this case, the second index remains the same and the first index assumes the values 0, 1, 2, ..., rectMatrix.length - 1. Therefore, the necessary code can be written as follows:

```
col = 2;
for (row = 0; row < rectMatrix.length; row++)
{
    //process rectMatrix[row][col];
}
```

You could adopt the above code to perform various row-processing tasks. For instance, the following code will find sum of all elements in column 1:

```
double sum;
...
col = 1;
sum = rectMatrix[0][col];
for (row = 1; row < rectMatrix.length; row++)
{
    sum = sum + rectMatrix[row][col];
}
```

Processing Entire Array Column by Column

Suppose you want to process all elements of the array column by column. The necessary code is

```
for (col = 0; col < rectMatrix[0].length; col++)
{
    // place any statements that needs to be executed
    //before processing the column with index value col

    for (row = 0; row < rectMatrix.length; row++)
    {
        //process rectMatrix[row][col];
    }
    // place any statements that needs to be executed
    //after processing column with index value col
}
```

Note 9.7 All the codes we have presented so far will work for ragged array as well. However, the segment of code presented in this subsection will work only if all the rows have the same number of columns. You may note that we have used `rectMatrix[0].length` in the outer loop and thus implicitly assume that every row has the same number of columns.

You could adapt the above code to find the sum of every column and store it in a array `colSum`. For instance, the following segment of code will find sum of all elements in column 0 and store it in `colSum[0]` and so on. Assume that required variables and arrays are properly declared.

```
for (col = 0; col < rectMatrix[0].length; col++)
{
    sum = rectMatrix[0][col];
    for (row = 1; row < rectMatrix.length; row++)
    {
        sum = sum + rectMatrix[row][col];
    }
    colSum[col] = sum;
}
```

The next segment of code can be used to determine the largest element in each row and each column:

```
//Largest element in each row
for (row = 0; row < matrix.length; row++)
{
```

```
    largest = matrix[row][0]; //assume that the first element
                              //of the row is the largest
    for (col = 1; col < matrix[0].length; col++)
        if (largest < matrix[row][col])
            largest = matrix[row][col];

    System.out.println("Largest element of row " + (row+1)
                                          + " = " + largest);
}

//Largest element in each column
for (col = 0; col < matrix[0].length; col++)
{
    largest = matrix[0][col]; //assume that the first
                                                element of
                            //the column is the largest
    for (row = 1; row < matrix.length; row++)
        if (largest < matrix[row][col])
            largest = matrix[row][col];

    System.out.println("Largest element of col " + (col+1)
                                          + " = " + largest);
}
```

Advanced Topic 9.11: Passing Arrays as Parameter in Methods

If a formal parameter is a two-dimensional array, then you need to include two pairs of square brackets without any specifications of size to distinguish it as a two-dimensional array. As in the case of one-dimensional array, if an actual parameter is a two-dimensional array, then you do not use the pair of square brackets. You use the array name to pass the reference.

The first method we present is the one for inputting the data into a two-dimensional array.

```
public static void getData(Scanner sc, double[][] arr)
{
    int row;
    int col;
    for (row = 0; row < arr.length; row++)
    {
    for (col = 0; col < arr[row].length; col++)
        arr[row][col] = sc.nextDouble();
    }
}
```

Another method that is quite useful in many situations is that of outputting, and hence we present one such method.

```
public static void print(double[] arr)
{
    int row;
    int col;

    for (row = 0; row < arr.length; row++)
    {
        for (col = 0; col < arr[row].length; col++)
            System.out.print(arr[row][col] + "\t");
        System.out.println();

    }
}
```

The following method takes two int arrays and deep copies the second array into the first array.

```
public static void copy(double[][] destination, double[][]
                                                        source)
{
    //works only if destination.length >= source.length
    //destination[row].length >= source[row].length
    //better option is let the copy method return destination

    int row;
    int col;

    destination = new double[source.length][];
      for(row = 0; row < source.length; row++)
    {
        destination[row] = new double[source[row].length];
        for(col = 0; col < source[row].length; col++)
            destination[row][col] = source[row][col];
    }

}
```

Observe that you specify `destination` and `source` as two-dimensional arrays by the type declaration `double[][]`.

TABLE 9.6 The 3 by 3 Window Centered at (r, c)

(r-1, c-1)	(r-1, c)	(r-1, c+1)
(r, c-1)	(r, c)	(r, c+1)
(r+1, c-1)	(r+1, c)	(r+1, c+1)

Sometimes you may be interested in only part of the array. Let us write a method to find the smallest value in a 3 by 3 window centered at (r, c) (see Table 9.6).

```
public static double minimum(double[][] arr, int r, int c)
{
    int row;
    int col;

    double min = arr[r-1][c-1];

    for (row = r-1; row <= r+1; row++)
        for (col = c-1; col <= c+1; col++)
            if (arr[row][col] < min)
                min = arr[row][col];
    return min;
}
```

Example 9.18

The following program processes 7 days traffic flow counted in the morning, lunch time, evening, and night at an intersection. The aim of this program is to compute the average traffic for location by day. Another data we would like to collect is the traffic flow at a location averaged over all days by the time of the day. Therefore, we create a class `TrafficStudy` with three data members:

```
private static final int ROW_SIZE = 7;
private static final int COLUMN_SIZE = 4;
private int[][] traffic = new int[ROW_SIZE][COLUMN_SIZE];
private int[] dailyAveTraffic = new int[ROW_SIZE];
private int[] timeAveTraffic = new int[COLUMN_SIZE];
```

The class requires at least three `public` operations:

```
public void collectData(Scanner sc);     // populate the array
public void analyzeData();               // compute averages
public String createTrafficReport();     // create traffic report
```

Of these, collectData and createTrafficReport are quite straightforward. Therefore, we list the code without any further explanation.

```
public void collectData(Scanner sc)
{
    int row;
    int col;

    for (row = 0; row < traffic.length; row++)
    {
        for (col = 0; col < traffic[row].length; col++)
            traffic[row][col] = sc.nextInt();
    }
}


public String createTrafficReport()
{
    int row;
    int col;

    String str;
    str =
    "\t\tMorning\t\tLunch\t\tEvening\tNight\t\tAverage\n";
    for (row = 0; row < traffic.length; row++)
    {
        str = str + "Day "+ row + " :\t\t";
        for (col = 0; col < traffic[row].length; col++)
            str = str + traffic[row][col] + "\t\t";
        str = str + dailyAveTraffic[row]+ "\n";

    }
    str = str + "Average"+ " :\t";
    for (row = 0; row < timeAveTraffic.length; row++)
        str = str + timeAveTraffic[row] + "\t\t";
    return str;
}
```

Now analyzeData needs to do the following. First, it must find the average along each row and store it in the array dailyAveTraffic. Therefore, we need to process row by row, and for each row we need to add all column values and divide the sum by traffic[row].length. Similarly, the average along each column needs to be computed and stored in the array timeAveTraffic. Therefore, we need to process column by column, and for each column we need to add all row values and divide the sum by traffic.length. Observe the use of static method Math.round. Thus, we have the following:

```
public void analyzeData()
{
    int row;
    int col;
    int sum;

    for (row = 0; row < traffic.length; row++)
    {
        sum = 0;
        for (col = 0; col < traffic[row].length; col++)
            sum = sum + traffic[row][col];
        dailyAveTraffic[row]
               = Math.round(sum / traffic[row].length);

    }

    for (col = 0; col < timeAveTraffic.length; col++)
    {
        sum = 0;
        for (row = 0; row < traffic.length; row++)
            sum = sum + traffic[row][col];
        timeAveTraffic[col]
               = Math.round(sum / traffic. length);
    }
}
```

The application program just needs to create an instance of TrafficStudy class and invoke the methods collectData, analyzeData, and create TrafficReport in sequence. Thus, we have the complete program listing.

```
import java.util.*;

/**
    Data analysis program for traffic study
*/
public class TrafficStudy
{
    private static final int ROW_SIZE = 7;
    private static final int COLUMN_SIZE = 4;
```

```
    private int[][] traffic = new int[ROW_SIZE][COLUMN_SIZE];
    private int[] dailyAveTraffic = new int[ROW_SIZE];
    private int[] timeAveTraffic = new int[COLUMN_SIZE];

    /**
        Stores data into a two-dimensional array
        @param sc a Scanner object
    */
    public void collectData(Scanner sc)
    {
        int row;
        int col;

        for (row = 0; row < traffic.length; row++)
        {
            for (col = 0; col < traffic[row].length; col++)
                traffic[row][col] = sc.nextInt();
        }
    }

    /**
        Analyze data
    */
    public void analyzeData()
    {
        int row;
        int col;
        int sum;

        for (row = 0; row < traffic.length; row++)
        {
            sum = 0;
            for (col = 0; col < traffic[row].length; col++)
                sum = sum + traffic[row][col];
            dailyAveTraffic[row]
                = Math.round(sum / traffic[row].length);

        }

        for (col = 0; col < timeAveTraffic.length; col++)
        {sum = 0;
            for (row = 0; row < traffic.length; row++)
```

```
                sum = sum + traffic[row][col];
                timeAveTraffic[col]
                        = Math.round(sum / traffic. length);
        }
    }

    /**
        Create study report as a String
        @return study report
    */
    public String createTrafficReport()
    {
        int row;
        int col;

        String str;
        str =
        "\t\tMorning\t\tLunch\t\tEvening\tNight\t\tAverage\n";
        for (row = 0; row < traffic.length; row++)
        {
            str = str + "Day "+ row + " :\t\t";
            for (col = 0; col < traffic[row].length; col++)
                 str = str + traffic[row][col] + "\t\t";
            str = str + dailyAveTraffic[row]+ "\n";

        }
        str = str + "Average"+ " :\t";
        for (row = 0; row < timeAveTraffic.length; row++)
            str = str + timeAveTraffic[row] + "\t\t";
        return str;
    }

}

import java.io.*;
import java.util.Scanner;

/**
    Application program for traffic study
*/
```

```
public class TrafficStudyApplication
{

    public static void main (String[] args) throws
                              FileNotFoundException, IOException
    {
        TrafficStudy trafficStudy = new TrafficStudy();
        Scanner ScannedInfo
            = new Scanner(new File("C:\\TrafficData. dat"));
        PrintWriter output
            = new PrintWriter(new FileWriter
                                ("C:\\TrafficData.out"));

        trafficStudy.collectData(ScannedInfo);
        trafficStudy.analyzeData();
        output.println(trafficStudy.createTrafficReport());
        output.close();
    }
}
```

Output

	Morning	Lunch	Evening	Night	Average
Day 0	574	389	697	215	468
Day 1	612	401	756	105	468
Day 2	655	399	809	206	517
Day 3	525	507	863	276	542
Day 4	634	472	742	241	522
Day 5	629	475	837	176	529
Day 6	598	487	787	281	538
Average	603	447	784	214	

Advanced Topic 9.12: Returning Arrays in Method Invocation

The following method creates a new array, then performs a deep copy of the rectangular subarray bounded by indices (`startRow, startCol`) and (`endRow, endCol`):

```
public static double[][] copy(double[][] source,
            int startRow, int startCol, int endRow, int endCol)
{
    int row;
    int col;
    double[][] destination = null;
    destination = new double[endRow-startRow+1][];
    for (row = 0; row < destination.length; row++)
```

```
    {
        destination[row] = new double[endCol-startCol+1];
        for (col = 0; col < destination[row].length; col++)
            destination[row][col]
                    = source[startRow + row][startCol+col];
    }
    return destination;
}
```

You need double[][] to indicate that the return type is a double two-dimensional array. However, in the return statement all you need is the array name.

The next example illustrates the use of two-dimensional array as a return type.

Example 9.19

The program creates an int array of size 5 by 4 and initializes using user input. Next, the program performs a deep copy of the rectangular subarray bounded by (1,2) and (4,4).

```
import java.io.*;
import java.util.Scanner;
import java.text.DecimalFormat;
/**
    Illustration of array as a parameter
*/
public class TwoDimArrayAsParameter
{

    /**
        Gets data from a scanner object and stores it in an
                                                       array
        @param sc input source; a scanner object
        @param arr the destination array name
    */
    public static void getData(Scanner sc, double[][] arr)
    {
    int row;
    int col;

    for (row = 0; row < arr.length; row++)
        {
        for (col = 0; col < arr[row].length; col++)
            arr[row][col] = sc.nextDouble();
        }
    }
```

```
/**
    copy part of source 2-dim. array to destination
                                            2-dim. array
    @param source array with data
    @param startRow starting row of source
    @param startCol starting column of source
    @param endRow ending row of source
    @param endCol ending column of source
    @return destination array
*/
public static double[][] copy(double[][] source,
    int startRow, int startCol, int endRow, int endCol)
{
    int row;
    int col;

    double[][] destination = null;
    destination = new double[endRow-startRow+1][];
    for(row = 0; row < destination.length; row++)
    {
        destination[row] = new
                              double[endCol-startCol+1];
        for(col = 0; col < destination[row].length; col++)
            destination[row][col]
                = source[startRow + row][startCol+col];
    }
    return destination;
}

/**
    print a two-dim. array
*/
public static void print(double[][] arr)
{
    int row;
    int col;

    for (row = 0; row < arr.length; row++)
    {
        for (col = 0; col < arr[row].length; col++)
            System.out.print(arr[row][col] + "\t");
        System.out.println();

    }
}
```

```
    public static void main (String[] args) throws IOException
    {
        double[][] sample = new double[5][4];
        Scanner ScannedInfo = new Scanner(System.in);
        DecimalFormat twoDecimalPlaces = new
                                 DecimalFormat("0.00");
        System.out.println("Enter 20 decimal values");
        getData(ScannedInfo, sample);

        System.out.println
                ("\nFive inputs per row and their sum
                                      follows:\n");
        print(sample);
        System.out.println
               ("\nSubarray (1,0) to (4,3) created is:\n");
        print(copy(sample,1,0,4,3));

    }
}
```

Output

```
Enter 20 decimal values

12.34 45.12 65.67 23.00 78.56 34.51 87.01 20.70 19.45 52.92 17.37 65.17
45.60 73.09 18.50 31.17 27.11 88.99 91.54 29.25

Five inputs per row and their sum are as follows:

12.34   45.12   65.67   23.0
78.56   34.51   87.01   20.7
19.45   52.92   17.37   65.17
45.6    73.09   18.5    31.17
27.11   88.99   91.54   29.25

Subarray (1,0) to (4,3) created is

78.56   34.51   87.01   20.7
19.45   52.92   17.37   65.17
45.6    73.09   18.5    31.17
27.11   88.99   91.54   29.25
```

Advanced Topic 9.13: Multidimensional Array

A two-dimensional array is an array of one-dimensional arrays. A three-dimensional array can now be defined as an array of two-dimensional arrays or an array of array of arrays. Thus, it is possible to define arrays of any dimension.

The syntax for declaring an n-dimensional array is

```
dataType[][]...[] arrayName;
```

The syntax for instantiating an n-dimensional array is

```
arrayName = new dataType[IE_1][IE_2]...[IE_n];
```

The above two statements can be combined as follows:

```
dataType[][]...[] arrayName= new dataType[IE_1][IE_2]...[IE_n];
```

where `IE_1, IE_2, ...,IE_n` are constant expressions that evaluate to nonnegative integer values. Individual elements of the array are accessed using the array name and `n` indices. Further, you need `n` nested loops to visit all the elements of an `n`-dimensional array.

To illustrate these concepts, we now revisit the traffic study introduced in Example 9.18. The study involved analyzing the traffic flow at a certain intersection 7 days with four readings per day. Thus, we created a two-dimensional array with 7 rows and 4 columns. Let us now modify our program so that study is carried out at 5 intersections.

Therefore, we use a three-dimensional array `weeklyTrafficData` to keep track of data. The following Java statement

```
int[][][] weeklyTrafficData = new int [5][7][4];
```

declares and instantiates a three-dimensional array `weeklyTrafficData`. Observe that the size of the first dimension is 5 and represents five intersections in the city. The size of the second dimension is 7 and represents 7 days of the week. The size of the third dimension is 4 and represents four readings. Thus, the array element `weeklyTrafficData[2][3][0]` denotes the first reading for day 3 at intersection 2.

The following nested repetition structure can be used to output the data:

```
for (loc = 0; loc < weeklyTrafficData.length; loc++)
{
    // perform initializations for an intersection
    for (day = 0; day < weeklyTrafficData[loc].length; day++)
    {
        // perform initializations for a day
        for (time = 0; time < weeklyTrafficData[loc][day].
                                              length; time++)
        {
            // process the array element
            // weeklyTrafficData[loc][day][time]
        }

        // perform additional steps for a day
    }

    // perform additional steps for an intersection
}
```

TABLE 9.7 Differences among a Variable and Arrays

	Variable	Arrays		
		One-Dimensional	Two-Dimensional	Three-Dimensional

Declaration
Instantiation
	`int rain;`	`int[] rain;`
		`rain = new`
		`int[2];`
			`int[][] rain;`
			`rain = new`
			`int[2][4];`
				`int[][][] rain;`
				`rain = new`
				`int[2][4][3];`

Declaration and instantiation		`int[] rain =`
		`new int[2];`
			`int[][]`
			`rain = new`
			`int[2][4];`
				`int[][][] = new`
				`int[2][4][3];`

Declaration and initialization	`int`
	`rain = 8;`
		`int[] rain =`
		`new {6, 8};`
			`int[][]`
			`rain =`
			`{{2,3,1,4},`
			`{5,7,8,6}}`
				`int[][] rain =`
				`{{{6,8,1},{7,2,1},`
				`{1,2,3},{4,7,8}},`
				`{{7,2,9},{1,9,2},`
				`{6,5,8},{5,6,8}}}`

Formal parameter / Return type specification	`int`	`int[]`	`int[][]`	`int[][][]`
Actual parameter / Variable in return statement	`rain` (`rain` is a variable)	`rain` (`rain` is a one-dimensional array)	`rain` (`rain` is a two-dimensional array)	`rain` (`rain` is a three-dimensional array)

We now summarize the major differences among a variable, a one-dimensional array, a two-dimensional array, and a three-dimensional array in Table 9.7.

`Vector` AND `ArrayList` CLASSES

In this chapter, you have seen arrays, a data structure that allow us to store and process items of the same type. As you have seen in Example 9.13, the restriction that an array must store items of the same type can be managed through the introduction of a superclass. However, arrays have their limitations. First, you must know the size of the array at the time of creation. Since an array size remains fixed, a predetermined number of elements alone can be stored in an array. Second, to insert an element at a specific index location you need to make room by shifting elements of the array. Removing an element from a specific index position in the array involves shifting elements of the array in the opposite direction.

Java provides two classes `Vector` and `ArrayList` to address these issues. `Vector` (`ArrayList`) objects can grow and shrink dynamically. As a user of the `Vector` (`Array List`) class, you need not be concerned with the size. Similarly, the `Vector` (`ArrayList`) allows you to insert at and delete from any position. Once again, as a user of the `Vector` (`ArrayList`) class, you need not be concerned with shifting the elements of a `Vector` (`ArrayList`) object. Internally, `Vector` (`ArrayList`) class is implemented using an array. While `Vector` is thread-safe, the `ArrayList` is not. The concept of thread-safety is far beyond the scope of this textbook and hence we do not attempt to explain. While it is enough to present either `ArrayList` or the `Vector` class, both are presented to illustrate the concept of an abstract data type (see Advanced Topic 9.14).

Table 9.8 lists selected constructors and methods of both the classes. Interested reader is invited to visit Java Collections Framework at `http://java.sun.com/javase/6/docs/api/index.html` to get a much broader view of the subject.

Self-Check

23. `Vector` (`ArrayList`) objects can ________ and ________ dynamically.
24. True or false: The `Vector` (`ArrayList`) allows you to insert at and delete from any position.

Wrapper Classes

Every component of a `Vector` (`ArrayList`) object is an `Object` reference. Therefore, primitive data type values cannot be directly assigned to a `Vector` element. This is made possible through wrapper classes. Corresponding to each primitive data type, there is a wrapper class. For example, corresponding to `int`, we have the `Integer` class. Table 9.9 lists all primitive data types and their corresponding wrapper classes.

TABLE 9.8 The `ArrayList<E>` and `Vector<E>` Classes

Constructor/Operation	Explanation
`public ArrayList()`	Constructs an empty `ArrayList` of size `10`
`public Vector()`	Constructs an empty `Vector` of size `10`
`public ArrayList(int capacity)`	Creates an `ArrayList` of size `capacity`
`public Vector(int capacity)`	Creates a `Vector` of size `capacity`
`public boolean add(E e)`	Adds the object `e` at the end
`public void add(int idx, E e)`	Inserts the object `e` at index location `idx`
`public boolean contains(Object obj)`	If the `ArrayList` (`Vector`) contains the object `obj`, then `true` is returned; else `false` is returned
`public E get(int idx)`	Returns element at location `idx`
`public int indexOf (Object obj)`	Returns the index of the first occurrence of the object `obj`. Returns `-1` if object is not in the `ArrayList` (`Vector`)
`public boolean is Empty()`	If the `ArrayList` (`Vector`) is empty, `true` is returned; else `false` is returned
`public int lastIndexOf (Object obj)`	Returns the index of the last occurrence of the object `obj`. Returns `-1` if object is not in the `ArrayList` (`Vector`)
`public E remove(int idx)`	Removes the element at `idx` and returns it
`public void set(int idx, E e)`	Replaces the element at `idx` by `e`
`public int size()`	Returns the number of elements in the `ArrayList` (`Vector`)

TABLE 9.9 Wrapper Classes

Primitive Type	Wrapper Class
`boolean`	`Boolean`
`byte`	`Byte`
`char`	`Character`
`short`	`Short`
`int`	`Integer`
`long`	`Long`
`float`	`Float`
`double`	`Double`

For the rest of this chapter, for any statement we make about Vector, there is a similar and equivalent statement involving ArrayList. Therefore, we explain various concepts through Vector class.

The Integer class can store one int value. Therefore, if you want to create a Vector to store int values, you need to create a Vector that can store Integer objects.

```
Vector<Integer> sampleVector = new Vector<Integer>();
```

In the above-mentioned statement, we used the type parameter Integer to create a Vector reference that can store Integer references. Therefore, sampleVector is called a parameterized Vector.

The following statement creates an Integer object with int value 10 and adds at the end of sampleVector:

```
sampleVector.add(10);
```

Note that the above-mentioned statement is in fact equivalent to the following:

```
sampleVector.add(new Integer(10));
```

This feature of the Java language, introduced in Java 5.0, is known as *auto-boxing*. Let us add four more integers into sampleVector.

```
sampleVector.add(21);
sampleVector.add(15);
sampleVector.add(31);
sampleVector.add(41);
```

The sampleVector now has five items as shown below:

```
[10, 21, 15, 31, 41].
```

Now index of 21 (or **new** Integer(21)) is 1. Similarly, both statements

```
sampleVector.contains(21)
```

and

```
sampleVector.contains(new Integer(21))
```

evaluate to true. However, the following expressions are not equivalent:

```
sampleVector.get(1) == 21                    //(1)
sampleVector.get(1) == new Integer(21)       //(2)
```

Note that in Line 1, Java compiler performs an *auto-unboxing* of the object returned by sam pleVector.get(1) and compares int 21 with int 21. Thus, Line 1 evaluates to true. In Case 2, such an unboxing is not necessary and thus the comparison is between two object references. Hence Case 2 evaluates to false. Auto-unboxing also takes place during numeric computation. For example, consider the following code to compute the sum of all values in the sampleVector:

```
sum = 0;
for (idx = 0; idx < sampleVector.size(); idx++)
    sum = sum + sampleVector.get(idx);
```

If you want to increment the second item by 5, you should first retrieve the item at 2. This can be accomplished by

```
sampleVector.get(2)
```

Next, we add 5 to it by

```
sampleVector.get(2) + 5          // (1)
```

and then replace the current item at location 2 by sampleVector.get(2) + 5. Thus, we have the following:

```
sampleVector.set(2, sampleVector.get(2) + 5); //(2)
```

Observe that there is an auto-unboxing performed in Line 1 and there is an auto-boxing done at Line 2. Similarly, adding element at location 3 to element at location 4 can be accomplished as follows:

```
sampleVector.set(4, sampleVector.get(3) + sampleVector.get(4));
```

Note that two auto-unboxing and one auto-boxing are performed during the execution of this statement.

The class Vector (ArrayList) is in the package java.util. Therefore, your program must import it using statement

```
import java.util.*;
```

or

```
import java.util.Vector; (import java.util.ArrayList;)
```

Example 9.20

In this example, we illustrate various methods of the Vector class. Since, these methods are also in the ArrayList class, the ArrayList class version can be obtained by replacing Vector by ArrayList. This example will also illustrate auto-unboxing and auto-boxing feature of the Java language.

```
import java.io.*;
import java.util.*;

/**
    Illustrate the Vector class
*/
```

```
public class VectorIllustrated
{
    public static void main(String[] arg)
    {
        int idx;
        int sum;

        Vector<Integer> sampleVector = new Vector<Integer>();

        sampleVector.add(10);
        sampleVector.add(21);
        sampleVector.add(15);
        sampleVector.add(31);
        sampleVector.add(41);

        System.out.println("After adding five "
                 + "elements to sampleVector");

        System.out.println("sampleVector: "
                           + sampleVector);
        System.out.println();

        System.out.println("indexOf(21) is "
                + sampleVector.indexOf(21));
        System.out.println("indexOf(new Integer(21)) is "
                + sampleVector.indexOf(new Integer(21)));
        System.out.println();
        System.out.println("sampleVector contains 21 is a "
                              + sampleVector.contains(21)
                                         +" statement");
        System.out.println(
              "sampleVector contains(new Integer(21)) is a "
                    + sampleVector.contains(new Integer(21))
                                          +" statement");
        System.out.println();
        System.out.println("sampleVector.get(1) == 21 is a "
                             + (sampleVector.get(1) == 21)
                                          +" statement");
        System.out.println(
              "sampleVector.get(1) == new Integer(21) is a "
                  + (sampleVector.get(1) == new Integer(21))
                                          +" statement");
        System.out.println();
```

```
        sampleVector.add(3, 77);
        System.out.println("After adding 77 "
                          + "at position 3");

        System.out.println("sampleVector: "
                           + sampleVector);

        System.out.println();

        sampleVector.remove(2);
        System.out.println("After removing item at 2 ");
        System.out.println("sampleVector: "
                           + sampleVector);
        System.out.println();
        System.out.println("Size of sampleVector is "
                              + sampleVector.size());

        sum = 0;
        for (idx = 0; idx < sampleVector.size(); idx++)
            sum = sum + sampleVector.get(idx);

        System.out.println();
        sampleVector.add(0, sum);
        System.out.println(
        "After inserting the sum of all numbers at index 0");

        System.out.println("sampleVector: "
                           + sampleVector);

        System.out.println();
        sampleVector.set(2, sampleVector.get(2) + 5);

        System.out.println(
        "After incrementing second item by 5");

        System.out.println("sampleVector: "
                           + sampleVector);

        System.out.println();
        sampleVector.set(4,
               sampleVector.get(3) + sampleVector. get(4));
```

```
        System.out.println(
        "Add third item to the fourth item");
        System.out.println("sampleVector: "
                                   + sampleVector);
    }
}
```

Output

```
After adding five elements to sampleVector
sampleVector: [10, 21, 15, 31, 41]

indexOf(21) is 1
indexOf(new Integer(21)) is 1

sampleVector contains 21 is a true statement
sampleVector contains(new Integer(21)) is a true statement

sampleVector.get(1) == 21 is a true statement
sampleVector.get(1) == new Integer(21) is a false statement

After adding 77 at position 3
sampleVector: [10, 21, 15, 77, 31, 41]

After removing item at 2
sampleVector: [10, 21, 77, 31, 41]

Size of sampleVector is 5

After inserting the sum of all numbers at index 0
sampleVector: [180, 10, 21, 77, 31, 41]

After incrementing second item by 5
sampleVector: [180, 10, 26, 77, 31, 41]

Add third item to the fourth item
sampleVector: [180, 10, 26, 77, 108, 41]
```

Example 9.21

In this example, we illustrate the use of `ArrayList` in place of an array. The `ArrayList` version of Example 9.13 is presented. For easy comparison, we have kept statements that warranted the required modification commented. Therefore, for the sake of clarity, other comments are omitted. Note that if we replace `ArrayList` by `Vector`, you get the `Vector` class version.

```
import java.util.*;
import java.io.*;

public class HeartlandCarsOfAmericaVector
{
    public static void main (String[] args) throws
                FileNotFoundException, IOException
    {

        /*

            Employee[] currentEmployees = new Employee[100];
        */


        ArrayList<Employee> currentEmployees
                                = new ArrayList <Employee>();


        char inputEmployeeType;
        String inputFirstName;
        String inputLastName;
        double inputBaseSalary;
        double inputPayPerHour;
        int inputSalesVolume;
        int inputHoursWorked;
        int noOfEmployees;
        int idx;

        Scanner ScannedInfo = new Scanner(
                    new File("C:\\Employee.dat"));
        PrintWriter outFile = new PrintWriter(
                new FileWriter("C:\\payroll.dat"));

        while (ScannedInfo.hasNext())
        {
            inputEmployeeType = ScannedInfo.next().
                                            charAt(0);
```

```
switch (inputEmployeeType)
{
    case 'F' :
    case 'f' :

    inputFirstName = ScannedInfo.next();
    inputLastName = ScannedInfo.next();
    inputBaseSalary = ScannedInfo.
                    nextDouble();
    inputHoursWorked = ScannedInfo.nextInt();

    /*

    currentEmployees[idx]= new
               FullTimeEmployee(inputFirstName,
                inputLastName, inputBaseSalary,
                             inputHoursWorked);
    */

    currentEmployees.add(new
               FullTimeEmployee(inputFirstName,
                inputLastName, inputBaseSalary,
                            inputHoursWorked));

    break;

    case 'P' :
    case 'p' :

    inputFirstName = ScannedInfo.next();
    inputLastName = ScannedInfo.next();
    inputPayPerHour = ScannedInfo.
                          nextDouble();
    inputHoursWorked = ScannedInfo.nextInt();

    /*
    currentEmployees[idx]= new
        PartTimeEmployee(inputFirstName,
        inputLastName, inputPayPerHour,
                     inputHoursWorked);
```

```
            */

            currentEmployees.add(new
                    PartTimeEmployee(inputFirstName,
                    inputLastName, inputPayPerHour,
                                inputHoursWorked));

            break;

            case 'S' :
            case 's' :

                inputFirstName = ScannedInfo.next();
                inputLastName = ScannedInfo.next();
                inputBaseSalary = ScannedInfo.
                                  nextDouble();
                inputSalesVolume = ScannedInfo.nextInt();

                /*

                    currentEmployees[idx] = new
                    SalesEmployee(inputFirstName,
                    inputLastName, inputBaseSalary,
                                inputSalesVolume);

                */

                currentEmployees.add(new
                    SalesEmployee(inputFirstName,
                    inputLastName, inputBaseSalary,
                                      inputSalesVolume));

            break;

            default:
                System.out.println("Check data file.");
             return;

        }
    }

    /*

    for (idx = 0; idx < noOfEmployees; idx++)

        */

    for (idx = 0; idx < currentEmployees.size() ; idx++)
    {
        /*
```

```
                outFile.println
                    (currentEmployees[idx].createPay Stub());


            */
            outFile.println(
                    currentEmployees.elementAt(idx).
                                    createPayStub());
        }
        outFile.close();
    }
}
```

A better solution is still possible! Introduce a new class, `EmployeeList`, which has an attribute that is either an array or `Vector` or `ArrayList` of `Employee` instances. Then use that class in the application program. See Programming Exercise 9.6. A similar approach is employed in Case Study 9.2.

Self-Check

25. The wrapper class corresponding to `int` is __________ and the wrapper class corresponding to `char` is __________.
26. The __________ method inserts the element in the `Vector` (`ArrayList`) and __________ method replaces the element in the `Vector` (`ArrayList`).

Advanced Topic 9.14: Abstract Data Types

Both `ArrayList` and `Vector` allow random access to all elements. You can access any element by specifying an integer index, you can replace any value by specifying an index, and you can insert a new item at a specified location using the index. Thus, both `ArrayList` and `Vector` seem to have many identical services (see Table 9.8). The reason is both `Vector` and `ArrayList` are concrete implementations of an *abstract data type (ADT)* list.

ADT is the very first attempt to distinguish the structure and operations on data from the data itself. For a better understanding of the concept, consider the following:

1. Contact information stored in your e-mail software
2. A set of index cards with various recipes arranged in alphabetical order
3. A deck of business cards of all the medical representatives in a doctor's office

Although the data maintained and the way it is maintained is different, all three items listed above have the same structure. In particular, observe the following commonalities among them:

1. Each is a collection of elements of certain type.
2. There is an inherent ordering of elements. Thus, there is a first element, there is a last element; given an element other than the last element, there is a "next" element; given an element other than the first element, there is a "previous" element.
3. Any element in the collection can be accessed randomly.
4. A new element can be added to the collection. Thus, a new item can be added as the first or the last item or anywhere in between.
5. An existing element can be removed from the collection.
6. It is possible to determine whether or not an item is currently in the collection.
7. An existing element can be updated or replaced by a new item.

A collection satisfying the above seven properties is commonly known as a *list*. Thus, list is an abstract concept. It can be implemented in many different ways. In particular, both `ArrayList` and `Vector` are two specific implementations of the ADT list.

CASE STUDY 9.2: MR. GRACE'S GRADE SHEET

Recall that Mr. Grace has decided to assign his letter grade based on class average. See Table 9.1 for his latest grading policy. To implement his new grading policy, he started out creating a new class `Course` with the following attributes:

`courseNumber : a String,`
`courseTitle : a String`
`term : a String`
`numberOfStudents :` an **`int`**
`courseAverage :` a **`double`**
`StudentList :` an array of `Student`

To simplify his work, Mr. Grace decided to have three `public` methods

1. `loadData` to read a file and load data into the `Course` class instance
2. `assignGrade` to assign grade to every student
3. `printGradeSheet` to write the grade sheet for the course in a file

instead of traditional get and set methods. (Pedagogically, this example illustrates the use of array of classes where class itself may contain an attribute that is an array. Our focus is on those issues. Further, by now the reader may know how to create get and set methods.) To simplify the method `assignGrade`, Mr. Grace decided to include a helper function `computeCourseAverage`.

The program listing is quite easy to follow. However, additional assistance is provided as the solution to Exercise 8.

```
import java.io.*;
import java.util.*;
/**
    Keeps information on all students in a course
*/
public class Course
{
    private static final int CLASS_SIZE = 25;          // Maximum
                                                       class size
    private String courseNumber;                       // Course
                                                          Number
    private String courseTitle;                        // Course
                                                           Title
    private String term;                               // Course
                                                           term
    private int numberOfStudents;
    private double courseAverage;
    private Student StudentList[] = new Student[CLASS_SIZE];
                                            // An array to keep
                                            student information


    /**
        stores data values
        @param sc a scanner object
    */
    public void loadData(Scanner sc)
    {
        Student st;
        int i = 0;


        courseNumber = sc.nextLine();
        courseTitle = sc.nextLine();
        term = sc.nextLine();
        while (sc.hasNext())
        {
            st = new Student();
            st.setStudentInfo(sc);
```

```
            StudentList[i] = st;
            i++;
        }
        numberOfStudents = i;
    }

    /**
        Computes the class average
    */
    private void computeCourseAverage()
    {
        int i;
        double sum = 0;

        for (i = 0; i < numberOfStudents; i++)
        {
            StudentList[i].computeGradeScore();
            sum = sum + StudentList[i].getGradeScore();
        }
        courseAverage = sum / numberOfStudents;
    }

    /**
        Assign grade to all students
    */
    public void assignGrade()
    {
        int i;
        double temp;

        computeCourseAverage();

        for (i = 0; i < numberOfStudents; i++)
        {
            temp = StudentList[i].getGradeScore();
            if (temp > 1.3 * courseAverage)
                StudentList[i].setGrade("A");
            else if (temp > 1.1 * courseAverage)
                StudentList[i].setGrade("B");
            else if (temp > 0.9 * courseAverage)
                StudentList[i].setGrade("C");
```

```
            else if (temp > 0.7 * courseAverage)
                StudentList[i].setGrade("D");
            else
                StudentList[i].setGrade("F");
        }

    }

    /**
        Prints grade sheet
        @param output a PrintWriter object
    */
    public void printGradeSheet(PrintWriter output)
    {
        int i;

        output.println("\t\t\t" + courseNumber);
        output.println("\t\t\t" + courseTitle);
        output.println("\t\t\t" + term);
        output.println("\t\t\tClass Average is " +
                                            courseAverage);
        for (i = 0; i < numberOfStudents; i++)
        {
            output.println(StudentList[i]);
        }

    }
}

import java.io.*;
import java.util.Scanner;

/**
    Application program that grades all students
*/
public class CourseGraded
{

    public static void main (String[] args)
            throws FileNotFoundException, IOException
```

```
    {
        Scanner scannedInfo
                    = new Scanner(new File("C:\\courseData.dat"));
        PrintWriter outFile
                    = new PrintWriter(new FileWriter
                                ("C:\\courseData.out"));

        Course crs = new Course();
        crs.loadData(scannedInfo);
        crs.assignGrade();
        crs.printGradeSheet(outFile);
        outFile.close();

    }
}
```

CourseData.dat file:

```
CSC 221
Java Programming
Fall 2008
Kim Clarke          70.50 69.85 90.25 100.0 81.75 100.0
Chris Jones         78.57 51.25 97.45 85.67 99.75 88.76
Brian Wills         85.08 92.45 67.45 71.57 50.92 72.00
Bruce Mathew        60.59 87.23 45.67 99.75 72.12 100.0
Mike Daub           56.60 45.89 78.34 64.91 66.12 70.45
```

CourseData.out file:

```
CSC 221
Java Programming
Fall 2008
Class Average is 81.4944
Kim Clarke        70.5  69.85 90.25 100.0 81.75 100.0 88.50 C
Chris Jones       78.57 51.25 97.45 85.67 99.75 88.76 90.04 B
Brian Wills       85.08 92.45 67.45 71.57 50.92 72.0  77.71 C
Bruce Mathew      60.59 87.23 45.67 99.75 72.12 100.0 83.94 C
```

REVIEW

1. An array is a named collection of contiguous storage locations that can store data items of the same type.
2. Each element of an array is referred to by the array name along with its position or index.

3. In the case of primitive data types, the array locations are initialized with default values. Thus, integral arrays are initialized by `0`, floating point arrays by `0.0`, and `boolean` arrays by `false`.
4. In the case of object references, the array locations are initialized by `null`.
5. In the case object references, the array locations need to be instantiated.
6. The length of an array is defined as the number of locations and this value is available in an attribute `length`.
7. During program execution if an array index becomes out of bounds, Java throws an `Array IndexOutOfBoundException` exception.
8. The assignment operator = and relational operators == and != can be used in the context of an array.
9. The assignment operator copies the reference of one array to the other. This form of copying is known as shallow copying.
10. To make a copy of an array object, memory has to be allocated. Further, an element-by-element copying using a repetition structure is required. This form of copying is known as deep copying.
11. The relational operators compare array references only.
12. A two-dimensional array is an array of one-dimensional arrays.
13. A three-dimensional array is an array of two-dimensional arrays or an array of array of arrays.
14. Insertion and deletion of an element in an arbitrary location of an array is not efficient.
15. In the case of a `Vector`, there is a default size of 10.
16. A `Vector` is thread-safe, whereas `ArrayList` is not.
17. Corresponding to each primitive data type, there is a wrapper class.
18. The wrapper class for `int` is `Integer`.
19. The wrapper class for `char` is `Character`.

EXERCISES

1. Mark the following statements as true or false:
   a. You must know the array size to create an array.
   b. Index can be of any numeric value.
   c. An array is created using new operator.
   d. You can store a `double` value in an `int` array.
   e. You can store an `int` value in a `double` array.
   f. The array `length` is the same as the upper bound of the array.
   g. The lower bound of all arrays is `0`.

h. A `boolean` array is never initialized by the system.

i. You can increase the size of an array inside your program.

j. The array name contains the reference of the array object.

k. If `cost` is a `double` array, then both `cost` and `cost[0] references` to the first item in the array `cost`.

l. If `cost` is a `double` array, then `cost[10]` refers to the ninth item in the array `cost`.

m. Let `costOne` and `costTwo` be two `double` arrays of the same size. Then `costOne = costTwo;` copies every element of array `costTwo` into array `costOne`.

n. Let `costOne` and `costTwo` be two `double` arrays of the same size. If `costOne.equals(costTwo)` is true, then `costOne == costTwo` is also true.

o. Let `cost` be a two-dimensional array having three rows and four columns. Then `cost[2][4]` is the eighth element of the array.

p. The default size of a `Vector` is 10.

q. The default size of an `ArrayList` is 10.

r. The size of a `Vector` is available in an attribute `length`.

2. Write Java statements that accomplish the following tasks. For each part, repeat it for a `Vector` and an `ArrayList`. If a certain task cannot be accomplished, explain.

   a. Declare an array named `priceList` to store `100` `int` values.

   b. Initialize the `10`th item of the `priceList` with `18.5`.

   c. Place the value `20.7` as the last item.

   d. Place the value `30.1` as the first item.

   e. Make the fifth item the sum of the fourth and the sixth items.

   f. Increment the second item by `12.9`.

   g. Print all values; seven items per row.

   h. Initialize the first and second values by `1.0`. All other values are initialized as the sum of the previous two values.

   i. Declare an array named `productList` to store `100` `String` values.

3. Consider the program segment presented in Example 9.9. For each part, if the answer is yes, rewrite the segment of code in Example 9.9.

   a. Is it possible to rewrite it using the enhanced loop statement?

   b. Is it possible to rewrite it so that the `if` statement is the last statement inside the block statement.

   c. Is it possible to rewrite it so that alternate lines have `7` and `11` items each.

4. Write method headings as specified:
   a. Method name: `trial`; two formal parameters: one-dimensional array of type `double`, `int`; return type: one-dimensional array of type `String`
   b. Method name: `tester`; two formal parameters: two-dimensional array of type `double`, one-dimensional array of type `char`; return type: two-dimensional array of type `double`
   c. Method name: `testing`; three formal parameters: one-dimensional array of type `int`, two-dimensional array of type `double`, two-dimensional array of type `String`; return type: `void`
   d. Method name: `test`; three formal parameters: two-dimensional array of type `int`, two-dimensional array of type `int`, one-dimensional array of type `double`; return type: one-dimensional array of type `char`
5. Write Java statements to invoke each of the methods in Exercise 4 if the methods are all `static` and belong to a class `GeneralUtil`. Show the necessary declarations and instantiations. However, you need not initialize any array.
6. Repeat Exercise 5. Assume that methods are members of a class `SpecialUtil` and they are not `static`.
7. Consider the following declarations:

```
double[][]    cost = new double[4][3];
int i, k, j;
```

What are the values stored in the array `cost` if each one of these segments are executed immediately after the above shown statements. If there is any error in the statement, indicate it.

a.
```
for (j = 0; j < cost.length; j = j + 2)
    for (k = 0; k < cost[0].length; k = k + 3)
        cost[j][k] = k * 10 + j;
```
b.
```
for (j = 0; j < cost.length; j = (j + 2) % cost.length)
    for (k = 0; k < cost[0].length;
        k = (k + 3) % cost[0].length, i++)
    cost[j][k] = i;
```
c.
```
for (k = 0;k < cost[2].length; k++)
    for (j = 1; j < k; j++)
        cost[j][k] = k/j;
```
d.
```
for (k = 0;k < cost[2].length; k++)
    for (j = cost.length; j > k; j++)
        cost[j][k] = k + j;
```

8. Explain each of the methods in the class `Course`, part of Mr. Grace's grade sheet program presented in Case Study 9.2.

9. Assume that the `loadData` method in the class `Course`, part of Mr. Grace's grade sheet program, is replaced with the following method:

```
public void loadData(Scanner sc)
{
    Student st;
    int i = 0;

    courseNumber = sc.nextLine();
    courseTitle = sc.nextLine();
    term = sc.nextLine();
    st = new Student();
    while (sc.hasNext())
    {
        st.setStudentInfo(sc);
        StudentList[i] = st;
        i++;
    }
    numberOfStudents = i;
}
```

   a. Is there a compilation error? Justify your answer.
   b. Is there a logical error? Justify your answer.

10. Assume that the `loadData` method in the class `Course`, part of Mr. Grace's grade sheet program, is replaced with the following method:

```
public void loadData(Scanner sc)
{
    int i = 0;

    courseNumber = sc.nextLine();
    courseTitle = sc.nextLine();
    term = sc.nextLine();
    while (sc.hasNext())
    {
        StudentList[i] = new Student();
        StudentList[i].setStudentInfo(sc);
        i++;
    }
    numberOfStudents = i;
}
```

a. Is there a compilation error? Justify your answer.

b. Is there a logical error? Justify your answer.

11. Consider the following segment of code:

```
public class HiThere
{
    public static void main (String[] args)
    {
        int[] test = new int[10];
        for (int k : test)
        {
            k = 10;
        }

        for (int k : test)
        {
            System.out.println("\t" + k);
        }

    }
}
```

a. Is there a compilation error? Justify your answer and correct it.

b. Is there a logical error? Justify your answer and correct it.

12. What is wrong with the approach mentioned in Advanced Topic 9.1 to address Common Programming Error 9.1. How is it different from the `loadData` method of Mr. Grace's grade sheet program?

PROGRAMMING EXERCISES

1. Given an array of double values, create another array of cumulative sums. For example, if 5.0, 6.5, 7.3, and 10.2 are the values, then their cumulative sums are 5.0, 11.5, 18.8, and 29.0.

2. One of the oldest approach to break a code is to perform a frequency count of letters. Write a program to perform a frequency count by reading the text from a file. Your program should output how many A's are there in the text, how many B's are there, and so on. Note that the program will not make any distinction between uppercase and lowercase letters.

3. Consider an `int` array `first` with possibly repeated values. Create a new array `second` that has each number in the `first` appear exactly once in their order of appearance. For example, if values in `first` are 10, 20, 6, 7, 10, 8, 5, 6, 4, 7, 1, then the `second` has 10, 20, 6, 7, 8, 5, 4, 1.

4. Consider an `int` array `first` with possibly repeated values. Write a program to perform the frequency count.

5. Based on the remainder obtained upon division by k, the integers in an array can be grouped into k disjoint sets. Write a program to print all groups.
6. Consider Example 9.21. Introduce a new class, `EmployeeList`, which has an attribute that is either an array or `Vector` or `ArrayList` of `Employees`. Then use that class to create the application program that implements the new policy of giving an extra $500.00 for the employee with maximum sales.
7. Redesign the classes of Example 9.16 so that you can eliminate the need for returning an array in methods.
8. Use `ArrayList` in place of arrays in both `Student` and `Course` classes of Mr. Grace's grade sheet.
9. A two-dimensional array with equal number of rows and columns filled with distinct integers is a magic square if the sum of the elements in each row, in each column, and in the two diagonals have the same value. Write a program to test whether or not a two-dimensional array is a magic square. Assume that user is supposed to enter numbers row by row. The program must do the following:
   a. Prompt user for enough number of integers
   b. Verify that the integers are distinct
   c. Test whether or not it is a magic square
10. A two-dimensional array with n rows and n columns filled with integers is a Latin square if each row and each column has all the numbers from 1 to n. A Latin square has a transversal, if all the elements in the diagonal are also distinct. Write a program to verify whether or not a square two-dimensional array is a Latin square. Assume that user is supposed to enter numbers column by column. The program must do the following:
    a. Prompt user for enough number of integers
    b. Test whether or not it is a Latin square
    c. If it is a Latin square, then test whether or not it has a diagonal
11. Write a program to verify whether or not a 9 by 9 grid is a sudoku. (*Hint*: In a sudoku grid, every row has all the numbers from 1 to 9, every column has all the numbers from 1 to 9, and every block has all the numbers from 1 to 9. Blocks are nonoverlapping 3 by 3 subarrays and there are nine blocks in a sudoku grid.
12. Modify Mr. Grace's grade sheet program so that the class average is computed for all the tests and printed as part of the grade sheet.
13. Create a class `PainterEstimater` to help painters determine the total surface area. The total surface is divided into various geometric shapes. Thus, `PainterEstimater` has an array of `GeometricFigure` objects. See Programming Exercise 14 of Chapter 7 for details on `GeometricFigure`.
14. Implement the following simple clustering algorithm. As a new point is entered by the user, check whether or not it lies within any cluster. If so, mark it with the cluster

number. If it does not fall in any existing cluster, start a new cluster with the new point as center and radius, r, that is specified by the user and is the same for all clusters.

15. Implement a slightly more sophisticated clustering algorithm. Implement all the steps in Programming Exercise 14 to determine the centers and number of clusters. Once all the cluster centers are determined, all the points are processed once more to determine the cluster they belong to. A point belongs to a cluster whose center is the closest.

ANSWERS TO SELF-CHECK

1. contiguous fixed-length, same type
2. array name, index
3. 0
4. integer expression
5. `double[] points;`
6. `Student[] classList;`
7. `points = new double[25];`
8. default
9. `points[2] = 23.5;`
10. `points[8] = 5 * points[6];`
11. `points.length`
12. `points.length - 1;`
13.
```
for (int i = 0; i < points.length; i++)
    points[i] = 50.67;
```
14. False
15.
```
for (int i = 0; i < points.length; i++)
    points[i] = 8.0 + i*5.;
```
16.
```
for (int i = 0; i < points.length; i++)
    points[i] = ScannedInfo.parseDouble();
```
17.
```
int final ITEMS_PER_LINE = 8;
for (int i = 0; i < points.length; i++)
{
    if (i % ITEMS_PER_LINE == 0 && i > 0)
        System.out.println();
    System.out.print(points[i]+ " ");
    }
```

18. ```
int sum = 0;
    for (int i = 0; i < points.length; i++)
        sum = sum + points[i];
    if (points.length > 0)
        average = sum / points.length;
    else
        average = 0.0;
```
19. ```
int maxIndex = 0;
for (int i = 1; i < points.length; i++)
        if (points[maxIndex] < points[i])
            maxIndex = i;
largestValue = points[maxIndex];
```
20. `double[][] = variance;`
21. `variance = new double [7][12];`
22. `variance[4][6] = 134.53;`
23. grow, shrink
24. True
25. `Integer`, `Character`
26. add, set

CHAPTER 10

Search and Sort

In this chapter you learn

- Fundamental concepts
  - Worst-case, average-case, and best-case time complexity; space complexity
- Programming skills
  - Adopt and use search algorithms such as linear search and binary search
  - Adopt and use sort algorithms such as selection sort, insertion sort, and bubble sort
  - Empirical method of measuring the performance of an algorithm and compare it with known time complexity measures

Searching and sorting are the two most common tasks in data processing. Can you imagine a telephone directory without names listed in sorted order? Sorting is also quite useful for ordinary folks such as Ms. Smart and Mr. Grace. If Ms. Smart needs to identify all sales personnel in the top `10` percentile, then she needs to sort the list of sales personnel based on their sales. Similarly, if Mr. Grace needs to identify all students in the class in the bottom `25` percentile, then he needs to sort the list of students based on their cumulative test scores. A sorting algorithm permutes or rearranges the elements of a collection either in an ascending or a descending order. Many sorting algorithms are known. In this chapter, you will be introduced to three different sorting algorithms: selection sort, insertion sort, and bubble sort. Once you have a sorted list, searching for an item becomes quite efficient. This chapter presents three search algorithms: linear search on an unsorted array, linear search on a sorted array, and binary search on a sorted array. All these algorithms are illustrated using `int` array for simplicity. However, once you master the algorithm, it is quite easy to adapt to another situation. In Case Study 10.1, Mr. Grace's grade sheet demonstrates how easy it is to adopt a sort algorithm to use in the case of array of objects.

Java has a class `java.util.Arrays` that contains many utility methods. However, we need certain utility methods that are not in `java.util.Arrays`. Therefore, we begin this chapter with a utility class of our own. To use this class, all that is required is to keep this class in the same folder that has your application program. Another option is to create a package and use it as explained in Chapter 6. There are four different utility methods. Three of them are overloaded methods `createAndSet`. These three methods let the user specify the size of the array and initialize it. The first `createAndSet` can be used to get data from the user. The second `createAndSet` initializes the array using a random number generator. The third `createAndSet` initializes the array using an arithmetic progression specified by the user. The only other method in our utility class is `printArray`. The `printArray` method lets the user specify the number of items per line and is the only output utility of the class.

```
import java.util.Random;
import java.util.Scanner;

/**
    This class contains many utility methods
    for array initialization and printing
*/
public class ArrayUtility
{
    private static Random randomGenerator
                            = new Random();

    /**
        Creates an int array and initializes with
        user input
        @param length the length of the array
        @param sc reference of the input Scanner
        @return reference of the int array of
        size length initialized with user input
    */
    public static int[] createAndSet(int length, Scanner sc)
    {
        int[] oneDim = new int[length];
        for (int i = 0; i < length; i++)
        {
            oneDim[i] = sc.nextInt();
        }
        return oneDim;
    }

    /**
        Creates an int array and initializes with
        random input
        @param length the length of the array
```

```
    @param limit is the largest random number
    @return reference of the int array of
    size length initialized with user input
*/
public static int[] createAndSet(int length, int limit)
{
    int[] oneDim = new int[length];
    for (int i = 0; i < length; i++)
    {
        oneDim[i] = randomGenerator.nextInt(limit);
    }
    return oneDim;
}

/**
    Creates an int array and initializes with
    user specified arithmetic progression
    @param length the length of the array
    @param first value for index 0
    @param inc the difference between any two
    adjacent index locations
    @return reference of the int array of
    size length initialized with user input
*/
public static int[] createAndSet(int length, int first, int
                                                        inc)
{
    int[] oneDim = new int[length];
    for (int i = 0; i < length; i++)
    {
        oneDim[i] = first + inc * i;
    }
    return oneDim;
}

/**
    output the array
    @param arr array to be printed
    @param itemsPerLine number of integers per line
*/
public static void printArray(int[] arr, int itemsPerLine)
{
    if (itemsPerLine <= 0)
        itemsPerLine = 10;
```

```
        for (int i = 0; i < arr.length; i++)
        {
            if (i % itemsPerLine == 0 && i > 0)
                System.out.println();
            System.out.print(arr[i]+ "\t");
        }
        System.out.println();
    }
}
```

SEARCH ALGORITHMS

Linear search algorithms can be written for a sorted and an unsorted array. In the case of an unsorted array, you need to compare every item in the array to determine whether or not a specified item is in the array. In the case of an ascending order sorted array, once an item that is greater than the one you are searching for is encountered, there is no need to search any further. In the case of sorted array, there is a better algorithm called binary search. In this chapter, linear search for unsorted array, linear search for sorted array, and binary search for sorted array are presented.

We shall use the term *search item* to denote the item we are searching for. If the search item is found, the method will return the index location, and if the search item is not found, the method will return `-1`.

Self-Check

1. _________ is a better algorithm than _________.
2. To use _________ the array must be in sorted order.

Linear Search

Linear search algorithm is probably the simplest of all search algorithms. Given a search item, the algorithm systematically checks the elements of the array for equality. Before presenting the program, let us consider some examples.

Example 10.1

In this example, we illustrate the behavior of the linear search algorithm using an unsorted array `L` with `8` elements as shown below:

0	1	2	3	4	5	6	7
7	10	8	21	35	17	26	3

Let the search item be `22`. Since `22` is not in the array, algorithm returns `-1`. However, if search item is `35`, the algorithm returns `4`.

TABLE 10.1 Linear Search: Search Item not Found

searchItem is 22		
i	**L[i]**	**L[i] == searchItem**
0	7	7 == 22 is **false**
1	10	10 == 22 is **false**
2	8	8 == 22 is **false**
3	21	21 == 22 is **false**
4	35	35 == 22 is **false**
5	17	17 == 22 is **false**
6	26	26 == 22 is **false**
7	3	3 == 22 is **false**
searchItem not found. **return** -1;		

TABLE 10.2 Linear Search: Search Item Found

searchItem is 35		
i	**L[i]**	**L[i] == searchItem**
0	7	7 == 35 is **false**
1	10	10 == 35 is **false**
2	8	8 == 35 is **false**
3	21	21 == 35 is **false**
4	35	35 == 35 is **true**
searchItem found. **return** 4;		

If the search item is in the array, then once it is found there is no need to search any more. However, if the search item is not in the array, then you need to compare every item of the array with the search item to conclude that search item is not in the array. Thus, if the search item is in the array, on average, you need to search only half the array and if the search item is not in the array, then you need to search the entire array.

Example 10.2

This example illustrates the linear search algorithm using the array L of Example 10.1 and search item 22. Assume that searchItem is a parameter variable of the method and thus searchItem is 22.

Let the index value be 0. Now L[0] is 7 and thus (L[0] == searchItem) is false. Increment the index by 1. Note that L[1] is 10 and thus (L[1] == searchItem) is false. Note that (L[i] == searchItem) is false for i = 2, 3, ..., 7. Since you have compared all items and the searchItem is not found, the method returns -1 (Table 10.1).

Example 10.3

This example illustrates the linear search algorithm using the array L of Example 10.1 and search item 35. As in Example 10.2, assume that searchItem is a parameter variable of the method and thus searchItem is 35.

Observe that (L[i] == searchItem) is false for i = 0, 1, 2, 3 and (L[4] == searchItem) is true. Therefore, the method stops comparing elements of the array and returns 4 (Table 10.2). The program and the test run are as follows:

```
import java.util.Scanner;
/**
    Linear search algorithm
```

```
*/
public class LinearSearch
{

    private int[] intData;

    /**
        Constructor
        @param inOutArray the array of integers
    */
    public LinearSearch(int[] inOutArray)
    {
        intData = inOutArray;
    }

    /**
        Searches the attribute inData for search item
        @param searchItem the search item
        @return index of search item if found and
        -1 if search item is not in the array
    */
    public int linearSearch(int searchItem)
    {
        for (int i = 0; i < intData.length; i++)
        {
            if (intData[i] == searchItem)
                return i;
        }
        return -1;
    }

}

import java.util.Scanner;
/**
    This is a test program for LinearSearch class
*/
public class LinearSearchApplication
{
    public static void main(String[] args)
    {
        int arraySize = 0;
        int numItems = 0;

        Scanner scannedInfo = new Scanner(System.in);
        System.out.print("Enter the number of integers : ");
```

```
        arraySize = scannedInfo.nextInt();
        System.out.println
                         ("Enter "+ arraySize +" integers\n");
        int[] testArray
            = ArrayUtility.createAndSet
                                     (arraySize, scannedInfo);
        System.out.println();
        System.out.print("Enter number of items per line : ");
        numItems = scannedInfo.nextInt();
        ArrayUtility.printArray(testArray, numItems);
        System.out.println();
        LinearSearch searchObject = new
                    LinearSearch(testArray);
        for (int i = 0; i < 4; i++)
        {
            System.out.print("Enter the item to be searched");
            int item = scannedInfo.nextInt();
            int index = searchObject.linearSearch(item);
            if ( index == -1)
                System.out.println
                      ("The item "+ item +" not found");
            else
                System.out.println
                   ("The item "+ item +" found at "+ index);
        }
    }
}
```

Output

```
Enter the number of integers: 20
Enter 20 integers

23   17   38   67   45   52   11   8    5    79
59   42   33   29   81   90   66   15   88   49

Enter number of items per line: 8

23      17      38      67      45      52      11      8
5       79      59      42      33      29      81      90
66      15      88      49

Enter the item to be searched 10
The item 10 not found
Enter the item to be searched 11
```

```
The item 11 found at 6
Enter the item to be searched 1
The item 1 not found
Enter the item to be searched 91
The item 91 not found
```

As you have seen in Example 10.2, in the case of an unsorted array, if the search item is not in the array, the entire array needs to be searched. This can be remedied, if the array is sorted. For example, if the array is sorted and the items are in ascending order, then once an item larger than the search item is encountered there is no need to search anymore.

Example 10.4

Consider the sorted array `L`. Let the search item be `22`. Assume that `searchItem` is a parameter variable of the method and thus `searchItem` is `22`.

0	1	2	3	4	5	6	7
7	10	18	21	35	45	48	50

Let the index value be `0`. Now `L[0]` is 7 and thus both `(L[i] == searchItem)` and `(L[i] > searchItem)` are `false`. Increment the index by `1`. Note that `L[1]` is `10` and thus `(L[i] == searchItem)` and `(L[i] > searchItem)` are `false`. Both `(L[i] == searchItem)` and `(L[i] > searchItem)` are `false` for `i = 2` and `i = 3` as well. Since `35` is greater than the search item, the `searchItem` is not in the array. The `boolean` expression `(L[i] > searchItem)` is `true` and search ends by returning `–1` (Table 10.3).

Example 10.5

Consider the sorted array `L` of Example 10.4. Let the search item be `21`.

Note that both `(L[i] == searchItem)` and `(L[i] > searchItem)` are `false` for `i = 0, 1, 2`. For `i = 3`, `(L[i] == searchItem)` is `true`. Therefore, the method returns `3` (Table 10.4).

TABLE 10.3 Sorted Array: Search Item Not Found

		searchItem is 22	
i	**L[i]**	**L[i] == searchItem**	**L[i] > searchItem**
0	7	7 == 22 is **false**	7 > 22 is **false**
1	10	10 == 22 is **false**	10 > 22 is **false**
2	18	18 == 22 is **false**	18 > 22 is **false**
3	21	21 == 22 is **false**	21 > 22 is **false**
4	35	35 == 22 is **false**	35 > 22 is **true**
		`searchItem` not found. **return** `-1;`	

TABLE 10.4 Sorted Array: Search Item Found

		searchItem is 21	
i	L[i]	L[i] == searchItem	L[i] > searchItem
0	7	7 == 21 is **false**	7 > 21 is **false**
1	10	10 == 21 is **false**	10 > 21 is **false**
2	18	18 == 21 is **false**	18 > 21 is **false**
3	21	21 == 21 is **true**	21 > 21 is **false**
		searchItem found. return 3;	

The `linearSearch` method for an array sorted in ascending order can be written as follows:

```
public int linearSearch(int searchItem)
{
    for (int i = 0; i < intData.length; i++)
    {
        if (intData[i] == searchItem)
            return i;
        if (intData[i] > searchItem)
            return -1;
    }
    return -1;
}
```

Self-Check

3. What is the minimum change you will have to make so that the `linearSearch` method for an array sorted in ascending order will become a `linearSearch` method for an array sorted in descending order?
4. Is the order of `if` statements appearing in the `linearSearch` method for an array sorted in ascending order important?

Binary Search

The binary search is the most efficient search algorithm on a sorted array. The array can be ordered in either an ascending or a descending manner. For the sake of discussion, assume that the array is sorted in an ascending order.

Most of you may be familiar with the following guessing game, quite often seen in popular game shows. There are two players. Player 1 picks a number between 1 and 1000 and keeps it as a secret. Player 2 has to guess the secret number correctly with minimum number of false guesses. Every time player 2 makes a guess, player 1 will say "higher" if the secret number is higher than what player 2 has guessed, "lower" if the secret number is lower than what player 2 has guessed, and "correct" if the secret number is the same as what player 2 has guessed. The best strategy in this situation is to guess the "middle value."

TABLE 10.5 The Guessing Game

Guess by Player 2	Reason for the Guess	Response by Player 1
500	500 is (1 + 1000)/2	Higher
750	750 is (501 + 1000)/2	Lower
625	625 is (501 + 749)/2	Higher
687	687 is (626 + 749)/2	Lower
656	656 is (626 + 686)/2	Higher
671	671 is (657 + 686)/2	Higher
679	679 is (672 + 686)/2	Lower
675	675 is (672 + 678)/2	Higher
677	677 is (676 + 678)/2	Higher
678	678 is (678 + 678)/2	Correct

Let us play the game once. Assume that player 1 has picked the number 678. Table 10.5 summarizes the game they played.

Notice that as the game starts, the secret number can be any number between 1 and 1000. The set of possible values is known as the *search space*. Thus, we begin with the search space [1, 1000]. Player 2 picked the middle value 500 as the first guess. Player 1 responded by "higher." Player 2 reasoned that the secret number has to be between 501 and 1000. Therefore, player 2 changed the lower limit from 1 to 501. Thus, the search space became [501, 1000]. Observe that the current search space is half (or less than half) of the previous search space. Player 2 guessed 750 and player 1 responded by "lower." Player 2 concluded that the number has to be between 501 and 749. Thus, the search space became [501, 749]. Once again, the current search space is half of the previous search space. The game went on like this, each time cutting down the search space into half. Finally, the search space became 1 and at that point, the guess matches the secret number. Let us summarize our observations as follows:

- Each time player 1 responded "higher," player 2 changed the lower limit of the search space to guess + 1.
- Each time player 1 responded "lower," player 2 changed the upper limit of the search space to guess − 1.
- The size of the search space becomes half with each guess.
- The search space ultimately becomes 1.

The binary search works quite similar to the above game.

Example 10.6

Consider the following sorted array:

0	1	2	3	4	5	6	7
7	10	18	21	35	45	48	50

Let the search item be `45`. Note that lower index is `0` and upper index is `7`. Thus, the middle index is `(0 + 7)/2 = 3`.

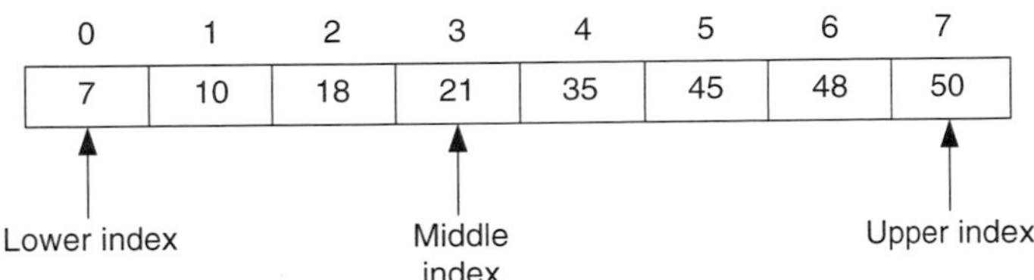

Note that the search item 45 is higher than the item at the middle index, namely 21. Therefore, the lower index is changed to middle index +1. Thus, lower index becomes 4. Similar steps are repeated until either the search item is found or the search space becomes zero.

Begin the next "iteration" by computing the middle index as (4 + 7)/2 = 5.

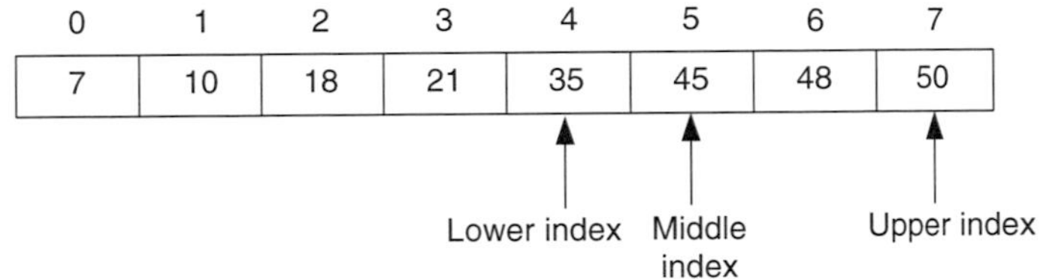

Observe that the search item and the item at middle index are the same. Therefore, the algorithm stops by returning the middle index, 5.

Table 10.6 summarizes our above discussion. For convenience, we use `lower`, `upper`, and `middle` to indicate lower, upper, and middle index values. We also use `L` as the array name.

Example 10.7

Consider the sorted array of Example 10.6. Let the search item be 56. The lower index is 0 and upper index is 7. Thus, the middle index is (0 + 7)/2 = 3.

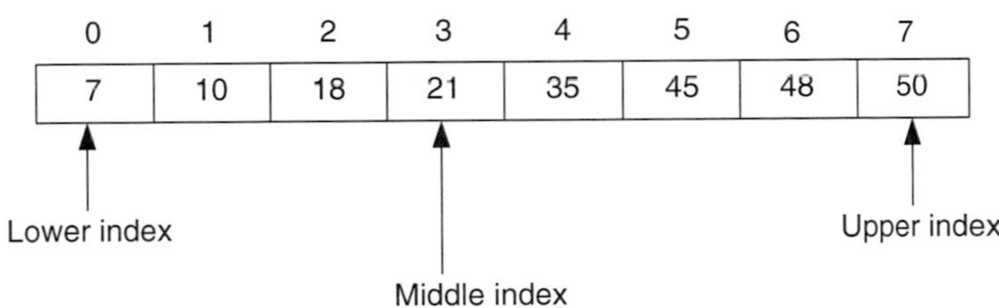

TABLE 10.6 Binary Search: Search Item Found

searchItem is 45					
lower	**upper**	**middle**	**L[middle]**	**Comparison**	**Action**
0	7	(0+7)/2 = 3	21	Higher	`lower = middle+1;`
4	7	(4+7)/2 = 5	45	Equal	**`return`** `middle;`
			return 5;		

TABLE 10.7 Binary Search: Search Item Larger than All Values

searchItem is 56					
lower	upper	middle	L[middle]	Comparison	Action
0	7	(0+7)/2 = 3	21	Higher	lower = middle + 1;
4	7	(4+7)/2 = 5	45	Higher	lower = middle + 1;
6	7	(6+7)/2 = 6	48	Higher	lower = middle + 1;
7	7	(7+7)/2 = 7	50	Higher	lower = middle + 1;
8	7	lower > upper and hence search space is 0. **return** -1;			
		return -1;			

The search item 56 is higher than the item at middle index. The lower index is changed to middle index +1. Thus, lower index becomes 4. Therefore, now the middle index is (4 + 7)/2 = 5.

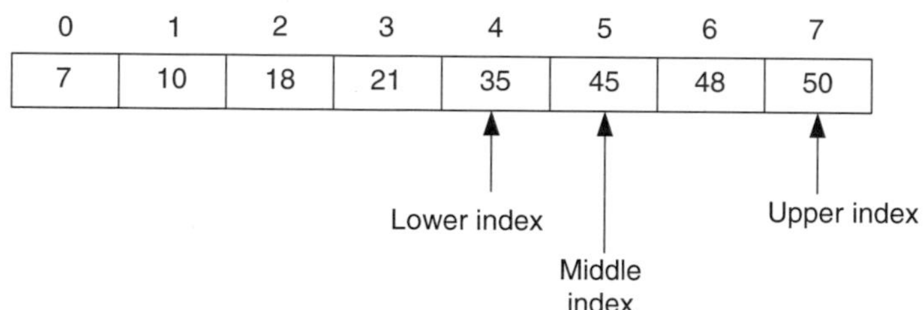

Once again, the search item is larger than the item at middle index. Thus, the lower index is again changed to middle index +1. Therefore, now the lower index is 6 and the middle index is (6 + 7)/2 = 6.

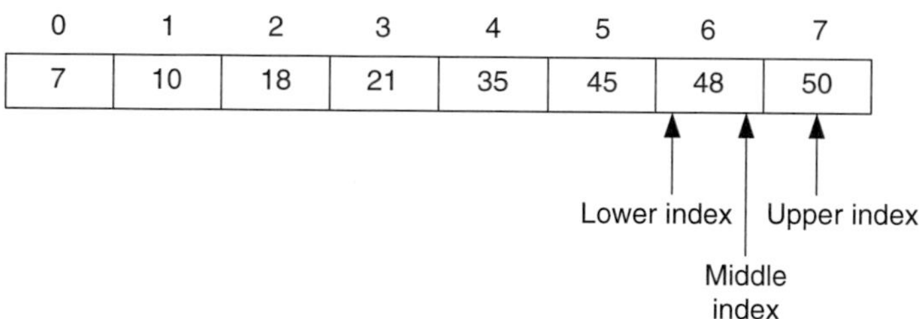

The search item is larger than the item at middle index. The lower index is changed to middle index +1. Therefore, the lower index is 7 and the middle index is (7 + 7)/2 = 7. Observe that search item is larger than the item at middle index. Therefore, lower index becomes middle index +1. That is, lower index is 8. Since lower index is larger than the upper index, the search space is zero. Therefore, the algorithm ends with the conclusion that search item is not in the array (Table 10.7).

Example 10.8

Consider the sorted array of Example 10.6. Let the search item be 15. The lower index is 0 and the upper index is 7. Thus, the middle index is (0 + 7)/2 = 3.

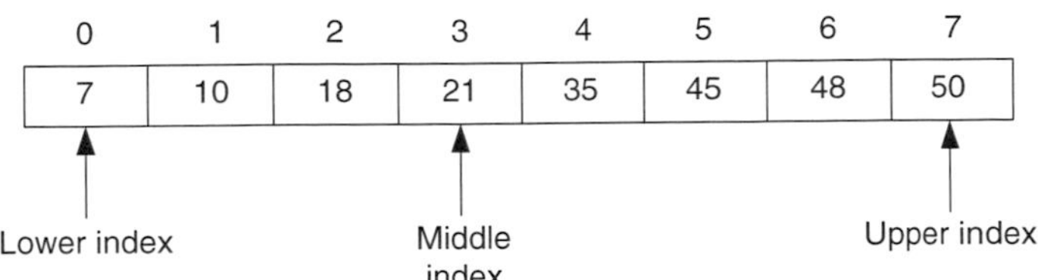

The search item `15` is lower than the item at middle index. The upper index is changed to middle index − `1`. Thus, new upper index is `2`. The middle index is `(0 + 2)/2 = 1`.

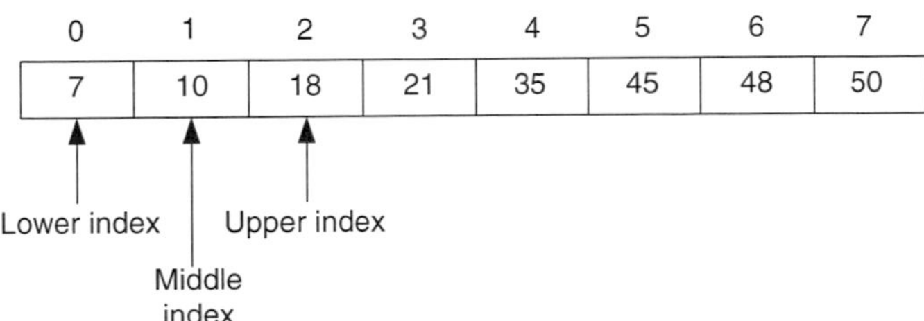

The search item `15` is larger than `10`, the item at middle index. Thus, the lower index is changed to middle index + `1`. The lower index becomes `2` and the middle index becomes `(2 + 2)/2 = 2`. Note that search item is smaller than the item at middle index. Therefore, upper index becomes middle index − `1`. The upper index is `1` and it is smaller than the lower index. Therefore, the search space is zero and the algorithm ends with the conclusion that search item is not in the array (Table 10.8).

The binary search program and the sample output are as follows:

```
import java.util.Scanner;
/**
    Binary search algorithm
*/
public class BinarySearch
{
  private int[] intData;

  /**
      Constructor
      @param inOutArray the array of integers
```

TABLE 10.8 Binary Search: Search Item Not Found

searchItem is 15					
lower	**upper**	**middle**	**L[middle]**	**Comparison**	**Action**
0	7	(0+7)/2 = 3	21	Lower	upper = middle-1;
0	2	(0+2)/2 = 1	10	Higher	lower = middle+1;
2	2	(2+2)/2 = 2	18	Lower	upper = middle-1;
2	1	lower > upper and hence search space is 0. **return** -1;			
		return -1;			

```
    */
    public BinarySearch(int[] inOutArray)
    {
        intData = inOutArray;
    }

    /**
        Searches sorted attribute inData for search item
        @param searchItem the search item
        @return index of search item if found and
        -1 if search item is not in the array
    */
    public int binarySearch(int searchItem)
    {
        int lower = 0;
        int upper = intData.length - 1;
        int middle = 0;

        while (lower <= upper)
        {
            middle = (lower + upper)/2;
            if (intData[middle] == searchItem)
                return middle;
            else if (intData[middle] < searchItem)
                lower = middle + 1;
            else
                upper = middle - 1;
        }
        return -1;
    }

}

import java.util.Scanner;
/**
    This is a test program for BinarySearch class
*/
public class BinarySearchApplication
{
    public static void main(String[] args)
    {
        int arraySize = 0;
        int numItems = 0;
        int start = 0;
```

```
        int inc = 0;

        Scanner scannedInfo = new Scanner(System.in);
        System.out.print("Enter the number of integers :");
        arraySize = scannedInfo.nextInt();
        System.out.print("\nEnter the first value of AP :");
        start = scannedInfo.nextInt();
        System.out.print("\nEnter the increment value of AP :");
        inc = scannedInfo.nextInt();
        int[] testArray
             = ArrayUtility.createAndSet(arraySize, start,inc);
        System.out.println();
        System.out.print("Enter number of items per line :");
        numItems = scannedInfo.nextInt();
        ArrayUtility.printArray(testArray, numItems);
        System.out.println();
        BinarySearch searchObject = new BinarySearch(testArray);
        for (int i = 0; i < 4; i++)
        {
            System.out.print("Enter the item to be searched ");
            int item = scannedInfo.nextInt();
            int index = searchObject.binarySearch(item);
            if ( index == -1)
                System.out.println
                      ("The item "+ item +" not found");
            else
                System.out.println
                      ("The item "+ item +" found at "+ index);
        }
    }
}
```

Output

```
Enter the number of integers : 32
Enter the first value of AP : 8
Enter the increment value of AP : 7
Enter number of items per line : 6

8      15     22     29     36     43
50     57     64     71     78     85
92     99     106    113    120    127
134    141    148    155    162    169
176    183    190    197    204    211
218    225
```

```
Enter the item to be searched 226
The item 226 not found
Enter the item to be searched 7
The item 7 not found
Enter the item to be searched 107
The item 107 not found
Enter the item to be searched 113
The item 113 found at 15
```

Self-Check

5. True or false: Binary search algorithm reduces the search space to approximately half after each comparison.
6. True or false: Linear search algorithm reduces the search space to approximately one-third after each comparison.

EFFICIENCY OF ALGORITHMS

There are two ways to measure the performance of an algorithm. First, and probably the most obvious way, is to write a program and run it using various test data and measure the time it takes to finish the computation. This we call the *empirical approach*. The second approach is mathematical in nature and in fact involves no programming at all. This we call the *analysis approach*. The aim of this section is to introduce both of these approaches.

Empirical Approach

Theoretically, one could run a program and measure the time it took using a stopwatch. However, this approach has two major flaws when it comes to computers. First, ordinary stopwatch is of no use because many small programs take only milliseconds to complete. Second, to execute the algorithm, first data has to be either read from a file or obtained interactively from the user. Similarly, the output needs to be displayed in the monitor or written in a file. All these activities take time and have no relevance to the performance of the algorithm we may be interested in. Therefore, we need a software stopwatch that can be started just before the start of the code that implements the algorithm and stops as soon as the last statement of the code is being executed.

We would like to measure the algorithm by computing the average time it took over many trials. Therefore, the `StopWatch` must have the capability to keep track of multiple start–stop sequences. Thus, we need an attribute `totalElapsedTime` to keep track of cumulative elapsed time between multiple start–stop sequences and another attribute `count` to keep track of the number of start–stop sequences. To compute the time elapsed between a start and a stop sequence, an attribute `startTime` is required. The `StopWatch` must have the following operations: clear, start, stop, and getAverageTime. Based on this analysis, we have the following class:

```
import java.util.Scanner;
/**
   StopWatch with multiple start-stop sequence
*/
public class StopWatch
{
    private long totalElapsedTime;
    private long startTime;
    private int count;

    /**
        Constructs a StopWatch with all
        attributes set to 0

    */
    public StopWatch()
    {
        clear();
    }

    /**
        Clears all attributes
    */
    public void clear()
    {
        totalElapsedTime = 0;
        startTime = 0;
        count = 0;
    }

    /**
        Starts the StopWatch
    */
    public void start()
    {
        startTime = System.nanoTime();
    }

    /**
        Stops the StopWatch
    */
```

```
    public void stop()
    {
        totalElapsedTime = totalElapsedTime +
                System.nanoTime() - startTime;
        count++;
    }

    /**
        Returns the average time elapsed
    */
    public long getAverageTime()
    {
        return totalElapsedTime / count;
    }

    /**
        Returns the total time elapsed
    */
    public long getTotalTime()
    {
        return totalElapsedTime;
    }

}
```

The following program uses the StopWatch class to measure the performance of linear search algorithm on a sorted array:

```
import java.util.Random;
import java.util.Scanner;
/**
    This a test program for LinearSearchSorted class
*/
public class LinearSearchSortedTiming
{
    private static Random randomGenerator = new Random();

    public static void main(String[] args)
    {
        int arraySize = 0;
        int trials = 0;
        int maxValue = 0;
        int item = 0;

        Scanner scannedInfo = new Scanner(System.in);
        System.out.print("Enter the number of integers :");
```

```
        arraySize = scannedInfo.nextInt();
        System.out.print("Enter the number of trials :");
        trials = scannedInfo.nextInt();
        int[] testArray
                = ArrayUtility.createAndSet(arraySize,7, 5);
        LinearSearchSorted searchObject
                        = new LinearSearchSorted(testArray);
        StopWatch timeKeeper = new StopWatch();

        maxValue = 7 + arraySize * 5;

        for (int i = 0; i < trials; i++)
        {
            item = randomGenerator.nextInt(maxValue);
            timeKeeper.start();
            searchObject.linearSearch(item);
            timeKeeper.stop();
        }

        long algorithmTime = timeKeeper.getTotalTime();

        System.out.println("Total Elapsed time for array of size "
                        + arraySize +" is :" + algorithmTime);
    }
}
```

Since each search takes very few milliseconds, we find the total time for arrays of sizes 1, 2, 3, 4, and 5 million. In all these cases, the number of searches is kept constant at 10,000. A similar study is conducted for binary search algorithm. The results on a Pentium 4, 1.70 GHz machine running under Windows XP operating system are summarized in Table 10.9 (see also Figure 10.1).

From Table 10.9, you can make the following observations:

1. Linear search takes quite a long time compared to binary search.
2. In the case of linear search, as the size of the array doubled, the time it took to search for 10,000 items approximately doubled. Similarly, as the array size tripled, the time

TABLE 10.9 Empirical Comparison of Linear and Binary Search

		Time in Milliseconds	
Array Length	**Number of Searches**	**Linear Search**	**Binary Search**
1 million	10,000	27,956	42
2 million	10,000	55,061	44
3 million	10,000	82,596	45
4 million	10,000	110,821	46
5 million	10,000	137,600	47

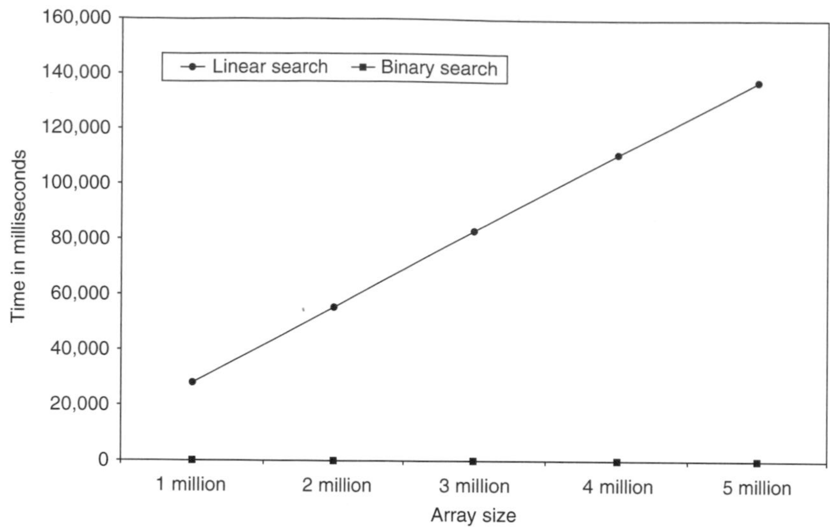

FIGURE 10.1 Empirical comparison of linear and binary search.

it took to search for 10,000 items approximately tripled and so on. The time it takes to find 10,000 items seems to be directly proportional to the size of the array. In other words, there seems to be a linear relationship between the size of the array and the time it takes to find 10,000 items.

3. In the case of binary search, as the size of the array doubled, the time it took to search for 10,000 items increased by a very small amount. Similar is the case as the array size tripled. Thus, the time it took to find 10,000 items seems to be an extremely slow growing function of the array length.

Self-Check

7. In the case of linear search, as the size of the array doubled, the computation time__________.
8. In the case of binary search, as the size of the array doubled, the computation time __________.

Analysis Approach

The purpose of this subsection is to introduce the concept of time complexity in a non-threatening fashion. Thus, the presentation in this subsection is intentionally simplified.

The analysis of an algorithm begins by identifying the most relevant operation. For example, in the case of matrix multiplication, multiplication is the most costly operation. Even though there are many additions involved, we identify multiplication as the relevant operation. In the case of search and sort algorithms, we identify comparison as the most relevant operation.

Once the most relevant operation is identified, you need to count the number of relevant operations required to solve a problem of size `n`. For example, consider the linear search algorithm. On an average, you may need `n/2` comparisons, where `n` is the size of the array.

Therefore, the time it takes to perform one linear search is directly proportional to the size of the array. We express this fact by saying linear search is an order n algorithm or the complexity of the linear search is O(n), pronounced as big-O n. Similarly, we can also measure the worst-case performance of linear search algorithm. If the search item is larger than any item in the list, the linear search algorithm has to compare all items even if the array is sorted. Thus, the algorithm has to perform n comparisons. Thus, the worst-case time complexity of the linear search algorithm is O(n). Yet another commonly considered measure is the best-case time complexity. If the search item is the first item in the array, only one comparison is all that is required. Therefore, irrespective of the array size, the number of comparisons is a constant. We say, the best-case time complexity is order 1 and we denote it by O(1).

A similar analysis for binary search algorithm can be done as follows. In the case of binary search, if the middle value is not the same as the search item, you need to consider only one half of the array. Therefore, if T(n) denotes the worst-case time complexity of binary search algorithm,

T(n) = time to compare one item + time complexity for an array of size n/2.

$T(n) = 1 + T(n/2)$.

Assume that $n = 2^k$. Then

$$
\begin{aligned}
T(n) &= T(2^k) = 1 + T(2^{k-1}) \\
&= 1 + 1 + T(2^{k-2}) = 2 + T(2^{k-2}) \\
&= 1 + 2 + T(2^{k-3}) = 3 + T(2^{k-3}) \\
&= 4 + T(2^{k-4}) \\
&\ldots \\
&= k + T(2^{k-k}) = k + T(2^0) = k + T(1).
\end{aligned}
$$

T(1) is the worst-case time required to perform binary search on an array of size 1. Therefore, T(1) = 1. Thus, T(n) = k + 1, where $n = 2^k$. Recall that in this case we could write $k = \log_2 n$. Therefore, $T(n) = \log_2 n + 1$. Thus, T(n) is a function of $\log_2 n$. In other words, the worst-case time complexity of the binary search algorithm is $O(\log_2 n)$. The average-case time complexity of the binary search is $O(\log_2 n)$ and the best-case time complexity of the binary search is O(1).

Self-Check

9. The linear search is a ________ algorithm.
10. The binary search is a ________ algorithm.

Advanced Topic 10.1: Levels of Complexity

You have already seen three levels of complexity: O(1), $O(\log_2 n)$, and O(n). You also know an O(1) algorithm takes less time than an $O(\log_2 n)$ algorithm. Similarly, an $O(\log_2 n)$ algorithm takes less time than an O(n) algorithm. Multiplying by n, we can conclude that an O(n) algorithm takes less time than an O($nlog_2n$) algorithm and an O($nlog_2n$) algorithm takes less time than an O(n^2) algorithm. Similarly, multiplying by n^k, we conclude that an O(n^k) algorithm takes less time than an O(n^klog_2n) algorithm and an O(n^klog_2n) algorithm takes less time than an O(n^{k+1}) algorithm.

As we consider the complexity, we are in fact concerned about the growth rate of the function; rather than the actual value of the function itself. Therefore, time complexity of a polynomial is the same as that of its leading term, for example, O($10n^2 + 8n - 15$) = O(n^2).

SORT ALGORITHMS

We begin this section with a fairly simple sort algorithm, selection sort. Similar to time complexity, we can also measure the space complexity. In this case, we are measuring the space (memory) requirements of the algorithm as a function of the input size. In the case of sort algorithms, we also pay attention to additional array requirements. A sort algorithm is said to be an *in-place* algorithm, if it does not require any additional arrays.

All the three algorithms presented in this chapter have the same time complexity, O(n^2). There are sort algorithms with time complexity O($nlog_2n$), which is very close to O(n) as log_2n is a very slowly increasing function. Merge sort and quick sort are two of the most commonly used algorithms with time complexity O($nlog_2n$).

It is a well-known fact that any sort algorithm based on comparison must have at least O($nlog_2n$) time complexity. Therefore, any algorithm with worst-case time complexity O($nlog_2n$) is known as an optimal sorting algorithm. Thus, both merge sort and quick sort are optimal sort algorithms. Similarly, binary search is an optimal search algorithm on sorted arrays.

Self-Check

11. Any sort algorithm based on comparison must have at least ________ time complexity.
12. The time complexity of an optimal sort algorithm is ________.

Selection Sort

The idea behind the selection sort is quite easy. First, find the smallest item and place it at the first position. From the remaining items in the list, find the smallest item and place it in the second place. Keep repeating these steps and eventually all items will be in the sorted order.

Example 10.9

This example demonstrates the selection sort. Consider an integer array L with six elements.

0	1	2	3	4	5
15	21	24	17	8	12

The algorithm begins by first finding the smallest element in the array. Note that 8 is the smallest item and it is at location 4. Since 8 is the smallest item, we need to keep it at location 0. However, location 0 currently has 15. Therefore, we swap the values at locations 4 and 0. The array L after swapping values at locations 4 and 0 is shown below. The location 0 is no longer considered as the part of L for the purpose of finding the smallest item, and hence is shaded.

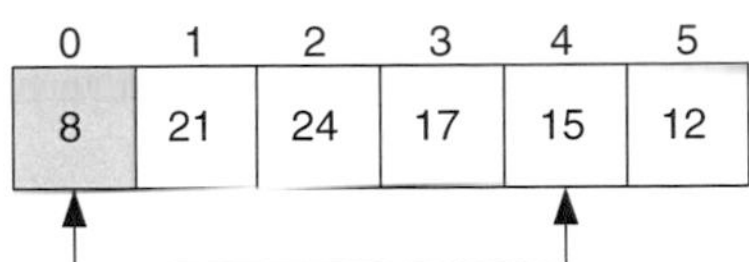

We now repeat the above steps. Find the smallest element in the array locations 1 through 5. Clearly, 12 is the smallest item and it is at location 5. We need to keep 12 at location 1. Therefore, we swap the values at locations 5 and 1. Once again, location 1 is no longer considered as part of the array as we search for the next smallest item, and hence shaded. Thus, we have the following:

0	1	2	3	4	5
8	12	24	17	15	21

The smallest value 15 is at location 4. Therefore, we swap the items at locations 4 and 2. Thus, the array L is as follows:

0	1	2	3	4	5
8	12	15	17	24	21

The smallest value 17 is at location 3. Note that in this case we swap the item at location 3 with the item at location 3. The array L after the swap can be visualized as follows:

0	1	2	3	4	5
8	12	15	17	24	21

The smallest item is 21. We swap the item at location 5 with the item at location 4. Observe that once we have placed five items in their respective locations, the last item is in its proper location. Therefore, the algorithm terminates. The array L is now sorted.

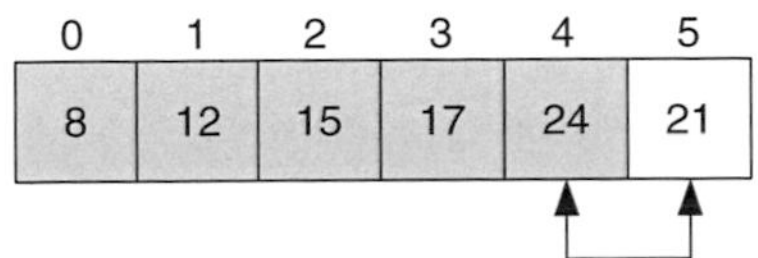

Table 10.10 summarizes the above discussion.

TABLE 10.10 Selection Sort

Starting Index i	Minimum Value	Index of Minimum Value	Action Taken
0	8	4	`swap(L[0], L[4]); i = i+1;`
1	12	5	`swap(L[1], L[5]); i = i+1;`
2	15	4	`swap(L[2], L[4]); i = i+1;`
3	17	3	`swap(L[3], L[3]); i = i+1;`
4	21	5	`swap(L[4], L[5]); i = i+1;`

The implementation is quite straightforward. We use two helper methods `findMin` and swap. They are intentionally kept as `private`.

```
import java.util.Scanner;
/**
    Selection sort algorithm
*/
public class SelectionSort
{


    private int[] intData;
    /**
        Constructor
        @param inOutArray the array to be sorted
    */
    public SelectionSort(int[] inOutArray)
    {
        intData = inOutArray;
    }


    /**
        Sorts the attribute inData
    */
    public void sort()
```

```
    {
        int minIndex;
        for (int i = 0; i < intData.length - 1; i++)
        {
            minIndex = findMin(i);
            swap(minIndex, i);
        }
    }

    /**
        Find the smallest value from the given location
        @param start location to begin finding minimum
        @return index of the minimum
        intData[start]...intData[intData.length-1]
    */
    private int findMin(int start)
    {
        int minLoc = start;
        for (int i = start + 1; i < intData.length; i++)
            if (intData[i] < intData[minLoc])
                minLoc = i;
        return minLoc;
    }

    /**
        interchange values between two locations of the array
        @param first one of the location to be interchanged
        @param second the other location to be interchanged
    */
    private void swap(int first, int second)
    {
        int temp = intData[first];
        intData[first] = intData[second];
        intData[second] = temp;
    }

}

import java.util.Scanner;
/**
    This a test program for SelectionSort class
*/
public class SelectionSortApplication
```

```
{
    public static void main(String[] args)
    {
        int arraySize = 0;
        int numItems = 0;
        Scanner scannedInfo = new Scanner(System.in);
        System.out.print
             ("Enter the number of items to sort: ");
        arraySize = scannedInfo.nextInt();
        System.out.println
              ("Enter "+ arraySize +" integers\n");
        int[] testArray
             = ArrayUtility.createAndSet
                         (arraySize, scannedInfo);
        System.out.println();
        System.out.print
              ("Enter number of items per line : ");
        numItems = scannedInfo.nextInt();
        System.out.println("\n The array after sort:\n");
        ArrayUtility.printArray(testArray, numItems);
        System.out.println();
        SelectionSort sortObject
                   = new SelectionSort(testArray);
        sortObject.sort();
        System.out.println
                     ("The array before sort:\n");
        ArrayUtility.printArray(testArray, numItems);
    }
}
```

Output

```
Enter the number of items to sort: 20
Enter 20 integers

7 21 34 17 45 5 8 31 5 59 42 30 29 62 1 55 47 20 9 54
Enter number of items per line : 8

The array before sort:

7   21  34  17  45  5   8  31
5   59  42  30  29  62  1  55
47  20  9   54
The array after sort:

1   5   5   7   8   9   17  20
21  29  30  31  34  42  45  47
54  55  59  62
```

The analysis of the selection sort is not that complicated. Assume that the number of items to be sorted is n. Then, you need n – 1 comparisons to determine the smallest item. Once the smallest item is placed at location 0, there are only n – 1 items in the unprocessed part of the array. Thus, n – 2 comparisons are required to find the second smallest number and so on. Therefore, the total number of comparisons is (n – 1) + (n – 2) + ··· + 1 = n(n – 1)/2 = 0.5n^2 + 0.5n. The growth rate of this expression is n^2. In other words, the worst-case time complexity of selection sort is O(n^2). As far as the space complexity is concerned, the algorithm requires an array of size n plus a few extra spaces. Thus, the space complexity is O(n). The algorithm requires no additional arrays and hence is an in-place algorithm.

Self-Check

13. Another way to perform selection sort is to find the largest item and place it at the ________ position.
14. The time complexity of selection sort algorithm is ________.

Insertion Sort

The idea behind the insertion sort is also quite easy to understand. Just like the selection sort, during the sorting process, insertion sort also considers the array as consisting of two parts. The first part is the sorted part of the array and the second part is the yet to be processed or the unprocessed part of the array. To begin with, the sorted part consists of just one item and all the remaining items are in the unprocessed part. We pick the first item from the unprocessed part and insert it in the sorted part so that the sorted part remains sorted. Keep repeating these steps until there is no more item in the unprocessed part of the array.

Example 10.10

This example demonstrates the insertion sort. Consider the same set of numbers we have used in Example 10.9. Thus, the integer array L with 6 elements is as follows:

0	1	2	3	4	5
15	21	24	17	8	12

First, we conceptually divide the array into two parts. The sorted part consists of one item, the element at index 0, and the yet to be processed part consists of all elements at index locations 1 through 5. Shading is used to distinguish the sorted part from the unprocessed part of the array.

0	1	2	3	4	5
15	21	24	17	8	12

Now, consider the first item in the yet to be processed part. Since 21 is greater than 15, the first two items are in sorted order. Therefore, nothing needs to be done, and the sorted part of the array is from index locations 0 to 1. The unprocessed part of the array is now from index locations 2 to 5.

0	1	2	3	4	5
15	21	24	17	8	12

The item to be processed is 24. Observe that 24 is greater than 21, and thus first three items are now sorted. The unprocessed part of the array is now from index locations 3 to 5.

0	1	2	3	4	5
15	21	24	17	8	12

The next item to be processed is 17. We need to insert 17 in the sorted part of the array. Therefore, we compare 17 with 24. Observe that 17 is smaller than 24. Move 24 to index location 3, while keeping the value 17 at a temporary variable temp. Thus, we have the following:

0	1	2	3	4	5
15	21		24	8	12

Next we compare 17 with 21. Again, 21 is larger than 17. So, we move 21 to index location 2.

0	1	2	3	4	5
15		21	24	8	12

Next we compare 17 with 15. Since 17 is larger than 15, we place 17 at index location 1. This marks the end of processing 17. The sorted part of the array is from index locations 0 to 3 and the unprocessed part of the array is from index locations 4 to 5.

0	1	2	3	4	5
15	17	21	24	8	12

The next item to be processed is 8. Since 8 is smaller than 24, we move 24 to index location 4.

0	1	2	3	4	5
15	17	21		24	12

Note that 8 is smaller than 21. Therefore, 21 is moved to index location 3.

0	1	2	3	4	5
15	17		21	24	12

Since 8 is smaller than 17, we move 17 to index location 2,

0	1	2	3	4	5
15		17	21	24	12

Observe that 8 is smaller than 15, and hence 15 is moved to index location 1.

0	1	2	3	4	5
	15	17	21	24	12

There are no more items to compare. So, we place 8 at index location 0. This marks the end of processing 8. The sorted part of the array is from index locations 0 to 4 and the unprocessed part of the array now consists of just one item.

0	1	2	3	4	5
8	15	17	21	24	12

The item to be processed is 12. Since 12 is smaller than 24, we move 24 to index location 5. By similar logic, we move 21 to index location 4, 17 to index location 3, and 15 to index location 2.

0	1	2	3	4	5
8		15	17	21	24

Note that 12 is not smaller than 8; therefore, we place 12 at index location 1. This marks the end of processing 12. The sorted part of the array is from index locations 0 to 5 and the unprocessed part of the array is empty (Table 10.11). Thus, we have the following sorted array:

0	1	2	3	4	5
8	12	15	17	21	24

The implementation is quite straightforward. We use two helper methods `isMoved` and `insert`. They are intentionally kept as `private`. Given an index location, say `i`, the method `insert(i)` will insert item at `i` in the sorted array `intData[0], ..., intData[i-1]`. This method uses the `boolean` method `isMoved` to move necessary data values to make room for insertion. Creation of the `InsertionSort` class and testing are quite similar to that of selection sort and is left as an exercise.

```
public void sort()
{
    for (int i = 1; i < intData.length; i++)
```

TABLE 10.11 Insertion Sort

Starting Index `i`, Unprocessed Part	Item to be Inserted	Index of the Item to Compare	Item to be Compared	Action Taken
1	21	0	15	none.
				i = i + 1;
2	24	1	21	none.
				i = i + 1;
3	17	2	24	temp = 17
				L[3] = 24
		1	21	L[2] = 21
		0	15	L[1] = 17
				i = i + 1;
4	8	3	24	temp = 8
				L[4] = 24
		2	21	L[3] = 21
		1	17	L[2] = 17
		0	15	L[1] = 15
				L[0] = 8
				i = i + 1;
5	12	4	24	temp = 12
				L[5] = 24
		3	21	L[4] = 21
		2	17	L[3] = 17
		1	158	L[2] = 15
		0	8	L[1] = 12
				i = i + 1;

```
        {
            insert(i);
        }
    }
    /**
        Insert the next item in the sorted part
        @param loc location of the next item
    */
    private void insert(int loc)
    {
        int temp = intData[loc];
        int i = loc - 1;
        while (i > -1 && isMoved(i,temp))
            i--;
        intData[i+1] = temp;
    }
    /**
```

```
        If item is smaller, move value at index is to index + 1
        @param index one of the location in sorted array
        @param item that is currently being processed
        @return true if moved false otherwise
    */
    private boolean isMoved(int index, int item)
    {
        boolean move = intData[index] > item;

        if (move)
            intData[index + 1] = intData[index];
        return move;
    }
```

The analysis of the insertion sort is quite similar to that of selection sort. Assume that number of items to be sorted is n. Then, you need 0 comparison to create a sorted part of size 1. To insert one item, you need to perform one comparison. Thus, to create a sorted part of size 2, you need 1 comparison. To insert third item, you need at most 2 comparisons, the size of the current sorted part, and so on. Thus, finally to insert the nth item, you need at most n − 1 comparisons. Therefore, the total number of comparisons is $1 + 2 + \ldots + (n - 1) = n(n - 1)/2 = 0.5n^2 + 0.5n$. The growth rate of this expression is n^2. In other words, the worst-case time complexity of insertion sort is $O(n^2)$. The space complexity is $O(n)$ and it is an in-place algorithm.

Self-Check

15. Another way to perform insertion sort is to keep the first part ________ and the second part ________.
16. The time complexity of insertion sort algorithm is ________.

Bubble Sort

Just as selection and insertion sort algorithms, bubble sort also divides the array into two parts: the sorted part and the unsorted part. The bubble sort algorithm compares two adjacent numbers and if they are not in order, then they are swapped to make them in order. Carrying out these steps for all adjacent pairs from the beginning to the end of the array has the effect of "bubbling" the largest item to the highest index position. Thus, the last item is the sorted part and all other elements in the beginning of the array are in the unprocessed part. Repeating these steps on the unprocessed part of the array will bubble the second largest item in the list to its proper location in the sorted order. Keep repeating these steps and eventually all items will be in the sorted order.

Example 10.11

This example demonstrates the bubble sort. Consider an integer array L with 6 elements.

0	1	2	3	4	5
15	21	24	17	8	12

The algorithm begins by comparing items at index locations 0 and 1. Since they are in sorted order, no action is taken.

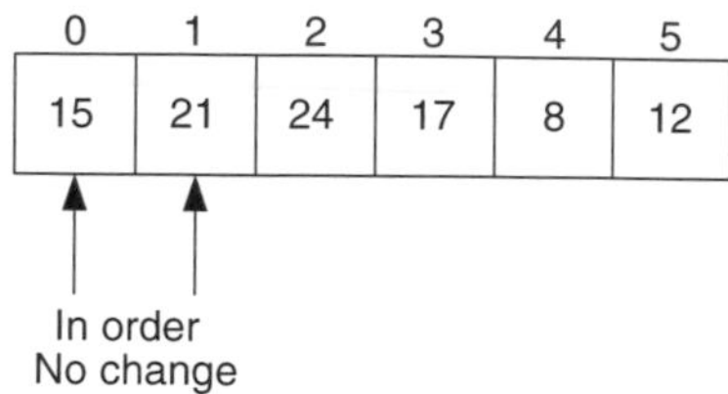

Next, items at index locations 1 and 2 are compared. They are in order and so there is no need for any action.

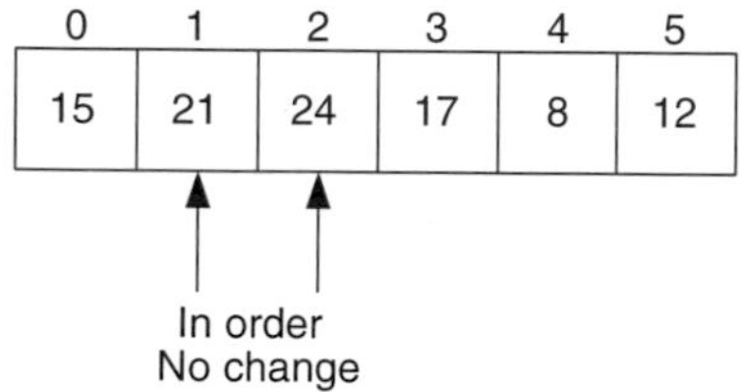

Comparing items at index locations 2 and 3, it is observed that they are not in order.

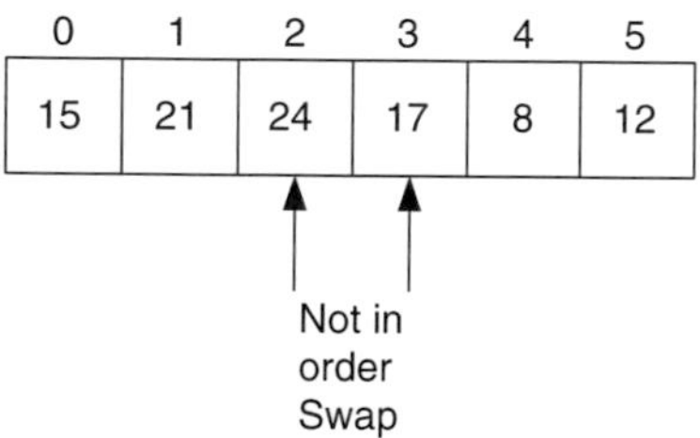

Therefore, we need to swap them. Thus, array changes as follows:

0	1	2	3	4	5
15	21	17	24	8	12

Next, index locations 3 and 4 are compared.

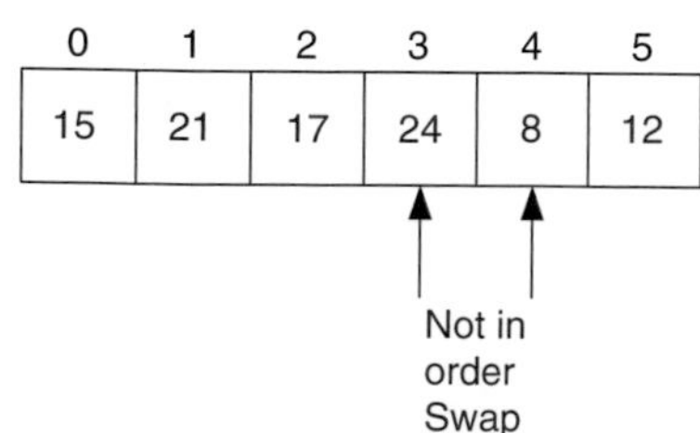

Since they are not in order, they are swapped.

0	1	2	3	4	5
15	21	17	8	24	12

Next, compare items at index locations `4` and `5`.

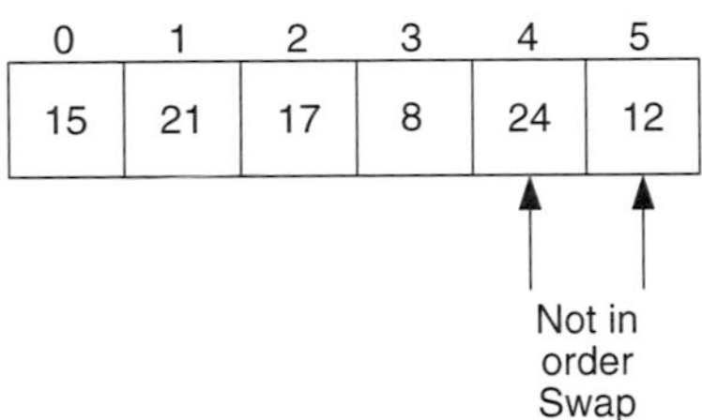

Since they are not in order, they are swapped. Note that there are no more pairs to compare. The largest item in the array, `24`, is at the highest index location. We consider location `5` as sorted part of the array. The array locations from index `0` to `4` are treated as unprocessed. In our illustrations, we use shading to indicate the sorted part of the array.

0	1	2	3	4	5
15	21	17	8	12	24

We repeat the above steps on the unprocessed part of the array. Thus, we start comparing items at locations `0` and `1`.

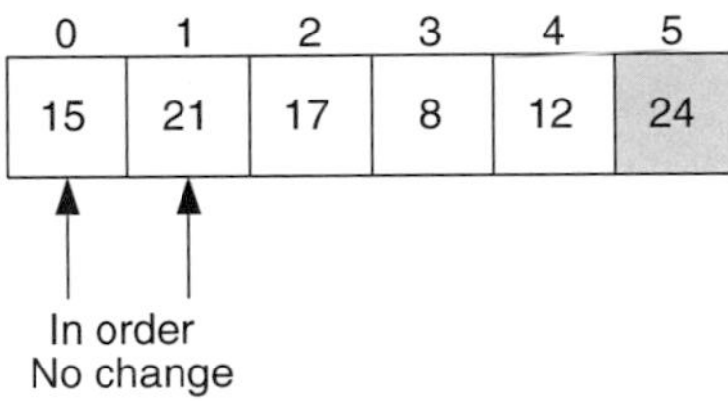

They are in order and as such no action is required. Next, items at index locations `1` and `2` are compared.

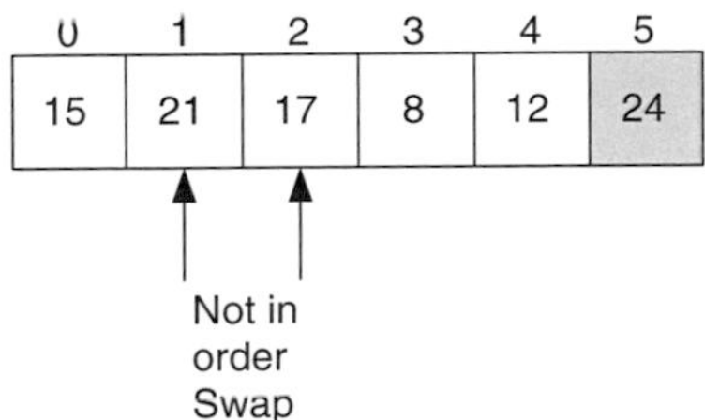

Since they are not in order, they are swapped.

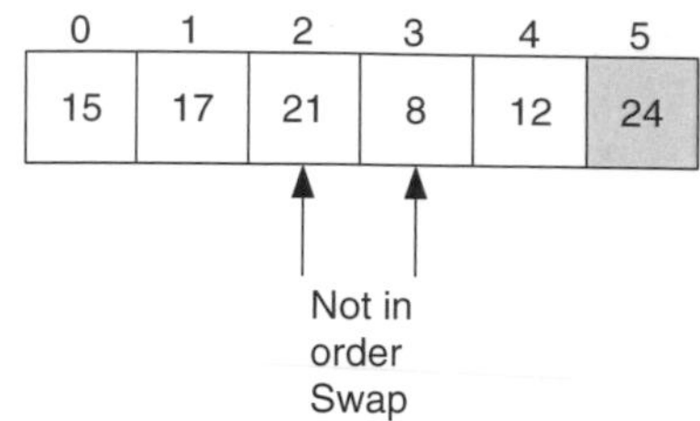

Items at locations 2 and 3 are compared and they are not in order. Hence, they are swapped.

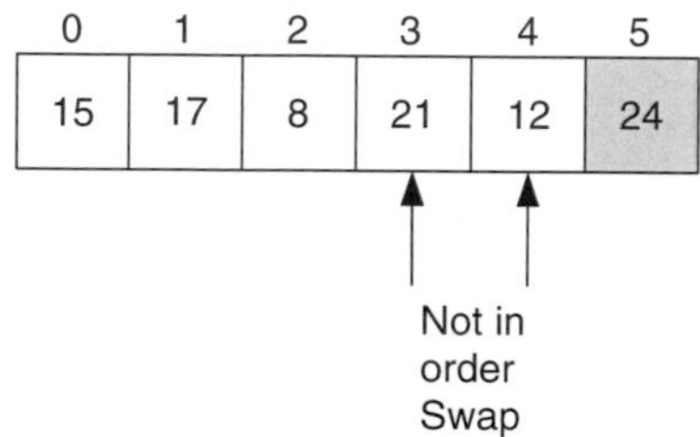

After comparing items at locations 3 and 4, they are swapped.

There are no more pairs to compare. The second largest item in the array, 21, is at the second highest index location. We consider locations 4 and 5 as sorted part of the array. The array locations from index 0 to 3 are treated as unprocessed.

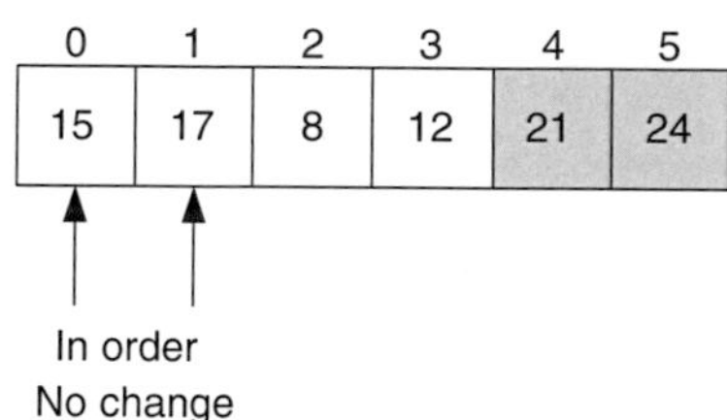

Once again, we start comparing items at locations 0 and 1. Since they are in order, no swapping is done. Next, items at locations 1 and 2 are compared.

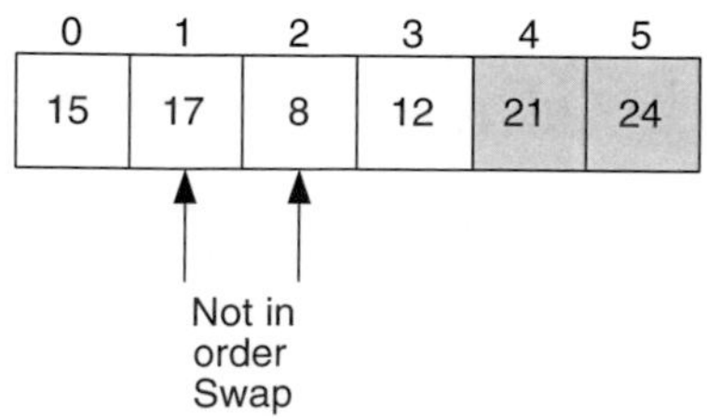

Since they are not in order, they are swapped.

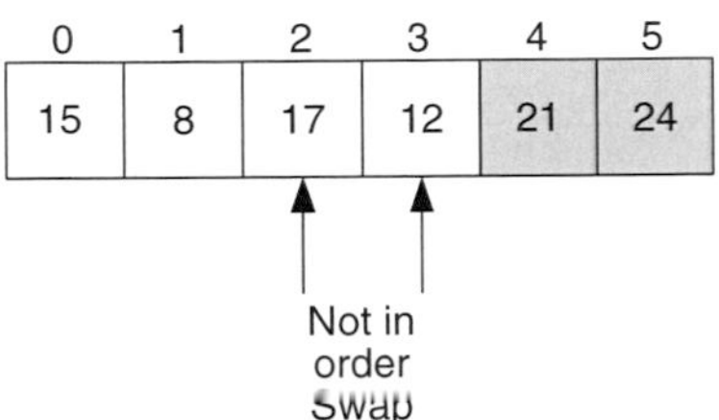

Items at locations `2` and `3` are not in order. Therefore, they are swapped. Note that `17` is now a processed item.

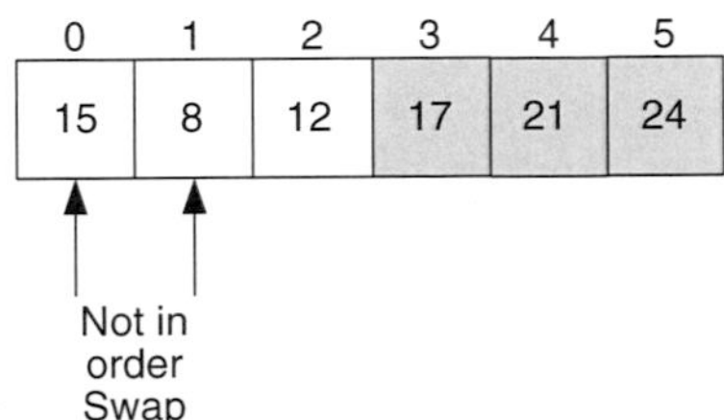

Compare items at index locations `0` and `1`. Note that they are not in order. Therefore, they are swapped.

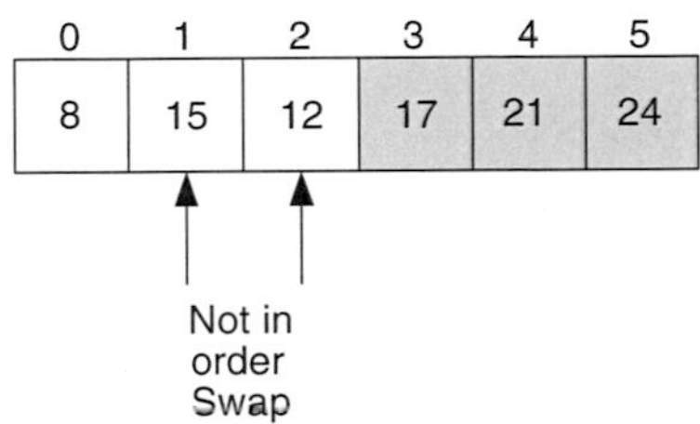

Note that items at locations `1` and `2` are not in order, and thus they are swapped. Observe that `15` is a processed item.

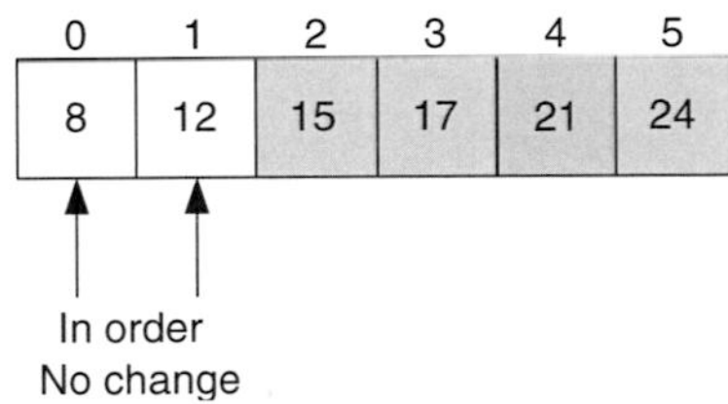

Once again, compare locations `0` and `1`. Since they are in order, no action is required and this makes `12` a processed item. Just as in the case of selection sort, there is no need to process the last item. Thus, algorithm terminates (Table 10.12).

The implementation is quite straightforward. We use two helper methods `placeMax` and `swap`. They are intentionally kept as `private`. Creation of the `BubbleSort` class and testing are quite similar to that of selection sort and is left as an exercise.

TABLE 10.12 Bubble Sort

Last Index of Unprocessed Part	Adjacent Indices	Values to be Compared	Comment and Action Taken
5	0, 1	15, 21	in order; no action
	1, 2	21, 24	in order; no action
	2, 3	24, 17	not in order; swapped
	3, 4	24, 8	not in order; swapped
	4, 5	24, 12	not in order; swapped
			24 is in the sorted part
4	0, 1	15, 21	in order; no action
	1, 2	21, 17	not in order; swapped
	2, 3	21, 8	not in order; swapped
	3, 4	21, 12	not in order; swapped
			21 is in the sorted part
3	0, 1	15, 17	in order; no action
	1, 2	17, 8	not in order; swapped
	2, 3	17, 12	not in order; swapped
			17 is in the sorted part
2	0, 1	15, 8	not in order; swapped
	1, 2	15, 12	not in order; swapped
			15 is in the sorted part
1	0, 1	8, 12	in order; no action
			array is sorted

```
import java.util.Scanner;
/**
    Bubble sort algorithm
*/
public class BubbleSort
{
    private int[] intData;

    /**
        Constructor
        @param inOutArray the array to be sorted
    */
    public BubbleSort(int[] inOutArray)
    {
        intData = inOutArray;
    }

    /**
        Sorts the attribute inData
    */
```

```
    public void sort()
    {
        for (int i = intData.length - 1; i > 0; i--)
        {
            placeMax(i);
        }
    }
    /**
        Bubble the largest value
        @param end location where unprocessed array ends
    */
    private void placeMax(int end)
    {
        for (int i = 0; i < end; i++)
            if (intData[i] > intData[i+1])
                swap(i, i+1);
    }
    /**
        interchange values between two locations of the array
        @param first one of the location to be interchanged
        @param second the other location to be interchanged
    */
    private void swap(int first, int second)
    {
        int temp = intData[first];
        intData[first] = intData[second];
        intData[second] = temp;
    }
}
```

The analysis of bubble sort algorithm is very similar to that of selection sort. Assume that the number of items to be sorted is n. Then, you need n – 1 comparisons to bubble the largest item. Once the largest item is placed at location n - 1, there are only n – 1 items in the unprocessed part of the array. Thus, n – 2 comparisons are required to bubble the second largest number and so on. Therefore, the total number of comparisons is $(n - 1) + (n - 2) + \ldots + 1 = n(n - 1)/2 - 0.5n^2 + 0.5n$. The growth rate of this expression is n^2. The worst-case time complexity of bubble sort is $O(n^2)$. The space complexity is O(n) and it is an in-place algorithm.

Self-Check

17. The time complexity of bubble sort algorithm is ________.
18. The space complexity of an in-place algorithm is ________.

CASE STUDY 10.1: MR. GRACE'S SORTED GRADE SHEET

Being a very popular teacher, Mr. Grace attracts many students to his course. Mr. Grace thought it would be great if he could sort the students based on their `gradeScore`. Thus, Mr. Grace modified his program to incorporate this feature. Mr. Grace looked into various sorting algorithms and decided to use selection sort. He realized that all he has to do is include the selection sort method in his `Course` class and invoke it as needed in his application program.

Although it is possible to compare students by inspecting their individual `gradeScore` values, it is better to have a `compareTo` method similar to one in the `String` class for the `Student` class. So, Mr. Grace decided to add a `compareTo` method in his `Student` class.

```
import java.util.*;
import java.text.DecimalFormat;
/**
    Keeps name, six test scores, grade score and letter grade
*/
public class Student
{
    private static final int ARRAY_SIZE = 6;

    private String firstName;
    private String lastName;
    private double testScores[] = new double[ARRAY_SIZE];
    private double gradeScore;
    private String grade;

    /**
        Compares implicit argument with explicit argument.
        negative int if implicit argument is smaller
        0 if they are equal up to two decimal places
        positive int if implicit argument is larger
        @param Student to compare
        @return -ve, 0 or +ve int if implicit parameter is <,
                                                           =, or >
    */
    public int compareTo(Student inStudent)
    {
        int diff;

        diff = (int)(this.gradeScore * 100
                                - inStudent.gradeScore * 100);
```

```
    return diff;
}

/**
    Loads student data to an instance of Student
    @param sca Scanner instance
*/
public void setStudentInfo(Scanner sc)
{
    firstName = sc.next();
    lastName = sc.next() ;

    for (int i = 0; i < testScores.length; i++)
        testScores[i] = sc.nextDouble() ;
}

/**
    Computes the sum of all test scores
    @return sum of all test scores
*/
private double computeSum()
{
    double sum = 0;

    for (double ts : testScores)
        sum = sum + ts ;

    return sum;
}

/**
    Computes the minimum of all test scores
    @return the minimum of all test scores
*/
private double findMin()
{
    double minimum = testScores[0];

    for (double ts : testScores)
        if (ts < minimum)
            minimum = ts;

    return minimum;
}
```

```
/**
    Computes the average test score ignoring the least score
*/
public void computeGradeScore()
{
    double adjustedTotal;

    adjustedTotal = computeSum() - findMin();
    gradeScore = adjustedTotal /( testScores.length - 1);
}

/**
    Create a String with all information on a student
    @return String with all information on a student
*/
public String createGradeReport()
{

    String str;
    DecimalFormat twoDecimalPlaces =
                          new DecimalFormat("0.00");

    str = firstName + "\t"+ lastName + "\t";
    if (str.length() < 10) str = str+ "\t";
    for (double ts : testScores)
        str = str + ts + "\t";
    str = str + twoDecimalPlaces.format(gradeScore);

    if (grade != null) str = str + "\t" + grade;
    return str;
}

/**
    Returns the average test score after ignoring the least
    @return average test score after ignoring the least
*/
public double getGradeScore()
{
    return gradeScore;
}

/**
    Returns the letter grade
    @return letter grade
*/
```

```
    public String getGrade()
    {
        return grade;
    }

    /**
        Mutator method for average test score
        @param inGradeScore new value of average test score
    */
    public void setGradeScore(double inGradeScore)
    {
        gradeScore = inGradeScore;
    }

    /**
        Mutator method for letter grade
        @param inGrade new value of letter grade
    */
    public void setGrade(String inGrade)
    {
        grade = inGrade;
    }

    /**
        toString method
        @return all information about student including letter
                                                        grade
    */
    public String toString()
    {
        return createGradeReport();
    }

}

import java.io.*;
import java.util.*;

/**
    Keeps information on all students
*/
public class Course
```

```
{
    private static final int CLASS_SIZE = 25;     // Maximum class
                                                                size
    private String courseNumber;                  // Course Number
    private String courseTitle;                   // Course Title
    private String term;                          // Course term
    private int numberOfStudents;
    private double courseAverage;
    private Student StudentList[] = new Student[CLASS_SIZE];
                    // An array to keep student information

    /**
        Sorts the attribute StudentList
    */
    public void sort()
    {
        int minIndex;
        for (int i = 0; i < StudentList.length - 1; i++)
        {
            minIndex = findMin(i);
            swap(minIndex, i);
        }
    }

    /**
        Find the smallest value from the given location
        @param start location to begin finding minimum
        @return index of the minimum
        intData[start]...intData[intData.length-1]
    */
    private int findMin(int start)
    {
        int minLoc = start;
        for (int i = start + 1; i < StudentList.length; i++)
            if (StudentList[i].compareTo(StudentList[minLoc])< 0)
                minLoc = i;
        return minLoc;
    }

    /**
        interchange values between two locations of the array
        @param first one of the location to be interchanged
        @param second the other location to be interchanged
    */
```

```
private void swap(int first, int second)
{
    Student temp = StudentList[first];
    StudentList[first] = StudentList[second];
    StudentList[second] = temp;
}

public void loadData(Scanner sc)
{
    Student st;
    int i = 0;

    courseNumber = sc.nextLine();
    courseTitle = sc.nextLine();
    term = sc.nextLine();
    while (sc.hasNext())
    {
        st = new Student();
        st.setStudentInfo(sc);
        StudentList[i] = st;
        i++;
    }
    numberOfStudents = i;
}

private void computeCourseAverage()
{
    int i;
    double sum = 0;

    for (i = 0; i < numberOfStudents; i++)
    {
        StudentList[i].computeGradeScore();
        sum = sum + StudentList[i].getGradeScore();
    }

    courseAverage = sum / numberOfStudents;
}

public void assignGrade()
{
    int i;
    double temp;
```

```
        computeCourseAverage();

        for (i = 0; i < numberOfStudents; i++)
        {
            temp = StudentList[i].getGradeScore();
            if (temp > 1.3 * courseAverage)
                StudentList[i].setGrade("A");
            else if (temp > 1.1 * courseAverage)
                StudentList[i].setGrade("B");
            else if (temp > 0.9 * courseAverage)
                StudentList[i].setGrade("C");
            else if (temp > 0.7 * courseAverage)
                StudentList[i].setGrade("D");
            else
                StudentList[i].setGrade("F");
        }

    }

    public void printGradeSheet(PrintWriter output)
    {
        int i;

        output.println("\t\t\t" + courseNumber);
        output.println("\t\t\t" + courseTitle);
        output.println("\t\t\t" + term);
        output.println("\t\t\tClass Average is " +
                                    courseAverage);
        for (i = 0; i < numberOfStudents; i++)
        {
            output.println(StudentList[i]);
        }

    }
}


import java.io.*;
import java.util.Scanner;

public class CourseGraded
{

    public static void main (String[] args)
    throws FileNotFoundException, IOException
```

```
    {
        Scanner scannedInfo
            = new Scanner(new File("C:\\courseData.dat"));
        PrintWriter outFile
            = new PrintWriter(new FileWriter
                                  ("C:\\courseData.out"));

        Course crs = new Course();
        crs.loadData(scannedInfo);
        crs.assignGrade();
        crs.sort();
        crs.printGradeSheet(outFile);
        outFile.close();

    }
}
```

CourseData.dat File:

```
CSC 221
Java Programming
Fall 2008
Kim Clarke          70.50 69.85 90.25 100.0 81.75 100.0
Chris Jones         78.57 51.25 97.45 85.67 99.75 88.76
Brian Wills         85.08 92.45 67.45 71.57 50.92 72.00
Bruce Mathew        60.59 87.23 45.67 99.75 72.12 100.0
Mike Daub           56.60 45.89 78.34 64.91 66.12 70.45
```

CourseData.out File:

```
CSC 221
Java Programming
Fall 2008
Class Average is 81.4944
Mike  Daub          56.6  45.89 78.34 64.91 66.12 70.45 67.28 D
Brian Wills         85.08 92.45 67.45 71.57 50.92 72.0  77.71 C
Bruce Mathew        60.59 87.23 45.67 99.75 72.12 100.0 83.94 C
Kim Clarke          70.5  69.85 90.25 100.0 81.75 100.0 88.50 C
Chris Jones         78.57 51.25 97.45 85.67 99.75 88.76 90.04 B
```

REVIEW

1. Searching and sorting are two most common tasks in data processing.
2. A sorting algorithm permutes or rearranges the elements of a collection in sorted order.
3. Once you have a sorted list, searching for an item becomes quite efficient.
4. Linear search algorithm can be written for a sorted or an unsorted array.

5. In the case of an unsorted array, you need to compare every item in the array to determine whether or not a specified item is in the array.
6. In the case of an ascending order sorted array, once an item that is greater than the one you are searching for is encountered, there is no need to search any further.
7. Binary search is the most efficient search algorithm on a sorted array.
8. The empirical approach to measure the performance of an algorithm is to write a program, run it using various test data, and measure the time it takes to finish the computation.
9. The analysis approach to measure the performance of an algorithm is mathematical in nature and in fact involves no programming at all.
10. The linear search is an `O(n)` algorithm.
11. The binary search is an `O(log`$_2$`n)` algorithm.
12. All three sorting algorithms presented in this chapter have `O(n`^{2}`)` time complexity.
13. There are sorting algorithms with `O(log`$_2$`n)` time complexity.
14. All three sorting algorithms presented in this chapter have `O(n)` space complexity.
15. A sort algorithm is said to be an in-place algorithm, if it does not require any additional arrays.
16. All three sorting algorithms presented in this chapter are in-place algorithms.

EXERCISES

1. Mark the following statements as true or false:
   a. In general, searching takes more time than sorting.
   b. The linear search algorithm is the best one possible on an unsorted array.
   c. Quick sort is one of the best algorithms for sorting.
   d. If someone asks you to pick an integer between 101 and 300, then your search space has 200 items.
   e. The analysis approach to determining the efficiency of an algorithm involves writing programs and so on.
2. Assume that array `L` has the following 10 items: 8, 13, 6, 14, 28, 29, 35, 12, 40, and 17. Illustrate the algorithm in a table format as presented in this chapter.
   a. Let the search item be 39. Apply linear search.
   b. Let the search item be 12. Apply linear search.
   c. Let the search item be 1. Apply linear search.
   d. Apply selection sort.
   e. Apply insertion sort.
   f. Apply bubble sort.
   g. Assume that array is sorted. Let the search item be 39. Apply binary search.

h. Assume that array is sorted. Let the search item be 12. Apply binary search.

i. Assume that array is sorted. Let the search item be 1. Apply binary search.

3. Assume that array `L` has the following 11 items: 7, 14, 2, 18, 39, 92, 88, 72, 41, 71, and 5. Illustrate the algorithm in a table format as presented in this chapter.

   a. Let the search item be 9. Apply linear search.

   b. Let the search item be 100. Apply linear search.

   c. Let the search item be 92. Apply linear search.

   d. Apply selection sort.

   e. Apply insertion sort.

   f. Apply bubble sort.

   g. Assume that array is sorted. Let the search item be 9. Apply binary search.

   h. Assume that array is sorted. Let the search item be 100. Apply binary search.

   i. Assume that array is sorted. Let the search item be 92. Apply binary search.

4. Consider the linear search method for a sorted array given in this chapter. Is the order of `if` statements important?

5. Consider the linear search method for a sorted array given in this chapter. Illustrate through three different examples how each of the three return statements are executed.

6. Assume that array `L` has the following 12 items: 2, 4, 7, 10, 14, 18, 22, 24, 29, 40, 45, and 57. What value of the search item will result in

   a. The maximum number of comparisons in the case of linear search algorithm?

   b. The minimum number of comparisons in the case of linear search algorithm?

   c. The maximum number of comparisons in the case of binary search algorithm?

   d. The minimum number of comparisons in the case of binary algorithm?

7. Give a sample data consisting of 10 values that will result in

   a. The maximum number of comparisons in the case of selection sort.

   b. The minimum number of comparisons in the case of selection sort.

   c. The maximum number of comparisons in the case of insertion sort.

   d. The minimum number of comparisons in the case of insertion sort.

   e. The maximum number of comparisons in the case of bubble sort.

   f. The minimum number of comparisons in the case of bubble sort.

8. What is the time complexity of an algorithm if the time complexity satisfies the following equations:

   a. $T(n) = T(n - 2) + 1$ and $T(1) = T(0) = 1$.

   b. $T(n) = T(n - 3) + 1$ and $T(2) = T(1) = T(0) = 1$.

   c. $T(n) = T(n/2) + 2$ and $T(1) = T(0) = 1$.

   d. $T(n) = T(n/3) + 2$ and $T(1) = T(0) = 1$.

PROGRAMMING EXERCISES

1. Modify the class `LinearSearch` so that searching can be done on a `Double` array.
2. Modify the class `LinearSearch` so that searching can be done on a `Double` `ArrayList`.
3. Modify the class `BinarySearch` so that searching can be done on a `Double` array.
4. Modify the class `BinarySearch` so that searching can be done on a `Double` `ArrayList`.
5. Modify the class `BinarySearch` so that searching can be done on an array of `Student`.
6. Modify the class `BinarySearch` so that searching can be done on an `ArrayList` of `Student`.
7. Modify the selection sort so that if `minIndex` and `i` are the same, swap is not necessary and as such no action is carried out.
8. Instead of finding the minimum value, selection sort can be modified to find maximum and minimum alternatively. Implement such a variation of the selection sort on an `int` array.
9. Perform an empirical comparison of selection sort with new selection sort described in Programming Exercise 8.
10. Perform an empirical comparison of selection sort, insertion sort, and bubble sort.
11. Write a program comparing the number of array location accessed by each of the sorting algorithms. (*Hint*: Let sorting method return this value. Memory access involves both reading from the location and writing back to the location.)
12. Create a class `EnglishFrenchDictionary` with the following three methods: `loadData`, `toEnglish`, and `toFrench`. `LoadData` is read in a pair of values from a file, where each pair is an English word and its equivalent French word. Test your program on a file. If an English word has no equivalent word in the `EnglishFrenchDictionary`, then the translation should print the word as it is.

ANSWERS TO SELF-CHECK

1. Binary search, linear search.
2. Binary search
3. Change `(intData[i] > searchItem)` to `(intData[i] < searchItem)`.
4. No
5. True
6. False
7. doubled
8. increased by a very small amount

9. $O(n)$
10. $O(\log_2 n)$
11. $O(n\log_2 n)$
12. $O(n\log_2 n)$
13. last
14. $O(n^2)$
15. unsorted, sorted
16. $O(n^2)$
17. $O(n^2)$
18. $O(n)$

CHAPTER 11

Defensive Programming

In this chapter you learn

- Java concepts
  - Exceptions, checked and unchecked exceptions, throwing and catching of exceptions, and user-defined exception classes
- Programming skills
  - Write robust programs by incorporating exception handling

INTRODUCTION

One of the most important qualities a program must possess is that of correctness. That is, if the input values satisfy the preconditions, the program will produce the output consistent with the specification. However, program correctness does not guarantee any behavior on the part of the program if one or more input values violate their respective preconditions. All of us know the importance of defensive driving. Anticipating the unexpected can definitely save us from many catastrophic accidents. Similarly, irrespective of whatever precautions we take, some unexpected event or data can lead to catastrophic failure of the entire software. To avoid such an eventuality, you need to cultivate the habit of defensive programming.

An exception is an abnormal condition that occurs during the program execution. Following are some of the common exceptions:

1. File not found or the failure to locate the input file specified.
2. Input mismatch such as the presence of a `boolean` literal when an `int` is expected.
3. Division by zero due to the presence of a denominator that is zero.

A reliable program must not only be logically correct but also include code to handle exceptional conditions.

During program execution, if an exception occurs, we say an exception is thrown. Once the exception is thrown, the program will terminate unless there is a code to handle the exception. Execution of the matching exception-handling code is called the catching of the exception.

Self-Check

1. An exception is an ________ that occurs during the program execution.
2. Once the exception is thrown, the program will ________ unless there is a code to handle the exception.

EXCEPTION AND ERROR

Recall that in Java, every class is a subclass of the `Object` class. One of the subclasses of the `Object` class is the `Throwable` class. The `Throwable` class has two subclasses, `Exception` class and `Error` class. Exceptions are thrown if an abnormal condition that can possibly be corrected occurs during the program execution. The presence of a `boolean` literal when an `int` literal is expected is an example of an abnormal condition that can be corrected. Errors are thrown if an abnormal condition that cannot be corrected occurs during the program execution. The system can no longer allocate additional memory requested by the program in an abnormal condition that cannot be corrected.

Self-Check

3. The `Throwable` class has two subclasses, ________ class and ________ class.
4. The presence of a `double` literal when an `int` literal is expected is an example of an ________.

Unchecked and Checked Exceptions

Java has a number of predefined exceptions and they are all subclasses of the `Exception` class (Figure 11.1). There are two types of exceptions in Java, *unchecked* and *checked exceptions.* All unchecked exceptions have `RuntimeException` class in their inheritance hierarchy. Most commonly occurring exceptions are unchecked exceptions and quite often they represent an error in the program such as passing an invalid argument during method invocation. `IndexOutOfBoundsException`, which you have seen in Chapter 9, is an example of unchecked exception. Observe that there is a logical error in this situation. In a typical program, there can be many unchecked exceptions and they can occur in many statements. Therefore, from a cost–reward perspective, it is not worth checking them. Java compiler does not require you to declare or `catch` an unchecked exception. Thus, as a programmer you have the option. You can handle an unchecked exception similar to a checked exception.

A checked exception does not have `RuntimeException` class in its inheritance hierarchy. `FileNotFoundException` and `IOException`, which you have seen in Chapter 5, are examples of a checked exception. A checked exception is not a logical error. Rather, it represents an invalid condition occurring in areas outside the control of the

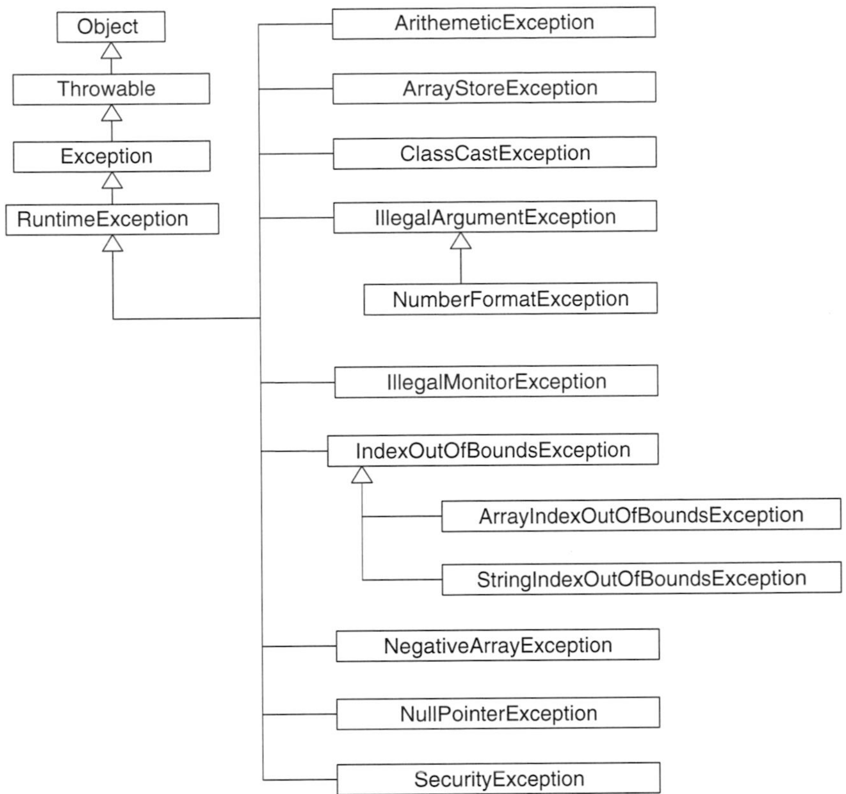

FIGURE 11.1 Part of Java's exception hierarchy.

program such as absence of files. A checked exception must be either declared or caught. In other words, the compiler checks for the declaration of the exception or the corresponding exception-handling code in your program. In fact, you have seen the declaration of exceptions in Chapter 5. In Example 5.1, you had declared two exceptions, `FileNotFoundException` and `IOException`, for the main method.

```
public static void main (String[] args) throws
          FileNotFoundException, IOException
```

The general syntax for declaring an exception is to include the method header with a `throws` clause that lists all exceptions. Recall that in Java, list items are always separated by a comma.

Table 11.1 lists some of the most commonly occurring exceptions.

Self-Check

5. Java has a number of predefined exceptions and they are all subclasses of the ________ class.
6. All unchecked exceptions have ________ class in their inheritance hierarchy.

TABLE 11.1 Some of the Most Commonly Occurring Exceptions

Class	Description	Checked?
`ArithmeticException`	Division by zero and other arithmetic exceptions	No
`ArrayIndexOutOfBoundsException` subclass of `IndexOutOfBoundsException`	Index is either negative or greater than or equal to array length	No
`FileNotFoundException`	File cannot be found	Yes
`IllegalArgumentException`	Invoking a method with illegal parameter	No
`IndexOutOfBoundsException`	Index is either negative or greater than or equal to the length	No
`InputMismatchException`	The token does not match the pattern for the expected type	No
`IOException`	All exceptions associated with I/O operations	Yes
`NullPointerException`	The reference variable has not been instantiated	No
`NumberFormatException` subclass of `IllegalArgumentException`	An illegal number format in the actual parameter	No
`StringIndexOutOfBoundsException` subclass of `IndexOutOfBoundsException`	Index is either negative or greater than or equal to string length	No

THROWING AND CATCHING OF EXCEPTIONS

An exception is thrown either by the system or by the user. For example, if there is an input mismatch, such as the expected input is an integer and instead the actual input happens to be a character, the system will `throw` the unchecked exception `InputMismatchException`. However, if the input value is out of range as far as the problem specification is considered, the system may not throw any exception. In such situations, it is the programmer's responsibility to provide the necessary code to throw appropriate exceptions.

As mentioned before, all checked exceptions need to be either declared or caught. If you do not `catch` an exception, Java virtual machine (JVM) will catch the exception and the program will terminate by printing the stacktrace.

To fully understand many of these concepts and what we mean by `stacktrace`, let us consider an example.

Example 11.1

Suppose you want to find the sum 1 + 1/2 + ⋯ + 1/n for a given input n. Clearly, n has to be a positive integer. Here is the class with corresponding application program and some sample test runs:

```
public class ComputeHarmonicOne
{
    /**
        Returns the sum 1/n + ...+ 1/2 + 1
        @params n a positive integer
        @return the sum 1/n + ...+ 1/2 + 1
    */
    public double harmonic(int n)
    {
        double sum = 1.0 / n;
        for (int k = n-1; k > 0; k--)
        {
            sum = sum + 1.0 / k;
        }
        return sum;
    }
}
import java.util.Scanner;
public class ComputeHarmonicOneApplication
{
    /**
        This application tests ComputeHarmonicOne
        class
    */
    public static void main(String[] args)
    {
        Scanner scannedInfo = new Scanner(System.in);
        System.out.print("Enter a positive integer : ");
        int number = scannedInfo.nextInt();
        System.out.println();
        ComputeHarmonicOne  computeHarmonic = new
                                ComputeHarmonicOne();
        double result = computeHarmonic.harmonic(number);
        System.out.println("The sum of first "+number+
                         " harmonic sequence is "+ result);
    }
}
```

Sample Run 1

```
Enter a positive integer : t
Exception in thread "main" java.util.InputMismatchException
```

```
        at java.util.Scanner.throwFor(Scanner.java:819)
        at java.util.Scanner.next(Scanner.java:1431)
        at java.util.Scanner.nextInt(Scanner.java:2040)
        at java.util.Scanner.nextInt(Scanner.java:2000)
        at ComputeHarmonicOneApplication.main
                            (ComputeHarmonicOneApplication.java:13)
```

Sample Run 2

```
Enter a positive integer : 0

The sum of first 0 harmonic sequence is Infinity
```

Sample Run 3

```
Enter a positive integer : -2

The sum of first -2 harmonic sequence is -0.5
```

From the sample runs of Example 11.1, the following points are worth mentioning:

1. In the case of sample run 1, the program encountered the `InputMismatchException` exception. Although the expected input was an integer, the actual input entered was a character. This led to throwing of the `InputMismatchException` exception. Since the user has not made any provisions to `catch` such an exception, the system used its default exception-handling technique of printing the `stacktrace`. The `stacktrace` not only identifies the type of exception thrown, but it also indicates the line (in this case, line 13 of the `main`) where an exception was encountered.
2. In the case of sample run 2, a division by zero has occurred. But Java proceeded without causing any problem. Note that no exception was thrown by the Java system. However, as a programmer, you can make provisions to `throw` and `catch` exceptions in such situations.
3. In the case of sample run 3, nothing illegal has happened. However, it does not make sense to compute the harmonic sequence sum of first `n` terms when `n` is negative. Once again, it is the programmer's responsibility to `throw` and `catch` exceptions.

Self-Check

7. All checked exceptions need to be either ________ or ________.
8. If a program does not `catch` an exception, JVM will `catch` the exception and the program will ________ by printing the ________

Throwing Exception

You can `throw` an exception using the `throw` statement. The syntax template for the `throw` statement is as follows:

```
throw exceptionInstance;
```

Note that throw is a keyword in Java. Throwing an exception involves creating an instance of the appropriate exception class and then using it in a throw statement.

Good Programming Practice 11.1

Use a helpful message as the actual parameter in the constructor of the appropriate exception class so that the user can take the most suitable action.

Example 11.2

```
public class ComputeHarmonicTwo
{
    /**
        Returns the sum 1/n + ...+ 1/2 + 1
        @params n a positive integer
        @return the sum 1/n + ...+ 1/2 + 1
    */
    public double harmonic(int n)
    {
        if ( n < 1)
            throw new IllegalArgumentException
                        ("Argument cannot be less than 1");
        double sum = 1.0 / n;
        for (int k = n-1; k > 0; k--)
        {
            sum = sum + 1.0 / k;
        }
        return sum;
    }
}
```

```
import java.util.Scanner;
public class ComputeHarmonicTwoApplication
{
    /**
        This application tests ComputeHarmonicTwo
        class
    */
    public static void main(String[] args)
    {
        Scanner scannedInfo = new Scanner(System.in);
        System.out.print("Enter a positive integer : ");
        int number = scannedInfo.nextInt();
```

```
            System.out.println();
            ComputeHarmonicTwo computeHarmonic = new
                                                   ComputeHarmonicTwo();
            double result = computeHarmonic.harmonic(number);
            System.out.println("The sum of first "+ number +
                          " harmonic sequence is "+ result);
        }
    }
```

Sample Run

```
Enter a positive integer : 0

Exception in thread "main"
                          java.lang.IllegalArgumentException:
                               Argument cannot be less than 1
   at ComputeHarmonicTwo.harmonic(ComputeHarmonicTwo.java:11)
   at ComputeHarmonicTwoApplication.main
                        (ComputeHarmonicTwoApplication.java:16)
```

Observe that with this modification, the program will no longer compute the harmonic sequence sum if n is not a positive integer. Further, inspect the stacktrace produced. The last line indicates that the error has occurred during the execution of Line 16 of the main. The line above indicates that the exception was thrown by Line 11 of the file ComputeHarmonicTwo.java and it occurred during the invocation of the method harmonic. Thus, stacktrace can clearly identify the statement, execution of which resulted in the throwing of an exception. Note that the message string "Argument cannot be less than 1" is also printed as part of the stacktrace. In fact, getMessage and printStackTrace are two methods of the Throwable class, the superclass of Exception class. Thus, those two methods are always available to the programmer.

Self-Check

9. Throwing an exception involves creating an instance of the appropriate __________ class and then using it in a __________ statement.
10. Two methods of the Throwable class are __________ and __________.

Catching Exception

To catch an exception, Java provides a try/catch/finally structure. The general syntax of the try/catch/finally structure is as follows:

```
try
{
    statementsTry
}
```

```
catch (ExceptionOne e1)
{
    statementsOne
}
[[catch (ExceptionTwo e2)
{
    statementsTwo
}]
.
.
.
[catch (ExceptionN eN)
{
    statementsN
}]]
[finally
{
    statementsFinally
}]
```

Note that `try`, `catch`, and `finally` are reserved words and `statementsTry`, `statementsOne`, `statementsTwo`, and so on can be one or more executable statements. The semantics of the `try/catch/finally` structure (Figure 11.2) can be explained as follows. If an exception is thrown inside a `try` block, execution of the statements in that `try` block stops immediately and control is transferred to the first matching `catch`, and then to the `finally` block. Note that unlike the method invocation, the control never returns to the `try` block. For example, the `statementsTry` is executed in sequence. If no exception is thrown, all `catch` blocks are skipped and `finally` block is executed. However, if an exception of the type `ExceptionT` is thrown during the execution of any of the statements in the `try` block and `ExceptionK` is the very first exception class among `ExceptionOne`, `ExceptionTwo`, ..., `ExceptionN` such that `ExceptionT` is either `ExceptionK` or a subclass of `ExceptionK`, then the `catch` block corresponding to `ExceptionK` followed by statements in `finally` block will be executed. Thus, the optional `finally` block will always be executed. If an exception is thrown, at most one `catch` block is being executed.

We now present a sequence of examples illustrating the `try/catch/finally` structure.

Example 11.3

This being the very first example on `try/catch/finally` structure, we apply the most simplistic technique. Even though two types of exceptions can be thrown, we use the fact that all exceptions are in fact subclasses of the `Exception` class; thus, we use just one `catch` block instead of multiple `catch` blocks. Further, since

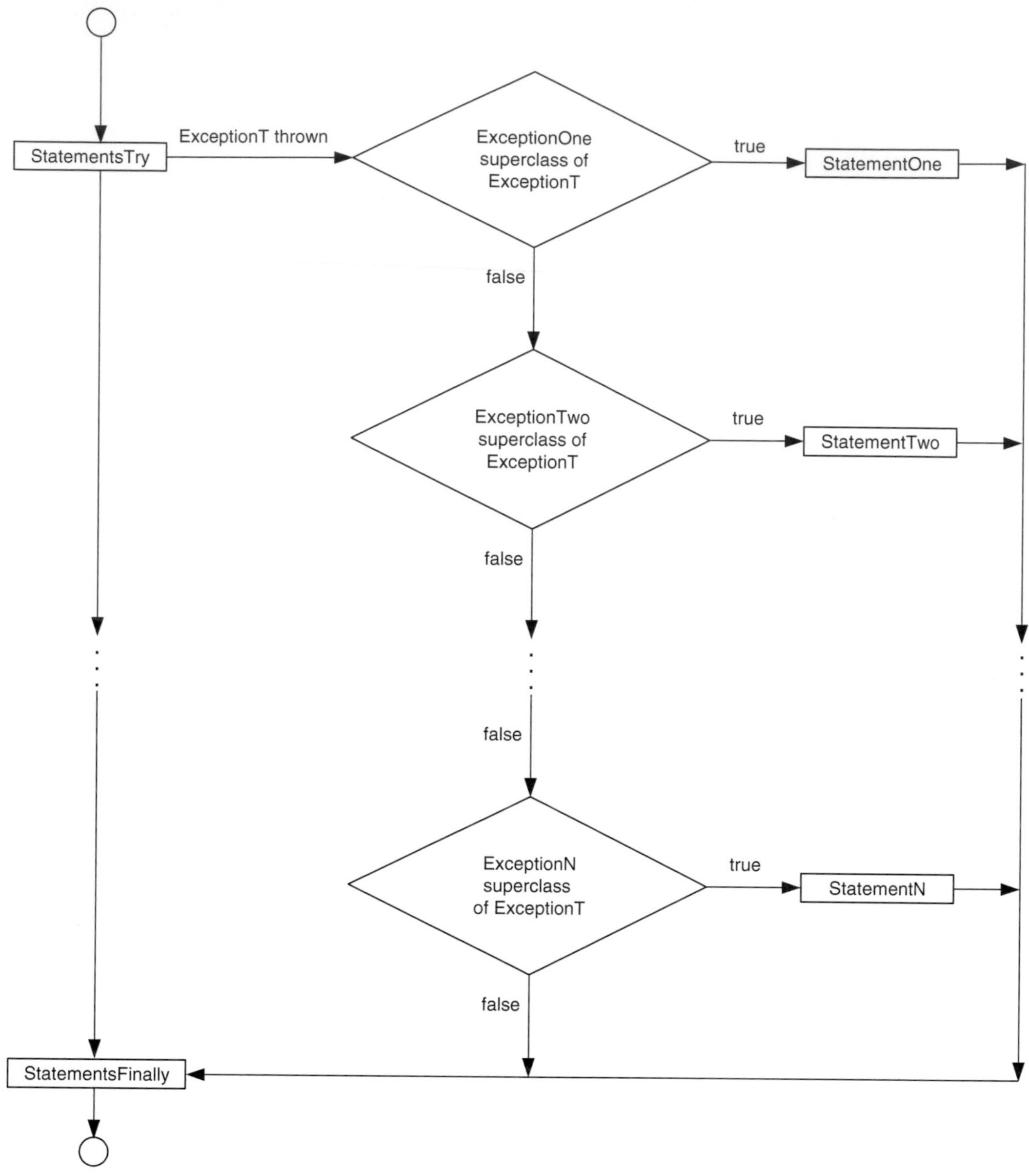

FIGURE 11.2 The `try/catch/finally` structure.

`finally` block is optional, we omit that also in this example. The user is given one chance to correct the abnormal condition that caused the exception.

```
import java.util.*;

/**
    Demonstration of try/catch blocks
    This version has one catch block
*/
```

```
public class TryCatchVersionOne
{

    public static void main(String[] args)
    {
        int number = 1;
        double result = 1.0;

        Scanner scannedInfo = new Scanner(System.in);
        ComputeHarmonicTwo computeHarmonic = new
                                ComputeHarmonicTwo();
        System.out.print("Enter a positive integer : ");

        try
        {
            number = scannedInfo.nextInt();
            System.out.println();
            result = computeHarmonic.harmonic(number);
        }
        catch (Exception e)
        {
            System.out.print
                    ("Enter a POSITIVE INTEGER please : ");
            scannedInfo.nextLine();
            number = scannedInfo.nextInt();
            System.out.println();
            result = computeHarmonic.harmonic(number);

        }
        System.out.println("The sum of first "+ number +
                    " harmonic sequence is "+ result);
    }
}
```

Sample Run 1

```
Enter a positive integer : t
Enter a POSITIVE INTEGER please : 3

The sum of first 3 harmonic sequence is 1.8333333333333333
```

Sample Run 2

```
Enter a positive integer : -3

Enter a POSITIVE INTEGER please : 5

The sum of first 5 harmonic sequence is 2.283333333333333
```

Example 11.4

In this example, we illustrate the use of multiple catch blocks. Two types of exceptions can be thrown, and therefore we use two separate catch blocks.

```
import java.util.*;

/**
    Demonstration of try/catch blocks
    This version has multiple catch blocks
*/
public class TryCatchVersionTwo
{
    public static void main(String[] args)
    {
        int number = 1;
        double result = 1.0;
        Scanner scannedInfo = new Scanner(System.in);
        ComputeHarmonicTwo computeHarmonic = new
                                    ComputeHarmonicTwo();
        System.out.print("Enter a positive integer : ");

        try
        {
            number = scannedInfo.nextInt();
            System.out.println();
            result = computeHarmonic.harmonic(number);
        }

        catch (IllegalArgumentException e)
        {
            System.out.println
                  ("Data entered a 0 or a negative integer");
            System.out.print
                     ("Enter a POSITIVE integer please : ");
            scannedInfo.nextLine();
            number = scannedInfo.nextInt();
            System.out.println();
            result = computeHarmonic.harmonic(number);

        }
        catch (InputMismatchException e)
        {
            System.out.println();
            System.out.println
                           ("Data entered is not an integer ");
```

```
            System.out.print
                       ("Enter a POSITIVE INTEGER please : ");
            scannedInfo.nextLine();
            number = scannedInfo.nextInt();
            System.out.println();
            result = computeHarmonic.harmonic(number);

        }
        System.out.println("The sum of first "+ number +
                                   " harmonic sequence is "+ result);
    }
}
```

Sample Run 1

```
Enter a positive integer : t
Data entered is not an integer
Enter a POSITIVE INTEGER please : 3
The sum of first 3 harmonic sequence is 1.8333333333333333
```

Sample Run 2

```
Enter a positive integer : -3
Data entered a 0 or a negative integer
Enter a POSITIVE integer please : 5
The sum of first 5 harmonic sequence is 2.283333333333333
```

Note 11.1 Having multiple `catch` blocks enable the programmer to write the `catch` block code specific to the exception. Thus, each type of exception can be handled differently.

Example 11.5

Recall that if an exception of the type `ExceptionT` is thrown during the execution of any of the statements in the `try` block and let `ExceptionK` be the very first exception class among `ExceptionOne`, `ExceptionTwo`, ..., `ExceptionN` such that `ExceptionT` is the same as `ExceptionK` or `ExceptionK` is in the inheritance hierarchy of `ExceptionT`, then the `catch` block corresponding to `ExceptionK` followed by `finally` block will be executed. Therefore, the `catch` blocks corresponding to exception subclasses must appear before the superclass. Otherwise, the `catch` block corresponding to the subclass will never be executed.

Consider Example 11.3. There are two `catch` blocks corresponding to two types of exceptions, `IllegalArgumentException` and `InputMismatchException`. They both are subclasses of the `Exception` class. Therefore, the `catch` block corresponding to `Exception` must not appear before the `catch` block corresponding to either one of the exceptions, IllegalArgumentException or InputMismatchException.

The following code is thus incorrect and can be corrected by placing the catch block corresponding to Exception as the last catch block:

```
import java.util.*;
/**
    Demonstration order of catch blocks
    The order of catch blocks is not correct
    Therefore version will not compile
*/
public class TryCatchVersionThree
{
    public static void main(String[] args)
    {
        int number = 1;
        double result = 1.0;
        Scanner scannedInfo = new Scanner(System.in);
        ComputeHarmonicTwo computeHarmonic = new
                                ComputeHarmonicTwo();
        System.out.print("Enter a positive integer : ");
        try
        {
            number = scannedInfo.nextInt();
            System.out.println();
            result = computeHarmonic.harmonic(number);
        }
        catch (Exception e)
        {
            System.out.print
                    ("Enter a POSITIVE INTEGER please : ");
            scannedInfo.nextLine();
            number = scannedInfo.nextInt();
            System.out.println();
            result = computeHarmonic.harmonic(number);
        }
        catch (IllegalArgumentException e)
        {
            System.out.println
                ("Data entered a 0 or a negative integer");
            System.out.print
                    ("Enter a POSITIVE integer please : ");
```

```
            scannedInfo.nextLine();
            number = scannedInfo.nextInt();
            System.out.println();
            result = computeHarmonic.harmonic(number);
        }
        catch (InputMismatchException e)
        {
            System.out.println();
            System.out.println
                        ("Data entered is not an integer ");
            System.out.print
                    ("Enter a POSITIVE INTEGER please : ");
            scannedInfo.nextLine();
            number = scannedInfo.nextInt();
            System.out.println();
            result = computeHarmonic.harmonic(number);
        }
        System.out.println("The sum of first "+number+
                        " harmonic sequence is "+ result);
    }
}
```

If you compile the above code, you may get an error message similar to the following:

```
C:\...\TryCatchVersionThree.java:32:exception
java.lang.IllegalArgumentException has already been caught
                        catch (IllegalArgumentException e)
                           ^

C:\...\TryCatchVersionThree.java:41:exception
java.util.InputMismatchException has already been caught
                        catch (InputMismatchException e)
                           ^

2 errors
```

Example 11.6

The aim of this example is to show the basic technique involved in allowing unlimited prompts to get correct input values. Starting with the program presented in Example 11.3, we have the following:

```
import java.util.*;

/**
    Demonstration of Try/Catch/Finally blocks
    This version allows unlimited data entry
```

```
*/
public class TryCatchFinally
{
    public static void main(String[] args)
    {
        int number = 1;
        double result = 1.0;
        boolean gotData = false;
        Scanner scannedInfo = new Scanner(System.in);
        ComputeHarmonicTwo computeHarmonic = new
                                    ComputeHarmonicTwo();
        System.out.print("Enter a positive integer : ");
        while (!gotData)
        {
            try
            {
                number = scannedInfo.nextInt();
                System.out.println();
                result = computeHarmonic.harmonic(number);
                gotData = true;
            }
            catch (IllegalArgumentException e)
            {
                System.out.println
                  ("Data entered a 0 or a negative integer");
                System.out.print
                      ("Enter a POSITIVE integer please : ");
            }
            catch (InputMismatchException e)
            {
                System.out.println();
                System.out.println
                        ("Data entered is not an integer ");
                System.out.print
                      ("Enter a POSITIVE INTEGER please : ");
            }
            finally
            {
                scannedInfo.nextLine();
            }
        }
```

```
        System.out.println("The sum of first "+number+
                     " harmonic sequence is "+ result);
    }
}
```

Output

```
Enter a positive integer : t

Data entered is not an integer
Enter a POSITIVE INTEGER please : -3

Data entered a 0 or a negative integer
Enter a POSITIVE integer please : 2.25

Data entered is not an integer
Enter a POSITIVE INTEGER please : A

Data entered is not an integer
Enter a POSITIVE INTEGER please : 5

The sum of first 5 harmonic sequence is 2.283333333333333
```

Self-Check

11. If no exception is thrown, all ________ blocks are skipped and ________ block is executed.
12. If an exception is thrown, at most one ________ block followed by ________ block is executed.

Advanced Topic 11.1: Design Options for `catch` Block

During the program execution, if an exception occurs within a `try` block, one of the `catch` block is executed. A `catch` block can be designed in three different ways:

1. *Complete handling.* Process the exception completely.
2. *Partial handling.* Process partially and then `throw` the same exception or a different exception.
3. *Minimal handling.* `throw` the same exception.

All the examples you have seen so far fall under complete handling. If the `catch` block is unable to completely rectify the abnormal condition, then one option is to perform whatever tasks possible at the current context and let the calling context take care of the rest of the tasks. In a sense, minimal handling is just a special case of partial handling. Therefore, in this section, we only provide an example of partial handling.

Example 11.7

In this example, the harmonic method does a partial handling of the exception and then throws the same exception so that the method main can handle the rest; in this case, reading the input again.

```
public class ComputeHarmonicThree
{
    /**
        Returns the sum 1/n + ... + 1/2 + 1
        @params n a positive integer
        @return the sum 1/n + ... + 1/2 + 1
    */
    public double harmonic(int n)
    {

        try
        {
            if (n < 1)
            throw new IllegalArgumentException
                         ("Argument cannot be less than 1");
        }
        catch (IllegalArgumentException e)
        {
            System.out.println("Illegal Argument Exception
                                              occurred");
            System.out.println("Partially handled.");
            System.out.print("Let the calling environment");
            System.out.println(" do the rest");

            throw e;
        }

        double sum = 1.0 / n;
        for (int k = n-1; k > 0; k--)
        {
        sum = sum + 1.0 / k;
        }
        return sum;
    }
}

import java.util.*;

/**
    Demonstration of try/catch blocks
```

```
    This version has one catch block
*/
public class CatchDesignOption
{
    public static void main(String[] args)
    {
        int number = 1;
        double result = 1.0;

        Scanner scannedInfo = new Scanner(System.in);
        ComputeHarmonicThree computeHarmonic
                              = new ComputeHarmonicThree();

        System.out.print("Enter a positive integer : ");

        try
        {
            number = scannedInfo.nextInt();
            System.out.println();
            result = computeHarmonic.harmonic(number);
        }
        catch (Exception e)
        {
            System.out.print
                      ("Enter a POSITIVE INTEGER please : ");
            scannedInfo.nextLine();
            number = scannedInfo.nextInt();
            System.out.println();
            result = computeHarmonic.harmonic(number);
        }
        System.out.println("The sum of first "+number+
                           " harmonic sequence is "+ result);
    }
}
```

Output

```
Enter a positive integer : -10

Illegal Argument Exception occurred
Partially handled.
Let the calling environment do the rest
Enter a POSITIVE INTEGER please : 10

The sum of first 10 harmonic sequence is 2.9289682539682538
```

Advanced Topic 11.2: User-Defined Exception Class

You can extend any exception class of Java to tailor your own needs. Recall that the `Exception` class is a subclass of `Throwable` class and therefore all exception classes inherit `getMessage` and `printStackTrace` methods of the `Throwable` class. In particular, the user-defined exception class will also inherit these two methods. Quite often, all you need is to provide the constructors for the exception class you create.

Example 11.8

In this example, we create a new exception class called `NegativeInteger Exception` by extending the `Exception` class and use it in the main.

```
public class NegativeIntegerException
{
    public NegativeIntegerException()
    {
        super("Non-negative integer is required");
    }
}

public class ComputeHarmonicFour
{
    /**
        Returns the sum 1/n + ...+ 1/2 + 1
        @params n a positive integer
        @return the sum 1/n + ...+ 1/2 + 1
    */
    public double harmonic(int n)
    {
        try
        {
            if (n < 1)
                        throw new NegativeIntegerException();
        }
        double sum = 1.0 / n;
        for (int k = n-1; k > 0; k--)
        {
            sum = sum + 1.0 / k;
        }
        return sum;
    }
}
```

```
import java.util.*;

/**
    Demonstration of try/catch blocks
    This version has multiple catch blocks
*/
public class UserDefinedIllustrated
{
    public static void main(String[] args)
    {
        int number = 1;
        double result = 1.0;
        boolean gotData = false;

        Scanner scannedInfo = new Scanner(System.in);
        ComputeHarmonicFour computeHarmonic
                               = new ComputeHarmonicFour();

        System.out.print("Enter a positive integer : ");
        while (!gotData)
        {
            try
            {
                number = scannedInfo.nextInt();
                System.out.println();
                result = computeHarmonic.harmonic(number);
                gotData = true;
            }
            catch (NegativeIntegerException e)
            {
                System.out.println
                   ("Data entered a 0 or a negative integer");
                System.out.print
                      ("Enter a POSITIVE integer please : ");
            }
            catch (InputMismatchException e)
            {
                System.out.println();
                System.out.println
                         ("Data entered is not an integer ");
                System.out.print
                      ("Enter a POSITIVE INTEGER please : ");
            }
            finally
```

```
                {
                    scannedInfo.nextLine();
                }
            }
                System.out.println("The sum of first "+number+
                        " harmonic sequence is "+ result);
        }
    }
```

Output

```
Enter a positive integer : -20

Non-negative integer is required
Enter a POSITIVE INTEGER please : 20

The sum of first 20 harmonic sequence is 3.597739657143682
```

Advanced Topic 11.3: Design Options for Exception Handling

There are three different choices for exception handling. They are as follows:

1. *Terminate the program.* In certain situations, the best option is indeed to terminate the program. Assume that your program is supposed to process data received from an external source. If the source is not feeding in the correct data, there is no reason to continue. Thus, the best choice is to terminate the program after sending an appropriate error message.
2. *Rectify and continue.* In many cases, it may be possible to rectify the error. A classical example is the user authorization. If you enter a wrong password, the system will not terminate. Rather, it will prompt the user for the correct password.
3. *Record and continue.* In this case, you may record the abnormal condition in a log file and proceed further. This option is most suitable in a batch-processing system.

REVIEW

1. One of the most important qualities a program must possess is that of correctness.
2. An exception is an abnormal condition that occurs during the program execution.
3. During program execution, if an exception occurs, we say an exception is thrown.
4. Once the exception is thrown, the program will terminate unless there is a code to handle the exception.
5. Execution of the matching exception-handling code is called the catching of the exception.
6. Java has a number of predefined exceptions and they are all subclasses of the `Exception` class.

7. There are two types of exceptions in Java, unchecked and checked exceptions.
8. All unchecked exceptions have `RuntimeException` class in their inheritance hierarchy.
9. Most commonly occurring exceptions are unchecked exceptions.
10. A checked exception does not have `RuntimeException` class in its inheritance hierarchy.
11. An exception is thrown either by the system or by the user.
12. All checked exceptions needs to be either declared or caught.
13. Throwing an exception involves creating an instance of the appropriate exception class and then using it in a `throw` statement.
14. The optional `finally` block will always be executed.
15. If an exception is thrown, at most one `catch` block is executed.
16. A `catch` block can be designed in three different ways: process the exception completely, process partially and then `throw` the same exception or a different exception, and `throw` the same exception.
17. `Exception` class is a subclass of `Throwable` class and therefore all exception classes inherit `getMessage` and `printStackTrace` methods of the `Throwable` class.

EXERCISES

1. Mark the following statements as true or false:
   a. In Java, `Exception` is a subclass of `Error`.
   b. One of the subclasses of `Exception` is `Throwable`.
   c. `PrintStackTrace` is a method of the `Throwable` class.
   d. Every checked exception must be either declared or caught in the program.
   e. If an exception occurs, at most one `catch` block is executed.
   f. The `finally` block is executed only if none of the `catch` block is executed.
   g. Once an exception is caught, it cannot be thrown.
   h. All unchecked exceptions are ignored by the system.
   i. As soon as an exception is thrown, the `try` block is exited.
   j. Once an exception is handled, the control goes back to the `try` block and starts executing the very next statement after the one that caused the exception.
   k. In Java, user-defined exception classes are allowed.
2. Fill in the blanks.
   a. An abnormal condition that can be corrected is an ________ and an abnormal condition that cannot be corrected is an ________.
   b. Two methods of the `Throwable` class are ________ and ________.

c. An exception can be either a __________ or an __________ one.

d. `Throwable` class has two subclasses __________ and __________.

e. Every user-defined exception must have __________ class in its inheritance hierarchy.

f. The keyword __________ is used to throw an exception and the keyword __________ is used to declare an exception, respectively.

3. Answer questions on the basis of the following error message:

   Exception in thread "main" `java.lang.ArrayIndexOutOfBoundsException:` 101 at `ArrayIndexOutOfBounds.main(ArrayIndexOutOfBounds.java:12)`

   a. Is the above error a syntax error?
   b. Is the above error an execution error?
   c. What is the exact reason for the above error?
   d. What is the significance of 101?
   e. What is the significance of 12?
   f. What is the significance of `ArrayIndexOutOfBounds.java`?
   g. What is the significance of `java.lang`?
   h. What is the significance of `ArrayIndexOutOfBoundsException`?

4. Answer the following questions on the basis of Example 11.6:

   a. What is the purpose of `finally` block?
   b. Can the order of `catch` blocks be interchanged? Justify the answer.
   c. Give examples of input that will throw each one of the exceptions caught in the program.
   d. Explain what happens if one of those exceptions is not caught.

PROGRAMMING EXERCISES

1. Redo Programming Exercise 1 of Chapter 5 by including necessary exception-handling code so that if the user enters anything other than a positive digit, the user is given unlimited number of chances to enter the correct data.
2. Redo Programming Exercise 4 of Chapter 5 by including necessary exception-handling code so that if the user enters anything other than an integer, the user is given unlimited number of chances to enter the correct data.
3. Redo Programming Exercise 13 of Chapter 5 by including necessary exception-handling code so that if the user enters a string with characters not in the telephone pad, the user is given unlimited number of chances to enter the correct data.
4. Redo Programming Exercise 2 of Chapter 4 by including necessary exception-handling code so that all three data values are positive and the user is given unlimited number of chances to enter the correct data.

5. Redo Programming Exercise 4 of Chapter 4 by including necessary exception-handling code so that if the data entered is not correct, the user is given exactly one chance to select from a list of choices.
6. Implement necessary exception-handling techniques so that if the fraction (see fraction calculator at the end of Chapter 6) is zero, system will prompt the user for a new fraction to perform inverse. Also, handle division by zero in the same way.
7. Implement necessary exception-handling code in Programming Exercise 2 of Chapter 7 so that both width and length are positive. Create your own exception class or classes.
8. Implement necessary exception-handling code in Programming Exercise 9 of Chapter 9 so that the user will be given exactly one more chance to enter the correct data. Use your own exception classes.

ANSWERS TO SELF-CHECK

1. abnormal condition
2. terminate
3. `Exception, Error`
4. exception
5. `Exception`
6. `RuntimeException`
7. declared, caught
8. terminate, `stacktrace`
9. `Exception, throw`
10. `getMessage, printStackTrace`
11. `catch, finally`
12. `catch, finally`

Appendix A: Operator Precedence

Operator	Operand Types	Operation	Level	Associativity
`( )`		Parentheses	0	
`.`		Member access	1	Left to right
`[]`		Array subscripting		
`(..)`		Parameters in message passing		
`++`	Numeric variable	Postincrement	1	
`--`		Postdecrement		
`++`	Numeric variable	Preincrement	2	Right to left
`--`		Predecrement		
`+`	Number	Unary plus	2	
`-`		Unary minus		
`!`	Logical	Logical NOT		
`~`	Integer	Bitwise NOT		
`new`		Object instantiation	3	
`(type)`		Type casting		
`*`	Number, number	Multiplication	4	Left to right
`/`		Division		
`%`		Modulus		
`+`	Number, number	Addition	5	
`+`	String, number Number, string	Concatenation		
`-`	Number, number	Subtraction		
`<<`	Integer, integer	Left shift	6	
`>>`		Right shift with sign extension		
`>>>`		Right shift with zero extension		
`<`	Number, number	Less than	7	
`<=`		Less than or equal		
`>`		Greater than		
`>=`		Greater than or equal		
`instanceof`	Object reference, class	Instance checking		
`==` `!=`	Any type, the same type	Equality operators	8	

(Continued)

<table>
<tr><th>Operator</th><th>Operand Types</th><th>Operation</th><th>Level</th><th>Associativity</th></tr>
<tr><td>&</td><td>Boolean, boolean</td><td>Logical AND (short circuit)</td><td>9</td><td></td></tr>
<tr><td></td><td>Integer, integer</td><td>Bitwise AND</td><td></td><td></td></tr>
<tr><td>^</td><td>Boolean, boolean</td><td>Logical XOR</td><td>10</td><td></td></tr>
<tr><td></td><td>Integer, integer</td><td>Bitwise XOR</td><td></td><td></td></tr>
<tr><td>|</td><td>Boolean, boolean</td><td>Logical OR (short circuit)</td><td>11</td><td></td></tr>
<tr><td></td><td>Integer, integer</td><td>Bitwise OR</td><td></td><td></td></tr>
<tr><td>&&</td><td>Boolean, boolean</td><td>Logical AND</td><td>12</td><td></td></tr>
<tr><td></td><td>Integer, integer</td><td>Bitwise AND</td><td></td><td></td></tr>
<tr><td>||</td><td>Boolean, boolean</td><td>Logical OR</td><td>13</td><td></td></tr>
<tr><td></td><td>Integer, integer</td><td>Bitwise OR</td><td></td><td></td></tr>
<tr><td>? :</td><td>Logical exp, exp, exp</td><td></td><td>14</td><td>Right to left</td></tr>
<tr><td>=, +=, -=,
*=, /=, %=,
<<=, >>=,
>>>=
&=, |=, ^=</td><td></td><td>Assignment and all compound operators</td><td>15</td><td></td></tr>
</table>

Appendix B: ASCII Character Set

	0	1	2	3	4	5	6	7	8	9
0	NUL	SOH	STX	ETX	EOT	ENQ	ACK	BEL	BS	HT
1	NL	VT	FF	CR	SO	SI	DLE	DC1	DC2	DC3
2	DC4	NAK	SYN	ETB	CAN	EM	SUB	ESC	FS	GS
3	RS	US	SP	!	"	#	$	%	&	'
4	(	)	*	+	,	-	.	/	0	1
5	2	3	4	5	6	7	8	9	:	;
6	<	=	>	?	@	A	B	C	D	E
7	F	G	H	I	J	K	L	M	N	O
8	P	Q	R	S	T	U	V	W	X	Y
9	Z	[	\	]	^	_	`	a	b	c
10	d	e	f	g	h	i	j	k	l	m
11	n	o	p	q	r	s	t	u	v	w
12	x	y	z	{	|	}	~	DEL		

The rows of the table are numbered 0–12 and columns are numbered 0–9. Row numbers stand for the rightmost digits of the character code 0–127 and column numbers stand for the left digits of the character code. For example, character A is in row 6, column 5. Thus, the character code of A is 65. Similarly, character g is in row 10, column 3. Therefore, the character code of g is 103.

Characters 0–31 and the character 127 are nonprintable. The description of abbreviations used in the above table is as follows:

NUL	Null character	VT	Vertical tab	SYN	Synchronous idle
SOH	Start of header	FF	Form feed	ETB	End of transmitted block
STX	Start of text	CR	Carriage return	CAN	Cancel
ETX	End of text	SO	Shift out	EM	End of medium
EOT	End of transmission	SI	Shift in	SUB	Substitute
ENQ	Enquiry	DLE	Data link escape	ESC	Escape
ACK	Acknowledge	DC1	Device control 1	FS	File separator
BEL	Bell	DC2	Device control 2	GS	Group separator

(Continued)

BS	Backspace	DC3	Device control 3	RS	Record separator
HT	Horizontal tab	DC4	Device control 4	US	Unit separator
LF	Line feed	NAK	Negative acknowledge	DEL	Delete

Character 32 is the space character produced by pressing the space bar. More details on ASCII character set can be found at `en.wikipedia.org/wiki/ASCII`.

Appendix C: Keywords

abstract : used in class and method definition. An abstract class cannot be instantiated. An abstract method declares a method without providing implementation.

assert : used to explicitly state an assumed condition.

boolean : used for data type declaration. This is one of the eight primitive data types in Java.

break : used to skip remaining statements in the current block statement and transfer control to the first statement following the current block.

byte : used for data type declaration. This is one of the eight primitive data types in Java.

case : used in a `switch` statement.

catch : used in connection with a `try` block.

char : used for data type declaration. This is one of the eight primitive data types in Java.

class : used for defining a class.

const : currently not used.

continue : used to skip remaining statements for the current iteration in a loop statement.

do : used in a `do` … `while` statement.

double : used for data type declaration. This is one of the eight primitive data types in Java.

else : used in an `if` … `else` statement.

enum : used to declare an enumerated type.

extends : used to specify the superclass in a subclass definition.

false : boolean literal.

final : used to specify a class, method, or variable cannot be changed.

finally : used in connection with a `try` block.

float : used for data type declaration. This is one of the eight primitive data types in Java.

for : used in a `for` statement.

goto : currently not used.

if : used in `if` and `if` … `else` statements.

implements : used to declare the interfaces implemented in the current class.

import : used to declare the packages need to be imported.

instanceof : used as a binary operator to determine whether or not an object is an instance of a class.

int : used for data type declaration. This is one of the eight primitive data types in Java.

interface : used to declare a special type of class in Java.

long : used for data type declaration. This is one of the eight primitive data types in Java.

native : used in method declaration to specify that the method is not implemented in Java but in another programming language.

new : used to create an instance of a class or array.

null : reference literal to indicate that the reference variable references no object.

package : used to declare a package.

private : used to declare a class, method, or variable has *private* access.

protected : used to declare a class, method, or variable has *protected* access.

public : used to declare a class, method, or variable has *public* access.

return : used to transfer the control back from the current method.

short : used for data type declaration. This is one of the eight primitive data types in Java.

static : used to declare an inner class, method, or variable has *static* context.

strictfp : used to restrict precision and rounding in floating point computations.

super : used to reference the object as an instance of its superclass.

switch : used in a `switch` control structure.

synchronized : used to acquire the `mutex` lock for an object during the execution of a `thread`.

this : used to reference the object.

throw : used to throw an instance of the exception.

throws : used in method declaration to specify exceptions not handled by the method.

transient : used to declare an instance variable is not part of the serialized form of an object.

true : boolean literal.

try : used in connection with a `try` block.

void : used in method declaration to indicate that the method returns nothing, or a void method.

volatile : used in variable declaration to indicate that the variable is modified asynchronously by other threads.

while : used in `while` and `do ... while` control structures.

Appendix D: Coding Conventions

A summary of Java coding conventions can be found at `http://java.sun.com/docs/codeconv/index.html`. This appendix presents a modified version that is appropriate for an academic programming course. In fact, we have followed these conventions throughout this textbook and this is just a summary of what we have been following.

COMMENTS

Java comments can be grouped into two sets. Within the first set, there are two types of comments. They are delimited by /* ... */ and //. The first type is used to comment multiple lines and the second type is used to comment a single line. These two types of comments are called implementation comments and are used to explain the implementation of the code. Use multiple line comments to describe files, methods, data structures, and algorithms. Use single line comments to further clarify a certain statement or method invocation.

The second set of comments, known as documentation comments, is unique to Java. They are delimited by /** ... */ and are used to describe the specification of the code. In fact, we have used this set of comments throughout this textbook and the reader must already be familiar with it. Java development kit (JDK) comes with a tool called `javadoc` that can create an HTML document similar to the official Java documentation that is available at `http://java.sun.com/javase/6/docs/api/` for Java libraries. In fact, Java online documentation is created using `javadoc`.

INDENTATION, BLANK LINE, AND BLANK SPACE

Use a blank line to separate

- Method definitions
- Instance variables from methods

Use a blank space to separate operands from operators wherever possible to improve readability.

Use indentation to show the logical structure of the program. For example, all statements in a block statement corresponding to an `if`, `if ... else`, `while`, `for`, and `do ... while` need to be indented. Java specification recommends four spaces of indentation.

NAMING CONVENTIONS

Class and interface: Nouns with alphabet alone are used. First letter of each word is in uppercase.

Method and instance variable, local variable, formal parameters: Nouns with alphabet alone are used. First letter of each internal word is in uppercase and the first letter is in lowercase.

Named constants: Uppercase letters alone are used. Each internal word is separated by the underscore character _.

Loop variables: Use single character such as `i`, `j`, `k` or meaningful names such as `row`, `col` (for column).

File name: Java source file must have the same name as the class and it must have `.java` extension.

BRACES

Curly braces are used to mark the beginning and end of blocks of code. Keep the braces aligned vertically.

DECLARATIONS

Keep one declaration per line. Initialize all local variables. Place all instance variable declarations together and before any method definition. Keep all local variable declarations together and at the beginning of a method before any executable statement.

EXECUTABLE STATEMENTS

Each executable statement must be placed in a single line. Control structures such as `if`, `if ... else`, `while`, `for`, and `do ... while` need to be organized to show the structure.

Appendix E: JDK and Documentation

The Java development kit (JDK) standard edition (SE) is free and is a collection of command line tools for developing Java software. It is available for various platforms at `http://java.sun.com/javase/downloads/index.jsp`. In this appendix, we give a brief introduction to some of the tools included in the JDK software.

`javac` is the *Java* compiler. It translates the Java source files into Java bytecode. Source file must obey the naming conventions outlined in Appendix D. If there are no syntax errors, the javac compiler will create a new file with extension `.class` having the same name as the source file. The javac command line is

```
javac [options] sourceFilesSeperatedBySpace
```

In fact, just entering

```
javac
```

will produce the following explanation for the usage of the command:

```
Usage: javac <options> <source files>
where possible options include:
  -g                          Generate all debugging info
  -g:none                     Generate no debugging info
  -g:{lines,vars,source}      Generate only some debugging info
  -nowarn                     Generate no warnings
  -verbose                    Output messages about what the
                                           compiler is doing
  -deprecation                Output source locations where
                                   deprecated APIs are used
  -classpath <path>           Specify where to find user class
                               files and annotation processors
  -cp <path>                  Specify where to find user class
                               files and annotation processors
```

```
-sourcepath <path>            Specify where to find input source
                                                         files
-bootclasspath <path>         Override location of bootstrap class
                                                         files
-extdirs <dirs>               Override location of installed
                                                    extensions
-endorseddirs <dirs>          Override location of endorsed
                                               standards path
-proc:{none,only}             Control whether annotation processing
                                          and/or compilation is done
-processor <class1>           Names of the annotation processors to
[,<class2>,<class3>...]             run; bypasses default discovery
                                                         process
-processorpath <path>         Specify where to find annotation
                                                    processors
-d <directory>                Specify where to place generated
                                                   class files
-s <directory>                Specify where to place generated
                                                  source files
-implicit:{none,class}        Specify whether or not to generate
                                     class files for implicitly
                                              referenced files
-encoding <encoding>          Specify character encoding used by
                                                  source files
-source <release>             Provide source compatibility with
                                             specified release
-target <release>             Generate class files for specific VM
                                                         ersion
-version                      Version information
-help                         Print a synopsis of standard options
-Akey[=value]                 Options to pass to annotation
                                                processors
-X                            Print a synopsis of nonstandard
                                                      options
-J<flag>                      Pass <flag> directly to the runtime
                                                        system
```

Thus, to compile `HiThere.java` file, all that is required is

```
javac HiThere.java
```

`java` is the Java interpreter. Java interpreter is used to run a Java application. Once the `.class` file is created, you can run the Java application as follows:

```
java className
```

Once again, you could issue the command `java` to see all the options.

The **`appletviewer`** tool can be used to run an applet without using a web browser.

The **`javadoc`** tool will create a set of HTML documents from a `.java` file. For example, to generate documentation for the `Circle` class presented in Chapter 7, the following command is required:

```
javadoc circle.java
```

The files created are shown below:

```
stylesheet.css, package-tree.html, package-summary.html, package-
list, package-frame.html, overview-tree.html, index-all.html, index.
html, help-doc.html, deprecated-list.html, constant-values.html,
Circle.html, allclasses-noframe.html, allclasses-frame.html.
```

You could open the file `index.html` using a web browser and see the documentation generated for the `Circle` class. The reader is encouraged to try this out.

Appendix F: Solution to Odd-Labeled Exercises

Solutions to odd-labeled components are provided for every exercise that has multiple components. For example, in Chapter solutions to Exercise 2a, 2c and 2e are provided, even though the exercise number is 2. If an exercise has no multiple components and is even-numbered then no solution is provided. Exercise 6 of Chapter 1 has no multiple parts and as such no solution is provided.

CHAPTER 1

1. a. False
   c. True
   e. True
   g. False
   i. False
   k. True
2. a. Microwave oven: Reheat, add 30 s
   c. Thermostat: Set temperature to start AC, set temperature to start heating
   e. Telephone: Dial a number, hang up
3. a. Class is a formal specification or template of a real world object.
   c. Keeping data along with the operation on them together as one unit is known as encapsulation.
   e. A client is an object that requests the service of another object.
   g. An interpreter is a software that translates each line of a program and executes it.
   i. A linker is a system software that links a user's bytecode program with other necessary precompiled programs to create a complete executable bytecode program.
4. a. 101001
   c. 40
   e. 68

  g. 63
  i. `HelloThere.java`
  k. Syntax error

5. Purpose of memory is to speed up the computation. Compared to the speed of the CPU, secondary storage devices are very slow.
7. Through the introduction of a Java virtual machine (JVM). Java is compiled to bytecode and JVM interprets the bytecode on the specific platform.
9. For functional requirement see solution to exercise 2. Only nonfunctional requirement is listed here.
   a. Microwave oven: Outside color must be white or silver.
   c. Thermostat: Must not weigh more than 8 oz.
   e. Telephone: Must not weigh more than 3 oz.
11.

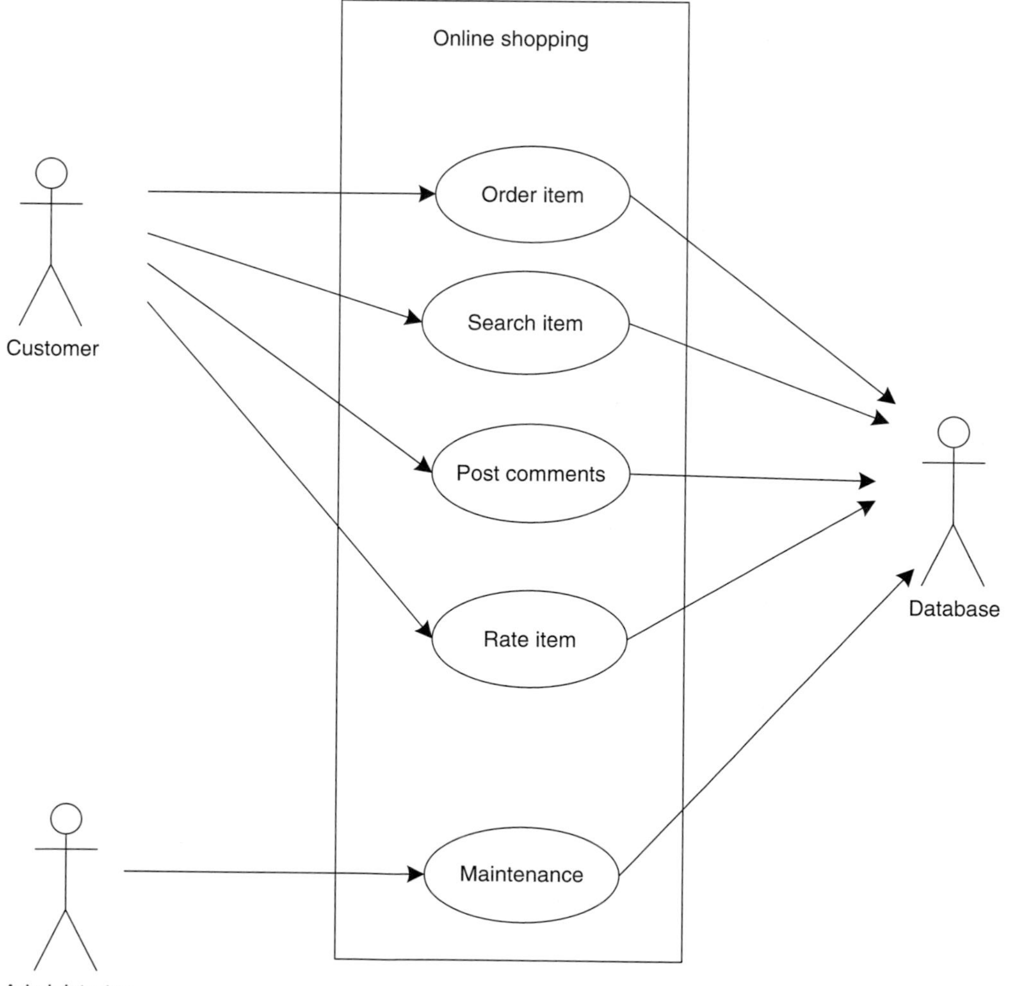

CHAPTER 2

1. a. True
   c. True
   e. False
   g. True
   i. False
   k. False
   m. False
   o. False
   q. True
   s. True
   u. True
2. a. Invalid; space not allowed
   c. Invalid; ? not allowed
   e. Invalid; first character cannot be a numeric
   g. Valid; not a good choice
   i. Invalid; % not allowed
   k. Valid
   m. Valid
   o. Invalid; – not allowed
   q. Invalid; / not allowed
   s. Valid
3. a. (i)
   c. (iii)
   e. (ii)
   g. (i)
4. a. `a = 7, b = 5, c = 9`
   c. `x = 10.7, y = 10.7`
   e. `a = 2, b = 10, x = 70.0`
   g. A double value (x) cannot be used to assign an `int` (b)
   i. `b = 9, x - 7.0, y = 2.0`
   k. `x = 11.0, y = 8.0`
   m. Division by zero error since `a = 0`
   o. `y = 2.0`
   q. `b = 20, c = 10`

5. a. Valid
   c. Use comma, not semicolon, `int a, b = 5, c;`
   e. String cannot initialize a char, `char space = ' ';`
6. a. Valid
   c. Valid
   e. Valid
   g. Valid
7. a. `double a, b;`
   c. `x = x - 17.5;`
   e. `k++;`
   g. `Scanner sc = new Scanner(System.in);`
      `k. = sc.nextInt();`
   i. `System.out.println("How are you");`
9. `b = 31, c = 1, d = -106, e = 106`
11. a. There is an extra " before)
    c. Subtraction from a String not allowed
    e. `b/c = 0`
    g. `1b = b + c`
12. a. There is an extra " before)
    c. Subtraction from a String not allowed
    e. `b/c = 0.7777777777777778`
    g. `16.0 = b + c`
13. a. `System.out.println("Good Morning America!");`
    c. `System.out.println("\"Good Morning America\" ");`
    e. `System.out.println("/Good Morning America\\");`
14. The syntax error corrected versions are as follows, wherein the commented lines are modified:

    a.
```
//public void class SyntaxErrOne
public class SyntaxErrOne
{
    //final ratio = 1.8;
    static final double ratio = 1.8;

    //static public main(String args())
    public static void main(String[] args)
    {
        //int b = c = 10;
```

```
        int b = 10, c = 10;
        //d = c + 5;
        int d = c + 5;
        //c = C - 10;
        c = c - 10;
        b = d/c;
        //ratio = b / d;
        //Cannot change the value of a final variable
        //system.out.println("b = + ", b);
        System.out.println("b = " +  b);

        //System.out.println("c = ", b);
        System.out.println("c = " + b);
        System.out.println("Ratio = " + b / d );
    }
}
```

c.
```
//void public class SyntaxErrorThree
public class SyntaxErrorThree
{
    //final int OFFSET 32;
    final static int OFFSET = 32;
    //void public main(String args())
    public static void main(String[] args)
    {
        //int b; c = 7; d;
        int b, c = 7, d;
        d = c + 5;
        //c = d - b;
        // b needs to be initialized.
        b = d/c;
        //c = b % OFFSET++;
        //c = b % (OFFSET + 1);
        //c++ = d + b;
        //left-hand side has to be a variable.
        c = d + b + 1;
        //system.out.println("B = + ", b);
        System.out.println("B =" + b);
        //System.out.println("C = ", c);
        System.out.println("C = "+ c);
        System.out.println("D " + d );
    }
}
```

15. a. `u -= 1;`

    c. `u *= (v/u - w);`

    e. `u %= v;`

16. a. `u = u - (v + w);`

    c. `v = v *( u++);`

    e. `u = u + (v % w);`

17. a. `Class Book;` Attributes: `title, author, publisher, price, isbn`; corresponding `get` and `set` methods.

    c. `Class House;` Attributes: `streetName, city, state, value, yearBuilt`; corresponding `get` and `set` methods.

    e. `Class Attraction;` Attributes: `Name, city, state, phoneNumber`; corresponding `get` and `set` methods.

18. a. Me and you are objects of `Person` class, ball currently in use is an object of `Ball` class.

    c. Jack and Mark are objects of `Person` class, Mark's dog is an object of `Dog` class.

    e. Joy is an object of `Person` class, the cycle he bought is an object of `Cycle` class, Cycle sport in an object of `Retailer` class.

CHAPTER 3

1. a. False

   c. True

   e. False

   g. False

   i. True

   k. True

2. a. Invalid; method `next` returns a String and is assigned to an `int` variable.

   c. Invalid; method `nextInt` has no parameter. The variable a must follow `int` as in (b).

   e. Invalid; wrong way to invoke `charAt`. See (g).

   g. Valid.

   i. Invalid; type mismatch. See (g).

   k. Invalid; missing the pair `()`.

   m. Invalid; `println` is a void method.

   o. Valid.

   q. Invalid; See (p).

   s. Valid.

   u. Invalid; wrong way to invoke the method setNumberOfShares.

3. a. (i)
   c. (ii)
   e. (ii)
   g. (ii)
4. a. Valid.
   c. Valid.
   e. Invalid; `println` is a void method.
   g. Invalid; `println` is a void method.
   i. Valid.
5. a. `public static void` is the order.
   c. `public void` or `public boolean`; `public` must appear first followed by either `boolean` or `void`.
   e. `private static String` is the order; missing `()`.
   g. Formal parameter list `int, int` must include parameters, such as `int a, int b`.
6. a. Replace `void` by `double`.
   c. Missing `()`.
   e. Omit `double` in the statement `double amount = amt;` in the body.
   g. Correct; `return quantity;` is all that is required.
   i. Replace `return;` by `return cost;` statement.
7. a.
   ```
   public String getProductName(){return productName;}
   public double getPrice(){return price;}
   public int getOnHand(){return onHand;}
   public boolean getIsBackOrder(){return isBackOrder;}
   public char getDeptCode(){return deptCode;}
   ```
   c.
   ```
   public String toString(){return "product Name is "+ productName + "price is " + price;}
   ```
   e.
   ```
   public void priceChange(double pct){price = (1 + pct) * price;}
   ```

CHAPTER 4

1. a. True
   c. False
   e. True
   g. True
   i. True

2. a. `Good Morning`
   c. `Good Night`
   e. No output. Syntax error: `else` has no matching `if`
   g. `Good`
   i. `Good`
   k. `ExcellentAcceptable`
   m. No output. Syntax error: `else` has no matching `if`
3. a. Syntax errors: There is no control variable `case default`, just `default`.
   c. Syntax error: Duplicate label not allowed.
   e. Syntax error: There is no `case all`.
4. a. `k = 8`
   `n = 0`
   `m = 16`
   c. `k = 8`
   `n = 9`
   `m = 24`
5. a. `k = 10`
   `n = 9`
   `m = 24`
   c. `k = 9`
   `n = 0`
   `m = 20`
6. a. `k = 8`
   `n = 0`
   `m = 16`
   c. `k = 8`
   `n = 0`
   `m = 16`

CHAPTER 5

1. a. False
   c. True
   e. False
   g. False
   i. True
   k. True

2. a. `x = 10, counter = 2`
   `x = 11, counter = 4`
   `x = 12, counter = 6`
   `x = 13, counter = 8`
   `x = 14, counter = 10`
   `x = 15, counter = 12`
   `x = 16, counter = 14`
   `x = 17, counter = 16`

   c. `x = 9, counter = 3`
   `x = 8, counter = 4`
   `x = 7, counter = 5`
   `x = 6, counter = 6`

   e. Syntax error: Compiler recognizes the empty statement as unreachable

   g. Syntax error: `break` outside the loop

3. a. `x = 10, counter = 0`
   `x = 9, counter = 1`
   `x = 8, counter = 2`
   `x = 7, counter = 3`
   `x = 6, counter = 4`

   c. `x = 10, counter = 0`
   `x = 10, counter = 1`
   `x = 10, counter = 2`
   `x = 10, counter = 3`
   `x = 10, counter = 4`
   `x = 10, counter = 5`
   `x = 10, counter = 6`
   `x = 10, counter = 7`
   `x = 10, counter = 8`
   `x = 10, counter = 9`

   e. Syntax error: Compiler recognizes the empty statement as unreachable

   g. `x = 30, counter = 0`
   `x = 27, counter = 3`
   `x = 24, counter = 6`

5. Any String that starts with `*` will work. For example, `***` could be used.

6. a.
```
Scanner scannedInfo = new Scanner(System.in);
int count = 1;
```

```
int sum = 0;
while (count <= 10)
{
    sum = sum + scannedInfo.nextInt();
    count++;
}
```

c.
```
Scanner scannedInfo = new Scanner(System.in);
int count = 1;
int sum = 0;
do
{
    sum = sum + scannedInfo.nextInt();
    count++;
} while (count <= 10);
```

7. a.
```
Scanner scannedInfo = new Scanner(System.in);

int count = 1;
int evensum = 0;
int oddsum = 0;
int value;
while (count <= 10)
{
    value = scannedInfo.nextInt();
    if (value % 2 == 0)
         evensum = evensum + value;
    else
         oddsum = oddsum + value;
    count++;
}
System.out.println(evensum + " " + oddsum);
```

c.
```
Scanner scannedInfo = new Scanner(System.in);
int count = 1;
int evensum = 0;
int oddsum = 0;
int value;
do
{
    value = scannedInfo.nextInt();
    if (value % 2 == 0)
        evensum = evensum + value;
```

```
    else
         oddsum = oddsum + value;
         count++;
} while (count <= 10);
System.out.println(evensum + " " + oddsum);
```

8. a.
```
x = 1, counter = 1
x = 1, counter = 1
x = 2, counter = 1
x = 3, counter = 1
x = 4, counter = 1
x = 5, counter = 1
x = 6, counter = 1
x = 7, counter = 1
x = 8, counter = 1
x = 9, counter = 1
```
c. Syntax error: The comma after `-5` has to be a semicolon

e.
```
x = 1, counter = -5
x = 2, counter = -5
x = 1, counter = -5
x = 0, counter = -5
x = -1, counter = -5
x = -2, counter = -5
x = -3, counter = -5
x = -4, counter = -5
x = -5, counter = -3
x = 2, counter = -3
x = 1, counter = -3
x = 0, counter = -3
x = -1, counter = -3
x = -2, counter = -3
x = -3, counter = -3
x = -4, counter = -3
x = -5, counter = -1
x = 2, counter = -1
x = 1, counter = -1
```

```
x = 0, counter = -1
x = -1, counter = -1
x = -2, counter = -1
x = -3, counter = -1
x = -4, counter = -1
x = -5, counter = 1
x = 2, counter = 1
x = 1, counter = 1
x = 0, counter = 1
x = -1, counter = 1
x = -2, counter = 1
x = -3, counter = 1
x = -4, counter = 1
x = -5, counter = 3
x = 2, counter = 3
x = 1, counter = 3
x = 0, counter = 3
x = -1, counter = 3
x = -2, counter = 3
x = -3, counter = 3
x = -4, counter = 3
x = -5, counter = 5
x = 2, counter = 5
x = 1, counter = 5
x = 0, counter = 5
x = -1, counter = 5
x = -2, counter = 5
x = -3, counter = 5
x = -4, counter = 5
```

g.

```
x = 1, counter = -5
x = 2, counter = -5
x = 1, counter = -5
x = 0, counter = -5
x = -1, counter = -5
x = -2, counter = -5
x = -3, counter = -5
```

```
x = -4, counter = -5
x = -5, counter = -4
x = 2, counter = -4
x = 1, counter = -4
x = 0, counter = -4
x = -1, counter = -4
x = -2, counter = -4
x = -3, counter = -4
x = -4, counter = -3
x = 2, counter = -3
x = 1, counter = -3
x = 0, counter = -3
x = -1, counter = -3
x = -2, counter = -3
x = -3, counter = -2
x = 2, counter = -2
x = 1, counter = -2
x = 0, counter = -2
x = -1, counter = -2
x = -2, counter = -1
x = 2, counter = -1
x = 1, counter = -1
x = 0, counter = -1
x = -1, counter = 0
x = 2, counter = 0
x = 1, counter = 0
x = 0, counter = 1
x = 2, counter = 1
x = 1, counter = 2
x = 2, counter = 3
x = 2, counter = 3
x = 1, counter = 3
x = 0, counter = 3
x = -1, counter = 3
x = -2, counter = 3
```

```
x = -3, counter = 3
x = -4, counter = 3
x = -5, counter = 4
x = 2, counter = 4
x = 1, counter = 4
x = 0, counter = 4
x = -1, counter = 4
x = -2, counter = 4
x = -3, counter = 4
x = -4, counter = 4
x = -5, counter = 5
x = 2, counter = 5
x = 1, counter = 5
x = 0, counter = 5
x = -1, counter = 5
x = -2, counter = 5
x = -3, counter = 5
x = -4, counter = 5
```

9. a. 100
   c. 34
   e. 7
   g. Infinite loop

CHAPTER 6

1. a. True
   c. False
   e. True
   g. True
   i. False
   k. False
   m. True
   o. True
   q. False
   s. True
   u. True
   w. False

2. a. Match.
   c. Match.
   e. `int`, `double`, `char` expected; only two parameters present.
   g. `q` is not a `char`. To be a `char` it has to be enclosed within a pair of parentheses.
   i. `"J"` is a `String`; `char` expected.
   k. Method has no formal parameters.
   m. Match.
   o. Match.
   q. Second parameter missing.
   s. Match.
3. a. `public double cashflow(int a, double b, char c)`
   c. `public int countVal(boolean a, long b, String c)`
   e. `public boolean getStatus()`
   g. `public Student()`
   i. `public Student(Student s)`
4. a. Syntax error: `int a, b` in formal parameter list must be `int a, int b`.
   c. Mismatch. 12.0 is `double`; `int` expected.
   e. x = 20, y = 40.
5. a. Corrected heading: `public static double trial(int a, int b, double c)`
   c. 7.75
   e. x = 20, y = 40, z = 18.2
6. a. Allowed as a `static` member
   c. 21.0
   e. `DataValues.testing(4, 5);`
7. a. You cannot have both `void` and `int` together. From the method name, it sounds like method is supposed to return new value of n. The method header can be

   `Public int nextValue (`**`int`**` n)`

   and you need the following return statement after if ... else statement :return;
   c. 22.
8. a. `Name()`, `Name(String)`, `Name(String, String)`, `Name(Name)`
   c.
```
public Name(Name n)
{
    if (this != n)
    {
        fName = n.fName;
        lName = n.lName;
    }
}
```

e.
```
public int compareTo(Name n)
{
     String str1 = fName+lName;
     String str2 = n.fName+n.lName;
     return str1.compareTo(str2);
}
```

9. a.
```
private double height;
private double weight;
```

c.
```
public Item(Item it)
{
     height = it.height;
     weight = it.weight;
}
```

e.
```
public int compareTo(Item it)
{
     final double ERR = 1.0E-14;
     int  val = 0;
     if (Math.abs(height*weight - it.height*it.weight) > ERR)
     if (height*weight > it.height*it.weight)
        val = 1;
     else
        val = -1;
     return val;
}
```

10. a. `public void copy(Stock obj)`

c. `obj`

e. Add the following Java statement as the last statement of the copy method

```
return this;
```

11. a. `public boolean equals(Stock obj)`

c. `obj`

e.
```
public boolean equals(Stock obj)
{
     if (this != obj)
     {
        if (numberOfShares != obj.numberOfShares)
             return false;
        if (!tickerSymbol.equals(obj.tickerSymbol))
             return false;
        if (dividend != obj.dividend)
```

```
            return false;
        }
        return true;
    }
```

12. a. 45
    c. 4
    e. 21
    g. 0

CHAPTER 7

1. a. False
   c. True
   e. False
   g. True
   i. False
   k. False
   m. True
2. a. `final`
   c. No
   e. `super`
   g. `super()`
3. i. `beta` of class `B`
   iii. `c = a;` is not correct
   v. `alpha` of `A`
   vii. `alpha` of `A`
4. a. Not a superclass/subclass relationship.
   c. Not a superclass/subclass relationship.
   e. Not a superclass/subclass relationship. However, they both can be subclasses of an `abstract` class vehicle.
   g. `Building` is a superclass of `House`. It can be defined as `abstract`.
   i. `Animal` is a superclass of `Dog`. It can be defined as `abstract`.
5. a. `BankAccount` is the superclass of the other two classes.
   c. `balance` is an attribute for the `BankAccount`.
   e. `ComputeInterest` can be the `abstract` method.
6. a. `testing()`
   c. `super.testing()`

e. `y.testing()` where `y` is a reference variable of the type `ClassOne`

g. Yes

i. No

7. a. `public abstract int trial(String s, double d);`

c. `public final static int trial(String s, double d);`

CHAPTER 8

1. a. True

c. True

e. False

g. False

i. False

k. False

m. False

2. a. Creating the application window, event-driven programming

c. `String`

e. `JTextComponent`

g. `ActionEvent`

i. HTML file

k. `actionPerformed`

m. 0, 0, 0

3. a. `super.setSize(300, 500);`

c. `Container conInterior = super.getContentPane();`

e. `conInterior.add(jBOk); //jBOk references a JButton`

g. `Font newFont = new Font("Courier New", Font.ITALIC, 48);`

i. `g.setColor(Color.blue); g.fillRect(350, 500, 300, 200);`

```
//g references the graphics context
```

5. a. `JTextField`

c. `JTextField`

e. `JButton`

7. `init`, `start`, `stop`, and `destroy`

9. `Serif`, `Monospaced Sanserif`, `Dialog`, `DialogInput`

CHAPTER 9

1. a. True

c. True

e. True

g. True

i. False

k. False

m. False

o. False

q. True

2. a.
```
int priceList[] = new int[100];
Vector<Integer> priceList = new Vector<Integer>(100);
ArrayList<Integer> priceList = new ArrayList<Integer>(100);
```

c.
```
priceList[priceList.length - 1] = 20.7;
priceList.add(20.7);
```

e.
```
priceList[4] = priceList[3] + priceList[5];
priceList.set(4, priceList.get(3) + priceList.get(5));
```

g.
```
int final ITEMS _ PER _ LINE = 7;
      ...
     for (index = 0; index < priceList.length; index++)
   {
       if (index % ITEMS_PER_LINE == 0 && index > 0)
             System.out.println();
       System.out.print(priceList[index]+ " ");
   }


   int final ITEMS_PER_LINE = 7;
     ...
   for (index = 0; index < priceList.size(); index++)
   {
       if (index % ITEMS_PER_LINE == 0 && index > 0)
             System.out.println();
       System.out.print(priceList.get(index)+ " ");
   }
```

i.
```
String productList[] = new String[100];
Vector <String> productList = new Vector<String>(100);
ArrayList <String>
ProductList = new ArrayList<String>(100);
```

3. a.
```
for (double p : itemPrice)
{
    System.out.println(p);
}
```

c.
```
int itemPerLine = 7;
int diff = 4;
 ...
boolean oddLine = true;
int count = 1;
for (index = 0; index < itemPrice.length; index++)
{
    System.out.print(itemPrice[index]+ " ");
    count = count + 1;
    if (oddLine)
    {
        if (count == itemPerLine)
        {
            count = 1;
            oddLine = false;
            itemPerLine = itemPerLine + diff;
            System.out.println();
        }
    }
    else
    {
        if (count == itemPerLine)
        {
            count = 1;
            oddLine = true;
            itemPerLine = itemPerLine - diff;
            System.out.println();
        }
    }
}
```

4. a.
```
public static String[] trial(double[] a, int[] b)
```
c.
```
public static void
    testing(int[] a, double[][] b, String[][] s)
```

5. a.
```
double[] a = new double[10];
int[] b = new int[30];
String[] str = GeneralUtil.trial(a, b);
```
c.
```
int[] a = new int[40];
double [][] b = new double[15][25];
String[]s = new String[15];
GeneralUtil.testing(a, b, s)
```

6. a.
```
SpecialUtil su = new SpecialUtil();
double[] a = new double[10];
int[] b = new int[30];
String[] str = su.trial(a, b);
```

   c.
```
SpecialUtil su = new SpecialUtil();
int[] a = new int[40];
double [][] b = new double[15][25];
String[]s = new string[15];
su.testing(a, b, s);
```

7. a. `cost[0][0] = 0.0`, and all Cost[2][0] = 2.0/other location unchanged
   c. `cost[1][2] = 2.0` and all other locations unchanged
9. a. There is no compilation error.
10. a. There is no compilation error.
11. a. There is no compilation error.

CHAPTER 10

1. a. False
   c. True
   e. False
2. a.

Linear Search: Search Item Not Found

	searchItem is 39	
i	**L[i]**	**L[i] == searchItem**
0	8	8 == 39 is **false**
1	13	13 == 39 is **false**
2	6	6 == 39 is **false**
3	14	14 == 39 is **false**
4	28	28 == 39 is **false**
	searchItem is 39	
i	**L[i]**	**L[i] == searchItem**
5	29	29 == 39 is **false**
6	35	35 == 39 is **false**
7	12	12 == 39 is **false**
8	40	40 == 39 is **false**
9	17	17 == 39 is **false**
	searchItem not found. **return** -1;	

c.

Linear Search: Search Item Not Found

searchItem is 1		
i	**L[i]**	**L[i] == searchItem**
0	8	8 == 1 is **false**
1	13	13 == 1 is **false**
2	6	6 == 1 is **false**
3	14	14 == 1 is **false**
4	28	28 == 1 is **false**
5	29	29 == 1 is **false**
6	35	35 == 1 is **false**
7	12	12 == 1 is **false**
8	40	40 == 1 is **false**
9	17	17 == 1 is **false**
searchItem not found. **return** -1;		

e.

Insertion Sort

Starting Index i, Unprocessed Part	Item to Be Inserted	Index of the Item to Compare	Item to Be Compared	Action Taken
1	13	0	8	none. i = i + 1;
2	6	1	13	temp = 6. L[2] = 13;
		0	8	L[1] = 8 L[0] = 6 i = i + 1;
3	14	2	13	none. i = i + 1;
4	28	3	14	none. i = i + 1;
5	29	4	28	none. i = i + 1;
6	35	5	29	none. i = i + 1;
7	12	6	35	temp = 12 L[7] = 35
		5	29	L[6] = 29
		4	28	L[5] = 28
		3	14	L[4] = 14
		2	13	L[3] = 13
		1	8	L[2] = 12 i = i + 1;
8	40	7	35	none. i = i + 1;
9	17	8	40	temp = 17 L[9] = 40
		7	35	L[8] = 35
		6	29	L[7] = 29
		5	28	L[6] = 28
		4	14	L[5] = 17 i = i + 1

g.

Binary Search

searchItem is 39

lower	upper	middle	L[middle]	Comparison	Action
0	9	(0+9)/2 = 4	14	Higher	lower = middle+1;
5	9	(5+9)/2 = 7	29	Higher	lower = middle+1;
8	9	(8+8)/2 = 8	35	Higher	lower = middle+1;
9	9	(9+9)/2 = 9	40	Lower	upper = middle-1;
9	8	lower > upper and hence search space is 0. **return** -1;			

return -1;

i.

Binary Search

searchItem is 1

lower	upper	middle	L[middle]	Comparison	Action
0	9	(0+9)/2 = 4	14	Lower	upper = middle-1;
0	3	(0+3)/2 = 1	8	Lower	upper = middle-1;
0	0	(0+0)/2 = 0	6	Lower	upper = middle-1;
0	-1	lower > upper and hence search space is 0. **return** -1;			

return -1;

3. a.

Linear Search: Search Item Not Found

searchItem is 9

i	L[i]	L[i] == searchItem
0	7	7 == 9 is **false**
1	14	14 == 9 is **false**
2	2	2 == 9 is **false**
3	18	18 == 9 is **false**
4	39	39 == 9 is **false**
5	92	92 == 9 is **false**
6	88	88 == 9 is **false**
7	72	72 == 9 is **false**
8	41	41 == 9 is **false**
9	71	71 == 9 is **false**
10	5	5 == 9 is **false**

searchItem not found. **return** -1;

c.

Linear Search: Search Item Found

searchItem is 92

i	L[i]	L[i] == searchItem
0	7	7 == 92 is **false**
1	14	14 == 92 is **false**
2	2	2 == 92 is **false**
3	18	18 == 92 is **false**
4	39	39 == 92 is **false**
5	92	92 == 92 is **true**

searchItem not found. **return** 5;

e.

Insertion Sort

Starting Index `i`, Unprocessed Part	Item to Be Inserted	Index of the Item to Compare	Item to Be Compared	Action Taken
1	14	0	7	`none.` `i = i + 1;`
2	2	1	14	`temp = 2.` `L[2] = 14`

		0	8	`L[1] = 7`
				`L[0] = 2`
				`i = i + 1;`

3	18	2	14	`none.` `i = i + 1;`
4	39	3	18	`none.` `i = i + 1;`
5	92	4	39	`none.` `i = i + 1;`
6	88	5	29	`temp = 88` `L[6] = 92`
		4	39	`L[5] = 88` `i = i + 1;`
7	72	6	92	`temp = 72` `L[7] = 92`
		5	88	`L[6] = 88`
		4	39	`L[5] = 72` `i = i + 1;`
8	41	7	92	`temp = 41` `L[8] = 92`
		6	88	`L[7] = 88`
		5	72	`L[6] = 72`
		4	39	`L[5] = 41` `i = i + 1;`

9	71	8	92	`temp = 71`
				`L[9] = 92`
				`L[8] = 88`
				`L[7] = 72`
				`L[6] = 71`
				`i = i + 1;`

10	5	9	92	`temp = 5` `L[10] = 92`
		8	88	`L[9] = 88`
		7	72	`L[8] = 72`
		6	71	`L[7] = 71`
		5	41	`L[6] = 41`
		4	39	`L[5] = 39`
		3	18	`L[4] = 18`
		2	14	`L[3] = 14`
		1	7	`L[2] = 7`
		0	2	`L[1] = 5` `i = i + 1`

g.

Binary Search

searchItem is 9					
lower	upper	middle	L[middle]	Comparison	Action
0	10	(0+10)/2= 5	39	Lower	upper = middle-1;
0	4	(0+4)/2 = 2	7	Higher	lower = middle+1;
3	4	(3+4)/2 = 3	14	Lower	upper = middle-1;
3	2	lower > upper and hence search space is 0. return -1;			
		return -1;			

i.

Binary Search

searchItem is 92					
lower	upper	middle	L[middle]	Comparison	Action
0	10	(0+10)/2 = 5	39	Higher	lower = middle+1;
6	10	(6+10)/2 = 8	72	Higher	lower = middle+1;
9	10	(9+10)/2 = 9	88	Higher	lower = middle+1;
10	10	(10+10)/2 = 10	92	Equal	return 10
		return 10;			

5. Let data be in an array L as follows: 3 7 9. If the search item is in L, the first return statement will be executed. For example, if the search item is 9, first return statement will be executed. If the search item is smaller than 9, but not in L, the second return statement will be executed. For example, if the search item is 8, the second return statement will be executed. If the search item is larger than 9, the third return statement will be executed. For example, if the search item is 10, the third return statement will be executed.
6. a. 58

   c. 58
7. a. Any 10 values: 5, 7, 16, 2, 22, 4, 9, 19, 27, 3

   c. Sorted in the reverse order: 67, 61, 57, 51, 49, 44, 22, 16, 7, 5

   e. Any 10 values: 5, 7, 16, 2, 22, 4, 9, 19, 27, 3
8. a. Change is linear; hence O(n).

   c. As n doubles, time increases by 2 units; hence $O(\log_2 n)$.

CHAPTER 11

1. a. False

   c. True

   e. True

   g. False

   i. True

   k. True

2. a. exception, error
   c. checked, unchecked
   e. `Throwable`
3. a. No
   c. There is no 101 array location
   e. The line that caused the exception
   g. The package name that contains the exception class `ArrayIndexOutOfBoundsException`
4. a. To perform some "clean-up" such as ignore the current input line
   c. −4, 2.5

Index

D

K

L

M

N

O

P

T

U

V

W

X